Clinical Chemistry in Diagnosis and Treatment

A Lloyd-Luke publication

Clinical Chemistry in Diagnosis and Treatment

Fifth Edition

Joan F. Zilva

MD (London), BSc (London), FRCP, FRCPath, DCC (Biochem)
Professor Emeritus in Chemical Pathology, University of London; Honorary Consulting Pathologist in the Riverside Health Authority, London; Formerly, Professor of Chemical Pathology, Charing Cross and Westminster Medical School, London

Peter R. Pannall

MB BCh (Witwatersrand), FFPath (SA), FRCPath, FRCPA, FAACB
Director in Clinical Chemistry, Queen Elizabeth Hospital, Adelaide; Clinical Senior Lecturer, University of Adelaide

Philip D. Mayne

MB BCh BAO (Dublin), BA (Mod), MSc (London), MRCPI, MRCPath
Senior Lecturer in Chemical Pathology, Charing Cross and Westminster Medical School; Honorary Consultant in Chemical Pathology, Westminster Hospital, London

Year Book Medical Publishers, Inc.
Chicago · London · Boca Raton

First published in Great Britain 1971 by
Lloyd-Luke (Medical Books) Ltd

Fifth edition first published in the United States 1988
by Year Book Medical Publishers, Inc.

Library of Congress Cataloging-in-Publication Data

Zilva, Joan F. (Joan Foster)
Clinical chemistry in diagnosis and treatment.
Includes bibliographies and index.
1. Chemistry, Clinical. I. Pannall, P. R. (Peter Ronald)
II. Mayne, Philip D. III. Title.
[DNLM: 1. Chemistry, Clinical. QY 90 Z69c]
RB40.Z54 1988 616.07′56 88–20453

ISBN 0–8151–9871–X

Typeset in 10/11 pt Plantin Compugraphic by Colset Private Limited, Singapore.
Printed and bound in Great Britain by Richard Clay Ltd, Bungay, Suffolk

Foreword to first edition

This book aims at giving, within a single cover, all the relevant biochemical and pathological facts and theories necessary to the intelligent interpretation of the analyses usually performed in departments of Clinical Chemistry or Chemical Pathology. The approach is firmly based on general principles and the authors have gone to great trouble to ensure that the development of ideas is logical and easy to follow. A basic knowledge of medicine and of elementary biochemistry is assumed but, given this, the reader should be able to understand even the more involved interrelationships without undue difficulty and should thereafter be in a strong position to apply this knowledge in medical practice.

I have appreciated very much the opportunity of reading this book during its preparation and feel that it represents a distinctly novel approach to the interpretation of biochemical data. In my opinion it could be read with advantage by all categories of reader mentioned in the Preface, and moreover I believe that they will enjoy the experience.

NF Maclagan

January, 1971

Preface to the fifth edition

During the 17 years since the first edition appeared our goals have remained unchanged. We are still aiming to explain, primarily to medical students and junior hospital staff, the clinical importance of understanding physiological and biochemical mechanisms before requesting tests from chemical pathology or clinical chemistry departments and when interpreting the results; we stress the importance of team work between clinicians and laboratory staff. We still state if there is no known explanation for the facts, or if the explanation given is only a working hypothesis. As before, the book is based on our own experience and we recommend tests and treatment that we have found to be useful, while mentioning that there may be other opinions, and giving references to these. Above all we continue to encourage students to reach their own conclusions by observing and thinking for themselves.

Since the fourth edition the original duo has become a triumvirate, and Dr Philip Mayne is now a coauthor. Naturally, other changes have been made, partly due to new advances and partly due to evolution of our ideas, often arising out of helpful comments and criticisms and of experience with the difficulties of students. As usual, all chapters have been revised – some more than others. With the advent of non-chemical methods for screening during pregnancy we believe that this subject no longer merits a chapter to itself, and have included it in that dealing with the reproductive system. We have introduced a chapter on the chemical pathology of the newborn infant.

We are again indebted to many people for helpful criticism. Amongst those to whom we are especially grateful are, in alphabetical order, Drs Stephen Absalom, Stephen Bangert, Clive Beng, Gurdeep Dhatt, Alan Grosset, Margaret Hancock, Stephen Holmes, Martyn Knapp, Patrick Little, Philip Nicholson and Maurice Wellby. Mr David Perry made helpful contributions to Chapters 16 and 23, and Dr Ilya Kovar to Chapter 17. Mrs Margaret Bicknell and Mrs Joanne Payne have again typed several drafts. Miss Denise Hoare, Staff Pharmacist of the Westminster Hospital Pharmacy, has checked the details of pharmaceutical preparations.

Sadly, Professor NF Maclagan, who trained both JFZ and PRP and who wrote the Foreword to the first edition, died just as this edition was ready to go to press. We would again like to pay tribute to his help and encouragement in our careers; we are especially grateful for the support he gave us at the beginning of this project, without which the book would not have been possible. He was also a valued friend, who will be much missed.

We are indebted to our new publishers for all their help and care in the preparation of this edition.

July, 1987

JFZ
PRP
PDM

Preface to the first edition

This book is intended primarily for medical students and junior hospital staff. It is based on many years practical experience of undergraduate and postgraduate teaching, and of the problems of a routine chemical pathology department. It is written for those who learn best if they understand what they are learning. Wherever possible, explanations of the facts are given: if the explanation is a working hypothesis (like, for instance, that for the ectopic production of hormones) this is stressed, and where no explanation is known, this is stated. Our experience of teaching and our discussions with students have led us to believe that many of them are willing to read a slightly longer book if it gives them a better understanding than a shorter one. Electrolyte and acid-base balance have been discussed in some detail because, in our experience, these are the most common problems of chemical pathology met with by junior clinicians and ones in which there are often dangerous misunderstandings. Some subjects, such as the porphyrias and conditions of iron overload, are discussed in greater detail than is necessary for undergraduate examinations: however, the incidence of these is high in some areas of the world, and an elementary source of reference seemed to be needed.

We have tried to stress the clinical importance of an understanding of the subject: by including in the chapter appendices some details of treatment, difficult to find together in other books, we hope that this one may appeal to clinicians as well as to students. Two chapters are included on the best use of the laboratory, including precautions which should be taken in collecting specimens and interpreting results. As we have stressed, pathologists and clinicians should work as a team, and full consultation between the two should be the rule. The diagnostic tests suggested are those which we have found, by experience, to be the most valuable. For instance, in the differential diagnosis of hypercalcaemia we find the steroid suppression test very helpful, phosphate excretion indices fallible and tedious to perform and estimation of urinary calcium useless.

Our own junior staff have found the drafts helpful in preparing for the Part I examination for the Royal College of Physicians, as well as for the primary examination for Membership of the Royal College of Pathologists, and our Senior Technicians are using it to study for the Special Examination for Fellowship in Chemical Pathology of the Institute of Medical Laboratory Technology. We feel that those studying for the primary examination for the Fellowship of the Royal College of Surgeons might also make use of it. It could provide the groundwork for study for the final examination for the Membership of the Royal College of

Pathologists in Chemical Pathology, and for the Mastership in Clinical Biochemistry.

Initially each chapter should be worked through from beginning to end. For revision purposes there are lists and tables, and summaries of the contents of each chapter. The short sub-indices in the Table of Contents should facilitate the use of the book for reference.

Appendix A lists some analogous facts which we hope may help understanding and learning. The list must be far from complete, and the student should seek examples for himself.

We wish to thank Professor NF Maclagan for his unfailing encouragement and his helpful advice and criticisms. We are also indebted to a great many other people, foremost among whom we should mention Dr JP Nicholson who has read and commented on the whole book, and Professor MD Milne, Professor DM Matthews, Mr KB Cooke and Dr BW Gilliver for helpful advice and criticism on individual chapters. Many registrars and senior house officers in the Department, in particular Dr Krystyna Rowland, Dr Elizabeth Small, Dr Nalini Naik and Dr Noel Walmsley, have been closely involved in the preparation of the book and have provided invaluable suggestions and criticisms. Students, too, have read individual chapters for comprehensibility, and we would particularly like to thank Mr BP Heather, Mr J Muir and Miss HM Merriman for helpful comments. Mr CP Butler of the Westminster Hospital Pharmacy was most helpful during the preparation of the sections on therapy. Mrs Valerie Moorsom and Mrs Marie-Lise Pannall, with the help of Mrs Brenda Sarasin and Miss Barbara Bridges, have borne with us during the typing of the drafts and the final transcript. The illustrations were prepared by Mr David Gibbons of the Department of Medical Photography and Illustration of the Westminster Medical School.

Finally, we would like to thank the publishers for their cooperation and understanding during the preparation of this book.

May, 1971

JFZ
PRP

Contents

Units in chemical pathology

Results in chemical pathology have been expressed in a variety of units. For example, concentrations of electrolytes were usually quoted in mEq/litre, those of protein in g/100 ml and of cholesterol in mg/100 ml. The units used might vary from laboratory to laboratory: calcium concentration might be expressed as mg/100 ml or mEq/litre; in Britain urea levels were expressed as mg/100 ml of *urea*, while in the United States it is usual to report mg/100 ml of *urea nitrogen*. This situation can be confusing and with patients moving, not only from one hospital to another, but from one country to another, dangerous misunderstandings have arisen.

Système International D'Unités (SI Units)

International standardization is obviously desirable; such standardization has long existed in many branches of science and technology.

The main recommendations for chemical pathology are as follows:

- if the molecular weight (MW) of the substance being measured is known, the unit of quantity should be the *mole* or a subunit of a mole.

$$\text{Number of moles (mol)} = \frac{\text{weight in g}}{\text{MW}}$$

In chemical pathology millimoles (mmol), micromoles (μmol) and nanomoles (nmol) are the most common units.

- the unit of volume should be the *litre*. Units of concentration are therefore mmol/litre, μmol/litre or nmol/litre.

Examples

Results previously expressed as mEq/litre

$$\text{Number of equivalents (Eq)} = \frac{\text{weight in g}}{\text{Equivalent weight}}$$

$$= \frac{\text{weight in g} \times \text{valency}}{\text{MW}}$$

- In the case of univalent ions, such as sodium and potassium, the units will be numerically the same. A sodium concentration of 140 mEq/litre becomes 140 mmol/litre.
- For polyvalent ions, such as calcium and magnesium (both divalent), the old units are divided by the valency. For example, a magnesium of 2.0 mEq/litre becomes 1.0 mmol/litre.

Results previously expressed as mg/100 ml. If results were previously expressed in mg/100 ml the method of conversion to mmol/litre is to divide by the molecular weight (to convert from mg to mmol) and to multiply by 10 (to convert from 100 ml to a litre). Thus effectively the previous units are divided by a tenth of the molecular weight. For example, the molecular weight of urea is 60, and of glucose 180. A urea value of 60 mg/100 ml and a glucose value of 180 mg/100 ml are both equivalent to 10 mmol/litre. The factor of 10 is, of course, only used for concentrations. The total amount of urea excreted in 24 hours in mg is numerically 60 times that in mmol.

Exceptions

- *Units of pressure* (for example mmHg) are expressed as pascals (or kilopascals – kPa). 1 kPa = 7.5 mmHg, so that a PO_2 of 75 mmHg is 10 kPa. Pascals are SI units.
- *Proteins*. Body fluids contain a complex mixture of proteins of varying molecular weights. It is therefore recommended that the gram (g) be retained, but that the unit of volume be the litre. Thus a total protein of 7.0 g/100 ml becomes 70 g/litre.
- 100 ml should be expressed as *decilitre* (dl).
- *Enzyme units* are not yet changed. Note that *the definition of international units for enzymes does not state the conditions of the reaction* (p. 309).
- Some constituents, such as some hormones, are still expressed in 'international' or other special units.

Different countries are still at different stages of implementation. We have adopted the following policy.

- If old and new units are numerically the same, we have given only the SI units (for example, sodium and potassium).
- We express protein concentrations only as g/litre.
- If it is generally accepted that SI units be adopted we have used these, with the equivalent old units in brackets.

A conversion table for some of the commoner results is listed opposite. Note that:

1 mol = 1000 mmol
1 mmol (10^{-3} mol) = 1000 μmol
1 μmol (10^{-6} mol) = 1000 nmol (nanomoles)
1 nmol (10^{-9} mol) = 1000 pmol (picomoles)

Some approximate conversion factors for SI units

	From SI units		To SI units	
Bilirubin	μmol/litre × 0.058	= mg/dl	mg/dl ÷ 0.058	= μmol/litre
Calcium				
Plasma	mmol/litre × 4	= mg/dl	mg/dl ÷ 4	= mmol/litre
Urine	mmol/24h × 40	= mg/24h	mg/24h ÷ 40	= mmol/24h
Cholesterol	mmol/litre × 39	= mg/dl	mg/dl ÷ 39	= mmol/litre
Cortisol				
Plasma	nmol/litre × 0.036	= μg/dl	μg/dl ÷ 0.036	= nmol/litre
Urine	nmol/24h × 0.36	= μg/24h	μg/24h ÷ 0.36	= nmol/24h
Creatinine				
Plasma	μmol/litre × 0.011	= mg/dl	mg/dl ÷ 0.011	= μmol/litre
Urine	μmol/24h × 0.11	= mg/24h	mg/24h ÷ 0.11	= μmol/24h
$P\text{O}_2$	kPa × 7.5	= mmHg	mmHg ÷ 7.5	= kPa
$P\text{CO}_2$	kPa × 7.5	= mmHg	mmHg ÷ 7.5	= kPa
Glucose	mmol/litre × 18	= mg/dl	mg/dl ÷ 18	= mmol/litre
Iron	μmol/litre × 5.6	= μg/dl	μg/dl ÷ 5.6	= μmol/litre
TIBC	μmol/litre × 5.6	= μg/dl	μg/dl ÷ 5.6	= μmol/litre
Phosphorus	mmol/litre × 3	= mg/dl	mg/dl ÷ 3	= mmol/litre
Proteins				
All serum	g/litre ÷ 10	= g/dl	g/dl × 10	= g/litre
Urine	g/litre × 100	= mg/dl	mg/dl ÷ 100	= g/litre
	g/24h		No change	
Urate	mmol/litre × 17	= mg/dl	mg/dl ÷ 17	= mmol/litre
Urea				
Plasma	mmol/litre × 6	= mg/dl	mg/dl ÷ 6	= mmol/litre
Urine	mmol/24h × 60	= mg/24h	mg/24h ÷ 60	= mmol/24h
5-HIAA	μmol/24h × 0.2	= mg/24h	mg/24h ÷ 0.2	= μmol/24h
HMMA	μmol/24h × 0.2	= mg/24h	mg/24h ÷ 0.2	= μmol/litre
Faecal 'fat'	mmol/24h × 0.3	= g/24h	g/24h ÷ 0.3	= mmol/24h

Abbreviations used in the book or in common use

ACP	Acid phosphatase
ACTH	Adrenocorticotrophic Hormone (corticotrophin)
ADH	Antidiuretic Hormone (Pitressin; arginine vasopressin; AVP)
AIDS	Acquired Immune Deficiency Syndrome
ALA	5-Aminolaevulinate
ALP	Alkaline Phosphatase
ALS	Aldolase
ALT	Alanine Transaminase (GPT)
AMS	α-Amylase
Anti-HB$_c$	Antibody to hepatitis B viral Core
Anti-Hb$_s$	Antibody to Hb$_s$Ag
APRT	Adenine Phosphoribosyl Transferase
APUD	Amine-Precursor Uptake and Decarboxylation
AST	Aspartate Transaminase (GOT)
AVP	Arginine Vasopressin (antidiuretic hormone; ADH)
BJP	Bence Jones Protein
BUN	Blood Urea Nitrogen (in mg/dl) $= \frac{28}{60} \times$ plasma urea in mg/dl (*or* 2.8 $\times$ plasma urea in mmol/litre)
CBG	Cortisol-Binding Globulin (Transcortin)
CC	Cholecalciferol
CK	Creatine Kinase (CPK)
CoA	Coenzyme A
CPK	Creatine Phosphokinase (CK)
CRF	Corticotrophin Releasing Factor
CSF	Cerebrospinal Fluid
DDAVP	1-Deamino-8-D-arginine vasopressin (desmopressin acetate)
1,25-DHCC	1,25-Dihydroxycholecalciferol
DIT	Diiodotyrosine
DNA	Deoxyribonucleic acid
DOC	Deoxycorticosterone
DOPA	Dihydroxyphenylalanine

DOPamine	Dihydroxyphenylethylamine
ECF	Extracellular Fluid
EDTA	Ethylene Diamine Tetracetate (Sequestrene)
EM Pathway	Embden–Meyerhof Pathway (glycolytic pathway)
ESR	Erythrocyte Sedimentation Rate
FAD	Flavine Adenine Dinucleotide
FFA	Free Fatty Acids (NEFA)
FMN	Flavine Mononucleotide
FSH	Follicle-Stimulating Hormone (follitropin)
FSH/LH-RH	Follicle-Stimulating and Luteinizing Hormone Releasing Hormone (Gn-RH)
FTI	Free Thyroxine Index
GDH	Glutamate Dehydrogenase (GMD)
GFR	Glomerular Filtration Rate
GGT	γ-Glutamyltransferase (γ-Glutamyltranspeptidase; γ-GT)
GH	Growth Hormone (somatotropin)
GMD	Glutamate Dehydrogenase (GDH)
Gn-RH	Gonadotrophin-Releasing Hormone (FSH/LH-RH)
GOT	Glutamate Oxaloacetate Transaminase (AST)
G-6-P	Glucose-6-Phosphate
G-6-PD	Glucose-6-Phosphate Dehydrogenase
GPT	Glutamate Pyruvate Transaminase (ALT)
GTT	Glucose Tolerance Test
HAV	Hepatitis A Virus
HBD	Hydroxybutyrate Dehydrogenase
HB_sAg	Hepatitis B Surface Antigen
25-HCC	25-Hydroxycholecalciferol
HCG	Human Chorionic Gonadotrophin
HDL	High-Density Lipoprotein
hGH	Human Growth Hormone
HGPRT	Hypoxanthine Guanine Phosphoribosyl Transferase
5-HIAA	5-Hydroxyindole Acetate
HIV	Human Immunodeficiency Virus
HMMA	4-Hydroxy-3-Methoxy-Mandelate (VMA)
HPL	Human Placental Lactogen
5-HT	5-Hydroxytryptamine (serotonin)
5-HTP	5-Hydroxytryptophan
ICD	Isocitrate Dehydrogenase
ICF	Intracellular Fluid
ICSH	Interstitial Cell-Stimulating Hormone (LH)
IDDM	Insulin-Dependent Diabetes Mellitus
Ig	Immunoglobulin
LCAT	Lecithin Cholesterol Acyl Transferase
LD	Lactate Dehydrogenase (LDH)
LDH	Lactate Dehydrogenase (LD)
LDL	Low-Density Lipoprotein
LH	Luteinizing Hormone
LH-RH	LH-Releasing Hormone (Gn-RH)

MEA	Multiple Endocrine Adenopathy (pluriglandular syndrome)
MIT	Monoiodotyrosine
MPS	Mucopolysaccharidosis
NAD	Nicotinamide Adenine Dinucleotide
NADP	Nicotinamide Adenine Dinucleotide Phosphate
NEFA	Non-Esterified Fatty Acids (FFA)
NIDDM	Non-Insulin-Dependent Diabetes Mellitus
5′-NT	5′-Nucleotidase (NTP)
NTP	5′-Nucleotidase (5′-NT)
17-OGS	17-Oxogenic Steroids
11-OHCS	11-Hydroxycorticosteroids (‘cortisol’)
17-OHCS	17-Hydroxycorticosteroids (17-oxogenic steroids)
OP	Osmotic Pressure
PBG	Porphobilinogen
PIF	Prolactin-Release Inhibitory Factor (prolactostatin)
PP factor	Pellagra Preventive Factor
PRPP	Phosphoribosyl Pyrophosphate
PTH	Parathyroid Hormone
RF	Releasing Factor (RH; liberin)
RH	Releasing Hormone (RF; liberin)
RNA	Ribonucleic Acid
RU	Resin Uptake (of T_3 or T_4)
SG	Specific Gravity
SGOT	Serum Glutamate Oxaloacetate Transaminase (AST)
SGPT	Serum Glutamate Pyruvate Transaminase (ALT)
SHBD	Serum Hydroxybutyrate Dehydrogenase (HBD)
SHBG	Sex-Hormone-Binding Globulin
T_3	Triiodothyronine
T_4	Thyroxine (tetraiodothyronine)
TBG	Thyroxine-Binding Globulin
TBPA	Thyroxine-Binding Prealbumin
TBW	Total Body Water
TCA cycle	Tricarboxylic Acid cycle (Krebs’ cycle; citric acid cycle)
TIBC	Total Iron-Binding Capacity (usually measure of transferrin)
TP	Total Protein
TPP	Thiamine Pyrophosphate
TRF	Thyrotrophin-Releasing Factor (TRH)
TRH	Thyrotrophin-Releasing Hormone (TRF)
TSH	Thyroid-Stimulating Hormone (thyrotrophin)
UDP	Uridine Diphosphate
UTP	Uridine Triphosphate
VLDL	Very Low-Density Lipoprotein
VMA	Vanillyl Mandelate (HMMA)
WDHA	Watery Diarrhoea, Hypokalaemia and Achlorhydria (Verner–Morrison syndrome)
Z-E syndrome	Zollinger–Ellison syndrome

1

The kidneys: renal calculi

The kidneys

The kidneys excrete waste products of metabolism and play an essential homeostatic role by adjusting the body water and solute balance. Normal function depends on:

- the integrity of the glomeruli and the tubular cells;
- a normal blood supply;
- normal secretion and feedback control of hormones acting on the kidney.

Other functions of the kidney, which will not be dealt with further in this chapter, are:

- the production of renin (p. 27);
- the conversion of 25-hydroxycholecalciferol to the active 1,25-dihydroxycholecalciferol (p. 176);
- the production of erythropoietin, a hormone stimulating erythropoiesis. The student is referred to textbooks of haematology for further details.

Passive filtration

About 200 litres of plasma ultrafiltrate enters the tubular lumina each day, mainly by glomerular filtration but also through the spaces between tubular cells ('tight junctions'). This flow of fluid from the blood depends on the fact that the hydrostatic pressure in the renal capillaries is higher than that in the lumen: any factor reducing this pressure gradient will lower the filtration rate.

Because this is a passive process, the filtrate contains diffusible constituents at almost the same concentrations as plasma. For example, at normal plasma concentrations about 30 000 mmol of sodium, 800 mmol of potassium, 300 mmol of free-ionized calcium, 1000 mmol (180 g) of glucose and 800 mmol (48 g) of urea would be filtered daily in 200 litres. The very large volume of filtrate allows adequate elimination of waste products such as urea, but unless the bulk of water and essential solutes were reclaimed, death from water and electrolyte depletion would occur within a few hours.

Proteins and protein-bound substances are filtered in only small amounts by normal glomeruli; most of that filtered is reabsorbed. The colloid osmotic pressure

(p. 34) of plasma is therefore slightly higher than that of tubular fluid and tends to oppose the filtration due to the hydrostatic pressure gradient. This osmotic effect is so weak that it can usually be ignored, but it should be remembered that over-enthusiastic intravenous infusion may dilute plasma proteins enough to cause an abnormally high filtration rate, and loss of some of the infused fluid (p. 39).

Tubular function

Changes in filtration rate alter the total amount of water and solute filtered, but not the composition of the filtrate.

Although about 200 litres of plasma is filtered, only about 2 litres of urine is formed each day. The composition of urine differs markedly from that of plasma (and therefore filtrate); the concentrations of individual constituents not only vary independently of each other, but also vary widely as physiological requirements alter. The reabsorption of about 99 per cent of the filtered volume, and adjustment of individual solutes, indicates that tubular cells have carried out selective *active* transport against physicochemical gradients. Active transport needs an energy supply, and is affected by cell death, enzyme poisons and hypoxia, which impair the production of adenosine triphosphate (ATP) by oxidative phosphorylation.

Transport of charged ions will tend to produce an electrochemical gradient which would inhibit further transport. This is minimized by two processes.

- *Isosmotic transport* occurs mainly in the proximal tubules and reclaims the bulk of filtered constituents essential for the body. Active transport of one ion leads to passive movement of an ion of the opposite charge in the same direction, along the electrochemical gradient. For instance isosmotic reabsorption of sodium (Na^+) depends on the availability of diffusible negatively charged ions (such as Cl^-). The process is 'isosmotic' because the active transport of solute causes equivalent movement of water in the same direction. Isosmotic transport also occurs in the distal part of the nephron, but is of less importance at that site.
- *Ion exchange* occurs mainly in the more distal parts of the nephron, and at this site is important for fine adjustment after bulk reabsorption has taken place. Ions of the same charge, usually cations, are exchanged and neither electrochemical nor osmotic gradients are created. There is therefore insignificant net movement of anion or water. For example, Na^+ may be reabsorbed in exchange for potassium (K^+) or hydrogen ion (H^+) secretion. Na^+ and H^+ exchange also occurs proximally, but that site is more important for bicarbonate reclamation than for fine adjustment (see Chapter 4).

Some other substances, such as phosphate and urate, are secreted into, as well as reabsorbed from, the lumen.

All body cells carry out both types of ion transport, but in most cells the pumps are uniformly distributed on the membrane surrounding the cell and solute passes into or out of the cell. In cells of the renal tubule, the intestine and many secretory organs, the pumps are located on the membrane on one side of the cell and pass solute between lumen and blood.

Waste products such as urea are not significantly handled by tubular cells. Almost all filtered urea is passed in the urine, although a small amount diffuses back passively with water. Its urinary concentration depends on the amount of water reabsorbed in excess of urea.

Reclamation in the proximal tubule

Over 70 per cent of the filtered *sodium* and free-ionized *calcium* and almost all the *potassium* is actively reabsorbed from the proximal tubules. Many inorganic anions follow the electrochemical gradient, and the reabsorption of sodium is limited by the availability of *chloride* – the most abundant diffusible anion in the filtrate (p. 97). *Bicarbonate* is almost completely recovered (though not strictly reabsorbed) following exchange of sodium and hydrogen ions (see Chapter 4). Specific active transport mechanisms result in almost complete reabsorption of *glucose, urate and amino acids*. There is incomplete *phosphate* reabsorption, and its presence in tubular fluid is important for buffering: inhibition of its reabsorption by parathyroid hormone (PTH) may occur here, or at a more distal site, and accounts for the hypophosphataemia of PTH excess (p. 174).

Isosmotic reabsorption of 70 to 80 per cent of filtered *water* from the proximal tubule depends on the solute transport described above.

Thus almost all the reutilizable nutrients and the bulk of electrolytes and water are reclaimed from the proximal tubules. Almost all the filtered metabolic waste products, such as urea and creatinine, which cannot be reused by the body, remain in the lumen.

Fine homeostatic adjustment of water and solute takes place distal to the proximal tubule. We will discuss renal handling of water, and then outline the control of solute by the distal tubule and collecting duct.

Water reabsorption: urinary concentration and dilution

Water is always reabsorbed *passively along an osmotic gradient*. However, *active* solute transport is necessary to produce this gradient. Two main processes are involved in water reabsorption:

- isosmotic reabsorption of water from the proximal tubule;
- differential reabsorption of water and solute from the loop of Henle, distal tubule and collecting duct.

Isosmotic reabsorption of water from the proximal tubule. The nephron as a whole reabsorbs 99 per cent of the filtered water, about 70 to 80 per cent (that is 140 to 160 litres a day) being returned to the body by the proximal tubules.

The proximal tubules pass through the renal cortex and their walls are freely permeable to water. We have seen that active solute reabsorption from the filtrate is accompanied by passive reabsorption of an osmotically equivalent amount of water. Blood flow is brisk in this area and solute and water are removed rapidly. Because water follows solute reabsorption, fluid entering the loop of Henle, though much reduced in volume, is still almost isosmotic and this process cannot adjust extracellular osmolality; it merely reclaims the bulk of filtered water and solute.

Differential reabsorption of water and solute from the loop of Henle, distal tubule and collecting duct. Normally between 40 and 60 litres of water enter the loops of Henle daily. Not only is this volume further reduced to about 2 litres, but, if changes in extracellular osmolality are to be corrected, the proportion of water reabsorbed must be varied according to need. At extremes of water intake urinary osmolality can vary from about 40 to about 1400 mmol/kg. (These figures should be compared with the normal for plasma, and therefore glomerular filtrate, of about 290 mmol/kg). Because the proximal tubules cannot dissociate water and solute reabsorption, the adjustment must occur between the end of the proximal tubule and the end of the collecting duct.

It is generally agreed that two mechanisms are involved.

- *Countercurrent multiplication* is an *active* process occurring in the *loop of Henle*, whereby high medullary osmolality is created, and urinary osmolality is reduced. This acts in the absence of antidiuretic hormone (ADH), and a dilute (hypoosmolal) urine is produced.
- *Countercurrent exchange* is a *passive* process, only occurring in the *presence of ADH*, whereby water without solute is reabsorbed from the *distal tubules* and *collecting ducts* into the *ascending vasa recta* along the osmotic gradient created by multiplication; by this means the urine is concentrated and the plasma diluted.

Countercurrent multiplication. The most generally held theory considers that this occurs in the loop of Henle, solute being actively pumped from the ascending to the descending limb while fluid is flowing through the loop. There is experimental evidence for a chloride pump at this site, but primary sodium transport would have the same effect.

Fluid entering the descending limb from the proximal tubule is almost isosmolal – that is, it is of the same osmolality as that in the general circulation. This is normally a little under 300 mmol/kg, and for ease of discussion we will use the figure 300 mmol/kg.

Suppose that the loop has been filled, no pumping has taken place, and the fluid in the loop is stationary. Osmolality throughout the loop and the adjacent medullary tissue would be about 300 mmol/kg.

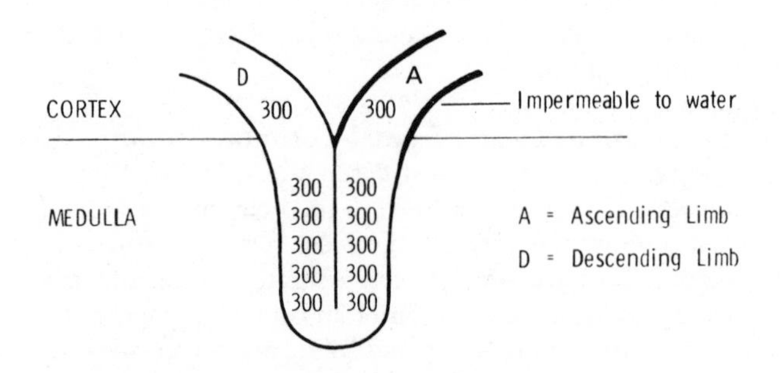

Suppose 1 mmol of solute per kg is pumped from the ascending limb (A) into the descending limb (D), the fluid column remaining stationary.

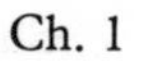

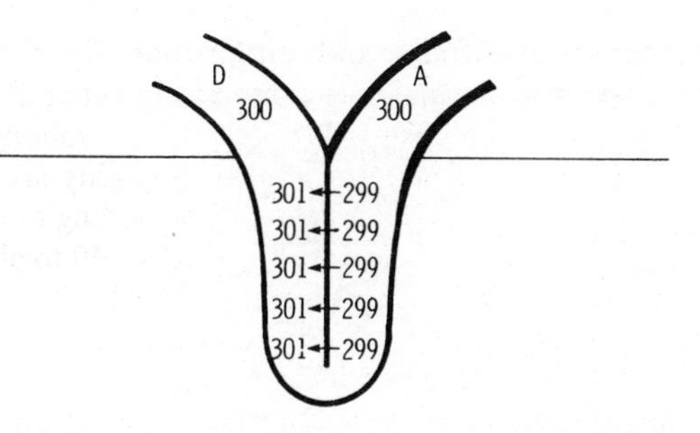

If this pumping were continued and there were no flow, limb D would become very hyperosmolal and limb A correspondingly hypoosmolal.

Let us now suppose that the fluid flows so that each figure 'moves two places'.

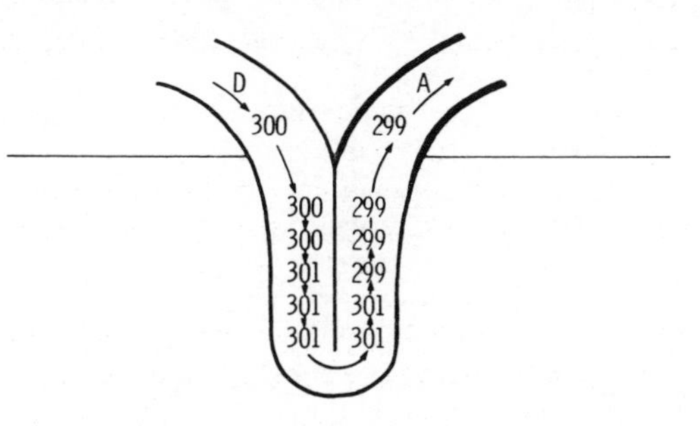

As this happens more solute is pumped from limb A to limb D.

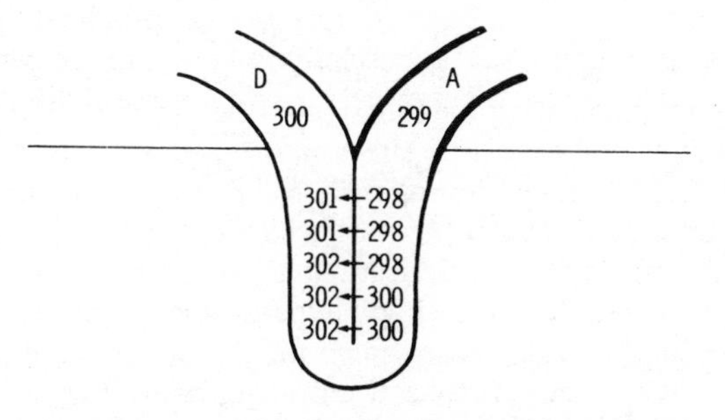

If the fluid again flows 'two places', then the situation will be:

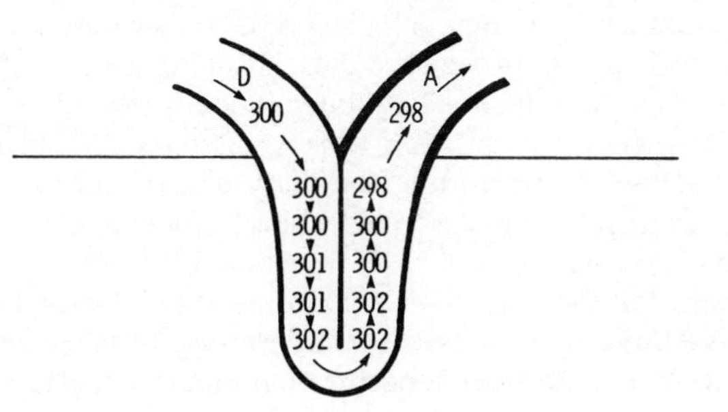

If these steps occur simultaneously and continuously, the wall of the *distal part of limb A being impermeable to water*, the consequences would be:

- increasing osmolality in the tips of the loops. Because the *walls of the loops are permeable* to water and solute, osmotic equilibrium would be reached with all the surrounding tissues and the deeper layers of the medulla including the plasma in the vasa recta, which therefore would also be of increasing osmolality;
- hypoosmolal fluid leaving the ascending limb.

The final result might be:

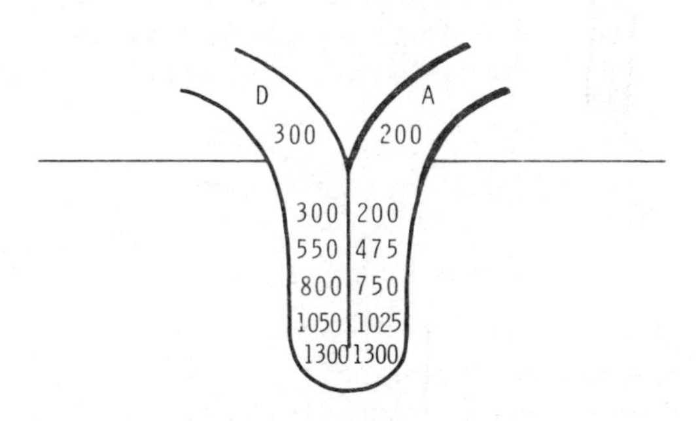

In the absence of ADH the walls of the distal tubules and collecting ducts are impermeable to water, no further change in osmolality occurs, and hypoosmolal urine would be passed.

Countercurrent exchange is essential, *together with multiplication*, for the *concentration of urine*. It can only occur in the presence of ADH, and depends on the 'random' apposition of collecting ducts and ascending vasa recta, a result of the close anatomical relations of *all* medullary constituents (Fig. 1.1) (apposition to descending vasa recta will have little effect on urinary osmolality). ADH actively alters the membranes of the cells lining the distal part of the tubule and collecting duct to increase their permeability to water, which then moves passively along the osmotic gradient created by multiplication; urine is thus concentrated as the collecting ducts pass into the increasingly hyperosmolal medulla. The increasing concentration of the fluid as it passes down the ducts would reduce the osmotic gradient if it did not meet even more concentrated plasma flowing in the opposite (countercurrent) direction. The gradient is thus maintained, and water can continue to be reabsorbed until the urine reaches the osmolality of the deepest layers (four or five times that of plasma). The low capillary hydrostatic pressure at this site, and the osmotic pull of plasma proteins, ensure that much of the reabsorbed water enters the vascular lumen. The diluted blood is carried towards the cortex and soon enters the general circulation, thus tending to reduce plasma osmolality.

The osmotic action of urea in the medullary interstitium may potentiate the countercurrent multiplication described above. Water reabsorption from the distal tubule under the influence of ADH increases the luminal urea concentration (the main reason for the rise in urinary osmolality). However, in the distal collecting ducts ADH not only increases the permeability of the cells to water, but also to urea: fluid of high urea concentration enters the deeper layers of the

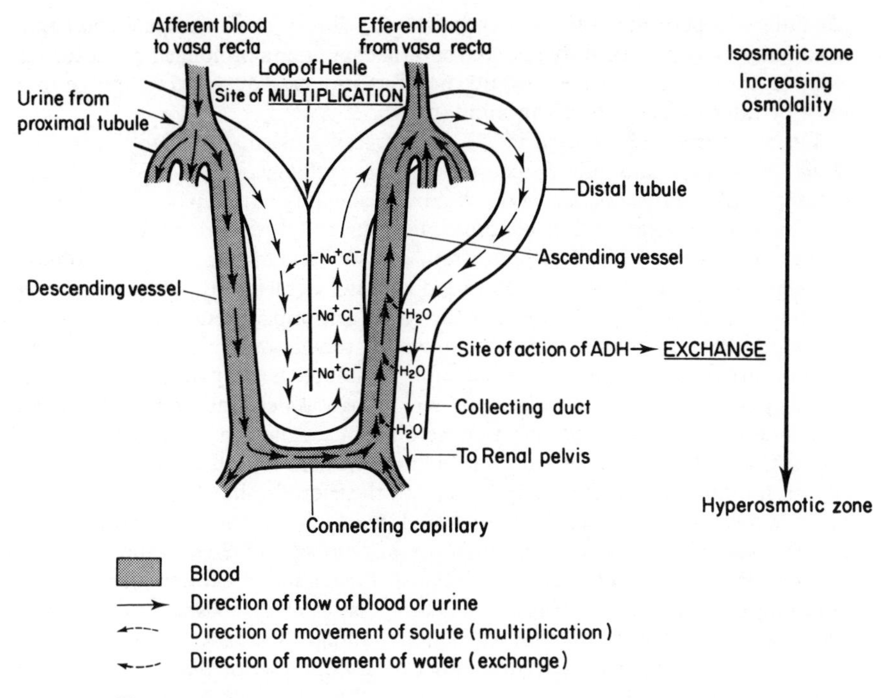

Fig. 1.1 The countercurrent mechanism.

medullary interstitium, and tends to draw water from the lower part of the permeable descending limb of the loop. This would concentrate sodium chloride in the fluid entering the ascending limb and reduce the amount of energy needed for pumping chloride.

Whatever the exact mechanism, both concentration and dilution of urine depend on active processes, which may be impaired if tubules are damaged.

Let us now look at the process in more detail as it works at extremes of water intake, bearing in mind that, if the very sensitive homeostatic mechanisms are functioning normally, even a small change in *plasma* osmolality is rapidly corrected.

Water load. A high water intake dilutes the extracellular fluid and the fall in osmolality reduces ADH secretion (p. 28). The walls of the collecting ducts are therefore impermeable to water, and *countercurrent multiplication* is acting *alone*, so that, as we have seen, a dilute urine will be produced. However, plasma osmolality would not be corrected unless the high medullary osmolality created by multiplication could be carried into the general circulation.

It has been found that during a maximal water diuresis the osmolality at the tips of the papillae may reach only about 600 mmol/kg, rather than the maximum of about 1400 mmol/kg. Increasing the circulating volume increases renal blood flow, and the *rapid flow in the vasa recta* 'washes out' medullary hyperosmolality, returning some of the solute, without extra water, to the circulation. Thus, not only is more water than usual lost in the urine, but more solute is 'reclaimed'.

Medullary hyperosmolality, and therefore the ability to concentrate the urine maximally, may only be fully restored several days after a prolonged water load is stopped. Results of a urine concentration test in a case of suspected polydipsia must be interpreted with this in mind (p. 72).

Water restriction. Water restriction, by increasing plasma osmolality, *increases ADH production* and allows countercurrent exchange. Reduced circulating volume results in *sluggish flow in the vasa recta*, allowing build-up of the medullary hyperosmolality produced from multiplication and therefore maximizing exchange. The reduced capillary hydrostatic pressure and increased colloid osmotic pressure, due to the haemoconcentration of water depletion (p. 29), ensure that much of the reabsorbed water enters the vascular compartment.

Osmotic diuresis. Water can only pass cell membranes passively, along an osmotic gradient created by reabsorption of solute: any filtered solute which cannot pass through the tubular cells will impair water reabsorption from the lumen. For example, *mannitol*, a sugar alcohol, cannot be transported across mammalian cell membranes, and is sometimes infused rapidly enough to reach a significant concentration in the plasma, and therefore in the glomerular filtrate. Some water leaves the proximal tubular lumen with sodium and other transportable solute by the normal mechanism, and the concentration, and therefore osmotic effect, of mannitol rises continuously throughout the proximal tubule, so opposing further water reabsorption. Thus less water than usual is reabsorbed proximally and a larger volume will enter the loop of Henle. Moreover, urine leaving the proximal tubules, although still isosmotic with plasma, contains a lower concentration of sodium than plasma, the difference being made up by mannitol: a lower sodium concentration than usual is therefore available for countercurrent multiplication in the loop, and, because of the consequent lower medullary osmolality, water reabsorption from the distal nephron is also impaired; a diuresis results. At physiological concentrations a little *urea* diffuses back in the proximal tubule and most *glucose* is actively reabsorbed. However, if these substances are filtered at high concentrations only a proportion can be reabsorbed and they too can act as osmotic diuretics (p. 29).

Homeostatic solute adjustment in the distal tubule and collecting duct

Sodium reabsorption in exchange for *hydrogen ion* occurs throughout the nephron. In the proximal tubule the main effect of this exchange is reclamation of filtered bicarbonate. Filtered bicarbonate has usually been reclaimed from fluid entering the distal tubule and collecting duct, and at these sites the process is more likely to be associated with net generation of bicarbonate to replace that lost in extracellular buffering, and so with fine adjustment of hydrogen ion homeostasis. Aldosterone stimulates the exchange. The possible mechanisms are discussed in Chapter 4.

Sodium reabsorption in exchange for *potassium* in the distal nephron is stimulated by aldosterone; the most important stimulus to aldosterone secretion is mediated *via* renal blood flow, and this method of reabsorption is part of the homeostatic mechanism controlling sodium and water balance (Chapter 2).

Potassium and hydrogen ions in tubular cells compete for secretion in exchange for sodium ions.

About 30 per cent of filtered free-ionized *calcium* and *magnesium* is not reclaimed from the proximal tubule. Most is reabsorbed at distal sites, possibly from the loop of Henle, and this reabsorption may be stimulated by parathyroid hormone and inhibited by loop diuretics such as frusemide (furosemide). Only about two per cent of filtered calcium appears in the urine.

The site of *urate* secretion has not been identified.

In summary

- *The very large daily volume of filtrate* allows *waste products to be excreted* at a rate equal to their production.
- Most of the filtered water, electrolytes and reusable metabolites are *reclaimed* from the *proximal tubule.*
- *Fine homeostatic adjustments* are made in the more *distal nephron,* often under hormonal control.

Chemical pathology of renal disease

Different parts of the nephron are in close anatomical association, and are dependent on a common blood supply. Renal dysfunction of any kind affects all parts of the nephron to some extent, although sometimes either glomerular or tubular dysfunction is predominant (see 'Renal circulatory insufficiency', p. 13 for an example of predominant glomerular dysfunction, and p. 16 for examples of the rarer predominant tubular dysfunction). The net effect of renal disease on plasma and urine depends on the proportion of glomeruli to tubules affected, and the number of nephrons involved. However, it may be easier to understand the consequences of renal disease if we start by considering hypothetical individual nephrons, first with a low glomerular filtration rate (GFR) and normal tubular function, and then with tubular damage but a normal GFR. *It must be stressed that these are hypothetical examples.*

Reduced GFR with normal tubular function

If the capacity of the proximal tubular cells to reabsorb solute, and therefore water, is normal a larger proportion than usual of the reduced filtered volume will be reclaimed by isosmotic processes. This will further reduce urinary volume.

In the subject with a low GFR there may also be a stimulus to antidiuretic hormone secretion; ADH, acting on the distal nephron, allows water to be reabsorbed in excess of solute, further reducing urinary volume, and increasing urinary osmolality well above that of plasma. This high urinary osmolality is mainly due to substances not actively handled by the tubules. For example, the urinary urea concentration will be well above that of plasma. This distal response will *only occur in the presence of ADH*; in its absence normal nephrons will form a dilute urine.

The total amounts of *urea and creatinine* excreted depend on the GFR. If the rate of filtration fails to balance that of production *plasma levels will rise.*

Phosphate and urate are released during cell breakdown. Plasma levels rise because less than normal is filtered. Most of the reduced amount reaching the proximal tubule can be reabsorbed and the capacity for distal secretion is impaired if the filtered volume is too low to accept the ions: these factors further contribute to *high plasma concentrations*.

A large proportion of the reduced amount of filtered sodium is reabsorbed by isosmotic mechanisms: less than usual is available for exchange with hydrogen and potassium ions.

This has two important results:

- *reduction of hydrogen ion secretion* throughout the nephron. Bicarbonate can only be reclaimed if hydrogen ion is secreted (p. 82), and *plasma bicarbonate levels will fall.*
- *reduction of potassium secretion* in the distal nephron, with potassium retention (potassium can still be reabsorbed proximally).

Only if the low GFR is accompanied by a low renal blood flow will systemic aldosterone secretion be maximal (p. 14); in such cases any sodium reaching the distal nephron will be almost completely reabsorbed in exchange for H^+ and K^+, and the urinary sodium concentration will be low.

Thus the findings in venous plasma and urine from the affected nephron will be:

Plasma

- high urea and creatinine concentrations;
- low bicarbonate concentration, with low pH;
- hyperkalaemia;
- hyperuricaemia and hyperphosphataemia.

Urine

- reduced volume;
- *only if renal blood flow is low* (stimulating aldosterone secretion), a low (appropriate) sodium concentration;
- *only if ADH secretion is stimulated*, a high (appropriate) urea concentration, and therefore a high osmolality.

Reduced tubular with normal glomerular function

Damage to the tubular cells impairs the adjustment of the composition and volume of the urine.

Impaired solute reabsorption from the proximal tubules reduces isosmotic water reabsorption at this site. Countercurrent multiplication may also be affected and hence the ability to respond to ADH is reduced. *A large volume of inappropriately dilute urine* is produced.

The tubules cannot secrete hydrogen ion and therefore *cannot reabsorb bicarbonate normally*, nor acidify the urine.

There is an impaired response to aldosterone of exchange mechanisms involving reabsorption of sodium, and the urine contains an *inappropriately high concentration of sodium* for the renal blood flow.

Potassium reabsorption from the proximal tubule is impaired and *plasma potassium levels may be low.*

Reabsorption of glucose, phosphate, magnesium, urate and amino acids is impaired. *Plasma phosphate, magnesium and urate levels may be low.*

Thus the findings in venous plasma and urine from the nephron would be:

Plasma

- normal urea and creatinine concentrations (normal glomerular function);
- due to *proximal or distal* tubular failure:
  low bicarbonate concentration with low pH;
  hypokalaemia;
- due to *proximal* tubular failure:
  hypophosphataemia, hypomagnesaemia and hypouricaemia.

Urine

- due to *proximal and/or distal* tubular failure:
  increased volume;
  pH inappropriately high for that of plasma.
- due to *distal* tubular failure:
  even if renal blood flow is low, an inappropriately high sodium concentration (inability to respond to aldosterone);
  even if ADH secretion is stimulated, an inappropriately low urea concentration, and therefore osmolality.
- due to *proximal* tubular failure:
  generalized aminoaciduria;
  phosphaturia;
  glycosuria.

Clinical syndromes of renal disease

There is a spectrum of conditions in which the proportion of tubular to glomerular dysfunction varies. The findings will depend on the relative contribution of each, and an attempt has been made to indicate this in Fig. 1.2. The dotted line indicates those findings which vary as these proportions change.

If the GFR falls, substances almost unaffected by tubular action, such as *urea and creatinine* will be retained; although their plasma concentrations start rising above the baseline for that individual soon after the clearance falls, they seldom rise above the reference range for the population until the GFR is below about 30 per cent of normal.

The degree of *potassium, urate and phosphate* retention depends on the balance between the degree of glomerular retention and the loss due to impairment of proximal tubular reabsorption. At the glomerular end of the spectrum so little is filtered that, despite failure of reabsorption, plasma levels rise; at the tubular end glomerular retention is more than balanced by impaired reabsorption of filtered potassium, urate and phosphate, and plasma levels may be normal or even low. Similarly the urine *volume* depends on the balance between the volume filtered,

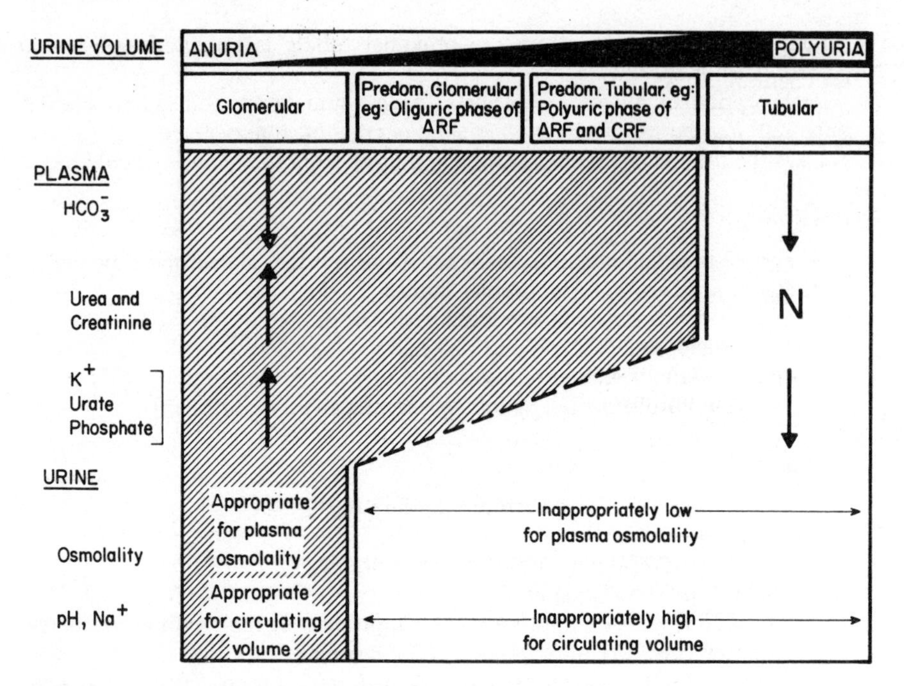

Fig. 1.2 Plasma and urinary findings in renal dysfunction. (ARF = acute & GRF = chronic, renal failure. Hatched area = results due to low GFR & clear those due to tubular dysfunction).

and the proportion reabsorbed by the tubules. At the glomerular end of the spectrum nothing is filtered and the patient is anuric: at the tubular end, although filtration is reduced, impairment of tubular reabsorption causes polyuria; since 99 per cent of filtered water is normally reabsorbed, a very small impairment of reabsorption will cause a large increase in urine volume.

Plasma levels of urea and creatinine depend largely on glomerular function; by contrast, *urinary concentrations* depend almost entirely on tubular action. However little is filtered at the glomerulus, the concentrations in the filtrate are those of a plasma ultrafiltrate. Any difference from these concentrations in urine passed is due to tubular activity. The more tubular function is impaired the nearer urine concentrations will be to those of plasma. Urinary concentrations *inappropriate to the state of hydration* suggest tubular damage, whatever the degree of glomerular dysfunction.

A low plasma *bicarbonate* concentration is a constant finding. Associated metabolic acidosis may aggravate the hyperkalaemia of glomerular dysfunction.

The plasma sodium concentration is not *primarily* affected by renal disease.

Acute oliguria

In the adult oliguria is defined as a urine output of less than 400 ml a day, or, in the short term, less than 15 ml an hour; it usually indicates a low GFR.

Acute oliguria with minimal renal damage, but with a low GFR, is due to mechanical factors reducing the hydrostatic pressure gradient between the renal capillaries and the tubular lumen.

A low intracapillary pressure is the commonest cause of this syndrome. It is known as *renal circulatory insufficiency* ('prerenal uraemia') and may be due to:

- intravascular depletion:
  of whole blood (haemorrhage);
  of plasma volume (due to gastrointestinal loss, or reduced intake);
- reduced pressure due to the vascular dilatation of 'shock'.

A rise in tubular luminal pressure as a primary cause of oliguria is rarer. The cause is usually but not always clinically obvious. It may be due to intra- or extrarenal obstruction to outflow ('postrenal uraemia').

- Intrarenal obstruction may be due to blockage of the tubular lumen by haemoglobin, myoglobin, and very rarely urate or calcium. Such causes of intrarenal obstruction as casts and oedema of tubular cells are usually the result of true renal damage.
- Extrarenal obstruction may be due to calculi, neoplasms, strictures or prostatic hypertrophy, any of which may cause sudden obstruction. The finding of a palpable bladder indicates urethral obstruction, and in the male is most likely to be due to prostatic hypertrophy, although there are other rarer causes.

Early correction of the cause of an acute reduction of GFR due to mechanical factors rapidly increases the urine output. The longer it remains untreated the greater the danger of ischaemic or pressure damage to renal tissue.

Acute oliguria due to renal damage often follows one of the above conditions, or may be due, for example, to:

- septicaemic damage;
- ingestion of a variety of poisons;
- acute glomerulonephritis.

Acute glomerulonephritis usually occurs in children. The history of a sore throat and the finding of red cells in the urine usually make the diagnosis obvious.

Septicaemia should be considered when the cause of oliguria is obscure.

The most difficult problem in the differential diagnosis of acute oliguria is to distinguish between renal circulatory insufficiency and the intrinsic renal damage which may have followed it.

Renal circulatory insufficiency is probably the commonest cause of a low GFR when tubular function is relatively normal, and is usually due to reduction in the circulating volume by haemorrhage or gastrointestinal loss of fluid. The many other causes of the 'shock' syndrome include myocardial infarction, acute intra-abdominal lesions (for example, rupture of an ectopic pregnancy, acute pancreatitis or perforation of a peptic ulcer) and intravascular haemolysis (including that due to mismatched blood transfusion). If renal blood flow is restored within a few hours the condition is reversible, but the longer it persists the greater the danger of ischaemic renal damage. Since most glomeruli are involved but tubular function is relatively normal, the findings in plasma and urine are those described on p. 9 as due to a low GFR in a single nephron. The patient will usually be hypotensive and possibly clinically volume depleted and, in addition to the laboratory findings listed above, there may be haemoconcentration (p. 29). Uraemia due to renal dysfunction is aggravated if there is increased protein breakdown, due either to tissue damage, or to the presence of blood in the gastrointestinal lumen or in large haematomas: intravenous amino acid infusion may have the same effect, because the urea in all these cases is derived by hepatic metabolism from the amino groups of amino acids. Increased tissue breakdown also aggravates hyperkalaemia, hyperuricaemia and hyperphosphataemia.

Acute oliguric renal failure often follows a period of reduced GFR due to renal circulatory insufficiency. The oliguria is due to glomerular damage and reduced cortical blood flow, aggravated by back-pressure on the glomeruli due to obstruction to tubular flow by oedema. At this stage, as indicated in Fig. 1.2, p 12, the plasma findings are at the glomerular end of the spectrum; the fact that the tubules are damaged is evident if the urinary concentrations are inappropriate – a fact which may occasionally be useful to differentiate acute oliguria due to true renal damage from that due to renal circulatory insufficiency. However, such laboratory tests are rarely necessary. In a patient with renal circulatory insufficiency urine output increases if normal renal blood flow is restored by correcting the systemic blood pressure and circulating volume. If the urine volume fails to rise, or if the patient is already normotensive and well hydrated, it is likely that there is renal damage. If laboratory tests are used, their limitations must be understood.

Aldosterone secretion will only be maximal if renal blood flow is still low; in such circumstances functioning tubules will respond appropriately by selectively reabsorbing sodium by distal exchange mechanisms. A urinary sodium concentration of less than about 30 mmol/litre, and certainly less than 20 mmol/litre, although not strictly normal if there is a very low renal blood flow, is usually taken to indicate that tubular, and therefore overall renal function is not significantly impaired. *Measurement of urinary sodium concentration cannot be used to test tubular function once renal blood flow has been restored and the stimulus to aldosterone secretion has therefore been removed*: in the absence of aldosterone the appropriate response of *normal* tubules is a reduction in differential sodium reabsorption, and the urinary sodium concentration will rise.

Measurement of *urinary osmolality* or other indicators of selective water reabsorption, such as urinary urea or creatinine concentrations, are even less valuable than assay of urinary sodium concentration, since *ADH secretion is not invariably stimulated.*

If renal damage is suspected, fluid should be given with caution, and only until volume depletion has been corrected: there is a danger of overloading the circulation if the glomerular membranes are unable to filter normally despite an adequate hydrostatic gradient.

During *recovery* oliguria is followed by *polyuria.* When cortical blood flow increases and as tubular oedema resolves, glomerular function recovers before that of the tubules. The findings gradually progress to the tubular end of the spectrum until they approximate to those for 'pure' tubular lesions; urinary output is further increased by the osmotic diuretic effect of the high load of urea, and the polyuria may cause water depletion. There may also be electrolyte depletion and the initial hyperkalaemia may be replaced by hypokalaemia. Mild acidosis (common to both glomerular and tubular lesions) persists until late. Finally, recovery of the tubules restores full renal function.

Chronic renal failure

Chronic renal failure may rarely follow an episode of acute oliguric renal failure. It is more usually the end result of a variety of conditions which include chronic glomerulonephritis, chronic obstructive uropathy, polycystic disease, renal artery stenosis, and any of the conditions listed on p. 16 as causes of tubular dysfunction. Sometimes there is no obvious precipitating factor.

In most cases of acute oliguric renal disease there is diffuse damage involving the majority of nephrons. If untreated, this would lead to early death, and a patient who survives long enough to develop chronic renal disease must have some functioning nephrons. Histological examination shows that not all nephrons are equally affected. Some may be completely destroyed, others may be almost normal, and yet others may have more damage to some parts of the nephron than to others. Some of the effects of chronic renal disease can be explained by this patchy distribution of damage: rarely acute renal disease exhibits the same picture (compare the effects of patchy pulmonary disease, p. 100).

Polyuric phase. At first glomerular function may be adequate to maintain plasma urea and creatinine levels within the reference range. As more glomeruli are involved the rate of urea excretion falls and cannot balance the rate of production: as a consequence the plasma urea, and therefore filtrate concentrations rise. This may cause an osmotic diuresis in functioning nephrons; in other nephrons the tubules may be damaged out of proportion to glomeruli. Both tubular dysfunction in nephrons with functioning glomeruli and the osmotic diuresis through intact nephrons contribute to polyuria. If the fluid lost is replaced by an equal increase in intake, urea excretion can continue through functioning glomeruli at a high rate: the increased excretion through normal glomeruli may balance the effects of reduced permeability of others, and a new steady state is reached at a higher level of plasma urea. At this stage the findings are near the tubular end of the spectrum in Fig. 1.2, p 12, the glomerular dysfunction being indicated by the high plasma urea level. If these subjects are kept well hydrated they may remain in a stable condition, with a moderately raised plasma urea, for years. Potassium levels are very variable, but tend to be raised.

Oliguric phase. If nephron destruction continues the condition approximates more and more to the glomerular end of the spectrum, until oliguria precipitates a steep rise in plasma urea and potassium concentrations. This stage, if untreated, is often terminal. Before assuming that this is the case, care should be taken to ensure that the sudden rise is not due to electrolyte and water depletion.

The diagnosis of chronic renal failure is usually obvious. Chemical estimations on urine are not helpful.

Incidental abnormal findings in renal failure

It is usual to estimate plasma urea or creatinine and electrolytes (especially potassium) to assess the severity and progress of renal failure. Other abnormalities, although not useful in diagnosis or assessment of renal dysfunction, may be misinterpreted if the cause is not recognized. Plasma *urate* levels rise in parallel with plasma urea, and a high level does not necessarily indicate primary hyperuricaemia (p. 383). Plasma *phosphate* levels also rise, and those of total *calcium* fall (p. 178). Hypocalcaemia should only be treated after correction of hyperphosphataemia (p. 189). After several years of chronic renal failure secondary hyperparathyroidism may cause decalcification of bone, with a high *alkaline phosphatase* activity (p. 183). Normochromic, normocytic *anaemia* is common, and does not respond to iron therapy.

Mild uraemia with normal urinary volume

A slight rise of plasma urea concentration is a common incidental finding especially in elderly subjects. It almost certainly indicates some degree of renal damage but, unless progressive, is unlikely to need treatment.

Congestive cardiac failure may impair renal circulation enough to cause mild uraemia.

Syndromes reflecting predominant tubular damage

A group of conditions initially affects tubules more than glomeruli, although eventually scarring involves the whole nephron and causes chronic renal failure. The patient may present complaining of polyuria and be found to have hypokalaemia, hypophosphataemia, hypomagnesaemia and hypouricaemia (the tubular end of the spectrum in Fig. 1.2, p. 12), and potassium supplementation may be necessary. These effects, as described on p. 11, are mostly due to the inability of the proximal tubule to reclaim the bulk of the filtrate. If there is detectable glycosuria, phosphaturia and non-selective aminoaciduria the condition is known as the acquired Fanconi syndrome.

Water and electrolyte depletion due to impaired tubular reabsorption may cause renal circulatory insufficiency. If intake is adequate to replace this loss plasma urea and creatinine concentrations are often normal.

Tubular cells may be damaged in several ways.

Precipitation of poorly soluble substances in or around them. Predisposing factors include:

- hypercalcaemia } the commonest causes in adults;
- hyperuricaemia }
- Bence Jones proteinuria.

Damage by exogenous toxins, which include:

- nephrotoxic drugs such as:
  - aminoglycosides;

amphotericin;
cisplatinum;
cyclosporin.

- other toxins, including heavy metals.

Among many inborn errors (discussed in the relevant chapters, and summarized on p. 377), the following may cause renal damage because of intracellular accumulation of a metabolite or other toxic substance:

- galactosaemia (galactose-1-phosphate);
- hereditary fructose intolerance (fructose-1-phosphate);
- Wilson's disease (copper);
- cystinosis (cystine);
- Fabry's disease (glycosphingolipids).

Infective damage in early pyelonephritis.

Vacuolation of tubular cells due to prolonged hypokalaemia.

Syndromes associated with isolated abnormalities of one aspect of tubular function are rarer. They are discussed in the relevant chapters.

Nephrotic syndrome

Increased glomerular permeability occurs in the *nephrotic syndrome*. All but the highest molecular weight plasma proteins can then pass the glomerulus and proteinuria of several grams a day results. The main effects are on *plasma proteins* and the subject is discussed more fully in Chapter 16. Uraemia only occurs in late stages of the disease, when many glomeruli cease to function.

Low plasma urea concentration

Occasionally the plasma urea concentration is found to be below 3 or even 1 mmol/litre. The causes of this finding are:

- **those due to an increased GFR (common):**

  pregnancy (the commonest cause in young women);
  overenthusiastic intravenous infusion (p. 41) (the commonest cause in hospital);
  'inappropriate' ADH secretion (p. 48).

- **those due to decreased synthesis:**

  use of amino acids for protein anabolism during growth, especially in children;
  low protein intake;
  very severe liver disease; } (rarer)
  inborn errors of the urea cycle (infants only). } (rarer)

Glomerular function tests

Plasma levels

Plasma urea and creatinine concentrations depend on the balance between their production and excretion.

Urea is derived in the liver from amino acids and therefore from protein, whether originating from the diet or tissues: the rate of production is accelerated by a high protein diet, by increased catabolism due to starvation, tissue damage or sepsis, or, for example, by absorption of amino acids and peptides from digested blood after haemorrhage into the gastrointestinal lumen. The normal kidney can excrete large amounts of urea, and in the presence of normal renal function only a very high protein diet, sustained for some time, or severe tissue damage can increase plasma concentrations above the reference range: in catabolic states glomerular function is often impaired due to circulatory factors, and this contributes more to the uraemia than increased production. A significantly elevated plasma urea concentration, certainly if it is above about 15 mmol/litre (blood urea nitrogen 42 mg/dl), can be taken to indicate impaired glomerular function. The probability is increased if there is a cause for renal dysfunction. Measurement of plasma creatinine may resolve any doubt in the remaining few cases.

Plasma *creatinine* is mostly derived from endogenous sources by tissue creatine breakdown. Its plasma concentration varies more than that of urea during the day because meat meals have a shorter effect on it. Sustained high meat diets and catabolic states probably affect the level of creatinine less than that of urea, and for this reason many laboratories prefer to use it for assessing renal function. However, not only is the estimation less precise than that of urea, but it is prone to serious analytical interference by such substances as bilirubin, acetoacetate and many drugs.

Whether plasma urea, creatinine or both are used as the first line investigation of glomerular function remains a matter of local choice. If the plasma concentration of either is significantly raised, and especially if it is rising, impaired glomerular function can safely be diagnosed. Changes reflect changes in the GFR, and progress can best be monitored using plasma levels alone.

Clearances

More than 60 per cent of glomeruli must be destroyed before either plasma urea or creatinine concentrations rise significantly, although in an individual they may rise above their baseline level while remaining within the population reference range.

Clearance tests should be more sensitive. They measure the volume of plasma which could theoretically be completely cleared of a substance per minute. For example:

$$\text{Creat. clearance (ml/min)} = \frac{\text{Urinary [creat.]} \times \text{Urine volume (ml)}}{\text{Plasma [creat.]} \times \text{Collection period (min)}}$$

Plasma and urinary concentrations must be expressed in the same units (for example, μmol/litre).

Only substances which are freely filtered by the glomerulus, but which are not acted on by the tubules will truly measure GFR. *Inulin* is thought to be such a substance: it is not produced in the body, and during clearance studies must therefore either be given by constant infusion to maintain plasma levels steady during the period of the test, or by a single injection followed by serial blood sampling to enable the concentration at the mid-point of the collection to be calculated. Similar considerations apply to the use of radiochromium labelled EDTA. Such exogenous clearances are not very practicable for routine use.

Endogenously produced substances are usually present at a fairly steady plasma concentration for the period of the test and blood need only be taken at the mid-point of the urine collection. Creatinine or, more rarely urea, clearances are used, but neither fulfils the necessary criteria. A little *urea* diffuses back from the tubules with water and the urea clearance is lower than that of inulin: however, the rate of protein catabolism, which may affect plasma urea levels, does *not* affect its clearance. A little *creatinine* is secreted by the tubules, and the creatinine clearance is higher than that of inulin by about the same amount as that of urea is lower. By custom creatinine clearances are more commonly performed, but there is little to choose in practice: although neither measures true GFR, both parallel it.

Precision and validity of clearances

Several factors render clearances more imprecise and inaccurate than plasma urea or creatinine concentrations.

- Every laboratory assay has an inherent imprecision. The combined imprecision of two assays is greater than that of one. Urine as well as plasma is assayed for clearance measurements.
- *The biggest error of any method depending on a timed urine collection is in the measurement of urine volume.* Even highly skilled staff and intelligent, highly motivated patients find accurate collection difficult; collection by personnel or patients who do not understand the concept of timing a urine collection (p. 450) yields very misleading results. The difficulties are increased in infants and young children, and in patients who have difficulty in bladder emptying, for example, because of prostatic hypertrophy. An added source of imprecision is introduced when the volume is measured. Unlike analytical imprecision, the probable magnitude of these errors cannot be calculated; it is likely to be much larger than those of laboratory assays. Most of these sources of error are minimized by increasing the collection period.
- Both creatinine and urea may be partly destroyed by bacterial action in old or infected urine. This error is *increased* by lengthening the collection period.

Clearance values will be equally low whether the reduced GFR is due to renal circulatory insufficiency, intrinsic renal damage, or 'post-renal' causes, and cannot distinguish between them.

Serial clearance studies are often performed on patients who are receiving potentially nephrotoxic drugs: in view of the imprecision, serial plasma levels are probably more reliable for this purpose. Since it is otherwise unlikely that treatment will be based on low clearances with normal plasma concentrations, we wonder if these tests are ever indicated.

Tubular function tests

The *water deprivation test* may be used if tubular damage is suspected. The ability to form a concentrated urine in response to fluid deprivation depends on normal tubular function (countercurrent multiplication) and on the presence of ADH. Failure of this ability is usually due to renal disease, but if there is any suspicion of diabetes insipidus the test can be repeated after giving the synthetic analogue of ADH, DDAVP (p. 72).

Proposed schemes for investigating patients with suspected renal disease, oliguria or polyuria are outlined on p. 70.

Biochemical principles of treatment of renal dysfunction

Oliguric renal dysfunction. The oliguria of volume depletion, which is often due to a reduced GFR only, should be treated with the appropriate fluid (Table 2.3, p. 54).

If oliguria is due to parenchymal damage the aims are:

- to restrict fluid and sodium, giving only enough fluid to replace losses (p. 30);
- to provide an adequate non-protein energy source using carbohydrate, with or without lipid, to minimize aggravation of uraemia and hyperkalaemia by increased endogenous catabolism;
- to prevent dangerous hyperkalaemia (p. 64).

Diuretics increase renal blood flow. Recovery may sometimes be hastened by giving an osmotic diuretic such as mannitol, or other diuretics such as frusemide (furosemide) or ethacrynic acid in large doses.

In chronic renal failure with polyuria the aim is to replace lost fluid and electrolytes. Sodium and water depletion may reduce the GFR further, and so aggravate the uraemia.

Haemodialysis or peritoneal dialysis removes urea and toxic substances from the plasma and corrects electrolyte balance by dialysing the patient's blood against fluid containing no urea, and appropriate concentrations of electrolytes, *free-ionized* calcium and other plasma constituents. The blood is either passed over a dialysing membrane before being returned to the body, or the folds of the peritoneum are used as a dialysing membrane, with their capillaries on one side, and suitable fluid infused into the peritoneal cavity on the other. The urea concentration should be reduced relatively slowly because of the danger of cellular overhydration if extracellular osmolality falls fast (p. 33). The urine output often falls after successful dialysis because the lower plasma urea level reduces the osmotic load on functioning nephrons. Dialysis is used in cases of acute renal failure until renal function improves, or as a regularly repeated procedure in suitable cases of chronic renal failure. It may also be used to prepare patients for transplantation, and to maintain them until the transplant functions adequately.

Renal calculi

Renal calculi are usually composed of products of normal metabolism which are present in normal glomerular filtrate, often at concentrations near their maximum solubility. Minor changes in urinary composition may cause precipitation of such constituents in the substance of the kidney (see causes of tubular damage, p. 16), or as crystals or calculi in the renal tract. Although this discussion deals with stone formation, crystalluria or parenchymal damage can occur under the same circumstances, and the treatment of all such conditions is the same.

Conditions favouring calculus formation

- **A high urinary concentration of one or more constituents of the glomerular filtrate,** due to:

  a low urinary volume, with normal renal function, due to restricted fluid intake or excessive fluid loss over a long period of time; this is particularly common in the tropics. It favours formation of most types of calculi, especially if one of the other conditions listed below is also present;
  an abnormally *high rate of excretion* of the metabolic product forming the stone, due either to a high plasma and therefore filtrate level, or to impairment of normal tubular reabsorption from the filtrate.

- **Changes in pH** of the urine, often due to bacterial infection, which favour precipitation of different salts at different hydrogen ion concentrations.

- **Urinary stagnation** due to obstruction to urinary outflow.

- **Lack of normal inhibitors.** It has been suggested that normal urine contains an inhibitor, or inhibitors, of calcium oxalate crystal growth absent in the urine of some patients liable to recurrent calcium stone formation.

Composition of urinary calculi

- Calcium-containing stones:
  calcium oxalate }
  calcium phosphate } with or without magnesium ammonium phosphate;
- Uric acid stones.
- Cystine stones.
- Xanthine stones.

Calculi composed of calcium salts

Between 70 and 90 per cent of all renal stones contain calcium. Precipitation is favoured by hypercalciuria, and the type of salt depends on urinary pH and on the availability of oxalate. Any patient presenting with calcium-containing calculi should have a plasma calcium estimation performed, and if the result is normal it should be repeated at regular intervals, to exclude primary hyperparathyroidism.

Hypercalcaemia causes hypercalciuria if glomerular function is normal, and estimation of urinary calcium in such cases does not help diagnosis. The causes and differential diagnosis of hypercalcaemia are discussed on p. 178.

In many subjects with calcium-containing renal calculi the plasma calcium level is normal. Any *increased release of calcium from bone*, as in actively progressing osteoporosis, (in which loss of matrix causes secondary decalcification) or in prolonged acidosis (in which ionization of calcium is increased), causes hypercalciuria; hypercalcaemia is very rare in such cases. One type of renal tubular acidosis (p. 94) both increases the load of calcium and, because of the relative alkalinity of the urine, favours precipitation in the kidney and renal tract: this is a rare cause.

Increased *oxalate* excretion favours the formation of the very poorly soluble calcium oxalate, even if calcium excretion is normal. The source of the oxalate may be dietary: the very rare inborn error, primary hyperoxaluria, should be considered if renal calculi occur in childhood.

We have already mentioned that alkaline conditions favour calcium phosphate precipitation: this type of stone is particularly common in *chronic renal infection* with urease-containing (urea-splitting) organisms, such as *Proteus vulgaris*, which convert urea to ammonia and bicarbonate.

A significant proportion of cases remain in which there is no apparent cause for calcium precipitation. The commonest cause of *hypercalciuria with normocalcaemia* is so-called *idiopathic hypercalciuria*, a name which reflects our ignorance of the aetiology of the condition. It is in such cases that estimation of the daily urinary calcium excretion may help.

Calcium-containing calculi are usually *hard, white and radiopaque*. Calcium phosphate may form 'staghorn' calculi in the renal pelvis, while calcium oxalate stones tend to be smaller and to lodge in the ureters, where they are compressed into a fusiform shape.

Treatment of calcium-containing calculi depends on the cause. Urinary calcium concentration should be reduced:

- by treating the primary condition, such as urinary infection or hypercalcaemia;
- if this is not possible, by reducing dietary calcium intake, and possibly decreasing its intestinal absorption by giving oral phosphate (p. 188);
- by reducing urinary calcium concentration by maintaining a high fluid intake day and night, unless there is glomerular failure. The concentration rather than the 24 hour output determines the tendency to precipitation.

Uric acid stones

About 10 per cent of renal calculi contain uric acid: these are sometimes associated with *hyperuricaemia*, with or without clinical gout.

In most cases no predisposing cause can be found. Precipitation is favoured in an *acid urine*.

Uric acid stones are usually *small, friable and yellowish-brown*, but can occasionally be large enough to form 'staghorn' calculi. They are *radiotranslucent*, but may be visualized by an intravenous pyelogram.

Treatment of hyperuricaemia is discussed on p. 385. If the plasma urate concentration is normal, fluid intake should be kept high and the urine alkalinized. A low purine diet may help to reduce urate production and therefore excretion.

Cystine stones

Cystine stones are rare. In normal subjects the urinary cystine concentration is well within its solubility. In homozygous cystinuria (p. 372) the solubility may be exceeded and the patient may present with *radiopaque* renal calculi. Like urate, cystine is more soluble in alkaline than acid urine and the principles of treatment are the same as for uric acid stones. Penicillamine can also be used to treat the condition (p. 373).

Xanthine stones

Xanthine stones are very uncommon and may be the result of the rare inborn error, xanthinuria (p. 386). They have *not* been reported during treatment with xanthine oxidase inhibitors such as allopurinol, which impair conversion of xanthine to urate.

A proposed scheme for the investigation of a patient with renal calculi is given on p. 73.

Summary

The kidneys

1. Normal renal function depends on a normal filtration rate and normal tubular function.
2. A low glomerular filtration rate (GFR) leads to:

- oliguria;
- uraemia and retention of other nitrogenous end-products including creatinine and urate, and of phosphate;
- a low plasma bicarbonate with metabolic acidosis;
- hyperkalaemia.

3. Tubular damage leads to:

- polyuria. The urine is inappropriately dilute and contains an inappropriately high sodium concentration in relation to the patient's state of hydration;
- a low plasma bicarbonate with metabolic acidosis;
- hypokalaemia;
- hypophosphataemia and hypouricaemia.

4. In most cases of renal disease impairment of glomerular and tubular function coexist. The clinical findings depend on the proportions of each, and on the total number of nephrons involved.
5. A low GFR without significant renal damage may be due to a reduced hydrostatic pressure gradient between the capillary plasma and the tubular lumen. This

is most commonly due to renal circulatory insufficiency, but may be caused by postglomerular obstruction.

6. In acute oliguric renal damage plasma findings cannot distinguish the condition from renal circulatory insufficiency.

7. The differentiation between the oliguria of renal circulatory insufficiency with relatively normal tubular function and of acute oliguric renal failure is best made on clinical grounds; if a laboratory test is felt to be necessary the urinary sodium concentration is the best indicator, but can only be interpreted if the renal blood flow was as low when the specimen was secreted.

8. In most cases plasma urea or creatinine levels reflect changes in renal clearance and assay of one or both is adequate to diagnose and to monitor glomerular dysfunction. Tubular function may be tested by assessing the concentrating capacity of the kidney.

9. Compared with plasma assays clearance tests are relatively imprecise and inaccurate.

Renal calculi

1. The formation of renal calculi is favoured by:

- a high concentration of the constituents of the calculi, whether due to oliguria or a high rate of excretion of the relevant substances;
- a urinary pH which favours precipitation of the constituents of the calculi;
- urinary stagnation.

2. 70 to 90 per cent of all renal calculi contain calcium. Calcium-containing stones are most commonly idiopathic in origin, but hypercalcaemia, especially that of primary hyperparathyroidism, should be excluded as a cause.

3. Uric acid stones account for about a further 10 per cent of renal calculi.

4. Rare causes are cystinuria, xanthinuria and hyperoxaluria.

Further reading

De Wardener H E. *The kidney: an outline of normal and abnormal function.* **5th ed.** Edinburgh and London: Churchill Livingstone, 1985.

Payne R B. Creatinine clearance: a redundant clinical investigation. *Ann Clin Biochem* 1986; **23**: 243–50.

2

Sodium and water metabolism

Water is an essential body constituent. Homeostatic processes ensure that:

- the *total water balance* is maintained within narrow limits;
- the *distribution of water* between the vascular, interstitial and cellular compartments is maintained.

Distribution depends on hydrostatic and osmotic forces acting across biological membranes.

Sodium is the most abundant extracellular cation and, with its associated anions, accounts for most of the osmotic activity of the extracellular fluid (ECF): it is important in determining water distribution across cell membranes. *Osmotic activity depends on concentration*, and so on the relative amounts of water and sodium in the extracellular fluid, rather than on the absolute quantity of either. An imbalance between the two causes hypo- or hypernatraemia, and therefore changes in osmolality: the clinical pictures which may be associated with these findings are due to the consequent movement of water. *If sodium and water are lost or gained in equivalent amounts the plasma sodium, and therefore osmolal, concentration is unchanged*: symptoms are then due to extracellular volume depletion or overloading. Of course, osmotic and volume disturbances often occur together.

Concentrations of associated anions such as chloride and bicarbonate usually alter at the same time as those of sodium.

Although transport of sodium, potassium and hydrogen ions across cell membranes is often interdependent, this chapter is separated from those on potassium and on hydrogen ion homeostasis for ease of discussion.

Total water and sodium balance

There are approximately 45 litres of water and 3000 mmol of osmotically active sodium in a 70 kg man. These total amounts depend on the balance between intake and loss. Water and electrolytes are taken in food and drink, and are lost in urine, faeces and sweat: in addition, about 500 ml of electrolyte-free water is lost daily in expired air.

Loss through the kidneys and gastrointestinal tract

The kidneys and intestine handle water and electrolytes in a very similar way. Net loss through both organs depends on the balance between the volume filtered proximally and that reabsorbed more distally. Any factor affecting either passive filtration or epithelial function may disturb this balance.

In addition to the approximately 200 litres of water and 30 000 mmol of sodium filtered by the kidney a further 10 litres of water and 1500 mmol of sodium enter the intestinal tract each day. The whole of the extracellular water and sodium could be lost by passive filtration in little more than an hour. Normally about 99 per cent of this initial loss is reabsorbed, and net daily losses amount to about 1.5 to 2 litres of water and 100 mmol of sodium in the urine, and 100 ml and 15 mmol in the faeces. It is not surprising that impairment of absorptive mechanisms causes such extreme disturbances of water and sodium balance.

Passive filtration. Most of the large amount of luminal water and electrolyte is derived from the plasma by ultrafiltration – in the kidney through the glomerulus, and in both organs through the so-called 'tight junctions' between the epithelial cells lining the lumen. Far from being 'tight', they are freely permeable to water and only slightly less so to small molecules and ions.

The hydrostatic gradient from plasma to lumen is the most important factor maintaining filtration in the kidney and in the resting small intestine: in the latter the postprandial flow of fluid into the lumen is greatly increased by the temporary increase in luminal osmolality due to partially digested food. Active secretion of digestive juices contributes only a small proportion of the total volume entering the lumen. Overenthusiastic fluid therapy, by increasing blood flow to the kidneys and intestine and by reducing capillary colloid osmotic pressure (dependent on plasma protein concentration) may cause inappropriate passive fluid loss (p. 39). A high luminal solute concentration may cause an osmotic diuresis (p. 8) or osmotic diarrhoea.

Bulk reabsorption. Solute transport is accompanied by an isosmotic water flow (p. 3). In the proximal renal tubules and the resting small intestine sodium and associated anions provide the bulk of absorbed solutes. After meals the products of digestion are the main solutes absorbed from the gut lumen. Epithelial cell dysfunction of either organ impairs this isosmotic reclamation of the bulk of the filtrate, and may cause an inappropriate diuresis or diarrhoea.

Fine adjustment. Fine adjustment of water and solute, often under hormonal control, occurs in the distal nephron and the large intestine. The effects of antidiuretic hormone (ADH) and aldosterone on the kidney seem to be the most important physiologically.

Loss in sweat and expired air

About 900 ml of water is lost daily in sweat and expired air: less than 30 mmol of sodium a day is lost in sweat. Although ADH and aldosterone have some effect on the composition of sweat, its volume is primarily controlled by skin temperature. Respiratory water loss depends on respiratory rate, and bears no relation to the body need for water. Normally losses in sweat and expired air are rapidly corrected by changes in renal and intestinal loss. However, because neither of the former can be controlled to meet sodium and water requirements, they may contribute considerably to abnormal balance when homeostatic mechanisms fail, or if there is gross depletion, whether due to poor intake or to excessive loss by other routes.

Control of sodium and water balance

Control of sodium balance

The most important factors controlling sodium balance are renal blood flow and the mineralocorticoid hormone aldosterone. This hormone controls loss, and there seems to be little, if any, control of sodium intake.

Aldosterone

Aldosterone is secreted by the zona glomerulosa of the adrenal cortex (p. 119). It affects sodium-potassium and sodium-hydrogen ion exchange across *all* cell membranes. We shall concentrate on its effect on renal tubular cells, but we must remember that it also affects loss in faeces, sweat and saliva, and the distribution of electrolytes in the body.

Aldosterone stimulates sodium reabsorption from the lumen of the distal nephron in exchange for potassium or hydrogen ion secretion. The net result is retention of more sodium than water, and loss of potassium and hydrogen ions. If the circulating aldosterone concentration is high, and tubular function is normal, the *urinary sodium concentration* is low.

Many factors have been implicated in the feed-back control of aldosterone secretion. Those such as local electrolyte concentrations in the adrenal gland and kidney are almost certainly of less physiological and clinical importance than the effect of the renin-angiotensin system.

The renin-angiotensin system

Renin is a proteolytic enzyme secreted by the juxtaglomerular apparatus, a cellular complex adjacent to the renal glomeruli. It splits a decapeptide, *angiotensin I*, from a circulating α_2-globulin known as renin substrate. Another peptidase, *angiotensin-converting enzyme*, located predominantly in the lungs, splits off a further two amino acid residues: the remaining octapeptide is the hormone *angiotensin II*, which has two important systemic actions:

- it acts directly on capillary walls, causing vasoconstriction, and so probably helps to maintain blood pressure;
- it stimulates the cells of the zona glomerulosa to synthesize and secrete aldosterone.

The most important stimulus to renin secretion seems to be reduced renal blood flow, possibly mediated by changes in the mean pressure in the afferent arterioles. Poor renal blood flow is often associated with an inadequate systemic blood pressure and the two effects of angiotensin II ensure that this is corrected:

- vasoconstriction may raise the blood pressure before the circulating volume can be restored;
- aldosterone stimulates sodium retention, which, as we shall see later, is usually followed by water retention and hence restoration of the circulating volume.

Thus aldosterone secretion responds, via renin, to a reduction in renal blood flow: since a low renal blood flow usually reflects a low systemic blood volume, we can make the simplified statement that *blood volume controls net sodium retention.*

Atrial natriuretic peptide(s)

A peptide hormone, or hormones, probably secreted in response to stimulation of atrial stretch receptors, may cause a natriuresis by increasing the GFR and by inhibiting renin and aldosterone secretion. Its importance in physiological control and in pathological states has not yet been elucidated.

In later discussion we will assume that the renin-aldosterone mechanism is of overriding importance in the control of sodium excretion.

Control of water balance

Both intake and loss of water are controlled by the osmotic gradient across cell membranes in hypothalamic centres. These centres, which are closely related anatomically, control *thirst* and secretion of *antidiuretic hormone* (ADH: arginine vasopressin); both thirst and ADH secretion are stimulated by flow of water out of cells caused by a relatively high extracellular osmolality (p. 33). Increased water intake due to thirst, and retention of relatively more water than solute due to the action of ADH on the collecting ducts, dilute extracellular osmolality. If osmolality inside cells is unchanged, an increase of only 2 per cent outside cells quadruples ADH output, and an equivalent fall almost completely inhibits it; this represents a change in plasma sodium concentration of only about 3 mmol/litre. In more chronic changes, when the osmotic gradient has been minimized by solute redistribution (p. 33), there may be little or no effect.

Severe hypovolaemia causes thirst and ADH secretion even if the plasma is hypoosmolal. Although these two responses help to correct total volume depletion, they are inappropriate to the osmolal state: since much of the retained water enters cells along the osmotic gradient, the correction of extracellular depletion is relatively inefficient (p. 39).

Normally plasma osmolality depends mostly on its sodium concentration. We have already pointed out that vascular volume controls sodium retention. We can now make the further *simplified* statement that the extracellular *sodium concentration controls water balance. However, sodium is not the only controlling factor if another osmotically active substance is circulating* (p. 33).

Sometimes the effectiveness of ADH is opposed by other factors; for example, during an osmotic diuresis the urine, although not hypoosmolal, will contain more water than sodium (p. 8). Patients being fed intravenously, or who, because of tissue damage, are breaking down more protein than usual, and so are producing more urea from the released amino acids, may become water depleted even if there is adequate ADH (p. 45): urinary osmolality will be high. Such non-ADH effects are usually relatively unimportant.

Assessment of sodium and water balance

Although it is easy to measure the water and sodium intake of patients receiving oral or intravenous liquid feeds, it is less so when a solid diet is being taken. Fortunately, accurate assessment is rarely needed in such cases.

Renal loss should also be easy to measure, but that in formed faeces, sweat and expired air ('insensible loss') is more difficult to assess and may be important when homeostatic mechanisms have failed, if there are abnormal losses by extrarenal routes, in unconscious patients and in infants (p. 45). The aim should be to ensure that, once such subjects are normally hydrated, they are kept 'in balance'. The doctor has the difficult task of imitating normal homeostasis without the help of the sensitive, interlinked mechanisms of the body.

Assessment of the state of hydration

Assessment of the state of hydration of a patient depends on observation of the clinical state and on laboratory evidence of haemoconcentration or haemodilution. Both these methods are crude, and quite severe disturbances can occur before they are obvious by clinical or laboratory methods.

Extracellular fluid is usually lost from the vascular compartment first and, unless the fluid is whole blood, depletion of water and small molecules leads to a rise in concentrations of large molecules and of blood cells: there is therefore a rise in the levels of all plasma protein fractions and of haemoglobin, and in the haematocrit reading *(haemoconcentration)*. Conversely, if there is plasma volume overloading these concentrations fall *(haemodilution)*. These findings may be affected by preexisting abnormalities of protein or red cell concentrations, and although changes are almost always more informative than single readings, both protein and haemoglobin levels may change following, for example, blood transfusion, or because of the primary disease.

It may be very difficult to assess hydration, but it is usually possible if the history and clinical and laboratory findings are all taken into account. Occasionally measurement of the urinary sodium concentration may help (p. 40).

Monitoring total fluid balance

By far the most important measurements in assessing changes in day-to-day fluid balance are those of intake and output. 'Insensible loss' is usually assumed to be about 900 ml per day, but this must be balanced against 'insensible' production of about 500 ml a day by metabolic processes. The *net* 'insensible loss' is therefore the difference between these two – about 400 ml a day. A normally hydrated patient, unable to control his own balance, should be given this basic volume of fluid daily with, in addition, the volume of urine, vomit, etc. measured during the preceding 24 hours. If he is thought to be volume depleted, fluid intake should be adjusted until normal hydration is restored. The intake for any day is thus calculated from the output during the previous day; this is satisfactory if the patient is normally hydrated before day-to-day monitoring is started.

A pyrexial patient may lose a litre or more in sweat, and if he is also overbreathing respiratory water loss can be considerable. In such cases an allowance of 400 ml for insensible loss may be inadequate.

Many very ill patients are incontinent of urine, and it may not be possible to measure even this volume. Although assessment of fluid balance may then be very difficult, every attempt must be made to measure fluid intake and loss accurately. In most circumstances carefully kept fluid charts are of more importance than frequent electrolyte estimations. *Inaccurate charting* is useless, and *may be dangerous*.

Distribution of water and sodium in the body

In mild disturbances of water and electrolyte balance the total amount in the body may be of less importance than their distribution within it.

Distribution of electrolytes

There are two main body fluid compartments in which different electrolytes contribute to osmolality. These are:

- **the intracellular compartment,** in which *potassium* is the predominant cation;
- **the extracellular compartment,** in which *sodium* is the predominant cation.

The extracellular fluid can be subdivided into:

interstitial fluid which is of very low protein concentration;
intravascular fluid (plasma) which contains protein in relatively high concentration.

Distribution of electrolytes between cells and interstitial fluid

The intracellular sodium concentration is less than a tenth of that in the extracellular fluid (ECF), while that of potassium is about thirty times as much. About 95 per cent of the osmotically active sodium is outside cells and about the same proportion of potassium is intracellular: energy is needed to maintain these differential concentrations.

Other ions tend to move across cell membranes at the same time as sodium and potassium. Hydrogen ion has already been mentioned. Magnesium and phosphate are predominantly intracellular, and chloride extracellular, ions. The distribution of these, and of bicarbonate, is often affected by those factors which affect that of sodium and potassium.

Distribution of electrolytes between plasma and interstitial fluid

The vascular endothelium is more freely permeable to small ions than the cell membrane. The protein concentration of plasma is relatively high, but that of interstitial fluid very low. The osmotic effect of the intravascular proteins is balanced by very slightly higher interstitial electrolyte concentrations; this difference is small, and for practical purposes plasma electrolyte levels can be assumed to be representative of those of the extracellular fluid as a whole.

Distribution of water

A little over half the body water is inside cells. About 15 to 20 per cent of the extracellular water is intravascular; the remainder constitutes the extravascular, extracellular interstitial fluid.

The distribution of water across biological membranes depends on the balance between the *in vivo* effective osmotic and the hydrostatic pressure differences between the fluids on each side. Correct interpretation of plasma electrolyte results depends on a clear understanding of these factors.

Osmotic pressure

Net movement of water across a membrane *permeable only to water* depends on the concentration *difference* of particles (ions or molecules) between the two sides. For any weight/volume, the larger the particle (the higher the molecular weight) the fewer there are in unit volume, and the less osmotic effect they will exert. However, if the membrane is freely permeable to smaller particles as well as to water, these smaller particles exert no osmotic effect and the larger ones become more important in affecting water movement. To explain water distribution in the body it is essential to understand the importance of these three factors:

- number of particles per unit volume;
- concentration gradient across the membrane;
- particle size relative to membrane permeability.

Units of measurement of osmotic pressure

Osmolar concentration can be expressed in two ways:

- in *osmolarity* expressed as mmol per *litre* of *solution*;
- in *osmolality* expressed as mmol per *kg* of *solvent*.

(The term milliosmole (mosmol) has been used to express osmolarity and osmolality: it has been recommended that this terminology be abandoned.)

If solute is dissolved in pure water at concentrations such as those in body fluids, osmolarity and osmolality will hardly differ. Plasma is, however, a complex solution, and contains large molecules such as proteins; the total volume of solution (water + protein) is greater than that of solvent (water only). The small molecules are dissolved only in water, and at a protein concentration of 70 g/litre the volume of water is about 6 per cent less than that of total solution (that is, the molarity will be about 6 per cent less than the molality). Many methods of measuring individual ions assess them in molarity (mmol/litre).

Measured plasma osmolality. Osmometers measure freezing point depression or vapour pressure, which depend on the *total* osmolality of the solution – the osmotic effect which would be exerted by the sum of all the dissolved molecules and ions across a membrane which, *unlike biological ones, is permeable only to water.* Table 2.1 shows that *sodium and its associated anions* (mainly chloride) contribute 90 per cent or more to this measured plasma osmolality, the effect of protein being negligible. The only major difference in composition between plasma and interstitial fluid is in protein content: thus *total* plasma osmolality is almost identical with that of the interstitial fluid surrounding cells.

Calculated osmolarity. It is the osmol*al*, not the osmol*ar* concentration which exerts an effect across cell membranes and which is controlled by homeostatic mechanisms. However, as we shall see, calculated plasma osmolarity is often at least as informative as measured plasma osmolality.

Although the measured osmolality of plasma should be higher than its osmolarity calculated by adding the molar concentrations of all the ions (because of its protein content), usually there is little difference between the two figures. This is because incomplete ionization of, for example NaCl to Na^+ and Cl^- reduces the osmotic effect by almost the same amount as the volume occupied by protein raises it. Calculated plasma osmolarity is then a valid approximation to true osmolality. However, if there is such gross *hyperlipidaemia* or *hyperproteinaemia* that protein or lipid contribute much more than 6 per cent to measured plasma volume, calculated osmolarity may be significantly lower than the true concentration in plasma water.

Many formulae of varying complexity have been proposed to calculate plasma osmolarity: however, since none of them can predict the osmotic *effect*, it seems

Table 2.1 Approximate contributions to plasma osmolality

	Osmolality (mmol/kg)	Per cent total
Sodium and anions	270	92
Potassium and anions	7	
Calcium (ionized) and anions	3+	
Magnesium and anions	1+	8
Urea	5	
Glucose	5	
Protein	Approximately 1	
Total	Approximately 292	

sensible to use the simplest which gives a close approximation to plasma osmolality, even if a more complicated one is very slightly more accurate. For this reason we find the formula given below satisfactory.

$$\text{Plasma osmolarity} = 2([Na^+] + [K^+]) + [\text{urea}] + [\text{glucose}] \text{ in mmol/litre}$$

The factor of 2 applied to the sodium and potassium concentrations allows for associated anions, and assumes complete ionization.

This calculation is *not* valid if:

- an unmeasured osmotically active solute, such as mannitol or alcohol is circulating. A significant difference between measured and calculated osmotic pressures, in the absence of hyperproteinaemia or hyperlipidaemia, may suggest alcohol or other poisoning. For example, a plasma alcohol concentration of 100 mg/dl contributes about 20 mmol/kg to osmolality;
- there is gross hyperproteinaemia or hyperlipidaemia.

In such cases the plasma sodium concentration may be misleading and osmolality should be measured.

Calculation of urinary osmolarity is not feasible because of the considerable variation in concentrations of different, sometimes unmeasured, solutes; its osmotic pressure can only be determined by measurement of its osmolality.

Plasma osmotic pressure: distribution of water across cell membranes

The hydrostatic pressure difference across cell membranes is negligible, and cell hydration depends on the osmotic difference between intra- and extracellular fluids. The cell membrane is freely permeable to water. However, unlike the theoretical semipermeable membrane, it is permeable to some solutes; different solutes diffuse, or are actively transported, across it at different rates, but always more slowly than water. In a stable state the total intracellular osmolality, due mostly to potassium and associated anions, equals that of the interstitial fluid, due mostly to sodium and associated anions, and there is no *net* movement of water into or out of cells. In some pathological states rapid changes of extracellular solute concentration affect cell hydration: slower changes may allow time for redistribution of solute and have little or no effect.

Because *sodium* and its associated anions account for at least 90 per cent of plasma osmolality in the normal subject, rapid changes of sodium concentration affect cell hydration; if there is no significant change in other solute, a rise causes cellular dehydration and a fall cellular overhydration.

Normal levels of *urea* and *glucose* contribute very little to measured plasma osmolality. However, concentrations 15-fold or more above normal can occur in severe uraemia and hyperglycaemia, and these solutes then make a significant contribution. Urea does diffuse into cells, but very much more slowly than water; although acute uraemia alters cell hydration, in chronic uraemia the osmotic *effect* of urea is reduced as the concentrations gradually equalize on the two sides of the membrane. Glucose is actively transported into many cells, but once there is rapidly metabolized: the intracellular concentration remains low and severe hyperglycaemia, whether acute or chronic, has a marked influence on cell hydration.

Although uraemia and hyperglycaemia can cause cellular dehydration, the contribution of normal urea and glucose concentrations to plasma osmolality is so small that reduced levels of these solutes, unlike those of sodium, do not cause cellular overhydration.

Rises in concentrations by a factor of more than about three of solutes such as calcium, potassium or magnesium, are incompatible with life for other than osmotic reasons: they therefore do not cause significant osmolality changes.

Substances not transported into cells, such as mannitol, can be infused to reduce cerebral oedema, and, like hypertonic glucose and urea, can also be used as osmotic diuretics (p. 8). Their concentrations, like that of alcohol, are included in osmolality measurements.

To interpret the potential consequences of changes in plasma osmolality we need to know the osmotic *effect* due to the concentration *difference* between intra- and extracellular fluid. Even if we could measure plasma osmolality with 100 per cent accuracy, we can only roughly gauge that inside cells by knowing the length of history, and the likely permeability of the cell membrane to the solute contributing most to the change in plasma osmolality. In almost all circumstances plasma osmolarity, calculated from sodium, potassium, urea and glucose concentrations, and interpreted with these other factors in mind, is at least as clinically valuable as measurement of plasma osmolality: it has the advantage that the solute responsible, and therefore its likely osmotic *effect*, is often identified.

Plasma colloid osmotic pressure: distribution of water across capillary walls

The distribution of water across capillary walls is little affected by electrolyte concentration, but is affected by the osmotic effect of *plasma proteins (the colloid osmotic, or oncotic, pressure)*.

Maintenance of blood pressure depends on the retention of intravascular fluid at a higher hydrostatic pressure than that of the interstitial fluid. Hydrostatic pressure therefore tends to force fluid into the extravascular space. In the absence of any opposing effective colloid osmotic pressure across capillary walls fluid would be lost rapidly from the vascular compartment.

Unlike the cell membrane, the capillary wall is permeable to small molecules; sodium therefore exerts almost no osmotic effect at this site. The smallest molecule present intravascularly at significant concentration which, because it cannot pass freely through the capillary wall, is in very low extravascular concentration is albumin (molecular weight 65 000). The plasma albumin concentration is therefore the most important factor opposing the net outward hydrostatic pressure: the higher molecular weight proteins, although present at much the same weight/volume as albumin, contribute much less to this effect because of their larger size. This effective colloid osmotic pressure across vascular walls is sometimes called the *oncotic* pressure.

Because proteins contribute negligibly to total plasma osmotic pressure (Table 2.1), measurement of plasma osmolality cannot be used to assess osmotic effects across capillary walls.

Interrelationship between sodium and water homeostasis

Let us now look more closely at the simplified statements made on p. 28 about the control of ADH and aldosterone secretion. We said that sodium, by its osmotic effect, controls ADH secretion, and therefore water balance, and that plasma volume, by its effect on renal blood flow, controls aldosterone secretion, and therefore sodium balance. The homeostasis of sodium and water is interdependent.

This simplified scheme is shown in Fig. 2.1. The diagonal line dividing the rectangle into triangles A and B indicates that changes in water balance (but not that of isosmotic fluid) also directly alter sodium concentration. The thirst mechanism is not shown here, but it should be remembered that an increase of extracellular osmolality, whether due to water depletion or sodium excess, not only reduces water loss by stimulating ADH release, but increases thirst and therefore intake, and that both these actions dilute the extracellular osmolality. Osmotic balance, and therefore cellular hydration, is rapidly corrected; if the primary abnormality is in sodium rather than water balance, cellular hydration is sometimes protected at the expense of aggravating abnormalities in total body volume.

We shall refer again to this simplified scheme when discussing pathological states. It does not allow for the effect of fluid shifts across cell walls on extracellular volume and concentrations, for changes in osmotically active solutes other than sodium, nor for redistribution of solute in chronic disturbances of extracellular osmolality.

Assessment of sodium status

As we have seen, the plasma sodium *concentration* is important because of its osmotic effect. Plasma sodium levels should be monitored while volume is being corrected to ensure that the distribution of fluid between cells and the extracellular compartment is optimal. The presence of other osmotically active solute should be taken into account. Measurement of total body sodium is not useful.

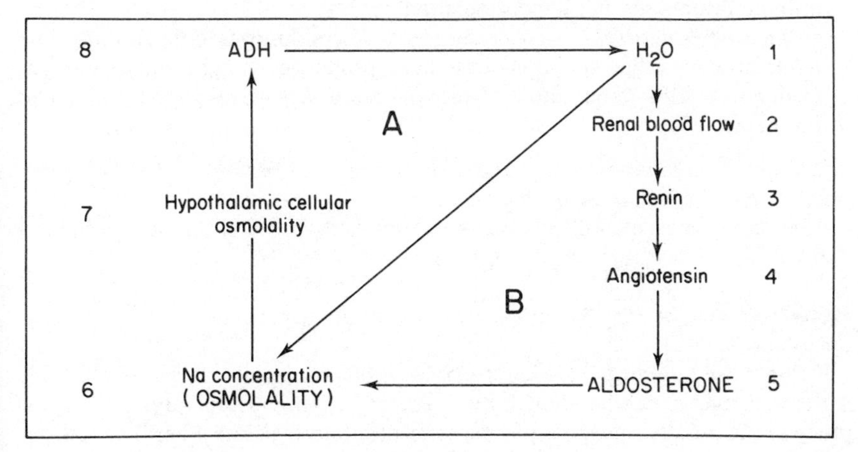

Fig. 2.1 Simplified cycle of sodium and water homeostasis.

Clinical features of water and sodium disturbances

The immediate clinical consequences of primary sodium disturbances depend on changes of extracellular osmolality, and hence of cellular hydration, and of primary water disturbances on changes in extracellular volume.

Clinical features of disturbances of sodium concentration

Measuring the plasma sodium concentration is a substitute for measuring plasma osmolality. Sodium levels *per se* are not important, but the osmotic gradient across cell membranes is. It is important to understand that the one does not always reflect the other.

If the concentration of sodium alters rapidly, and that of other extracellular solutes remain the same, most of the clinical features are due to an osmotic difference across cell membranes. Gradual changes, which allow time for redistribution of diffusible solute such as urea, and so for equalization of osmolality without major shifts of water, may produce little effect.

Hyponatraemia *may* reflect extracellular hypoosmolality, and may therefore cause cellular overhydration. However:

- *artefactual* hyponatraemia may be due to taking blood from the limb into which fluid of low sodium concentration is flowing, and may not reflect the sodium concentration in the general circulation (p. 448);
- *appropriate* hyponatraemia may be due to, for example, acute uraemia, hyperglycaemia, infusion of amino acids or mannitol, or high alcohol or other solute concentrations, all of which increase extracellular osmolality: the consequent homeostatic dilution of *total* extracellular solute concentration towards normal causes hyponatraemia which reflects partial or complete compensation of hyperosmolality, not hypoosmolality;
- *'pseudohyponatraemia'* may be due to gross hyperlipidaemia (such as that due to intravenous infusion of lipid) or to hyperproteinaemia. The sodium concentration in plasma water, and therefore the osmolality at cell walls may then be normal (p. 32).

Infusion of sodium-containing fluids with the aim of 'correcting' artefactual, appropriate or pseudohyponatraemia is dangerous.

When hyponatraemia reflects true hypoosmolality cerebral cellular overhydration may cause *headache, confusion, fits and even death.*

Hypernatraemia may be *artefactual* if sodium heparin is used as an anticoagulant (p. 449).

If this cause has been excluded *hypernatraemia always reflects hyperosmolality,* with the danger of cellular dehydration. Severe hypernatraemia cannot be appropriate to the osmolal state: the plasma concentrations of solute other than sodium are very low (Table 2.1, p. 32), and, for example, although hyperglycaemia can increase plasma osmolality by as much as 55 in 280 mmol/kg (about 20 per cent), complete absence of glucose would only reduce it by about 5 mmol/kg (about 2 per cent).

The clinical effects of cerebral cellular dehydration are *thirst*, *mental confusion* and later *coma*.

Hyponatraemia is a much commoner finding than hypernatraemia and is often appropriate. If artefactual causes are excluded, the incidence of true hypoosmolality is probably not much higher than that of hypernatraemia, which always reflects hyperosmolality.

Disturbances of sodium and water metabolism

Disturbances of sodium and water balance are most commonly due to excessive losses from the body, usually of gastrointestinal fluid. We shall see that these losses may be inappropriately increased if the volume or composition of infused fluid is incorrect: repletion will then be relatively ineffective, and the increased gastro-intestinal excretion may aggravate the clinical abnormalities.

More rarely the primary defects are excessive or deficient secretion of aldosterone or ADH.

Water and sodium deficiency

Apart from loss of solute-free water in expired air, water and sodium are always lost from the body together. An imbalance between the degree of their deficiencies is relatively common, and may be due to the composition of the fluid lost or to that of the fluid given to replace it.

The initial effects depend on the composition of the fluid lost *compared with that of plasma*.

- *Isosmolar volume depletion* results if the *sodium concentration is similar; changes in plasma sodium concentration are then rare.*
- *Predominant sodium depletion is usually the result of inappropriate treatment*, since no secretion has a significantly higher sodium concentration than that of plasma. *Hyponatraemia* results.
- *Predominant water depletion* results if the *sodium concentration is much lower. Hypernatraemia* indicates loss of relatively more water than sodium, even if there is little evidence of volume depletion.

Subsequent effects depend on the efficiency of homeostatic mechanisms and on the availability and composition of fluid replacement.

We believe that the use of the term 'dehydration' should be abandoned, and the type of fluid lost specified. It is often used to describe all the conditions listed above, although the clinical findings are very different. The consequent confusion may, by leading to inappropriate treatment, be dangerous.

Isosmolar volume depletion

Causes of isosmolar fluid loss

The sodium concentrations of all small intestinal secretions, and of the urine passed when tubular function is grossly impaired, are between 120 and 140 mmol/litre. Clinical conditions causing approximately isosmolar loss are therefore:

- small intestinal fistulae, including new ileostomies;
- small intestinal obstruction and paralytic ileus, in which the fluid accumulating in the gut lumen has, like urine in the bladder, been lost from the ECF;
- severe renal tubular damage with minimal glomerular dysfunction (for example, the recovery phase of acute oliguric renal failure, or polyuric chronic renal failure).

Intestinal losses are much more likely to produce severe volume depletion than is renal tubular disease.

Results of isosmolar fluid loss

Hypovolaemia reduces renal blood flow and causes renal circulatory insufficiency with *oliguria, uraemia* and the other changes described on p. 10. Sodium and water are lost in almost equivalent amounts, and the plasma sodium concentration is usually normal: for this reason the patient may not complain of thirst despite volume depletion.

Haemoconcentration confirms considerable loss of fluid other than blood, although its absence does not exclude such loss. Apart from the possibility of pre-existing abnormalities of haemoglobin and plasma protein concentrations, in shock increased capillary permeability may allow albumin to diffuse more freely than usual into the interstitial fluid and cause hypoalbuminaemia which is not due to haemodilution.

A fall in blood pressure on standing (*postural hypotension*) is a relatively early sign of volume depletion. Severe hypovolaemia causes *hypotension* even in the supine position.

Changes produced by homeostatic mechanisms

The homeostatic response depends on renal tubular function, and does not occur if the depletion is due to tubular damage.

The student should refer to Fig. 2.1a. The diagonal line has been omitted, because loss of *isosmotic* fluid has no direct effect on plasma sodium concentration.

The reduced intravascular volume impairs renal blood flow and stimulates renin and therefore aldosterone secretion (Steps 1 to 5). There is selective sodium reabsorption (Steps 5 to 6) and a *low urinary sodium concentration*.

The tendency of the retained sodium to increase plasma osmolality stimulates ADH secretion (Steps 6 to 8) and water is reabsorbed in equivalent amounts to sodium: this tends to correct volume and keep the plasma sodium concentration normal. Occasionally severe volume depletion stimulates enough ADH secretion,

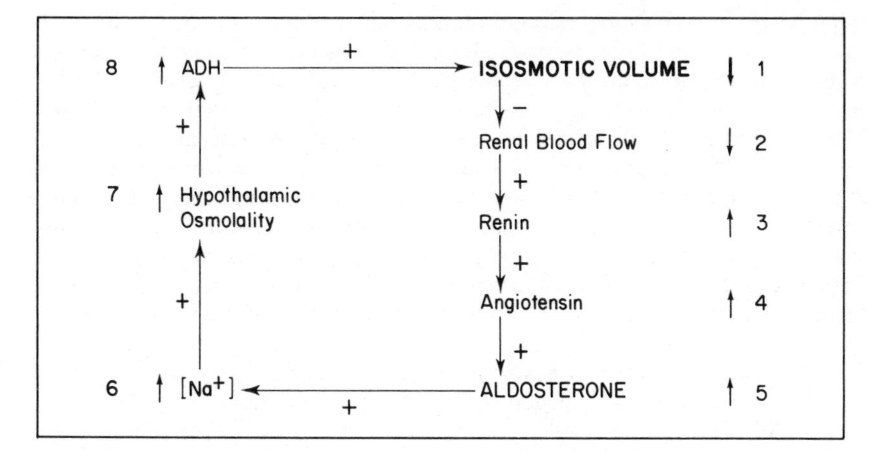

Fig. 2.1a Homeostatic correction of isosmotic volume depletion.
Lettering in **bold** indicates the primary abnormality. Arrows indicate the direction of change.

and therefore water retention, to cause mild hyponatraemia: much of this water will move from the depleted and now slightly hypoosmolar ECF into the relatively well-hydrated cells.

Even complete water and sodium retention could not correct extrarenal losses which exceed those of a normal urine output. Sodium and water must be replaced in adequate amounts to provide 'substrate' for the kidneys. However, it is important to understand that excessive replacement, especially with fluid of inappropriate concentration, has undesirable side effects. Replacement of isosmolar loss with fluid of low sodium concentration is the commonest cause of predominant sodium depletion, and is discussed below. Here we will describe the consequences of volume replacement with protein-free fluid.

Effects of intravenous volume replacement

Those unable to absorb adequate amounts of oral fluid because of gastrointestinal loss usually need intravenous replacement. The following discussion applies to such cases, and assumes normal renal function.

Correction of presenting hypovolaemia can be monitored by clinical observation and measurement of urine output. There is the danger that overcorrection will increase intestinal loss of fluid, which may accumulate in the already distended bowel of intestinal obstruction or paralytic ileus.

To explain this fluid loss due to overinfusion we should remember that passive filtration depends on the capillary hydrostatic pressure and is opposed by the capillary colloid osmotic pressure. Infusion of protein-free fluid increases the hydrostatic gradient, and reduces the opposing osmotic gradient by diluting plasma proteins. The necessary increase in glomerular filtration during correction of hypovolaemia is inevitably accompanied by a less desirable intestinal loss and this will be aggravated by overcorrection: 'wastage' in the urine occurs for the same reasons but, although uncomfortable for the patient, is of less clinical importance. The unwanted effect of reduced effective osmolality would be minimized if the infused fluid included molecules of about the size of albumin; this would also prevent inappropriate loss from the vascular compartment into the interstitial fluid. Unfortunately infused albumin has a very short intravascular half-life: most other colloid preparations have some disadvantages.

It is difficult to assess clinically when volume repletion is complete, but not excessive. Measurement of urinary sodium concentration may help. If, during the period of known hypovolaemia, it was less than about 30 mmol/litre, tubular function is adequate and further measurements can be used to assess the presence of circulating aldosterone. A very low urinary sodium concentration suggests that renal blood flow is still low enough to stimulate maximal renin secretion, and infusion should be increased: a urinary sodium concentration much higher than 30 mmol/litre *in a patient with adequate tubular function* suggests overcorrection and the need to slow the infusion. Urinary sodium may sometimes need to be monitored in this way until intestinal obstruction is relieved. In cases in which all losses can be measured, further maintenance of normal balance can be based on *accurate* fluid balance charts.

Predominant sodium depletion

Effects of the composition of the infused fluid

No body secretion has a sodium concentration significantly higher than plasma and predominant sodium depletion is almost always due to intravenous infusion. The composition of the fluid is even more important than the volume. The effect of the absence of colloid in electrolyte solutions has already been discussed.

Patients with isosmolar depletion, or postoperatively, are often infused with fluid such as 'dextrose saline' containing about 30 mmol/litre of sodium. Although the glucose in the infused fluid renders it isosmolar despite the low sodium concentration, the glucose is metabolized after infusion, and both plasma sodium and osmolality are diluted by the remaining hypoosmolar fluid. Homeostatic mechanisms involving ADH tend to correct this hypoosmolality, but may be prevented from doing so if the infused volume is high.

The reader should refer to Fig. 2.1b. In this case the hypoosmolar fluid directly dilutes the plasma sodium concentration, as shown by the – sign on the diagonal line.

The mechanisms which tend to correct hypoosmolality (and therefore to safeguard cell hydration) involve inhibition of ADH secretion in response to hypoosmolality (Steps 1 to 6, and 6 to 8). *The excess water is lost in the urine* until restoration of normal plasma osmolality again stimulates normal ADH secretion.

This would tend to restore osmolality at the expense of volume. The loss of fluid *should* stimulate aldosterone secretion (Steps 1 to 5); sodium would then be retained and the consequent restoration of osmolality would hasten the return to normal ADH secretion: if an adequate volume of isosmolar fluid were provided, the balance of both sodium and water would be corrected. However, *if the volume is maintained by replacing the urinary volume with effectively hypotonic fluid*, hypoosmolality (as shown by hyponatraemia) persists, and sodium depletion is

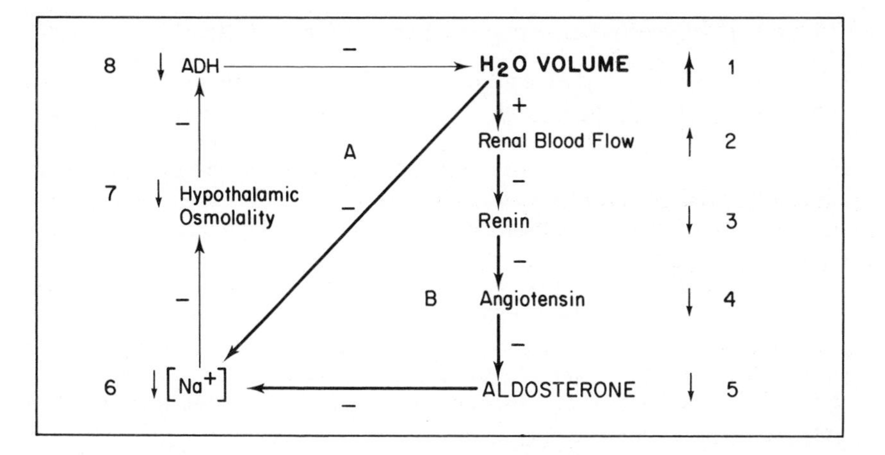

Fig. 2.1b Infusion of hypotonic fluid as a cause of predominant sodium depletion. Lettering in **bold** indicates the primary abnormality. Arrows indicate the direction of change.

aggravated. As shown in Fig. 2.1b, restoration of the plasma volume inhibits aldosterone secretion and sodium is lost in the urine despite hypoosmolality. The net effect is restoration of circulating volume by infusion, but sodium depletion and cellular overhydration due to hypoosmolality.

The clinical signs are those of hypoosmolality (p. 36).

The laboratory findings are:

- hyponatraemia;
- passage of a large volume of dilute urine due to inhibition of ADH secretion;

and if fluid intake is excessive:

- haemodilution;
- a low plasma urea concentration due to the high GFR (excessive intravenous infusion is one of the commonest causes of this finding);
- a high urinary sodium concentration, due to inhibition of aldosterone secretion.

It is, therefore, not surprising that postoperative hyponatraemia is so common. In subjects with normal renal function the mechanisms described above rapidly correct both sodium and water balance when the infusion is stopped, and no serious harm is done. However, by alternating 'dextrose saline' and isotonic saline, we find that most patients can maintain relatively normal plasma sodium and urea levels, while passing an adequate volume of urine: the danger of overloading the circulation is minimal if renal function is normal. Infusion should be stopped as soon as the patient can take oral fluids.

Infusion of 'dextrose saline' into seriously ill patients, with impairment of homeostatic mechanisms, is even more likely to cause hyponatraemia.

Failure of homeostatic mechanisms for sodium

Aldosterone deficiency is a much rarer cause of sodium depletion than that described above. It initiates homeostatic reactions which tend to maintain osmolality at the expense of volume.

The least rare of the conditions causing aldosterone deficiency is *Addison's disease*, which is discussed more fully on p. 126.

Loss of sodium in excess of water reduces plasma osmolality and cuts off ADH secretion (Triangle A, Fig. 2.1c). Water is lost until osmolality is corrected as shown by the + sign on the diagonal line, and the patient becomes hypovolaemic.

Aldosterone secretion cannot increase significantly in response to the consequent rise in renin-angiotensin secretion (Triangle B). Osmolality is therefore maintained by water loss, and the plasma sodium concentration does not at first fall below the reference range: later hypovolaemia stimulates ADH secretion, and hyponatraemia may occur.

The clinical signs are:

- those of volume depletion (p. 38);
- *later* those of hypoosmolality (p. 36).

The laboratory findings are:

- haemoconcentration (due to fluid depletion);
- renal circulatory insufficiency with mild uraemia, due to volume depletion;
- *later* hyponatraemia;
- an inappropriately high urinary sodium concentration in the face of volume depletion, due to aldosterone deficiency.

Predominant water depletion

Predominant water depletion is due to loss of water in excess of that of sodium. It is

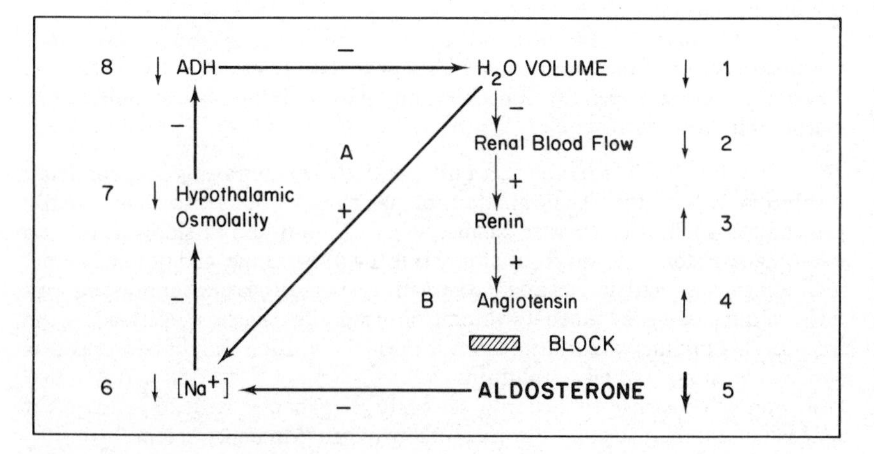

Fig. 2.1c Effect of aldosterone deficiency.
Lettering in **bold** indicates the primary abnormality. Arrows indicate the direction of change.

usually due to loss of fluid with a lower sodium concentration than that of plasma, or to deficient water intake. The sodium concentrations in *sweat, gastric juice* and *diarrhoea stools* are about half those of plasma; in *diabetes insipidus due to ADH deficiency*, during an *osmotic diuresis*, or in *nephrogenic diabetes insipidus*, in which the renal tubules are unresponsive to ADH, urine of low sodium concentration is passed. *Because hyperosmolality due to predominant water loss causes thirst, laboratory abnormalities are only found if water is unavailable or cannot be taken in adequate amounts.*

The causes of predominant water loss are:

- *water deficiency in the presence of normal homeostatic mechanisms.*
  Excessive water loss:
    loss of excessive amounts of sweat;
    loss of gastric juice;
    loss of fluid stools of low sodium concentration, usually in infantile gastroenteritis;
    excessive respiratory loss;
    loss of fluid from extensive burns.
  Deficiency of water intake:
    Inadequate water supply, or mechanical obstruction to its intake.
- *failure of homeostatic mechanisms for water retention.*
  Inadequate response to thirst, for example in comatose patients and in infants, or because the thirst centre is damaged.
  Significant water depletion is very rare if the response to thirst is normal and water is available;
  Diabetes insipidus due to deficiency of ADH;
  Overriding of ADH action by osmotic diuresis;
  Nephrogenic diabetes insipidus due to failure of tubular cells to respond normally to ADH.

In most clinical states associated with predominant water depletion more than one of these factors is contributory.

Predominant water depletion with normal homeostatic mechanisms. The reader should refer to Fig. 2.1d. The direct effect of water depletion on plasma sodium concentration is shown by the + sign on the diagonal line.

Triangle B. Immediate effects.

- *loss of water in excess of sodium* increases the plasma sodium concentration, and hence osmolality (shown by the + sign on the diagonal line);
- *reduction of circulating volume reduces renal blood flow* and stimulates aldosterone secretion (Steps 1 to 5). Sodium is retained and hypernatraemia is aggravated (Steps 5 to 6).

Triangle A. Compensatory effects.
Increased plasma osmolality stimulates:

- *thirst, increasing water intake* if water is available and if the patient can respond to it (not shown on the diagram);

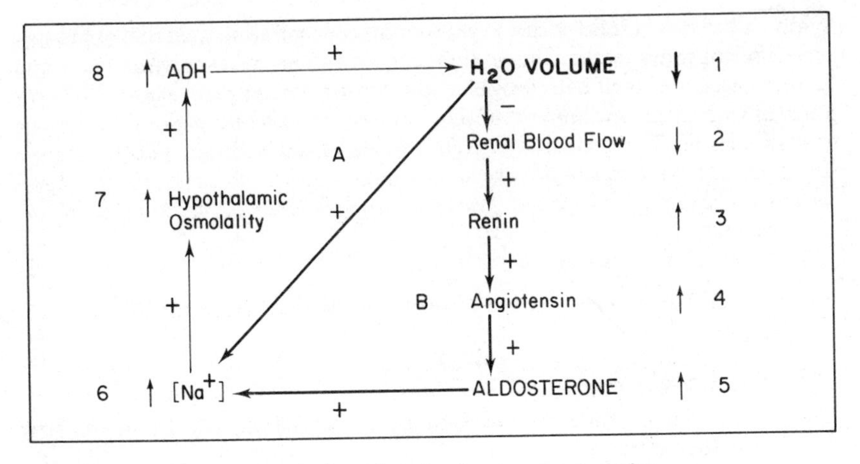

Fig. 2.1d Homeostatic correction of predominant water depletion.
Lettering in **bold** indicates the primary abnormality. Arrows indicate the direction of change.

- *ADH secretion* (Steps 6 to 8). Urinary volume falls (Steps 8 to 1) and water loss through the kidney is minimized.

If an adequate amount of water is available depletion is rapidly corrected. If there is an inadequate intake to replace the loss the mechanisms depicted in Triangle A are impaired and homeostasis is ineffective. *In water depletion hypernatraemia may be found before clinical signs of volume depletion are detectable.*

The clinical signs are those of:

- hyperosmolality (p. 37);
- oliguria due to ADH secretion;
- *later*, there may be signs of volume depletion (p. 38).

The laboratory findings are:

- hypernatraemia;
- haemoconcentration due to fluid depletion;
- mild uraemia, due to volume depletion and hence low GFR;
- a high urinary osmolality and urea concentration due to the action of ADH;
- a low urinary sodium concentration in response to high aldosterone levels stimulated by the low renal blood flow.

Failure of homeostatic mechanisms for water retention. These syndromes are relatively rare. The *thirst centre* may occasionally be so damaged, usually by invasion by tumour, as to cause water depletion. Abnormalities of secretion of, or response to, ADH are slightly less uncommon.

Diabetes insipidus is the syndrome associated with impairment of ADH secretion. It may be due to pituitary or hypothalamic damage caused by head injury, or by invasion of the region by tumour. It may be idiopathic in origin.

Hereditary nephrogenic diabetes insipidus is a rare inborn error of renal tubular function in which ADH levels are high but the tubules cannot respond to it: the

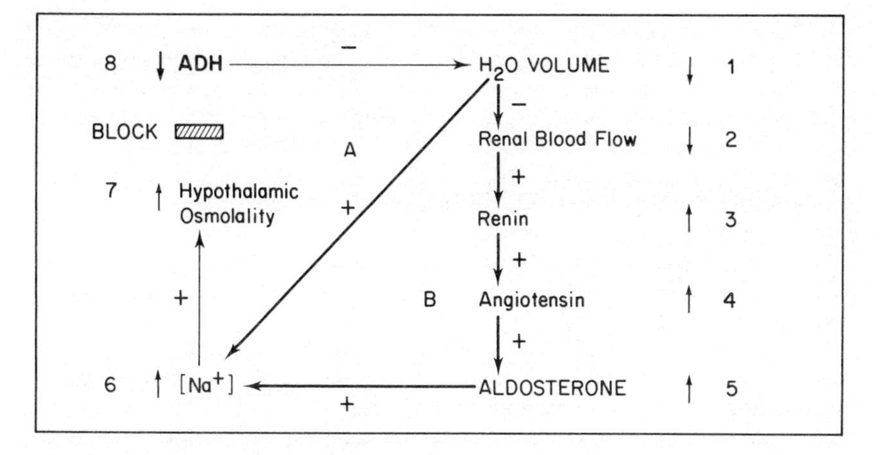

Fig. 2.1e Results of ADH deficiency.
Lettering in **bold** indicates the primary abnormality. Arrows indicate the direction of change. No compensation possible.

newborn infant passes a large volume of urine and rapidly becomes hyperosmolar and volume depleted. Urinary loss is difficult to assess at this age, and the cries of thirst may be misinterpreted.

Some drugs, such as lithium, may cause the clinical picture of nephrogenic diabetes insipidus.

The reader should refer to Fig. 2.1e and to the immediately preceding section.

The first stage is identical with that described in the last section. The mechanisms previously depicted in Triangle A are not functioning, compensation cannot occur, and it may not be possible for intake to replace loss.

Clinical signs and findings which differ from the description in the previous section are due to ADH deficiency. These are:

- *polyuria*, not oliguria;
- *a dilute*, not concentrated, urine.

The diagnostic procedure is described on p. 71.

The unconscious patient. The syndrome of water depletion with hypernatraemia is found most commonly in unconscious or confused patients or in infants with gastroenteritis or pneumonia. In such subjects there is usually more than one cause of water depletion.

1. *Pyrexia* increases the loss of hypotonic sweat.
2. *Overbreathing* due to pneumonia, acidosis or brain stem damage increases water loss in expired air.
3. *Osmotic diuresis* overrides the effect of ADH and causes further water loss. This diuresis may be due to hypertonic intravenous infusions, for example to provide nutrient (glucose or amino acids), to tissue damage and hence increased production of urea from protein, or to glycosuria in diabetic coma.
4. *Diabetes insipidus* may be caused by head injury.
5. *The subject cannot respond to hyperosmolality by drinking.*

In the presence of factors 3 and 4 a high urinary volume contributes to water depletion and does *not* indicate 'good' hydration: the extent of the loss may not be appreciated if the subject is incontinent. Homeostatic mechanisms may not be able to respond to hyperosmolality (3, 4 and 5). Hypernatraemia is found before the more usual signs of volume depletion are detectable: it is dangerous because of the resulting cellular dehydration. Monitoring cumulative fluid balance of patients at risk of developing hypernatraemia helps to detect water depletion early enough to allow preventive measures to be taken (p. 69).

In unconscious or confused patients with extensive *burns* the water in exuded ECF evaporates, and some of the electrolyte is reabsorbed into the circulation, aggravating hyperosmolality.

Hyperosmolar saline should never be used as an emetic in cases of poisoning. Movement of water into the gut along the osmotic gradient, and absorption of some of the sodium, can cause marked hypernatraemia: the vomiting or unconscious patient cannot respond to hyperosmolality by drinking. Death has occurred as a consequence of this practice.

Water and sodium excess

An excess of water and sodium is usually rapidly corrected. Syndromes of excess are usually associated with impaired homeostatic mechanisms.

Volume excess with no change in osmotic difference across cell walls causes *hypertension* and *cardiac failure* due to overloading of the heart. If the plasma albumin concentration is low the effect of reduced colloid osmotic pressure is added to that of increased hydrostatic pressure, and water passes into the interstitial fluid causing *oedema*. Laboratory findings are characteristic of *haemodilution* and, unless there is glomerular dysfunction, plasma urea levels tend to be low.

Secondary aldosteronism: oedema. Any of the conditions in which aldosterone secretion is stimulated by a low renal blood flow could, strictly, be called secondary aldosteronism. The use of the term is usually confined to the description of those conditions in which, because the initial abnormality is not corrected, long-standing hyperaldosteronism itself produces effects.

Aldosterone is secreted following stimulation of the renin-angiotensin system by a low renal blood flow. This may be due to local abnormalities in renal vessels or to a reduced circulating volume.

Secondary aldosteronism may be the result of the following conditions.

- Redistribution of extracellular fluid, leading to a reduced plasma volume despite a normal or high total extracellular fluid volume. Such conditions are due to a reduced plasma colloid osmotic pressure, and are therefore associated with low plasma albumin concentrations. *Oedema is present.* Persistent hypoalbuminaemia may be due to:

  chronic liver disease;
  nephrotic syndrome;
  protein malnutrition.

- Damage to renal vessels, reducing renal blood flow. These conditions are *rarely associated with oedema*:
    - essential hypertension;
    - malignant hypertension;
    - renal hypertension, such as that due to renal artery stenosis.
- Cardiac failure, in which two factors may reduce renal blood flow:
    - a low cardiac output results in poor renal perfusion;
    - high venous hydrostatic pressure in congestive failure may cause redistribution, with *oedema*.
    - Aldosterone metabolism is also impaired in cardiac failure.

The mechanisms in Fig. 2.1f are brought into play. The diagonal line has been omitted because increased *isotonic* fluid should have no direct effect on the plasma sodium concentration.

- Reduced renal blood flow stimulates aldosterone secretion (Steps 1 to 5);
- Sodium retention stimulates ADH secretion (Steps 5 to 8) and therefore *water retention* (Steps 8 to 1).

These processes tend to restore the intravascular volume, but if there is hypoalbuminaemia or congestive cardiac failure more fluid passes into the interstitial fluid when the retained water redilutes albumin and raises hydrostatic pressure; the cycle restarts. The circulating volume can only be maintained by further fluid retention, and a vicious cycle leads to its accumulation in the interstitial compartment with oedema. If the low renal blood flow is due to a narrowing of blood vessels hypertension rather than oedema results.

This cycle should stimulate parallel water and sodium retention, with normonatraemia. Mild hyponatraemia is, however, common in oedematous secondary aldosteronism. This may be due to stimulation of ADH secretion which is inappropriate to the osmolality when fluid leaves the intravascular for the inter-

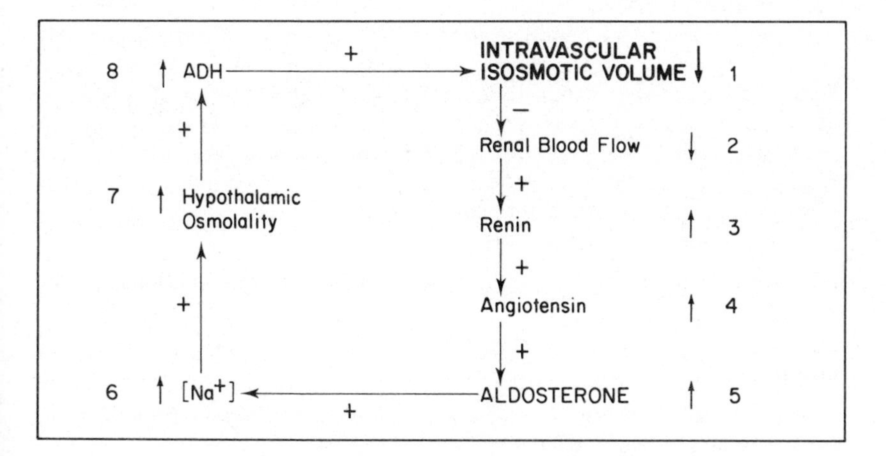

Fig. 2.1f Effects of secondary aldosteronism.
Lettering in **bold** indicates the primary abnormality. Arrows indicate the direction of change.

stitial compartment. Whatever the cause of the hyponatraemia, it is important to realize that it co-exists with an *increase* in total body sodium (Table 2.2, p. 52), and that administration of sodium with the aim of correcting the hyponatraemia may be dangerous.

Hypokalaemia is less common in secondary than in primary hyperaldosteronism, perhaps because the low GFR reduces the amount of sodium reaching the distal tubule (p. 10). However, potassium depletion is more readily precipitated by loop diuretics than in subjects without hyperaldosteronism.

The clinical features found in these cases are those of the primary condition.

The laboratory findings are:

- normonatraemia or mild hyponatraemia;
- a low urinary sodium concentration;
- those due to the primary abnormality, such as hypoalbuminaemia.

Predominant excess of water

Predominant water overloading may occur in circumstances in which normal homeostasis has failed:

- in *glomerular dysfunction*, if fluid of low sodium concentration has been replaced in excess of that lost. Fluid balance must be carefully controlled in such cases;
- if there is *'inappropriate' ADH secretion*. Hormone secretion is defined as inappropriate if it continues when it should be cut off by feedback control, in this case by low plasma osmolality. ADH, or a peptide with ADH-like activity, can, like many other peptide hormones, be synthesized by malignant cells in non-endocrine tissues (ectopic secretion; Chapter 22). Inappropriate secretion of ADH, probably from the pituitary or hypothalamus, and therefore not ectopic, is very common in most illnesses. The finding of a urine of higher osmolality than plasma, despite plasma hypoosmolality, is evidence of such inappropriate secretion.
- during intravenous administration of the posterior pituitary hormone *oxytocin* 'Syntocinon' to induce labour. Oxytocin has an antidiuretic effect similar to that of the chemically closely related ADH (p. 108). If 5 per cent glucose, or dextrose saline, is used as a carrier the glucose is metabolized and the net effect is retention of solute-free water. Death from acute hypoosmolality has resulted after prolonged infusion: the oxytocin should be given in the least possible volume of isotonic saline, and a careful watch kept on fluid balance and plasma sodium concentrations.

If we refer to Fig. 2.1g we can see how excessive water intake is normally corrected.

Triangle B. Immediate effects:

- *excess water* tends to *lower plasma sodium concentration* (indicated by the – sign on the diagonal line);
- *increased renal blood flow* cuts off aldosterone secretion, increasing urinary

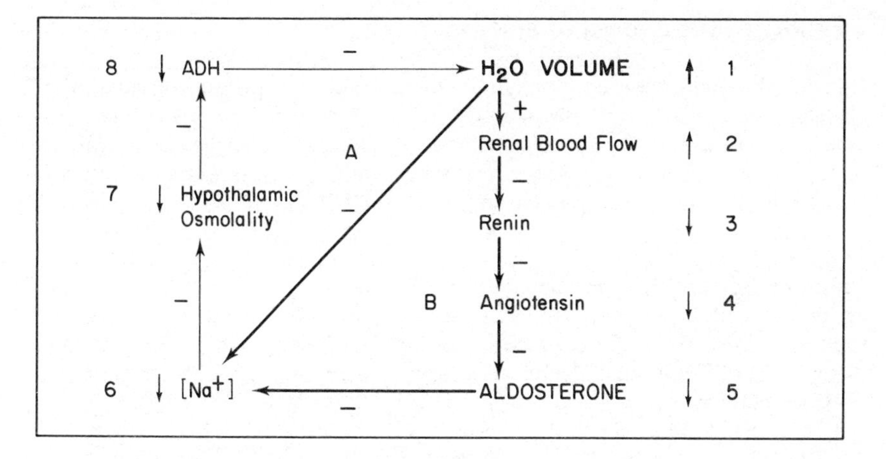

Fig. 2.1g Homeostatic control of water excess.
Lettering in **bold** indicates the primary abnormality. Arrows indicate the direction of change.

sodium loss and therefore further decreasing its plasma concentration (Steps 2 to 6).

Triangle A. Compensatory effects:

- *ADH is cut off* (Steps 6 to 8) and a large volume of *dilute urine* is passed (Steps 8 to 1).

If glomerular function is impaired, if there is inappropriate ADH secretion, or if oxytocin is being given, the mechanisms depicted in Triangle A are impaired and compensation cannot occur.

The clinical consequences are:

- those of volume excess;
- if overhydration is rapid, those of hypoosmolality.

In patients with inappropriate ADH secretion the onset of hypoosmolality is usually gradual; if there has been redistribution of other diffusible solute across cell membranes clinical symptoms may be absent despite very severe hyponatraemia. By contrast, administration of oxytocin can reduce osmolality within a few hours, with the serious danger of cellular overhydration.

The findings are:

- haemodilution;
- hyponatraemia.

If there is glomerular dysfunction, there will be uraemia. In the syndrome of inappropriate ADH secretion the GFR is high and plasma urea concentrations tend to be low.

Predominant excess of sodium

Predominant sodium excess is rare, but is usually due to inappropriate secretion of aldosterone or other corticosteroids in *primary hyperaldosteronism (Conn's syndrome)*, or in *Cushing's syndrome*. Cushing's syndrome is described more fully on p. 123. In these syndromes sodium retention stimulates that of water, minimizing changes in plasma sodium concentration.

Primary hyperaldosteronism (Conn's syndrome). About half the cases are due to a single benign adenoma of the adrenal cortex, about 10 per cent are multiple, and most of the remaining cases are associated with bilateral nodular hyperplasia of the adrenal glands. Aldosterone secretion is not subject to normal feedback control.

The reader should refer to Fig. 2.1h.

- Aldosterone excess causes *urinary sodium retention (Steps 5 to 6).*
- *The rise in plasma sodium* concentration stimulates ADH secretion (Steps 6 to 8) and water retention (Steps 8 to 1).
- *Water retention tends to restore plasma sodium concentrations to normal* (indicated by – sign on diagonal line).
- The mechanisms depicted in Triangle B are ineffective, aldosterone secretion cannot be cut off in response to a fall in angiotensin concentration, and its continued action causes sodium (and water) retention at the expense of potassium loss; *hypokalaemia is common.*

Typical clinical features are:

- those of volume excess. Patients are hypertensive but rarely oedematous;
- those of hypokalaemia (p. 63).

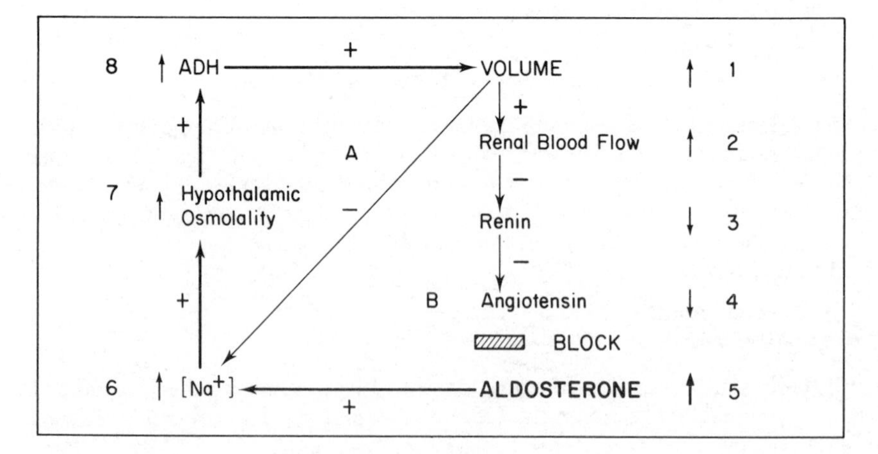

Fig. 2.1h Effects of primary hyperaldosteronism.
Lettering in **bold** indicates the primary abnormality. Arrows indicate the direction of change.

Laboratory findings are:

- hypokalaemia, due to aldosterone excess;
- a high plasma bicarbonate concentration (for explanation see p. 58).
- a plasma sodium concentration in the upper reference range, or just above it;
- a low urinary sodium concentration in the early stages. Later sodium excretion may rise, possibly because of hypokalaemic tubular damage (p. 64).

Hypokalaemic alkalosis is a common finding in potassium depletion from other causes and primary hyperaldosteronism is a rare condition. However, the association of these findings in a patient without an obvious reason for potassium loss, and with hypertension suggests the diagnosis of primary hyperaldosteronism. The investigative procedure described on p. 74 should be followed. If the cause does not soon become obvious, the finding of high plasma aldosterone with low renin concentrations (Fig. 2.1h) confirms the diagnosis; in secondary aldosteronism *both* are high (Fig. 2.1f, p. 47).

Urinary sodium estimation

Estimation of daily loss of sodium in urine and other fluids with a view to quantitative replacement is not only unnecessary, but may be dangerous. *Urinary sodium excretion is not related to body sodium content but mainly to renal blood flow.* Replacement of the small amount of sodium lost in hypovolaemic patients will be inadequate, and of the large amount in volume-expanded patients excessive. Cell hydration must be protected by giving fluid containing the *proportion* of sodium to water which will maintain *plasma concentrations* normal; the total *volume* of this fluid must be determined by the state of volume repletion.

Estimation of the urinary sodium concentration in a random specimen may be of value to monitor volume requirements (p. 40). More rarely it may help to differentiate renal circulatory insufficiency from intrinsic renal damage (p. 14).

The clinical significance of plasma sodium concentrations

Table 2.2 summarizes the situations in which hypernatraemia and hyponatraemia may be found and indicates some aspects of the clinical picture which may help to differentiate the causes. The two most important points to remember are:

Hypernatraemia almost invariably indicates water depletion.

Hyponatraemia is more commonly due to water excess than to sodium depletion: if this is so, the body content of sodium may be normal, but is more often high. In the primary depletion of Addison's disease the patient is likely to have both clinical and laboratory evidence of volume depletion, with mild to moderate uraemia.

Table 2.2 Hypothetical numerical examples to illustrate conditions associated with normal and abnormal plasma sodium concentrations.

	ECF litres	ECNa mmol	Plasma [Na] mmol/litre	Features (see text)	Principles of treatment
Normal subject	*20*	*2800*	*140*		
Normonatraemia due to isosmotic changes in volume					
Na and H_2O loss eg diarrhoea	15↓	2100↓	*140*	Volume depletion	Treat cause Isosmolar saline
Na and water gain eg non-oedematous 2° aldosteronism	30↑	4200↑	*140*	Volume excess	Treat cause Restrict Na and H_2O
Hyponatraemia due to *relative* H_2O excess					
H_2O excess eg Inapprop. ADH	25↑	*2800*	112↓	Volume excess Hypoosmolality	Restrict H_2O
Na depletion eg Post-op. infusion of low [Na] fluid	*20*	2240↓	112↓	Hypoosmolality	Give at least isosmolar saline
Na and H_2O excess eg oedematous 2° aldosteronism	30↑	3360↑	112↓	Volume excess Hypoosmolality	Restrict Na and H_2O
Na and H_2O loss eg late Addison's disease	15↓	1680↓	112↓	Volume depletion Hypoosmolality	Treat cause Give at least isosmolar saline
Hypernatraemia due to *relative* Na excess					
H_2O depletion eg unconscious patient	18↓	*2800*	155↑	Hyperosmolality	Give hypoosmolar saline *slowly*
Na excess (very rare) Excessive intake, usually in infants	*20*	3100↑	155↑	Hyperosmolality	Remove Na by dialysis

Values are related to those of a hypothetical 'normal subject' (line 1). These 'normal' values are in italics throughout. Shifts of fluid across cell membranes, along the osmotic gradient, would slightly reduce changes in plasma sodium concentrations. *Note especially the relatively slight water depletion associated with hypernatraemia*.

Plasma sodium concentrations down to about 125 mmol/litre occur in many ill patients and, unless there is evidence of volume depletion, require no treatment. In states in which solutes other than sodium contribute significantly to plasma osmolality, mild hyponatraemia may be appropriate.

The student should read the indications for electrolyte estimation on p. 74.

Biochemical basis of treatment of sodium and water disturbances

It cannot be stressed too strongly that treatment should not be based on plasma sodium concentrations alone. Hyponatraemia *per se* only rarely needs to be treated. Assessment of the clinical history, findings indicating haemoconcentration or haemodilution, and the plasma urea concentration help to unravel the cause of the low sodium concentration and to determine whether its correction is indicated. Rarely it may be useful to know the urinary sodium concentration or osmolality. Hypernatraemia, on the other hand, should always be treated by *slow* infusion of hypotonic fluid.

Table 2.2 uses hypothetical values to illustrate the various combinations of disturbances of sodium and water metabolism and their treatment. Solutions available for intravenous use are listed on pp. 54 and 55. These rules can only be a guide to treatment in often complicated clinical situations. They may, however, help to avoid some of the more dangerous errors of electrolyte therapy.

If homeostatic mechanisms are impaired, and especially if there is glomerular damage, normal hydration should be maintained according to the principles outlined on p. 30.

Summary

1. Homeostatic mechanisms for sodium and water are interlinked. Potassium and hydrogen ions often take part in exchange mechanisms with sodium.
2. Aldosterone secretion is the most important factor affecting body sodium content.
3. Aldosterone secretion is controlled by the renin-angiotensin mechanism which responds to changes in renal blood flow.
4. Antidiuretic hormone (ADH) secretion is the most important factor affecting body water excretion.
5. ADH secretion is controlled by plasma osmolality which normally depends mainly on the plasma sodium concentration.
6. Distribution of fluid between cells and extracellular fluid depends on the osmotic difference between intra- and extracellular fluids. Changes in this are usually due to changes in extracellular sodium concentrations.
7. Distribution of fluid between the vascular and interstitial compartments depends on the balance between the hydrostatic pressure and the plasma colloid osmotic pressure; the latter depends mainly on the albumin concentration.
8. Clinical effects of disturbances of water and sodium metabolism are due to:

- changes in extracellular osmolality, dependent mainly on sodium concentration. In pathological states urea, glucose and ingested solutes can be important;
- changes in circulating volume.

Further reading

Anonymous. Atrial natriuretic peptide. *Lancet* 1986; **2**: 371–2.

Tan S Y, Mulrow P J. Aldosterone in hypertension and edema. In: Bondy P K and Rosenberg L E eds. *Metabolic Control and Disease*, Philadelphia: W B Saunders, 1980; **8th ed**: 1501–33.

Weitzman R, Vorherr H, Kleeman C R. Water metabolism and the neurohypophyseal hormones. In: Bondy P K and Rosenberg L E eds. *Metabolic Control and Disease*. Philadelphia: W B Saunders, 1980: **8th ed**: 1241–1323.

Penney M D, Walters G. Are osmolality measurements clinically useful? *Ann Clin Biochem* 1987; **24**: 566–71.

Twigley A J, Hillman K M. The end of the crystalloid era? A new approach to peri-operative fluid administration. *Anaesthesia* 1985; **40**: 860–71.

Zilva J F. Fluid and electrolyte disturbances and their management. In: Ellis H ed. *Intestinal Obstruction*. New York: Appleton-Century-Crofts 1982: 23–37.

Table 2.3 Some electrolyte-containing fluids for intravenous infusion.

	Na	K	Cl	HCO_3	Glucose	Ca	Approximate osmolarity × plasma
	←		mmol/litre			→	
					(g/dl)		
Saline							
'Normal' (physiological 0.9%)	154	–	154	–	–	–	× 1
Twice 'normal' (1.8%)	308	–	308	–	–	–	× 2
Half 'normal' (0.45%)	77	–	77	–	–	–	× 0.5
Fifth 'normal' (0.18%)	31	–	31	–	–	–	× 0.2
'Dextrose' saline							
2.5% 0.45%	77	–	77	–	140 (2.5)	–	× 1
4% 0.18%	31	–	31	–	222 (4.0)	–	× 1
5% 0.45%	77	–	77	–	278 (5.0)	–	× 1.5
Sodium bicarbonate							
1.4%	167	–	–	167	–	–	× 1
2.74%	327	–	–	327	–	–	× 2
8.4%*	1000	–	–	1000	–	–	× 6
Complex solutions							
Ringer's	147	4.2	156	–	–	2.2	× 1
Hartmann's	131	5.4	112	29†	–	1.8	× 1
Na lactate, sixth molar	167	–	–	167†	–	–	× 1

*Most commonly used bicarbonate solution. Note marked hyperosmolarity. Only use if strongly indicated.
†as lactate

Table 2.4 Composition of some amino-acid solutions (all hyperosmolar)

	Na	K	Cl	Ca	PO_4	Mg	Nitrogen
	←		mmol/litre			→	g/litre
Synthamin 9	73	60	70	—	30	5	9.1
Synthamin 14*	73	60	70	—	30	5	14.0
Synthamin 17	73	60	70	—	30	5	16.5
Vamin 9	50	20	55	2.5	—	1.5	9.4
Vamin 9 glucose**	50	20	55	2.5	—	1.5	9.4
Vamin 14*	100	50	100	5	—	8	13.5
Vamin 18	—	—	—	—	—	—	18.0

*available electrolyte-free
**provides 1700 kJ (406 kcal)/litre, from glucose

Table 2.5 Electrolyte-free glucose (dextrose) solutions for intravenous administration (mostly used as an energy source)

Dextrose per cent	Glucose mmol/litre	Osmolarity	Approximate kJ/litre	(Approximate kcal/litre)
5	277	Isosmolar	860	(205)
10	555	Hyperosmolar	1720	(410)
20	1110	Hyperosmolar	3440	(820)
50	2770	Hyperosmolar	8600	(2050)

Table 2.6 Other hyperosmolar solutions (for inducing an osmotic diuresis or for reducing cerebral oedema)

	per cent	mmol/litre	Approximate osmolarity × plasma
Mannitol	10	550	× 2
	20	1100	× 4

3

Potassium metabolism: diuretic therapy

The total amount of potassium in the body is about 3000 mmol. Since about *98 per cent* of this is *intracellular*, the plasma potassium concentration is a poor indicator of the total amount in the body. This disadvantage is more apparent than real, because the low *extracellular (plasma) concentrations of potassium*, like those of calcium and magnesium, affect neuromuscular activity and cardiac action, and are of immediate importance in therapy, whatever the amount of intracellular potassium; both severe hypokalaemia and severe hyperkalaemia are dangerous, and must be treated. However, an attempt should be made to estimate the overall situation, and so to anticipate therapeutic needs.

Factors affecting the plasma potassium concentration

The large amount of intracellular potassium ion provides a reservoir for the extracellular compartment, and changes in water balance have little direct effect on the plasma potassium concentration, as they do on that of sodium. The hyperkalaemia often found when there is volume depletion is the result of renal retention (p. 10) rather than of haemoconcentration.

Potassium enters and leaves the extracellular compartment by three main routes:

- the intestine;
- the kidney:
    - the glomerulus;
    - the tubular cells;
- through the membranes of *all* other cells.

The intestine

Potassium is absorbed throughout the small intestine. Dietary intake replaces net urinary and faecal loss, and amounts to 100 mmol a day or less.

Potassium leaves the extracellular compartment in all intestinal secretions, usually at concentrations near or a little above those in plasma. A total of about 60 mmol a day is lost into the intestinal lumen; most of this is absorbed, together with the dietary intake, and less than 10 mmol a day is present in formed faeces. As in the case of sodium, excessive intestinal potassium loss in diarrhoea stools, in ileostomy fluid or through other fistulae is derived more from the fluid entering the intestinal lumen from the body than from dietary intake. However, prolonged starvation can cause potassium depletion, sometimes with hypokalaemia.

The kidney

Glomerular filtrate. Potassium is filtered by the glomerulus at almost the same concentration as in plasma. Because of the very large volume of the filtrate about 800 mmol (about a quarter to a third of the body content) would be lost daily if there were no tubular regulation. The net loss, although very variable, is only about 10 per cent of this.

The tubules. Potassium is normally almost completely *reabsorbed* in the *proximal tubule* and so renal tubular dysfunction can cause potassium depletion.

Potassium is secreted in the *distal tubule* and *collecting duct* in exchange for *sodium*. Hydrogen ions compete with potassium, and *aldosterone stimulates both exchanges*. If the proximal tubules are functioning normally potassium loss in the urine depends on three factors:

- the amount of *sodium available* for exchange. *This depends on the filtration rate, filtered sodium load*, and *sodium reabsorption in the proximal tubule and loop of Henle*. Reabsorption in the loop is inhibited by many diuretics (p. 65).
- the relative *amounts of hydrogen and potassium ions* in the cells of the distal tubule and collecting ducts, and on the *ability to secrete* H^+ *in exchange for* Na^+ (inhibited during treatment with carbonate dehydratase inhibitors and in some types of renal tubular acidosis).
- the circulating *aldosterone* level. This is increased following fluid loss with volume contraction (which usually accompanies intestinal loss of potassium, p. 27) and in almost all conditions requiring diuretic therapy.

The cell membrane

Potassium is the predominant intracellular cation and is continually lost from the cell down a concentration gradient. This loss is opposed by the 'sodium pump' on the cell surface. Sodium is pumped out of the cell in exchange for potassium and hydrogen ions.

A small shift of potassium out of cells causes a significant rise in plasma levels, whether *in vivo* or *in vitro*; in the latter case the artefactual hyperkalaemia due to haemolysed or old specimens may be misinterpreted (p. 449). Usually the shift of potassium across cell membranes is accompanied by a shift of sodium in the opposite direction, but the *percentage* change in extracellular sodium levels will be much less than that of potassium. A simplified example will demonstrate this. Let us assume extra- and intracellular volumes to be equal, the plasma sodium to be 140 mmol/litre and potassium 4 mmol/litre, with reversal of these concentrations inside cells. A one-to-one exchange of 4 mmol/litre of sodium for potassium across the cell membrane would double the plasma potassium concentration (clinically a very significant change) while only reducing the plasma sodium concentration from 140 to 136 mmol/litre. Thus, redistribution between cells and ECF has a more significant effect on plasma potassium than sodium concentrations.

There is *net loss* of potassium from the cell:

- if *potassium is lost from the ECF* and replenished from the cell;
- if the *sodium pump is inefficient*, as it is in hypoxia, or if glucose metabolism is impaired in severe diabetes;
- *in acidosis*, when potassium is displaced from cells by hydrogen ions.

There is *net gain* of potassium by the cell:

- if the *activity of the sodium pump is increased*, as it is after the administration of glucose and insulin: this effect may be used to treat hyperkalaemia. It is the main cause of the change from hyperkalaemia to hypokalaemia during treatment of diabetic coma. Catecholamines have been said to have a similar effect on the pump, and to contribute to the hypokalaemia sometimes found after the stress of myocardial infarction; however, hypokalaemia is not a prominent finding in phaeochromocytoma, in which catecholamine levels are very high;
- *in alkalosis*. Induction of alkalosis can also be used to treat hyperkalaemia.

Interrelationship between hydrogen and potassium ions

The extracellular hydrogen ion concentration affects the entry of potassium into all cells: changing the relative proportions of K^+ and H^+ in distal renal tubular cells affects urinary loss of potassium. In acidosis increased loss of potassium from the cells into the ECF, coupled with reduced urinary secretion of the ion, causes hyperkalaemia: in alkalosis hypokalaemia is due both to net increased entry of potassium into cells and to increased urinary loss.

Because the relationship between K^+ and H^+ is reciprocal, changes in K^+ balance affect that of hydrogen ion.

In absorptive cells, including those in the renal tubules, the 'sodium pump' is situated on only one, probably the non-luminal, surface. In the kidney the sodium derived from the luminal fluid is pumped through the cell in exchange for K^+ or H^+. If K^+ is lost from the ECF loss from cells follows down the increased concentration gradient, so reducing the amount of intracellular K^+; unless the loss is due to renal tubular cell damage, sodium is reabsorbed in exchange for less K^+ and more H^+ than usual. Despite extracellular alkalosis, an acid urine is formed. H^+ within the tubular cell is formed by the carbonate dehydratase (CD) mechanism (p. 82).

$$H_2O + CO_2 \underset{CD}{\rightarrow} H^+ + HCO_3^-$$

As more H^+ is secreted into the urine the reaction is accelerated, and more HCO_3^- is generated and passes into the ECF, accompanied by the reabsorbed sodium. Chronic potassium depletion is therefore usually accompanied by a *high plasma [HCO_3^-] and extracellular alkalosis*. If other causes for a raised plasma bicarbonate concentration, such as severe chronic obstructive airways disease, are absent, and especially if factors known to cause potassium depletion, such as diuretic therapy, are present, this finding is a sensitive indicator of potassium depletion, even with normokalaemia. *The combination of hypokalaemia and a high plasma [HCO_3^-] is more likely to be due to K^+ depletion, which is common, than primarily to metabolic alkalosis, which is rare.* These are the changes of *chronic* potassium depletion.

After *acute* loss the slight lag in potassium release from cells may result in more severe hypokalaemia for the same degree of depletion than in the chronic state: bicarbonate levels are less likely to be raised, because it takes several days before enough retained bicarbonate accumulates to be detectable.

Theoretically, potassium excess could cause intracellular alkalosis and extracellular acidosis, with a low plasma bicarbonate concentration. However, *the combination of hyperkalaemia and a low plasma [HCO_3^-] is more likely to be due to metabolic acidosis, which is common, than primarily to potassium excess, which is extremely rare.* In *respiratory* acidosis the plasma bicarbonate concentration is often high (p. 95), but that of plasma potassium will also tend to be high normal.

These situations are summarized in Table 3.1.

Abnormalities of plasma potassium concentrations

In any clinical situation no single factor accounts for the changes in plasma potassium concentration. For example, in conditions associated with intestinal potassium loss, concomitant volume loss causes secondary hyperaldosteronism; similarly, most conditions requiring diuretic therapy are associated with

Table 3.1 Interrelationships of plasma potassium and bicarbonate concentrations.

Plasma [K^+]	Plasma [HCO_3^-]	Most likely cause	Examples of clinical causes
Low N or ↓	↑	Chronic K^+ loss	Diuretic treatment Chronic diarrhoea (eg purgative takers)*
↓ or ↓↓	N or ↓	Acute K^+ loss	Severe acute diarrhoea Fistulae etc
↓	↓	Respiratory alkalosis	Overtreatment on respirator Hysterical overbreathing
↑	↓	Metabolic acidosis	Renal dysfunction* Diabetic ketoacidosis
↑	↑	Respiratory acidosis	Bronchopneumonia
↑	N	Acute K^+ load	Excess K^+ treatment

*It must be stressed that this is *only a guide*. For example, severe diarrhoea may cause so much bicarbonate loss that plasma [HCO_3^-] is low despite potassium depletion. Similarly, renal tubular lesions deplete both bicarbonate and potassium.

hyperaldosteronism. This aggravates urinary loss and may also increase entry of potassium into all body cells. However, if volume and sodium depletion are very severe, the reduced total amount of filtered sodium means that less is available for exchange with potassium in the distal nephron and aldosterone cannot produce maximal effects: hypokalaemia will then be 'unmasked' when the patient is rehydrated. This effect should be anticipated.

Hypokalaemia

Hypokalaemia is usually the result of potassium depletion, although, if the rate of loss of potassium from cells into the extracellular compartment equals or exceeds that from the ECF, potassium depletion may not cause hypokalaemia. By contrast, it can occur without depletion if there is a shift into cells, as in alkalosis, after taking a large amount of glucose and in the rare condition, familial periodic paralysis.

Misleading temporary hypokalaemia may occur for a few hours after ingestion of any of those oral diuretics which cause potassium loss. The finding of a higher value in a later specimen does not usually indicate a 'laboratory error'.

The causes of hypokalaemia may be classified as follows:

1. Predominantly due to **loss of potassium from the body.**

(a) Predominantly due to loss from the ECF in *intestinal secretions.*

- Prolonged vomiting;
- Diarrhoea;
- Loss through intestinal fistulae.

Intestinal loss is usually aggravated by the secondary hyperaldosteronism consequent on fluid loss. The resultant inappropriately high urinary loss may sometimes contribute more to the potassium depletion than the original pathology. The following points should be noted.

- The concentration of potassium in fluid from a *recent ileostomy* and in *diarrhoea stools* be may 5 to 10 times that of plasma, but a prolonged drain of any intestinal secretion causes depletion, especially if urinary loss is increased by secondary hyperaldosteronism.
- Habitual purgative takers may present with hypokalaemia and are often reluctant to admit to the habit.
- Rare, large *mucus-secreting villous adenomas of the intestine* may cause considerable potassium loss.

(b) Predominantly due to *loss from the ECF into urine.*

(i) Increased activity of sodium: potassium exchange mechanisms in the distal renal tubule.

- *Secondary hyperaldosteronism* often aggravates other causes of potassium depletion.
- *Cushing's syndrome and steroid therapy.* Patients secreting excess of, or on prolonged therapy with, glucocorticoids tend to become hypokalaemic due to the mineralocorticoid effect on the distal renal tubule.

- *Synacthen or ACTH therapy and ectopic ACTH secretion*, which stimulate cortisol secretion (p. 121).
- *Primary hyperaldosteronism* (p. 50).
- *Bartter's syndrome* is a very rare condition in which hyperplasia of the renal juxtaglomerular apparatus (p. 27) increases renin, and therefore aldosterone secretion. The biochemical findings are those of primary hyperaldosteronism, but the patient is normotensive.
- *Carbenoxolone therapy.* Carbenoxolone has been used to accelerate healing of peptic ulcers. It potentiates the action of aldosterone, and can cause hypokalaemia.
- *Liquorice and tobacco* contain glycyrrhizinic acid, which also has an aldosterone-like effect. Overindulgence in liquorice-containing sweets, or habitual tobacco chewing, can cause hypokalaemia.

(ii) Excess available sodium for exchange in the distal renal tubule.

- *Prolonged infusion of saline.*
- *Diuretics inhibiting the 'sodium pump' in the loop of Henle* (p. 65). The increased distal sodium: potassium exchange is aggravated by secondary hyperaldosteronism.

(iii) Decreased renal sodium: hydrogen ion exchange, favouring sodium: potassium exchange.

- *Carbonate dehydratase inhibitors* (p. 66).
- *Renal tubular acidosis* (p. 93).

(iv) Reduced proximal tubular potassium reabsorption.

- *Renal tubular dysfunction* (for example, the recovery phase of acute oliguric renal failure).
- *'Fanconi syndrome'* (p. 16) (hypercalcaemia is a common cause).

2. Predominantly due to **reduced potassium intake**.

- *Chronic starvation.* If water and salt intake are also reduced, secondary hyperaldosteronism may aggravate the hypokalaemia.

3. Predominantly due to **redistribution** within the body. Loss into cells.

- *Glucose and insulin therapy.* This may be used to treat severe hyperkalaemia.
- *Increased secretion of catecholamines*, such as may occur after the stress of myocardial infarction, may be a cause of hypokalaemia (p. 58).
- *Familial hypokalaemic periodic paralysis* (very rare). In this condition episodic paralysis is associated with entry of potassium into cells.

4. Loss from ECF by **more than one route**.

(a) Into cells and urine.
Alkalosis.

(b) Into cells, urine and intestine.
Pyloric stenosis with alkalosis. The losses into urine and cells are probably more important causes of the hypokalaemia than the loss in gastrointestinal secretion.

Hyperkalaemia

Potassium results must not be reported if the specimen is haemolysed, or if the plasma has not been separated from cells within a few hours after the blood was taken, *whether refrigerated or not* (p. 449). *Pseudohyperkalaemia* due to *in vitro* leakage of potassium from cells into plasma is often misinterpreted, sometimes with dangerous consequences.

True hyperkalaemia is most common when the rate of potassium entering the ECF from cells is greater than its rate of excretion from the body. The causes of hyperkalaemia are:

1. Predominantly due to **gain of potassium by the body**.

Gain by the ECF from the *intestine or by the intravenous route.*

- *Overenthusiastic potassium therapy, especially in patients with impaired glomerular function.*
- *Failure to stop potassium therapy* when depletion has been corrected and the cause removed.

2. **Impaired renal secretion of potassium.**

(a) Decreased activity of sodium: potassium exchange mechanisms in the distal renal tubule.

- *Hypoaldosteronism* (as in Addison's disease).
- *Captopril and enalapril* (used to treat hypertension), inhibit conversion of angiotensin I to angiotensin II, and therefore aldosterone secretion (p. 27); they may cause hyperkalaemia, especially if glomerular function is impaired.
- *Diuretics acting on the distal tubule* by antagonizing aldosterone, or by direct inhibition of the 'sodium pump' (p. 66).

(b) Too little sodium available for exchange in the distal tubule.

- *Renal glomerular dysfunction.* Hyperkalaemia is usually aggravated by the concomitant acidosis and release of potassium from cells.
- *Sodium depletion.*

3. Predominantly due to **redistribution** of potassium within the body.

Gain of potassium by ECF from cells.

- *Severe tissue damage.*
- *Familial hyperkalaemic periodic paralysis (exceedingly rare).* Episodic paralysis is associated with release of potassium from cell.

4. Gain by ECF by **more than one route**.

Reduced renal excretion despite gain by ECF from cells.

- *Acidosis.*
- *Hypoxia* causes impairment of the 'sodium pump' in all cells, with a gain of potassium by the extracellular fluid. Such impairment in the distal nephron causes potassium retention. If hypoxia is severe, lactic acidosis aggravates hyperkalaemia.

Diabetic ketoacidosis. In poorly controlled diabetes mellitus potassium enters the ECF from cells, because normal action of the 'sodium pump' depends on the energy supplied by glucose metabolism. While glomerular function is relatively normal urinary potassium loss is increased, causing body depletion: despite this loss in the urine rarely keeps pace with gain from the cells and mild hyperkalaemia is common. As the condition progresses two other factors contribute to hyperkalaemia:

- volume depletion causing a reduced GFR;
- keto-, and sometimes lactic, acidosis.

All these factors are reversed during insulin and fluid therapy. As potassium enters cells extracellular levels fall and the depletion is revealed. *Plasma potassium levels must be monitored during therapy*, and potassium given as soon as concentrations begin to fall.

Measurement of urinary and intestinal losses

Pure urinary or intestinal loss as a cause of hypokalaemia is very rare. Measurement of such losses with a view to quantitative replacement may lead to even more dangerous errors of therapy than in the case of sodium. Exchanges across cell walls cannot be measured. If gain of potassium by the ECF from cells is faster than loss from it into urine and the intestinal lumen, replacement of measured loss could endanger the patient's life by aggravating hyperkalaemia; if loss from the ECF into cells is predominant, therapy based on urinary excretion may be inadequate. Moreover, high urinary potassium excretion may be appropriate when, for instance, trauma has damaged many cells. It would be as rational to replace urinary glucose quantitatively in an uncontrolled diabetic as to use urinary potassium measurements in the same way. The student should remember this analogy when considering the value of measuring daily losses of any constituent, especially to control enteral and parenteral feeding (p. 271). *It is the plasma potassium levels that are important*, and in rapidly changing states frequent estimation of these is the only safe way of assessing therapy. In chronic depletion the plasma bicarbonate level may help to indicate the state of cellular repletion.

The diagnostic use of urinary potassium estimations to determine the primary cause of depletion is also more often misleading than helpful. Most extrarenal potassium loss is associated with volume depletion, and therefore with secondary hyperaldosteronism; a high urinary excretion is not proof of a primary renal cause. A urinary potassium excretion of less than about 20 mmol a day can only be expected in the *well-hydrated* hypokalaemic patient with extrarenal losses, in whom aldosterone secretion is inhibited: a low potassium excretion therefore confirms extrarenal loss, but a high one does not prove that it is the primary cause.

Clinical features of disturbances of potassium metabolism

The clinical features of disturbances of potassium metabolism are due to changes in the extracellular concentration of the ion.

Hypokalaemia, by interfering with neuromuscular transmission, causes *muscular weakness, hypotonia and cardiac arrhythmias*, and may precipitate digoxin toxicity. It may also aggravate paralytic ileus and hepatic encephalopathy.

Intracellular potassium depletion causes extracellular alkalosis (p. 58). This reduces the ionization of calcium salts (p. 173) and in long standing depletion of gradual onset the presenting symptom may be muscle *cramps* and even *tetany*. This syndrome is accompanied by high plasma bicarbonate levels.

Prolonged potassium depletion causes vacuolation of renal tubular cells; the renal tubular damage may cause polyuria and further potassium depletion.

Severe hyperkalaemia always carries the danger of cardiac arrest. Both hypo- and hyperkalaemia cause characteristic changes in the electrocardiogram.

Treatment of potassium disturbances

Abnormalities of *plasma* potassium should be corrected whatever the state of the total body potassium balance. However, an attempt should be made to assess the latter so that sudden changes in plasma potassium (for instance, during treatment of diabetic coma) can be anticipated. Treatment should be controlled by frequent plasma potassium estimations.

Hyperkalaemia. Treatment of hyperkalaemia is based on three principles. The first two are most useful in severe hyperkalaemia.

Very severe hyperkalaemia can cause cardiac arrest. Calcium and potassium ions have opposing actions on cardiac muscle, and the immediate danger can be minimized by infusing calcium salts (usually as the gluconate, p. 68). This has no effect on the plasma potassium concentration, but, because it has a rapid effect on cardiac sensitivity, allows time to institute measures to reduce hyperkalaemia.

Plasma potassium can be reduced within an hour by increasing the rate of entry into cells. Glucose and insulin speed up glucose metabolism and the action of the 'sodium pump'. Induction of alkalosis by infusion of bicarbonate also increases the rate of entry into cells (p. 69). For practical reasons, these treatments, which involve intravenous infusion, cannot be continued indefinitely, but they allow long-term treatment to be instituted.

In moderate hyperkalaemia a slower acting method can be used. Potassium can be removed from the body at a rate higher than, or equal to, that at which it is entering the extracellular fluid by giving ion-exchange resins by the oral or rectal route. These are not absorbed and exchange potassium for sodium or calcium ions. Plasma potassium is lowered at the expense of body depletion, and the potassium may have to be replaced when the cause of the hyperkalaemia has been removed.

Hypokalaemia. *Mild* hypokalaemia can be treated with *oral* potassium supplements, which must continue to be given until plasma potassium *and bicarbonate* levels return to normal. These levels should be monitored regularly. Undertreatment is more common than overtreatment. The normal subject loses about

60 mmol of potassium daily in the urine, much larger amounts being excreted during diuretic therapy. By the time hypokalaemic alkalosis is present the total deficit is probably several hundred mmol. A patient with hypokalaemia should be given *at least* 80 mmol a day: much more may be needed if plasma levels fail to rise.

In *severe* hypokalaemia, particularly if the patient is unable to take oral supplements, *intravenous potassium* should be given cautiously (p. 68). Diarrhoea not only reduces absorption of oral potassium supplements, but may itself be aggravated by them; this too may be an indication for intravenous therapy.

It is sometimes suggested that hypokalaemia could be treated by a high intake of fruit or of fruit juice. Table 3.2 shows that if it were possible to ingest the quantities required, the consequent diarrhoea might well be self-defeating. The relatively high cost is only a minor disadvantage.

Table 3.2 Potassium content of fruit and fruit juice

	Approximate K^+ content (mmol)	Approximate amount containing 50 mmol	Price for 50 mmol relative to 'Slow K' in UK. January 1987
Tomato juice	82 per litre	610 ml	10
Orange juice	30 per litre	1700 ml	25
Grapefruit juice	30 per litre	1700 ml	24
Lemon juice	43 per litre	1160 ml	35
Orange cordial (*undiluted*)	16 per litre	3100 ml	25
Bananas	8 per banana	6 bananas	22
Slow K' (Ciba)	8 per tablet	6 tablets	1

Diuretic therapy

In oedematous states fluid accumulation is associated with an excess of sodium in the body, *even if there is hyponatraemia* (p. 48). Diuretics act by inhibiting sodium reabsorption in the renal tubule and secondarily by causing water loss. They may be used to treat hypertension as well as oedema. All diuretics tend to affect potassium balance, and this effect should be anticipated, especially in patients with secondary hyperaldosteronism or renal dysfunction.

Diuretics can be divided into three main groups.

1. *Those inhibiting the pump in the loop of Henle*, and therefore water reabsorption: the increased sodium load on the distal tubule and collecting ducts increases sodium: potassium exchange at this site (which is stimulated by the accompanying hyperaldosteronism). In our experience long-term diuretic therapy almost invariably causes significant *potassium depletion*, and sometimes symptomatic hypokalaemia, even if potassium supplements are given (although some authors have questioned this). *A high plasma bicarbonate concentration* is common in long-continued use of such diuretics. This is because loss of potassium from the ECF

causes an increased $Na^+ : H^+$ exchange in the distal nephron. High plasma bicarbonate levels are a more sensitive indicator of cellular K^+ depletion than plasma potassium levels.

The *thiazide* group of diuretics act at the junction of the loop and the distal tubule (sometimes called the 'cortical diluting segment'). *Frusemide* (furosemide; 'Lasix'), *bumetanide* ('Burinex') and *ethacrynic acid* ('Edecrin') are true 'loop diuretics', and inhibit the pump in the ascending limb.

2. Those either directly *inhibiting aldosterone* or inhibiting the exchange mechanisms in the *distal tubule and collecting duct*. These cause *potassium retention* and may lead to hyperkalaemia, especially if glomerular function is impaired: potassium supplements should *not be* used. Potassium-retaining diuretics include:

- *Spironolactone* ('Aldactone'), a competitive aldosterone antagonist.
- *Amiloride* ('Midamor') } Inhibitors of the $Na^+ : K^+$ exchange
- *Triamterene* ('Dytac') } mechanism in the renal tubule.

This group of diuretics is often used, together with those causing potassium loss, when hypokalaemia cannot be controlled by potassium therapy, or to potentiate sodium loss by inhibiting reabsorption at more than one site. Used alone they have only a weak diuretic action.

3. Carbonate dehydratase inhibitors such as acetazolamide ('Diamox') (p. 94) inhibit sodium reabsorption in the proximal tubule and so act as diuretics. They tend to cause hypokalaemia, but are now rarely used for their diuretic effect because of the danger of acidosis, and because they are relatively ineffective.

Summary

1. Changes in plasma potassium concentrations are the net result of exchanges between ECF and cells, kidney and intestine.
2. In any clinical situation many factors are involved, and monitoring of plasma potassium levels is the only safe guide to treatment.
3. Because hydrogen and potassium compete for exchange with sodium ions in the renal tubule and at other cell membranes, disturbances of hydrogen ion homeostasis and potassium balance often coexist. A raised plasma bicarbonate (TCO_2) level may indicate intracellular potassium depletion.
4. Clinical manifestations of disturbances of potassium balance are due to its action on neuromuscular transmission and on the heart.
5. Diuretics fall into two main groups:

- those inhibiting the 'sodium pump' in the loops of Henle and diluting segments, causing potassium depletion;
- those antagonizing aldosterone, either directly or indirectly, by affecting $Na^+ : K^+$ transport mechanisms, causing potassium retention.

Further reading

Brown R S. Extrarenal potassium homeostasis. *Kidney Int* 1986; **30**: 116–27.
Lant A F. Modern diuretics and the kidney. *J Clin Pathol* 1981; **34**: 1267–75.
Diuretics. In: *British National Formulary*, London: British Medical Association and Pharmaceutical Society of Great Britain, 1987; **13**: 71.

Treatment of hypokalaemia

Potassium-containing preparations

One gram of potassium chloride contains 13 mmol of potassium.

For oral use

1. Potassium Effervescent Tablets BP. 6.5 mmol K^+ per tablet (as bicarbonate and acid tartrate).
2. 'Slow K' (Ciba). 8 mmol K^+ per tablet (as chloride).
3. 'Sando-K' (Sandoz). 12 mmol K^+ per tablet (as bicarbonate and chloride).
4. 'Kloref-S' (Cox-Continental). 20 mmol K^+ per sachet (mostly as chloride).
5. 'Kloref' tablets (Cox-Continental). 6.7 mmol K^+ (mostly as chloride and bicarbonate).

For intravenous use

These preparations should only be used in serious depletion, or when oral potassium cannot be taken or retained. In most cases oral potassium treatment is preferable. Intravenous potassium should be given with care, especially in the presence of poor glomerular function and the following rules should be observed.

1. Unless the deficit is unequivocal and severe, intravenous potassium should not be given if there is oliguria.
2. Potassium in the intravenous fluid should not usually exceed a concentration of 40 mmol/litre.
3. Intravenous potassium should not usually be given at a rate of more than 20 mmol per hour. In very severe depletion this dose may have to be exceeded; in such cases frequent plasma potassium estimations must be performed.

Strong potassium chloride solution BP
20 mmol of potassium chloride in 10 ml.
WARNING. This should *never* be given undiluted. It should be added to a full bag of other intravenous fluid. 10 ml added to 500 ml of fluid gives a concentration of 40 mmol/litre.

Potassium chloride and dextrose intravenous infusion BP

5 per cent glucose (dextrose) with:
10 mmol/litre of potassium and chloride.
20 mmol/litre of potassium and chloride.
40 mmol/litre of potassium and chloride.

Treatment of hyperkalaemia

Emergency treatment

Calcium chloride (or gluconate). 10 ml of a 10 per cent solution is given intravenously with ECG monitoring. This treatment antagonizes the effect of hyperkalaemia on heart muscle, but does not alter potassium concentrations.

WARNINGS.

- Calcium should never be added to bicarbonate solutions, because calcium carbonate is poorly soluble in water.
- If too much calcium is given hypercalcaemia may cause cardiac arrest.

Glucose 50 g with 20 units of soluble insulin by intravenous injection starts to lower plasma potassium levels within about 30 minutes by increasing entry into cells. If the situation is less urgent 10 units of soluble insulin may be added to a litre of 10 per cent glucose.

If acidosis is present, bicarbonate may be used as an alternative to glucose and insulin; *40 ml of 8.4 per cent* sodium bicarbonate (40 mmol of bicarbonate) may be injected over 5 minutes. This solution is hyperosmolar.

Long-term treatment

Sodium or calcium polystyrene sulphonate ('Resonium-A'; 'Kayexalate' or 'Calcium Resonium' Winthrop) 45 to 60 g a day by mouth in 15 g doses, *or* 30 g in methyl cellulose solution as a retention enema retained for 9 hours. These remove potassium from the body.

Biochemical investigation of renal, water and electrolyte disorders

Monitoring fluid balance

Maintenance and inspection of *accurate* ward fluid balance charts is as important as measurement of daily plasma electrolyte concentrations.

Fluid imbalance may develop so gradually in those patients who are unconscious, or who have abnormal losses, that it may pass unnoticed if individual *daily* charts are relied upon, especially if insensible losses are ignored. Maintenance of a record of *cumulative* fluid balance is a more sensitive way of detecting a trend, which may then be corrected before serious abnormalities develop. This is especially important in patients at risk of predominant water depletion, which is less clinically obvious than the more common isosmolal volume depletion (p. 44), and which may not be noticed until dangerous hypernatraemia has developed.

In the example given below the minimum of *400 ml has been allowed for insensible loss*; calculated losses are therefore more likely to be under- than overestimated, and the danger of fluid overload by replacement of the stated volume is minimal.

	Intake (ml)	Measured output (ml)	Total output (minimum ml)	Daily balance (ml)	Cumulative balance (ml)
Day 1	2000	1900	2300	−300	− 300
Day 2	2000	2000	2400	−400	− 700
Day 3	2100	1900	2300	−200	− 900
Day 4	2200	2000	2400	−200	−1100

This shows how insidiously a serious deficit can develop over a few days.

The volume of fluid infused should be based on the calculated cumulative balance and on clinical evidence of the state of hydration, and its composition adjusted to maintain the plasma electrolyte concentrations normal. If the cumulative balance figures are plotted on the same graph as plasma sodium and urea values, potential hypernatraemia and renal circulatory insufficiency can often be detected and treated before they become serious and clinically obvious.

If the ambient temperature is high, or if the patient is pyrexial, hyperventilating, or has large unmeasured intestinal losses, as in intestinal obstruction, more than 400 ml should be allowed for insensible loss.

Investigation of acute oliguria

Correct management of acute oliguria depends on distinguishing between renal and extrarenal causes.

1. Estimate plasma electrolytes (especially potassium) at once. Knowledge of the urea and/or creatinine concentrations is of less immediate therapeutic importance. Treat dangerous hyperkalaemia (p. 68).

2. Exclude post-renal causes. Prostatic hypertrophy is the commonest cause of oliguria

in the adult male with a palpable bladder.

3. Examine the urine for protein and casts: their presence suggests significant renal damage. Microbiological investigation may reveal urinary tract infection.

4. *If the patient is oedematous, perhaps due to congestive cardiac failure, do not continue with steps 5 and 6.* Fluid administration is contraindicated, whether or not there is renal damage.

5. Rehydrate the patient, watching for signs of incipient overhydration, and measure the urine output accurately. If it increases rapidly, and the plasma urea/creatinine concentrations start to level off or fall within 24 hours, the cause is likely to be prerenal. If, after rehydration, the oliguria persists and the plasma urea continues to rise, renal damage is likely and fluid balance should be maintained as described on p. 30.

6. In the unlikely event that the cause is still in doubt, estimate the urinary sodium concentration. This is only useful if the specimen is collected when the patient is still volume depleted or shocked, since only under these conditions is aldosterone secretion likely to be maximal (p. 14). If the concentration is below 30 mmol/litre, and certainly if it is below 20 mmol/litre, significant renal damage is unlikely and rehydration can safely be continued.

7. Plasma potassium concentrations must be measured at least daily and treatment adjusted appropriately.

Investigation of polyuria

A. Confirm polyuria and exclude obvious causes

1. Take a history.

- Try to distinguish between true polyuria (a high 24 hour urinary volume) and frequency (a normal 24 hour volume but abnormally frequent micturition).
- Is the patient taking diuretics? This cause may easily be forgotten.
- If there is a recent history of oliguric renal dysfunction the patient has probably entered the polyuric phase, and should be treated accordingly.

2. Is there a cause for an osmotic diuresis such as:

- gross glycosuria due to diabetes mellitus or infusion of glucose (dextrose)?
- severe tissue damage, leading to a high urea load?
- infusion of amino acids?

If so investigate and treat, or change the composition of the infusion.

3. If fluid, especially of low sodium concentration, is being infused, adjust accordingly.

4. Is there an obvious cause for diabetes insipidus, such as a head injury?

B. Distinguish between polyuria appropriate to a high intake (see also A.3) and that due to failure of homeostatic mechanisms.

1. Volume depletion suggests failure of homeostasis.
2. Accurately monitor fluid intake and output.

- A *negative balance* suggests *failure of homeostasis.*
- A *positive balance* suggests that *polydipsia is primary.*

The patient should be carefully watched for secret fluid ingestion ('hysterical polydipsia') or even addition of fluid to the urine.

3. Estimate plasma urea/creatinine and electrolytes.

- *Very high plasma urea/creatinine levels suggest renal failure.* Treat and monitor accordingly (p. 20).
- *Low or low normal plasma urea/creatinine levels, especially if there is mild hyponatraemia, suggest that polydipsia is the primary cause* and that the polyuria is appropriate. Question the patient about his intake in relation to true thirst. Patients with true 'hysterical polydipsia' may give misleading answers: they rarely complain of nocturia.
- *High normal or slightly high urea/creatinine levels* (urea up to about 15 mmol/litre or 90 mg/dl; urea nitrogen 42 mg/dl), *especially if the plasma sodium concentration is high normal or high, suggest impairment of the tubular ability to concentrate.* Severe hypokalaemia may suggest that this is a cause of tubular dysfunction.

4. If the plasma urea/creatinine concentration is slightly high, cautiously increase fluid intake. A fall in urea in response to rehydration suggests that glomerular damage is minimal. Seek a cause for tubular damage.

- Take a drug history.
- Send urine for microbiological examination to detect pyelonephritis.
- Estimate plasma calcium and urate concentrations.
- Consider inborn errors of metabolism in infants (Chapter 18).
- Exclude the other causes listed on p. 16.

5. If it is still not possible to distinguish between polydipsia, tubular impairment and diabetes insipidus, *contact the laboratory* to arrange a water deprivation test, if necessary with DDAVP.
6. Treat mild tubular impairment with a high fluid intake.
7. In cases of diabetes insipidus, seek a neurological opinion.

Water deprivation test

Restriction of water intake for some hours should stimulate ADH secretion (p. 27). Solute free water is reabsorbed from the collecting ducts and a concentrated urine is passed. Maximal water reabsorption is impaired if either:

- the countercurrent multiplication mechanism is impaired;
- ADH activity is low.

If the feedback mechanism is intact ADH levels are already high, and administration of exogenous hormone will not improve the renal concentrating power: if diabetes insipidus is the cause, and tubular function is normal, such administration will convert it to normal.

Warning. *The test should not be performed if the patient is volume depleted or has even mild hypernatraemia.* In such cases the finding of a low urine to plasma osmolality ratio (see below) is diagnostic, and further fluid restriction may be dangerous. If the test is necessary, *it must be stopped if the patient becomes distressed and the plasma osmolality (or sodium concentration) rises to high or high normal levels*: urine and blood specimens should be collected at once. *If the apparent 'distress' does not coincide with a high normal or high plasma osmolality it is likely to be psychological in origin.*

Procedure

Always contact your laboratory *before* starting the test, both to ensure efficient and speedy analysis, and to check local variations in the protocol.

The patient is allowed no food or water after 18.00 h on the night before the test. *He must be in hospital, and kept under observation* during the period of fluid restriction (see above).

On the day of the test:

07.00 h. The bladder is emptied. Blood is collected and plasma osmolality measured. If this is low normal or low water depletion is unlikely, and polyuria is probably due to an appropriate response to a high intake. If it is high normal or high the test should be stopped and the osmolality of the urine measured.

08.00 h. Blood and urine are again collected and the plasma and urinary osmolality measured. If the urinary osmolality exceeds 850 mmol/kg neither significant tubular disease nor diabetes insipidus are present and the test may be stopped: if the osmolality of urine is below 850 mmol/kg, and that of plasma is normal, fluid restriction should be continued and the estimations repeated at hourly intervals. The test should be stopped as soon as urinary concentration occurs.

Failure to concentrate three consecutive urine specimens indicates either tubular disease or diabetes insipidus, and the differential diagnosis is usually clear on clinical grounds; the test may then be stopped. If the diagnosis of diabetes insipidus is considered, continue with the DDAVP test.

DDAVP test. DDAVP (1-**D**eamino-8-**D**-**A**rginine **V**aso**p**ressin; desmopressin acetate) is a potent synthetic analogue of vasopressin.

4 μg DDAVP is injected intramuscularly. If there is tubular dysfunction there will be no improvement, but in diabetes insipidus the urinary concentration will increase.

Note. After prolonged overhydration, usually due to hysterical polydipsia, concentrating power, even after administration of exogenous hormone, is impaired. This is due to washing out of medullary hyperosmolality (p. 8). *Unless the patient is volume depleted or there is plasma hyperosmolality*, the test should be repeated after several days of relative water restriction. *The patient should be kept under careful observation*, on the one hand for signs of genuine distress associated with a rise in plasma osmolality, and on the other for surreptitious drinking.

We feel that measurement of urinary specific gravity is too inaccurate to be useful, and should no longer be carried out. If an osmometer is unavailable estimation of the urinary urea concentration may help. Values of about 90 to 100 times that of plasma indicate good concentrating ability.

Investigation of the patient with renal calculi

1. If the stone is available, send it to the laboratory for analysis.
2. Exclude *hypercalcaemia* and *hyperuricaemia*.
3. If the *plasma calcium is normal*, collect a 24-hour specimen of urine for *urinary calcium estimation: acid must be added to the specimen before analysis (to keep calcium in solution).*
4. *If all these tests are negative*, and especially if there is a family history of calculi, screen the urine for *cystine*. If the qualitative test is positive the 24-hour excretion of cystine should be estimated (p. 373).
5. *If the urine is alkaline despite metabolic acidosis* the diagnosis of *renal tubular acidosis* is likely (p. 93).
6. In children a *low plasma urate* and *high urinary xanthine* suggest xanthinuria. If these values are normal the 24-hour excretion of oxalate should be determined to exclude primary hyperoxaluria.

Investigation of electrolyte disturbances

Electrolyte estimations provide the numerical bulk of the workload and are responsible for a large part of the expenditure of most chemical pathology departments. In the last two chapters we have outlined conditions in which plasma sodium or potassium concentrations may be abnormal, and have stressed that an abnormal level, especially of sodium may not be of clinical significance, while normal levels do not guarantee normal balance.

Before requesting any test it is worth considering whether a high, normal or low result will aid diagnosis or treatment. For technical reasons, sodium and potassium are estimated simultaneously by the laboratory, but their clinical value should be considered separately: potassium results are more often useful than those of sodium. This mental discipline may help to avoid some common and dangerous therapeutic errors which arise from a misunderstanding of the underlying pathophysiological processes.

Plasma sodium should be estimated regularly:

- *in unconscious or confused patients and in infants losing fluid,* because of the danger of hyperosmolality due to hypernatraemia;
- *in the patient in diabetic coma or precoma,* because of the danger of sustained hyperosmolality due to hypernatraemia despite successful control of plasma glucose levels;
- *in the volume depleted patient,* or *those with abnormal losses,* to help diagnosis and to indicate the type of replacement fluid.

Plasma potassium (and bicarbonate) should be estimated regularly in any patient in whom there is a possible cause for abnormal levels, because they, unlike those of sodium must always be treated:

- *in patients with abnormal losses* from the gastrointestinal tract, or kidneys (especially due to *diuretic, steroid or ACTH therapy*);
- *in patients on potassium therapy*;
- *in patients with renal failure*;
- *in patients with diabetic ketoacidosis or hyperosmolar coma or precoma.*

The indications for electrolyte estimation are few compared with the numbers usually requested. In the fully conscious, normally hydrated, patient with no abnormal losses plasma sodium estimation rarely helps. Mild hyponatraemia is common (p. 52), but treatment in such subjects is usually contraindicated. In the absence of renal failure potassium estimation is also unhelpful in such subjects.

Investigation of hypokalaemia

Hypokalaemic alkalosis

Diuretic therapy is by far the commonest cause of hypokalaemic alkalosis in hypertensive patients. Primary hyperaldosteronism and ectopic ACTH production are very rare.

The following procedure will almost always elucidate the cause without the need for expensive and time-consuming hormone assays.

1. Exclude obvious causes of potassium loss, such as diarrhoea. Prolonged fasting may occasionally cause hypokalaemia.

2. Take a careful drug history, with special reference to potassium-losing diuretics, purgatives, or steroid, ACTH or tetracosactrin (for example 'Synacthen') therapy; rare causes of a steroid-like effect are ingestion of carbenoxolone or liquorice.

3. Look at the plasma sodium value. If it is high or high normal the hypokalaemia is likely to be due to steroids or a steroid-like effect. If it is low normal or low another cause is probable.

4. If not clinically contraindicated, stop all therapy known to affect potassium loss. This

is, in any case, necessary if hormone assays are later indicated, because their results are affected by such treatment.

5. Give adequate oral potassium supplements until both the plasma potassium *and the bicarbonate* ($T\text{CO}_2$) concentrations are normal.

6. Stop supplements and continue to monitor plasma potassium levels for several days. Most cases remain normokalaemic and need no further investigation.

If the plasma potassium level falls again without an obvious cause, or if adequate supplementation fails to correct the hypokalaemia, and if there is no evidence of renal tubular damage, blood specimens should be taken under carefully controlled conditions for hormone assays. If primary hyperaldosteronism is suspected, *contact your laboratory before taking blood* for both renin (or angiotensin) and aldosterone assays: the diagnosis can only be made if the aldosterone is high, *and* the renin(angiotensin) activity low.

If ectopic ACTH secretion is suspected, measure plasma cortisol, and if the concentration is high, ACTH. ACTH and therefore cortisol secretion is stimulated by stress, and moderately high plasma levels of ACTH are not necessarily due to ectopic hormone secretion.

Hypokalaemic acidosis

Hypokalaemic acidosis is relatively rare, and the cause is usually obvious.

1. Loss of potassium through *fistulae in the proximal small intestine* may be accompanied by significant bicarbonate loss.

2. Hypokalaemic acidosis is a complication of transplantation of the *ureters into the ileum or colon* (p. 93).

3. If neither of these causes is present, the diagnosis of renal tubular acidosis should be considered (p. 93).

Investigation of hyperkalaemia

1. Exclude artefactual hyperkalaemia due to haemolysis or delayed separation of plasma from blood cells (p. 449).

2. Is there a cause for cell damage, such as hypoxia or severe trauma?

3. What are the plasma urea and bicarbonate concentrations? Severe renal glomerular dysfunction is one of the commonest causes of hyperkalaemia. Acidosis, whether metabolic (with a low plasma bicarbonate concentration) or respiratory (with a normal or high plasma bicarbonate concentration), may cause mild hyperkalaemia even if glomerular function is normal. A cause for respiratory acidosis is usually obvious clinically.

4. What drugs is the patient taking?

- Is he taking *potassium supplements*? It is surprisingly common for such supplementation to be continued once the need for it has passed and for this cause of hyperkalaemia to be forgotten.
- Is he taking *potassium-retaining diuretics?*
- Is he taking *captopril* or *enalapril?*

5. The combination of hypotension, skin pigmentation, hyponatraemia with hyperkalaemia and mild uraemia is common in many serious illnesses, and is only rarely due to *Addison's disease*. However, this diagnosis should not be forgotten. Is there pigmentation of the *mucous membranes*? If there is any doubt, estimate the plasma cortisol concentration. It should be very high *at any time of day* in an ill patient: if it is low, or even normal, proceed as on p. 154.

4

Hydrogen ion homeostasis: blood gas levels

About 50 to 100 *milli*moles of hydrogen ions are released from cells into 15 to 20 litres of extracellular fluid each day. Despite fluctuations in the rate of release during the day, homeostatic mechanisms keep the extracellular hydrogen ion concentration between about 35 and 45 *nano*moles/litre (40 nmol/litre = pH 7.40). Control of hydrogen ion balance depends ultimately on H^+ secretion from the body, mainly into the urine; *renal impairment causes acidosis.*

Aerobic metabolism of the carbon skeletons of organic compounds converts hydrogen, carbon and oxygen to water and carbon dioxide (CO_2). It does not directly affect hydrogen ion balance, but the CO_2 is an essential component of the extracellular buffering system. Control of CO_2 depends on normal *lung function.*

Hydrogen ions are released during *metabolism of amino acids*, or by *incomplete metabolism of carbon skeletons.*

- Conversion of amino nitrogen to urea, or of the sulphydryl groups of some amino acids to sulphate, releases equimolar amounts of hydrogen ions. *Subjects on a high protein diet pass an acid urine*, but although this source of hydrogen ions may aggravate pre-existing acidosis, it is rarely of clinical importance.
- *Anaerobic carbohydrate metabolism* yields lactate (p. 206) and *anaerobic metabolism of fatty acids and of ketogenic amino acids yields acetoacetate (p. 204): these processes release equimolar amounts of* H^+, directly or indirectly. In pathological lactic acidosis or ketoacidosis the rate of these reactions is so rapid that the compensatory capacity is exceeded, and blood pH falls significantly.

Many anabolic processes, including gluconeogenesis, use hydrogen ions.

Acidosis is more common than alkalosis because metabolism produces hydrogen and not hydroxyl ions.

Definitions

An *acid* can dissociate to produce hydrogen ions (protons: H^+): *a base* can accept hydrogen ions. Table 4.1 includes examples of acids and bases of importance in hydrogen ion balance.

An *alkali* dissociates to produce hydroxyl ions (OH^-). Because OH^- is not a primary product of metabolism, alkalis are of little importance in the present discussion.

Table 4.1 Some weak acids and their conjugate bases

	Acid			Conjugate base	
Carbonic acid	H_2CO_3	$\leftrightarrow H^+$	+	HCO_3^-	Bicarbonate ion
Dihydrogen phosphate	$H_2PO_4^-$	$\leftrightarrow H^+$	+	HPO_4^{2-}	Monohydrogen phosphate ion
Ammonium ion	NH_4^+	$\leftrightarrow H^+$	+	NH_3	Ammonia
Lactic acid	$CH_3CHOHCOOH$	$\leftrightarrow H^+$	+	$CH_3CHOHCOO^-$	Lactate ion
Acetoacetic acid	CH_3COCH_2COOH	$\leftrightarrow H^+$	+	$CH_3COCH_2COO^-$	Acetoacetate ion
3-hydroxy-butyric acid	$CH_3CHOHCH_2COOH$	$\leftrightarrow H^+$	+	$CH_3CHOHCH_2COO^-$	3-hydroxy-butyrate ion

A *strong acid* is mostly dissociated in aqueous solution, and so yields many hydrogen ions: hydrochloric acid, an example of a strong acid, is almost entirely in the form H^+Cl^- in water. Although, by comparison, the examples given in Table 4.1 are *weak acids*, yielding relatively few hydrogen ions, even very small changes in pH are important.

Buffering is the process by which a strong acid (or base) is replaced by a weaker one, with a consequent reduction in the number of free hydrogen ions (H^+); the 'shock' of the hydrogen ions is taken up by the buffer, and the change in pH after addition of acid is less than it would be in the absence of the buffer.

For example:

H^+Cl^-	+ $NaHCO_3$	$\leftrightarrow$	H_2CO_3	+	$NaCl$
Strong acid	**Buffer**		Weak acid		Neutral salt

pH is a measure of hydrogen ion activity. It is $\log_{10}$ of the reciprocal of the hydrogen ion concentration ($[H^+]$) in mol/litre. The $\log_{10}$ of a number is the power to which 10 must be raised to produce that number.

Thus

$$\log 100 = \log 10^2 = 2, \text{ and } \log 10^7 = 7.$$

Let us suppose $[H^+]$ is 10^{-7} (0.000 000 1) mol/litre

Then $\log [H^+] = -7$

But $$pH = \log \frac{1}{[H^+]} = -\log [H^+] = 7$$

Since at pH 6 $[H^+] = 10^{-6}$ (0.000 001) mol/litre (1000 nmol/litre) and at pH 7 $[H^+] = 10^{-7}$ (0.000 000 1) mol/litre (100 nmol/litre) a change of *one pH unit represents a ten-fold change in $[H^+]$*. This is much more than is immediately obvious from the difference in pH values. Although changes of this magnitude do not occur in the cells or in extracellular fluid during life, in pathological conditions the blood pH can change by 0.3 of a unit: because 0.3 is log 2, a *decrease of pH by 0.3* (for example from 7.4 to 7.1) represents a *doubling of $[H^+]$* from 40 to 80 nmol/litre. Again the use of the pH notation makes a very significant change in $[H^+]$ appear deceptively small. (Compare the situation if the plasma sodium concentration had changed from 140 to 280 mmol/litre).

Urinary pH is much more variable than that of blood: $[H^+]$ can increase 1000-fold (a fall of 3 pH units).

The Henderson–Hasselbalch equation. A buffer replaces a strong acid by a weaker one, so minimizing the change in pH when H^+ is added to a system. A weak acid and its conjugate base form a *buffer pair* (Table 4.1). The Henderson–Hasselbalch equation expresses the relationship between pH and a buffer pair: in aqueous solution the pH is determined by the concentration *ratio* of the acid to its conjugate base.

The bicarbonate pair, which is of great biological importance, can be used as an example. Carbonic acid (H_2CO_3) dissociates into H^+ and HCO_3^- until equilibrium is reached (in this case very much in favour of H_2CO_3), when the ratio of the two forms will remain constant (dissociation constant, K). We can therefore write:

$$K[H_2CO_3] = [H^+] \times [HCO_3^-]$$

(that is, at equilibrium, the concentration of H_2CO_3 is K times that of the product of $[H^+]$ and $[HCO_3^-]$).

Transposing

$$[H^+] = K \frac{[H_2CO_3]}{[HCO_3^-]}$$

Although some laboratories express results in terms of hydrogen ion concentration, the pH notation, which will be used in this edition, is more common.

$$pH = \log \frac{1}{[H^+]}$$

Taking reciprocals and logarithms in the equation for $[H^+]$ given above (when taking logs, multiplication becomes addition):

$$\log \frac{1}{[H^+]} = \log \frac{1}{K} + \log \frac{[HCO_3^-]}{[H_2CO_3]}$$

$$\log \frac{1}{K} \text{ is called pK}$$

Therefore

$$pH = pK + \log \frac{[HCO_3^-]}{[H_2CO_3]}$$

This is an example of the Henderson–Hasselbalch equation which is valid for any buffer pair. It is important to remember that the pH depends on the *ratio* of the concentrations of base (in this case $[HCO_3^-]$) to acid (in this case $[H_2CO_3]$).

It is not possible to measure the very low carbonic acid concentration directly, but it is in equilibrium with dissolved CO_2; if the carbon dioxide concentration is inserted in the equation in place of $[H_2CO_3]$, the overall dissociation constant is now that for the sum of the two reactions:

$$K_1[H_2CO_3] = [H^+] \times [HCO_3^-]$$
$$\text{and } K_2[CO_2] \times [H_2O] = [H_2CO_3]$$

This combined constant is usually written as K′ and the pK′ is about 6.1. The Henderson–Hasselbalch equation for the bicarbonate system then becomes:

$$pH = 6.1 + \log \frac{[HCO_3^-]}{[CO_2]}$$

In practice the partial pressure of CO_2 gas (PCO_2) in blood is measured, and the concentration in solution in plasma is derived by multiplying PCO_2 by the solubility constant for carbon dioxide. If the PCO_2 is expressed in kilopascals (kPa) this constant is 0.23: if it is in mmHg it is 0.03.

Therefore, if PCO_2 is expressed in kPa, the equation becomes:

$$\boxed{pH = 6.1 + \log \frac{[HCO_3^-]}{PCO_2 \times 0.23}}$$

We shall use this form in the rest of the chapter.

Hydrogen ion homeostasis

- *Hydrogen ions can be incorporated into water.*
This is the normal mechanism during oxidative phosphorylation.
During the conversion of H_2CO_3 to CO_2 and water, H^+ is also incorporated into water.

$$H^+ + HCO_3^- \leftrightarrow H_2CO_3 \leftrightarrow CO_2 + H_2O$$

As this is a reversible reaction H^+ will only continue to be inactivated in this way if CO_2 is removed. This results in bicarbonate depletion.
- *Buffering of hydrogen ions is a temporary measure.* The H^+ has not been excreted from the body, and the production of the weak acid of the buffer pair does cause a small change in pH (see the Henderson–Hasselbalch equation). If H^+ is not completely neutralized, or eliminated from the body, and if production continues, buffering power will eventually be so depleted that the pH will change significantly.
- *Hydrogen ions can be lost from the body only through the kidney and the intestine.* This mechanism is coupled with generation of bicarbonate ion (HCO_3^-). In the kidney this is the method by which secretion of excess H^+ ensures regeneration of buffering capacity.

Control systems

Carbon dioxide and hydrogen ions are among the potentially toxic products of aerobic and anaerobic metabolism respectively. Although most CO_2 is lost through the lungs some is converted to bicarbonate, thus providing a buffering system: inactivating one toxic product provides a means of minimizing the effects of the other.

The Henderson–Hasselbalch equation for any buffer pair is:

$$pH = pK + \log \frac{[base]}{[acid]}$$

A buffer pair is most effective at maintaining a pH near its pK. The optimum pH of the extracellular fluid is about 7.4, but the pK′ of the bicarbonate system is 6.1. Despite this apparent disadvantage bicarbonate is the most important buffer in the body. Not only does it account for over 60 per cent of the blood buffering capacity, but the bicarbonate system is central to all the other important homeostatic mechanisms for dealing with hydrogen ions; these include buffering by haemoglobin (which provides most of the rest of the blood buffering capacity) and secretion of hydrogen ions by the kidney.

Aerobic metabolism provides a plentiful supply of CO_2, the denominator in the equation:

$$pH = 6.1 + \log \frac{[HCO_3^-]}{PCO_2 \times 0.23}$$

The control of CO_2 by the respiratory centre and lungs

The partial pressure of CO_2 in plasma is normally about 5.3 kPa (40 mmHg). Maintenance of this level depends on the balance between production by metabolism and loss through the pulmonary alveoli.

The sequence of events is as follows:

- inspired oxygen is carried from the lungs to tissues by haemoglobin;
- the tissue cells use the oxygen for aerobic metabolism, and some of the carbon in organic compounds is oxidized to CO_2;
- CO_2 diffuses along a concentration gradient from the cells into the extracellular fluid and is returned by the blood to the lungs, where it is eliminated in expired air;
- the rate of respiration, and therefore the rate of CO_2 elimination, is controlled by chemoreceptors in the medullary respiratory centre which respond to the $[CO_2]$ (or pH) of the plasma and cerebrospinal fluid. If the PCO_2 rises much above 5.3 kPa (or if the pH falls) the rate of respiration rises. Normal lungs have a very large reserve capacity for CO_2 elimination.

Thus, not only is there a plentiful supply of CO_2, but the normal respiratory centre and lungs can maintain its concentration within narrow limits, so controlling the denominator in the Henderson–Hasselbalch equation.

Some diseases of the lungs, or abnormalities of respiratory control, primarily affect the PCO_2

The control of bicarbonate by the kidneys and erythrocytes

The renal tubular cells and erythrocytes use some of the CO_2 retained by the lungs to form bicarbonate. Under physiological conditions the erythrocyte mechanism makes fine adjustments to the plasma bicarbonate concentration in response to changes in PCO_2 in the lungs and tissues: the kidneys play the major role in maintaining the circulating bicarbonate concentration.

The carbonate dehydratase system

Carbonate dehydratase (carbonic anhydrase: CD) catalyses the first reaction in the chain:

$$CO_2 + H_2O \underset{CD}{\rightarrow} H_2CO_3 \rightarrow H^+ + HCO_3^-$$

Not only do erythrocytes and renal tubular cells have a high concentration of CD, but they also have a means of removing one of the products, H^+; thus both reactions continue to the right, and HCO_3^- will be produced. One of the reactants, water, is freely available, and one of the products, H^+, is being removed; HCO_3^- production would therefore be accelerated either by a rise in the intracellular concentration of the other reactant, CO_2, or by a fall in the intracellular concentration of the other product HCO_3^-.

In the normal subject, at a plasma $P\text{CO}_2$ of 5.3 kPa (a CO_2 concentration of about 1.2 mmol/litre, p. 103) erythrocytes and renal tubular cells maintain the extracellular bicarbonate concentration at about 25 mmol/litre. The extracellular ratio of $[HCO_3^-]$:$[CO_2]$ (both in mmol/litre) is just over 20:1. It can be calculated from the Henderson–Hasselbalch equation that, with a pK′ of 6.1, this ratio represents a pH very near 7.4. An increase of intracellular $P\text{CO}_2$, or a decrease in intracellular $[HCO_3^-]$ accelerates the production of HCO_3^-, and minimizes changes in the *ratio*, and therefore changes in pH.

Bicarbonate generation by the erythrocytes (Fig. 4.1)

Haemoglobin is an important blood buffer.

$$pH = pK + \log \frac{[Hb^-]}{[HHb]}$$

However, it only works effectively in cooperation with the bicarbonate system.

Erythrocytes lack aerobic pathways and therefore produce little CO_2. Plasma CO_2 diffuses into the cell along a concentration gradient, where carbonate dehydratase catalyses its reaction with water to form carbonic acid. As H_2CO_3 dissociates, much of the H^+ is buffered by haemoglobin. The concentration of HCO_3^- in the erythrocyte rises, and it diffuses into the extracellular fluid along a

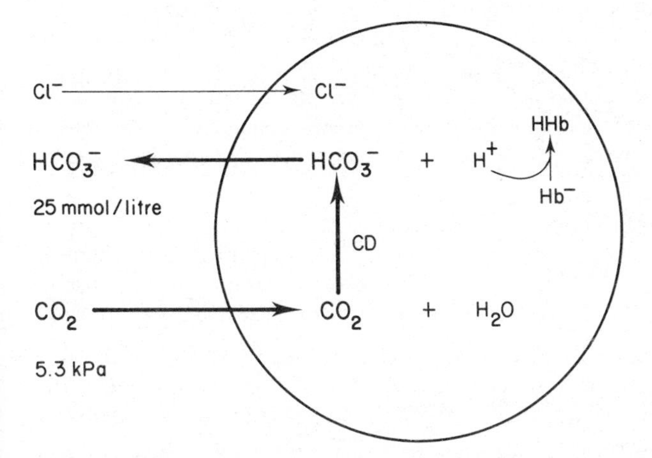

Fig. 4.1 Generation of bicarbonate by the erythrocyte.

concentration gradient; electrochemical neutrality is maintained by diffusion of chloride in the opposite direction into the cell (the 'chloride shift').

Under *physiological conditions* the higher $P\text{CO}_2$ in blood leaving tissues stimulates erythrocyte HCO_3^- production, and the lower $P\text{CO}_2$ in blood leaving the lungs slows it down; the arteriovenous difference in the ratio $[HCO_3^-]:[CO_2]$, and therefore the pH, is kept relatively constant, but there is little effect on overall HCO_3^- 'balance'.

The kidneys

Carbonate dehydratase is also of central importance in the renal mechanisms involved in hydrogen ion homeostasis. The hydrogen ion is secreted from the renal tubular cell into the lumen, where it is buffered by constituents of the glomerular filtrate. Unlike haemoglobin in the erythrocyte these buffers are constantly being replenished by continuing glomerular filtration. *For this reason, and because most of the excess H^+ can only be eliminated from the body by the renal route, the kidneys are of major importance in compensating for chronic acidosis.* Without them haemoglobin buffering capacity would soon become saturated.

Two renal mechanisms control $[HCO_3^-]$ in the extracellular fluid:

- *bicarbonate 'reabsorption'*, the predominant mechanism in maintaining the steady state. The CO_2 driving the carbonate dehydratase mechanism in the renal tubular cell is derived from filtered bicarbonate, and there is no *net* loss of hydrogen ions;
- *bicarbonate generation*, a very important mechanism for correcting acidosis, in which the levels of CO_2 or $[HCO_3^-]$ affecting the carbonate dehydratase reaction in the renal tubular cell reflect those in the extracellular fluid. There *is* net loss of hydrogen ions.

Bicarbonate 'reabsorption' (Fig. 4.2). Normal urine is almost bicarbonate-free. An amount equivalent to that filtered by the glomeruli is returned to the body by the tubular cells.

The luminal surfaces of the renal tubular cells are impermeable to bicarbonate, which therefore cannot be reabsorbed directly. It must first be converted to CO_2 in the tubular lumen, and an equivalent amount of CO_2 is converted to bicarbonate within the tubular cell. The mechanism depends on the action of carbonate dehydratase within the tubular cell, and on H^+ secretion from the cell into the lumen in exchange for the sodium filtered with the bicarbonate. The small letters in the explanation below refer to those on Fig. 4.2, in which the sequence of events is depicted.

(a) Bicarbonate is filtered through the glomeruli at plasma concentration (normally about 25 mmol/litre).

(b) Filtered bicarbonate combines with H^+ secreted by tubular cells to form H_2CO_3.

(c) The H_2CO_3 dissociates to CO_2 and water. In the proximal tubules this reaction is catalysed by carbonate dehydratase on the luminal membrane of the tubular cells: in the distal nephron, where the pH is usually lower, it probably dissociates spontaneously.

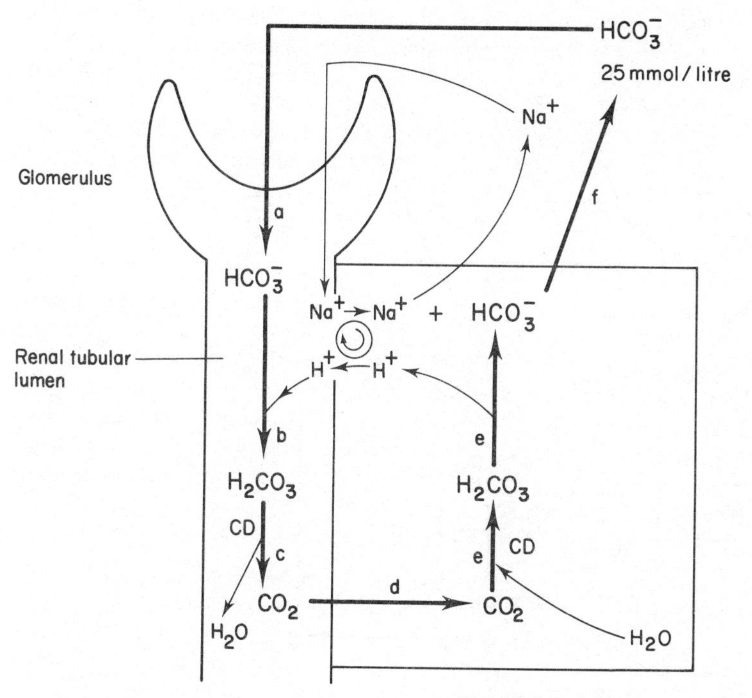

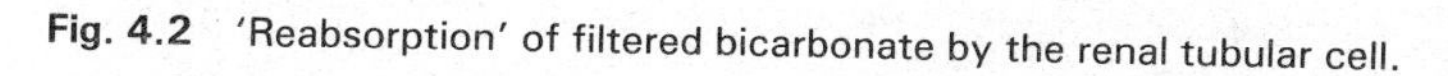
Fig. 4.2 'Reabsorption' of filtered bicarbonate by the renal tubular cell.

(d) As the luminal $P\text{CO}_2$ rises, CO_2 diffuses into tubular cells along a concentration gradient.

(e) As the intracellular $[CO_2]$ rises, carbonate dehydratase catalyses its combination with water to form carbonic acid again, which dissociates into H^+ and HCO_3^-.

(f) As H^+ is secreted (and so starts the reactions from (b) again) the intracellular concentration of HCO_3^- rises, and the bicarbonate diffuses into the extracellular fluid accompanied by the sodium reabsorbed in exchange for H^+.

This self-perpetuating cycle reclaims buffering capacity which would otherwise be lost to the body by glomerular filtration. The secreted H^+ is derived from cellular water, and is reincorporated into water in the lumen. Because there is no net change in hydrogen ion balance and no net gain of bicarbonate, *this mechanism cannot correct an acidosis* but *can maintain a steady state.*

Bicarbonate generation (Fig. 4.3). The mechanism in the renal tubular cell for generating bicarbonate is identical to that of bicarbonate 'reabsorption', but there *is* net loss of H^+ from the body as well as a net gain of HCO_3^-; this mechanism is therefore well suited to correcting acidosis.

The carbonate dehydratase mechanism may be stimulated by a rise in $P\text{CO}_2$ or a fall of $[HCO_3^-]$ within the tubular cell. In this case the rise in $[CO_2]$ is the indirect result of a rise in extracellular $P\text{CO}_2$. The renal tubular cell, unlike the erythrocyte,

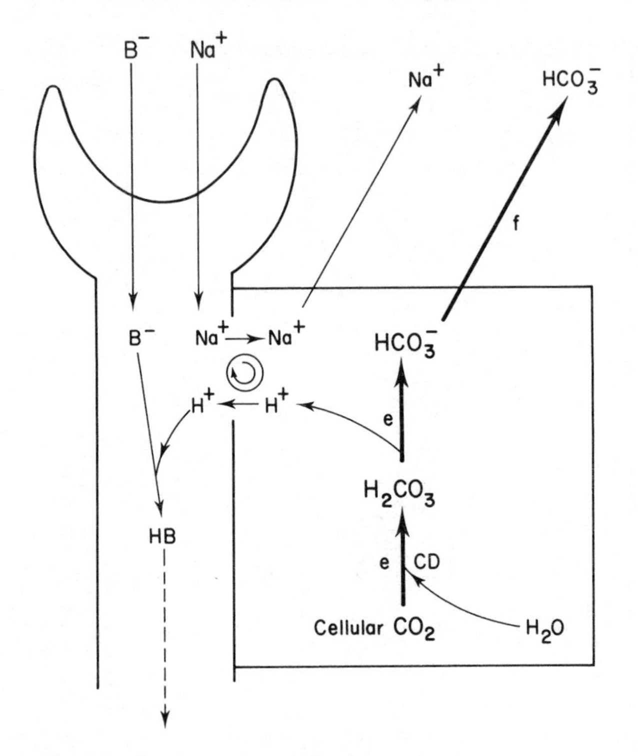

Fig. 4.3 Net generation of bicarbonate by the renal tubular cell. (B^- = non-bicarbonate base)

is constantly producing CO_2 by aerobic pathways: this CO_2 diffuses out of the cell into the extracellular fluid down a concentration gradient. An increase in extracellular $P\text{CO}_2$, by reducing the gradient, slows this diffusion, and intracellular $P\text{CO}_2$ rises. Conversely, reduction of extracellular [HCO_3^-], by increasing the gradient for this anion, increases loss of HCO_3^- from the cell.

Normally almost all filtered bicarbonate is 'reabsorbed' by the mechanism described above: once the luminal fluid is bicarbonate-free continued secretion of H^+ and generation of HCO_3^- depend on the presence of other filtered buffers (B^- in Fig. 4.3): in their absence the luminal acidity would increase so much that further H^+ secretion would be inhibited. These buffers, unlike bicarbonate, do not form compounds capable of diffusing back into tubular cells: nor is H^+ incorporated into water. *There is net loss of* H^+ *in urine* as HB. The bicarbonate formed in the cell is derived from cellular CO_2, not luminal bicarbonate, and therefore represents a *net gain in bicarbonate*. Whenever a mmol of H^+ is secreted into the urine a mmol of HCO_3^- passes into the extracellular fluid with sodium.

This mechanism is very similar to that in erythrocytes, but unlike the red cell, the renal tubular cell is exposed to a relatively constant $P\text{CO}_2$. Bicarbonate generation coupled with hydrogen ion secretion becomes very important in acidosis, when it is stimulated by a fall in extracellular [HCO_3^-] (metabolic acidosis) or a rise in extracellular $P\text{CO}_2$ (respiratory acidosis).

Extra- and intracellular buffers other than bicarbonate and haemoglobin. The plasma concentration of phosphate is only about 1 mmol/litre and this ion does not contribute significantly to blood buffering: the higher concentration in bone and inside cells is of more importance. Plasma proteins play a very minor part in hydrogen ion homeostasis: intracellular proteins have an important local role in buffering.

Urinary buffers

Urinary buffers other than bicarbonate are involved in the bicarbonate generation linked to H^+ secretion. The two most important of these are phosphate and ammonia.

The phosphate buffer pair. At pH 7.4 most of the phosphate in plasma, and therefore in the glomerular filtrate, is in the form of monohydrogen phosphate (HPO_4^{2-}), and this can accept H^+ formed by the carbonate dehydratase mechanism to become dihydrogen phosphate ($H_2PO_4^-$); bicarbonate generation can continue. The pK of this pair is about 6.8.

$$\mathrm{pH} = 6.8 + \log \frac{[HPO_4^{2-}]}{[H_2PO_4^-]}$$

Phosphate is normally the most important buffer in the urine because its pK is relatively near that of the glomerular filtrate, and because the concentration increases 20-fold to near 25 mmol/litre as water is reabsorbed from the tubular lumen.

Even in mild acidosis bone releases more phosphate ions than at normal pH, and the need for increased urinary H^+ secretion is linked with increased buffering capacity in the glomerular filtrate due to the increase in phosphate.

More monohydrogen phosphate is converted to dihydrogen phosphate as H^+ is added to tubular fluid until, at a pH below 5.5, most is in the acid form. Bicarbonate generation is very important in severe acidosis, and at low pH urinary phosphate cannot maintain the essential continued hydrogen ion secretion. The predominant urinary anion is chloride, which, because hydrochloric acid is almost completely ionized in aqueous solution, cannot act as a buffer.

The role of ammonia. As the urine becomes more acid it can be shown to contain increasing amounts of ammonium ion (NH_4^+); urinary ammonia probably allows H^+ secretion, and therefore bicarbonate formation, to continue after other buffers have been depleted.

The enzyme glutaminase, which is present in renal tubular cells, catalyses the hydrolysis of the terminal amino group of glutamine ($GluCONH_2$) to form glutamate ($GluCOO^-$) and the ammonium ion.

$$H_2O + GluCONH_2 \rightarrow GluCOO^- + NH_4^+$$

Ammonia and the ammonium ion form a buffer pair

$$\mathrm{pH} = 9.8 + \log \frac{[NH_3]}{[NH_4^+]}$$

but the pK of the system is about 9.8 and at pH 7.4 and below the equilibrium is overwhelmingly in favour of NH_4^+. NH_3 can diffuse out of the cell into the tubular lumen much more rapidly than NH_4^+, and if the luminal fluid is acid, will be retained there by avid combination with H^+ derived from the carbonate dehydratase mechanism. This allows hydrogen ion to be excreted as ammonium chloride: thus, in severe acidosis, bicarbonate formation can continue even when phosphate buffering power has been exhausted, and dissociation of NH_4^+ in the cell is maintained by removal of NH_3 into the luminal fluid.

However, H^+ is also liberated within the tubular cell during dissociation of NH_4^+. There would seem to be no advantage in buffering one secreted hydrogen ion in the lumen if, at the same time, another is produced within the cell. The fate of the glutamate ($GluCOO^-$) produced at the same time as the ammonium ion provides a possible explanation. After further deamination to 2-oxoglutarate it can be converted to glucose; *gluconeogenesis uses an equivalent amount of* H^+ *to that of* NH_4^+ *produced from glutamine.* Thus, the H^+ liberated into the cell is probably incorporated into glucose (Fig. 4.4).

As usual the net result is gain of HCO_3^-.

Both glutaminase activity and the rate of gluconeogenesis have been shown to be increased in acidosis.

Bicarbonate formation in the gastrointestinal tract (Fig. 4.5)

Intestinal mucosal cells form bicarbonate by the carbonate dehydratase mechanism. The bicarbonate may either pass into the extracellular fluid or the intestinal lumen; in either case the mechanism can only continue if H^+ is pumped in the opposite direction. Electrochemical neutrality is maintained by one of two mechanisms:

- Na^+ exchange for H^+, by a mechanism which is the opposite of that in the renal tract;
- passage of Cl^- with H^+.

Acid secretion by the stomach. The parietal cells of the stomach secrete H^+ into the lumen together with Cl^-. As H^+ Cl^- enters the gastric lumen, the gain of bicarbonate by the extracellular fluid accounts for the postprandial 'alkaline tide'. In the normal subject this is rapidly corrected by bicarbonate secretion, mainly by the pancreas, as food continues to pass down the intestinal tract. *This mechanism explains the metabolic alkalosis due to pyloric stenosis* (p. 97).

Sodium bicarbonate secretion by pancreatic and biliary cells Duodenal fluid is alkaline because of the high bicarbonate concentration of secretions entering it from the common bile duct. Sodium bicarbonate secretion by pancreatic and biliary cells, in response to stimulation by secretin, occurs by the reverse process of sodium bicarbonate reabsorption by renal tubular cells (p. 82). The pancreatic and biliary mechanisms are accelerated by the local rise in $P\text{CO}_2$ when H^+ is pumped into the extracellular fluid, and reacts with the HCO_3^- generated by gastric parietal cells. This is analogous to the stimulation of renal bicarbonate formation by the rise in luminal $P\text{CO}_2$. *Loss of large amounts of duodenal fluid may cause bicarbonate depletion* (p. 92).

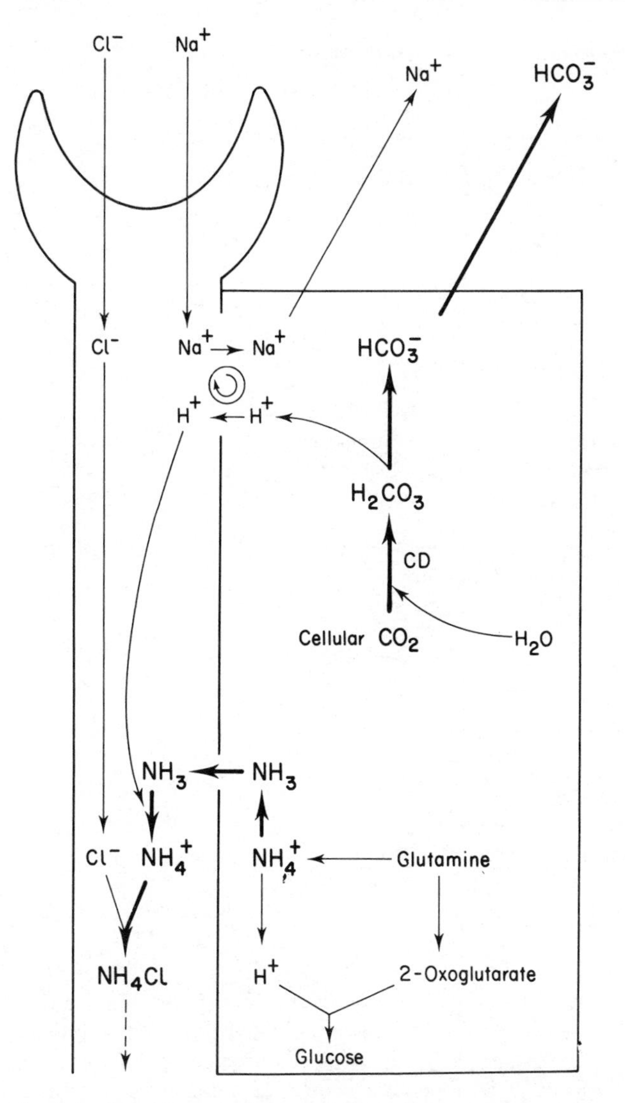

Fig. 4.4 Role of ammonia in renal bicarbonate generation.
(Modification reproduced by kind permission from Williams DL, Marks V, eds. *Biochemistry in clinical practice*, London: Heinemann Medical Books, 1983.)

Bicarbonate secretion and chloride reabsorption by intestinal cells. As fluid passes down the intestinal tract, bicarbonate enters and chloride leaves the lumen by a reversal of the gastric mucosal mechanism. Thus the gastric loss of chloride and gain of bicarbonate are finally corrected. *Preferential reabsorption of urinary chloride by this mechanism after ureteric transplantation into the ileum, ileal loops or the colon explains the hyperchloraemic acidosis which this operation may cause* (p. 93).

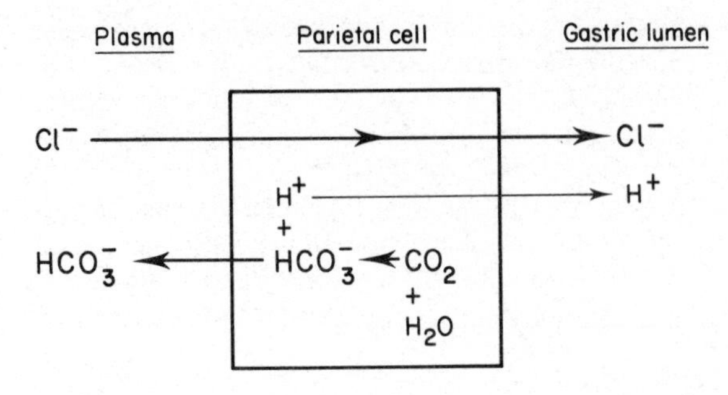

Net effect on plasma - Gain of bicarbonate and loss of chloride

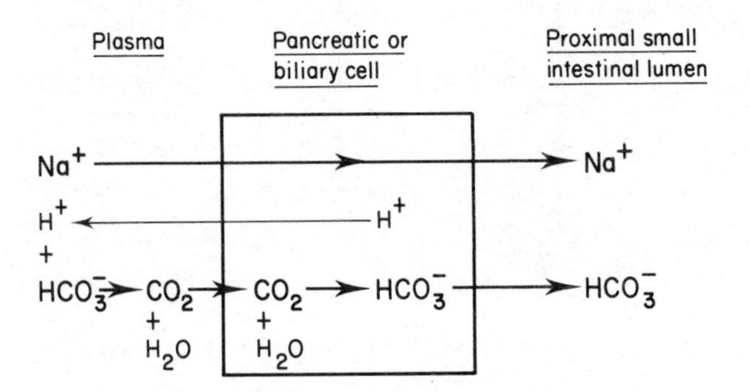

Net effect on plasma - Loss of sodium bicarbonate

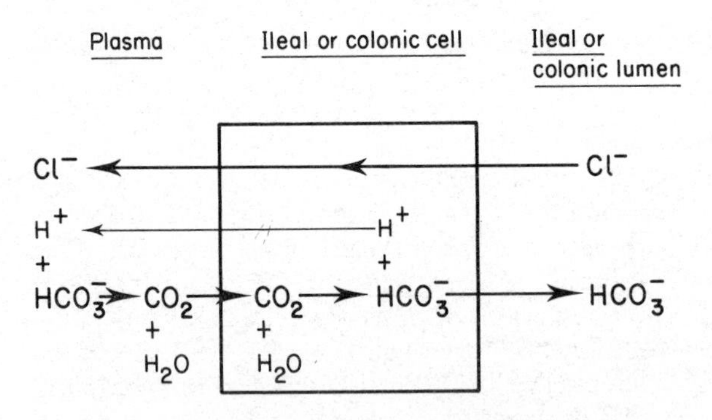

Net effect on plasma - Loss of bicarbonate and gain of chloride

Fig. 4.5 Possible mechanisms for gastrointestinal handling of bicarbonate.

In summary. CO_2 is of central importance in hydrogen ion homeostasis. Despite the apparently unfavourable pK of the bicarbonate buffer system, the tendency of H_2CO_3 to form volatile CO_2, the partial pressure of which can be controlled by the respiratory centre and lungs to about 5.3 kPa, and the ability of the renal tubular cells and erythrocytes to maintain the $[HCO_3^-]$ at 25 mmol/litre at this $P\text{CO}_2$, enable the pH to be kept above the pK of the system. Renal or pulmonary disease may impair control of extracellular pH. Inhibition of carbonate dehydratase activity (p. 94) decreases bicarbonate formation by erythrocytes and renal tubular cells, and bicarbonate reabsorption from the glomerular filtrate, and causes bicarbonate depletion.

Disturbances of hydrogen ion homeostasis

Disturbances of hydrogen ion homeostasis involve the bicarbonate buffer pair. In 'respiratory' disturbances abnormalities of CO_2 are primary, while in so-called 'metabolic' disturbances $[HCO_3^-]$ is affected early and changes in CO_2 are secondary.

Acidosis

Acidosis occurs if there is a fall in the ratio $[HCO_3^-]$: $P\text{CO}_2$ in the extracellular fluid.

$$\text{pH} = 6.1 + \log \frac{[HCO_3^-]}{P\text{CO}_2 \times 0.23}$$

In *metabolic (non-respiratory) acidosis* the primary abnormality in the bicarbonate buffer system is a *reduction in* $[HCO_3^-]$.

In *respiratory acidosis* the primary abnormality in the bicarbonate buffer system is a *rise in* $P\text{CO}_2$.

In either metabolic or respiratory acidosis the ratio of $[HCO_3^-]$:$P\text{CO}_2$, and therefore the pH, can be corrected by a change in concentration of the other member of the buffer pair in the same direction as the primary abnormality. This *compensation* may be partial or complete. The compensatory change in a metabolic acidosis is a reduction in $P\text{CO}_2$: in respiratory acidosis it is a rise in $[HCO_3^-]$.

In a fully compensated acidosis the pH is normal. However, the levels of the other components of the Henderson–Hasselbalch equation are abnormal. *All parameters can only return to normal if the primary abnormality is corrected.*

Metabolic acidosis

The primary abnormality in the bicarbonate buffer system in metabolic acidosis is a reduction in $[HCO_3^-]$ which, by reducing the ratio $[HCO_3^-]$: $[CO_2]$, causes a fall in pH. The bicarbonate may be lost in the urine or gastrointestinal tract, its generation may be impaired, or it may be used in buffering H^+ more rapidly than it can be generated. If the number of negatively charged ions (in this case HCO_3^-) is reduced, electrochemical neutrality might be maintained by replacing them with an equivalent number of other anion(s), or by loss of an equivalent number of cation(s). If metabolic acidosis is classified in this way many of the associated findings can be explained. It also allows critical evaluation of the rare need to estimate chloride.

One negative charge balances one positive charge, and some substances are multivalent (have more than one charge per mole): by contrast, if the molecular weight is divided by the valency the charges on each resulting *equivalent* will be the same as those on an equivalent of any other chemical. Most of the ions in the following discussion are monovalent, and hence the number of mmol is numerically the same as that of mEq: however, the latter notation should be used when calculating ion balance.

In the normal subject over 80 per cent of *plasma anions* is accounted for by *chloride and bicarbonate*: the remaining 20 per cent or so (sometimes referred to as '*unmeasured anion*') is made up of protein, and the normally low concentrations of, for example, urate, phosphate, sulphate, lactate and other organic anions. The protein concentration remains relatively constant, but the levels of other unmeasured anions can vary considerably in disease.

Sodium and potassium provide over 90 per cent of *plasma cation* concentration in the normal subject; the balance includes low concentrations of calcium and magnesium, which vary very little even in disease (p. 34).

The difference between the total concentration of *measured* cations, sodium and potassium, and that of *measured* anions, chloride and bicarbonate, is sometimes called the 'anion gap': it is normally about 15 to 20 mEq/litre. If we represent unmeasured anion as A^-, we can express the normal situation as:

$$\begin{array}{ccccccccccc} [Na^+] & + & [K^+] & = & [HCO_3^-] & + & [Cl^-] & + & [A^-] & \\ 140 & + & 4 & = & 25 & + & 100 & + & 19 & \text{mEq/litre} \end{array}$$

The anion gap due to $[A^-]$ is 19 mEq/litre.

In the following examples the plasma $[HCO_3^-]$ will be assumed to have fallen by 10 mmol(mEq)/litre.

Increase in $[A^-]$. In *renal glomerular dysfunction*, with relatively normal tubular function, bicarbonate generation is impaired because the amount of sodium available for exchange with H^+, and the amount of filtered buffer anion, B^- (Fig. 4.3, p. 84), available to accept H^+ are both reduced (p. 10). These buffer anions contribute to the unmeasured anion (A^-). For each mEq of buffer anion retained, one mEq fewer H^+ can be secreted, and therefore one mEq fewer HCO_3^- is generated. The retained A^- therefore replaces HCO_3^-. There is no change in chloride in uncomplicated cases. Abnormal figures are in **bold** type.

$$\begin{array}{ccccccccccc} [Na^+] & + & [K^+] & = & [HCO_3^-] & + & [Cl^-] & + & [A^-] & \\ 140 & + & 4 & = & \mathbf{15} & + & 100 & + & \mathbf{29} & \text{mEq/litre} \end{array}$$

In this example, as the $[HCO_3^-]$ has fallen from 25 to 15, $[A^-]$ has risen from 19 to 29 mEq/litre. The anion gap, entirely due to $[A^-]$, has risen by the same amount.

If renal bicarbonate generation is so impaired that it cannot keep pace with its peripheral utilization the pH will fall.

$$\mathbf{pH}\downarrow = 6.1 + \log \frac{\mathbf{[HCO_3^-]\downarrow}}{P\text{CO}_2 \times 0.23}$$

Compensation occurs as the respiratory centre responds to the acidosis. CO_2 is lost through the pulmonary alveoli and the pH returns towards normal. *In the partially or fully compensated case the PCO_2 is low.*

The cause of the low $[HCO_3^-]$ is usually obvious if plasma urea is estimated.

Correction can only occur when the *GFR increases* (for example, by correction of volume depletion). Treatment of the acidosis of irreversible glomerular dysfunction, except by dialysis, is usually contraindicated because it is dangerous to give sodium salts of, for example, bicarbonate, if the ability to excrete sodium is impaired: it is, in any case, rarely necessary.

Increase in a single anion other than chloride (X^-). Such anions are:

- acetoacetate and 3-hydroxybutyrate in **ketoacidosis;**
- lactate in **lactic acidosis.**

In both these syndromes the rise in $[X^-]$ is due to overproduction rather than underexcretion: in both there is simultaneous production of equimolar amounts of H^+. The reduction of $[HCO_3^-]$ results from its use in buffering the H^+ which accompanies the X^-.

$$HCO_3^- + H^+ \rightarrow H_2CO_3 \rightarrow CO_2 + H_2O$$

and X^- replaces the lost HCO_3^-.

$$\begin{array}{ccccccccccccc} [Na^+] & + & [K^+] & = & [HCO_3^-] & + & [Cl^-] & + & [A^-] & + & [X^-] & \\ 140 & + & 4 & = & \mathbf{15} & + & 100 & + & \underbrace{19 \quad + \quad \mathbf{10}}_{\text{'Anion gap'}} & & & \text{mEq/litre} \end{array}$$

In this example, as the $[HCO_3^-]$ has fallen from 25 to 15 mEq/litre it has been replaced by 10 mEq/litre of X^-. The anion gap of 29 mEq/litre is the sum of $[A^-]$ and $[X^-]$.

In the uncomplicated case there is no change in chloride concentration. Because the renal and erythrocyte mechanisms are functioning normally, bicarbonate formation and reabsorption from the glomerular filtrate are complete.

The falling extracellular $[HCO_3^-]$ accelerates its generation in renal tubular cells and erythrocytes (Figs 4.1, p. 81 and 4.3, p. 84), and only when this generation fails to keep pace with utilization does $[HCO_3^-]$ become abnormal. The urine becomes acid as the rate of H^+ secretion, linked with HCO_3^- generation, increases.

By this ingenious mechanism H^+ is incorporated into water after being buffered by HCO_3^-; it is therefore inactivated near the site of production.

Water is used by the renal tubular cells to replete HCO_3^-; H^+ is again liberated, but this time at a site where it can be eliminated from the body into the urine. Water is a carrier for hydrogen, transporting it in a non-toxic form from the site of production to the site of secretion (Fig. 4.6).

The CO_2 derived from buffering by bicarbonate is lost through the lungs as the respiratory centre responds to the acidosis; because of the large reserve capacity of the lungs, *the blood PCO_2 is never raised if respiratory function is normal.* The respiratory centre continues to respond to the low pH until the PCO_2 falls enough to correct the ratio of $[HCO_3^-]:PCO_2$; the acidosis is then *compensated.* Full correction by the kidneys is only possible if the rate of H^+ production is reduced to a level at which generation can keep pace with utilization.

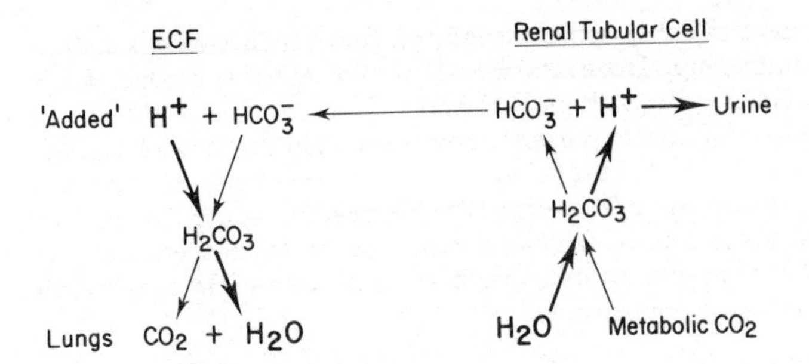

Fig. 4.6 Hydrogen ion 'shuttle' between site of buffering and the kidneys.

The diagnosis of the underlying disorder is usually obvious in this type of acidosis.

The commonest cause of *ketoacidosis* is uncontrolled diabetes mellitus (p. 214); occasionally starvation ketosis can be severe enough to cause mild acidosis.

The commonest cause of *lactic acidosis* is impairment of aerobic metabolism by a reduced tissue blood flow in the *shocked, hypotensive patient*. Drugs, such as *phenformin* or an overdosage of *salicylates* may also interfere with lactate metabolism and cause lactic acidosis (pp. 207 and 99). A drug history should be taken if a low [HCO_3^-] is found for no immediately obvious reason.

Lactic acidosis can complicate ketoacidosis if tissue perfusion is impaired due to volume depletion. In both keto- and lactic acidosis such depletion may reduce the GFR, and acidosis may be aggravated by impairment of glomerular function. It is usually impossible to determine the relative contributions made by these different factors to the fall in [HCO_3^-], and treatment rarely depends on such knowledge: it should always be directed at improving tissue perfusion by rehydration and by other measures designed to restore normal blood pressure. Diabetic ketoacidosis should be treated as outlined on p. 216. Bicarbonate infusion may sometimes be needed in very severe acidosis (pH below 7), but should be used with caution because of the danger of causing hyperosmolality and hypokalaemia (pp. 54 and 58). Respiratory dysfunction, which will impair compensatory CO_2 elimination, should also be treated.

Loss of a mixture of anions and cations. In this group electrochemical neutrality is maintained by loss of cation (sodium) and anion (bicarbonate) in equivalent amounts, and the anion gap is unaffected. The losses are variable, and the effects depend on fluid intake, but the following is an example of a possible pattern of findings.

$$[Na^+] + [K^+] = [HCO_3^-] + [Cl^-] + [A^-]$$
$$\mathbf{130} + 4 = \mathbf{15} + 100 + 19 \text{ mEq/litre}$$

Loss of intestinal secretions. Duodenal fluid, with a bicarbonate concentration about twice that of plasma, is alkaline. If the rate of loss through small intestinal fistulae exceeds that of the renal ability to regenerate HCO_3^-, the plasma [HCO_3^-] may fall enough to cause acidosis. The initial plasma electrolyte concentrations depend on the composition of the lost secretion, but are soon affected by the composition of the replacement fluid. Renal generation of HCO_3^- is accelerated by the falling concentration, and only very profuse intestinal loss lowers the plasma $[HCO_3^-]$ significantly. However, if volume depletion is severe enough to reduce the GFR, impairment of renal mechanisms may precipitate acidosis.

The *diagnosis* is usually obvious on clinical evidence.

Treatment should aim to restore the circulating volume by administering appropriate fluid (p. 54). If an adequate GFR is maintained in this way the kidneys will correct the acidosis.

Generalized renal tubular dysfunction may cause a loss of a similar mixture of ions in the urine. Renal HCO_3^- reabsorption and generation are impaired due to damage to H^+ secreting mechanisms. *The cause is suggested by polyuria.*

Increase in $[Cl^-]$. In the cases discussed so far $[Cl^-]$ is relatively unchanged. For example, in renal glomerular dysfunction, and in keto- or lactic acidosis, the reduction in $[HCO_3^-]$ is compensated by a rise in unmeasured anion, and $[Cl^-]$ is unaffected in uncomplicated cases. In intestinal loss, or generalized tubular dysfunction, $[Cl^-]$ is variable, but usually near normal. The combination of a low $[HCO_3^-]$ and a high $[Cl^-]$ is known as 'hyperchloraemic acidosis' and is rare; it can usually be predic ed on clinical grounds. The 'anion gap' in such cases is normal.

$$[Na^+] + [K^+] = [HCO_3^-] + [Cl^-] + [A^-]$$
$$140 + 4 = \mathbf{15} + \mathbf{110} + 19 \text{ mEq/litre}$$

- Such a combination would be expected if *HCO_3^- were lost in a one-to-one exchange for Cl^-*. This occurs if the *ureters are transplanted into the ileum or colon*, usually after cystectomy for carcinoma of the bladder. If chloride-containing fluid enters the lumen at this level of the intestinal tract, the cells exchange some of the chloride for HCO_3^- (Fig. 4.5, p. 88). Bicarbonate depletion may occur if urine is delivered into the ileum, ileal loops or colon. Large doses of oral bicarbonate are needed to prevent hyperchloraemic acidosis.
- Hydrogen ion secretion, and therefore bicarbonate production, are affected by tubular disease of any kind. However, if the tubular ability to handle H^+ is an isolated lesion, and other functions are relatively unimpaired, hyperchloraemic acidosis results.

In the normal tubule most filtered sodium is reabsorbed with chloride. The rest is exchanged for secreted H^+ or K^+. If H^+ secretion is impaired, and yet the same amount of sodium is reabsorbed, Na^+ must be accompanied by Cl^- or exchanged for K^+. *This type of hyperchloraemic acidosis is therefore often accompanied by hypokalaemia* – an unusual finding in acidosis, which usually causes hyperkalaemia (p. 59).

The two causes of this type of acidosis are:

- renal tubular acidosis;
- administration of carbonate dehydratase inhibitors.

Renal tubular acidosis. Renal tubular acidosis may be caused by many of the factors leading to generalized tubular dysfunction and listed on p. 16. It is more commonly due to an acquired lesion than to an inherited defect. By contrast with generalized tubular dysfunction the ability to secrete ammonia when the urinary pH falls very low is retained.

Because there is no primary glomerular lesion the plasma urea and creatinine concentrations are often normal. However, in acidosis free-ionized calcium is released from bone more rapidly than usual (p. 173), and in untreated cases calcium precipitation in the kidney may cause nephrocalcinosis; the consequent fibrosis may eventually lead to chronic renal failure. Increased bone breakdown also partly explains the phosphaturia often found in renal tubular acidosis.

There are two forms of the syndrome.

- The least rare, sometimes called '*classical renal tubular acidosis*', is due to a *distal tubular defect*. The urinary pH cannot fall much below that of plasma, even in severe acidosis. The distal luminal cells are probably abnormally permeable to H^+; this impairs the ability of the distal nephron to build up a high gradient of $[H^+]$ between the tubular lumen and cells: the re-entry of H^+ into the distal tubular cells inhibits carbonate dehydratase activity at that site, but proximal HCO_3^- reabsorption is normal. The inability to acidify the urine normally can be demonstrated by giving an ammonium chloride load (p. 105).
- In the rarer form a defect in the carbonate dehydratase mechanism *impairs bicarbonate reabsorption in the proximal tubule.* Loss of bicarbonate may cause systemic acidosis, but *the ability to form an acid urine when acidosis becomes severe is retained*: the response to ammonium chloride loading is therefore normal.

Acetazolamide therapy. Acetazolamide is used to treat glaucoma because, by inhibiting carbonate dehydratase activity in the eye, it reduces formation of aqueous humour. Inhibition of the enzyme in renal tubular cells and erythrocytes impairs H^+ secretion and HCO_3^- formation (Figs. 4.2, 4.3 and 4.4, pp. 83, 84 and 87). Hyperchloraemic acidosis sometimes complicates treatment.

Table 4.2 Changes in concentrations of ions balancing fall in $[HCO_3^-]$ in metabolic acidosis

	Measured ions		Unmeasured anions		
	$[Na^+]$	$[Cl^-]$	Mixture $[A^-]$	Single $[X^-]$	'Anion gap'
Glomerular dysfunction	N	N	↑	—	↑
Keto- and lactic acidosis	N	N	—	↑ Lactate or ketoacid	↑
Intestinal loss or renal tubular dysfunction	←———	———	VARIABLE	———	———→
Ureteric transplantation, renal tubular acidosis, acetazolamide	N	↑	N	N	N —

The last group is the rarest

It must be stressed that these are the changes in uncomplicated cases. In many clinical situations there is more than one abnormality

The changes in anion and cation concentrations which coincide with the low $[HCO_3^-]$ of metabolic acidosis are summarized in Table 4.2.

The few indications for chloride assay are discussed on p. 106.

To summarize the plasma findings in metabolic acidosis:

- *$[HCO_3^-]$ always low;*
- P_{CO_2} usually low (compensatory change);
- pH low (uncompensated or partly compensated) or normal (fully compensated);
- $[Cl^-]$ unaffected in most cases; raised after ureteric transplantation, giving acetazolamide, or in renal tubular acidosis.

Tests which may help to elucidate the cause of metabolic acidosis are:

- plasma urea estimation;
- plasma glucose estimation;
- tests for ketones in urine or plasma.

Respiratory acidosis

The findings in respiratory acidosis differ significantly from those in non-respiratory disturbances.

The primary defect is CO_2 retention, usually due to impaired alveolar ventilation (p. 99). The consequent *rise* in P_{CO_2} is the constant finding in respiratory acidosis. As in the metabolic disturbance, the acidosis is accompanied by a fall in the ratio of $[HCO_3^-]$: P_{CO_2}.

$$\mathbf{pH}\downarrow = 6.1 + \log \frac{[HCO_3^-]}{\mathbf{P_{CO_2}}\uparrow \times 0.23}$$

Compensatory changes in $[HCO_3^-]$ are initiated by the acceleration of the carbonate dehydratase mechanism in erythrocytes and renal tubular cells by the high P_{CO_2}; $[HCO_3^-]$ generation is speeded up, tending to compensate for the raised P_{CO_2} (Figs. 4.1, p. 81 and 4.3, p. 84). The urine becomes more acid as more H^+ is secreted.

In **acute respiratory failure** (for example due to bronchopneumonia or status asthmaticus) both the erythrocyte and renal tubular mechanisms increase the rate of HCO_3^- generation as soon as the P_{CO_2} rises. In the short term there is not time for renal production to contribute much to the plasma levels, nor for haemoglobin buffering power to become saturated, and a relatively high proportion of the slight rise in plasma $[HCO_3^-]$ is derived from erythrocytes. This degree of compensation is rarely adequate to prevent a fall in pH.

In **chronic respiratory failure** (for example due to chronic obstructive airways disease) the renal tubular mechanism is of prime importance. Haemoglobin buffering power is of limited capacity, but, so long as the glomerular filtrate provides an adequate supply of sodium for exchange with H^+ and of buffers to accept H^+, tubular cells continue to generate bicarbonate until the ratio $[HCO_3^-]$: P_{CO_2} is normal. In stable chronic respiratory failure the pH may be normal despite a very high P_{CO_2}, with a plasma bicarbonate level as high as twice normal.

The blood findings in respiratory acidosis are:

- $P\text{CO}_2$ always raised.

In *acute* respiratory failure:

- pH low;
- $[HCO_3^-]$ high normal, or slightly raised.

In *chronic* respiratory failure:

- pH normal or low, depending on chronicity (time for compensation to occur);
- $[HCO_3^-]$ raised.

Mixed metabolic and respiratory acidosis

If CO_2 retention occurs with metabolic acidosis, some of the extra bicarbonate generated as a compensatory response to the high $P\text{CO}_2$ is used to buffer acid other than H_2CO_3. The rise in $[HCO_3^-]$ is impaired, more CO_2 is produced during buffering, and the pH falls more than during CO_2 retention alone. This combination is common in the 'respiratory distress syndrome' of the newborn and after cardiac arrest in adults, when tissue hypoxia causes lactic acidosis. In adults or children it may also be due to coexistence of, for example, respiratory disease with ketoacidosis or renal failure. In such cases the plasma $[HCO_3^-]$ may be high, normal or low, depending on the relative contributions from respiratory and metabolic components.

Alkalosis

Alkalosis occurs if there is a rise in the ratio $[HCO_3^-]$: $P\text{CO}_2$ in the extracellular fluid.

$$\text{pH} = 6.1 + \log \frac{[HCO_3^-]}{P\text{CO}_2 \times 0.23}$$

In *metabolic alkalosis* the primary abnormality in the bicarbonate buffering system is a *rise in* $[HCO_3^-]$. There is little compensatory change in $P\text{CO}_2$.

In *respiratory alkalosis* the primary abnormality is a *fall in* $P\text{CO}_2$. The compensatory change is a fall in $[HCO_3^-]$.

The primary products of metabolism are hydrogen ions and CO_2, not hydroxyl ions and HCO_3^-; alkalosis is therefore less common than acidosis.

The presenting clinical symptom of alkalosis may be tetany, which occurs despite a normal plasma total calcium concentration; this is due to a reduced amount of free-ionized calcium in a relatively alkaline medium (p. 173).

Metabolic alkalosis

A primary rise in plasma $[HCO_3^-]$ may occur in three situations:

- **Generation of bicarbonate by the kidney in potassium depletion.** This is discussed on p. 58.

- **Administration of bicarbonate.** Ingestion of large amounts of bicarbonate to treat indigestion, or during intravenous bicarbonate infusion. Usually the cause and treatment are both obvious.

- **Generation of bicarbonate by the gastric mucosa** when hydrogen and chloride ions are lost in *pyloric stenosis or by gastric aspiration*. This causes *hypochloraemic alkalosis*.

Pyloric stenosis. Secretion of H^+Cl^- into the gastric lumen is accompanied by the generation of equivalent amounts of HCO_3^- by the parietal cells (Fig. 4.5 p. 88). Normally this is followed by secretion of HCO_3^- into the duodenum and, if there is free communication between the stomach and duodenum, the loss of secreted HCO_3^- in vomitus counteracts the effect of its generation by the gastric mucosa. Vomiting usually causes little disturbance of hydrogen ion balance, the effects being due to volume and electrolyte loss.

The vomiting of pyloric stenosis also causes volume depletion. However, marked obstruction between the stomach and duodenum reduces loss of HCO_3^-, Na^+ and K^+. Loss of H^+ stimulates its continued production by the carbonate dehydratase of gastric mucosal cells, with generation of equivalent amounts of HCO_3^-. H^+ continues to be lost and if duodenal secretion and renal excretion of bicarbonate are insufficient to correct for the rise in plasma $[HCO_3^-]$ alkalosis may result.

$$\mathbf{pH}\uparrow = 6.1 + \log \frac{\mathbf{[HCO_3^-]}\uparrow}{P\text{CO}_2 \times 0.23}$$

Compensation for the alkalosis of pyloric stenosis is relatively ineffective. Although acidosis stimulates the respiratory centre, alkalosis cannot usually depress it sufficiently to bring the pH to normal: respiratory inhibition not only leads to CO_2 retention, but also causes hypoxia, which can override the inhibitory effect of alkalosis on the centre (p. 99). Thus CO_2 may not be retained in adequate amounts to compensate for the rise in plasma $[HCO_3^-]$.

Correction of the alkalosis of pyloric stenosis depends on the rate of urinary bicarbonate loss exceeding that of gastric production. The rising extracellular $[HCO_3^-]$, derived from gastric mucosal cells, inhibits the formation of H^+ and HCO_3^- in renal tubular cells, and diminished renal secretion of H^+ reduces HCO_3^- 'reabsorption' (Fig. 4.2 p. 83). The resulting urinary HCO_3^- loss may correct for the increased generation by the gastric mucosa.

Two factors may impair this capacity to lose HCO_3^- in the urine and so may aggravate the alkalosis.

- *A reduced GFR* due to volume depletion limits the total amount of HCO_3^- which can be lost.
- If there is severe hypochloraemia due to gastric Cl^- loss, the *chloride concentration in the glomerular filtrate may be reduced.* Isosmotic sodium reabsorption in the proximal tubule depends on passive reabsorption of chloride along the electrochemical gradient. A reduction in available chloride limits isosmotic reabsorption, and more sodium becomes available for exchange with H^+ and K^+. The urine becomes inappropriately acid, and H^+ secretion stimulates inappropriate HCO_3^- reabsorption. The increased K^+ loss aggravates the hypokalaemia due to alkalosis.

Thus, vomiting due to pyloric stenosis may cause:

- hypochloraemic alkalosis;
- hypokalaemia;
- mild uraemia and haemoconcentration due to volume depletion.

The hypokalaemia may only become apparent when the plasma volume has been repleted. It should be anticipated.

Pyloric stenosis is usually treated before severe hypochloraemic alkalosis develops. Nevertheless, the typical changes in bicarbonate and chloride may indicate the diagnosis. The plasma chloride concentration may be as much as 80, rather than the usual 40, mmol/litre lower than that of sodium.

Treatment of the biochemical disorder of pyloric stenosis. If renal function is normal, water and chloride should be replaced by infusing a large volume of at least isosmolar saline. The restoration of the GFR and the correction of hypochloraemia will enable the kidney to correct the alkalosis. Potassium should be added if the plasma potassium concentration is low normal or low (p. 68).

Respiratory alkalosis

A primary fall in $P\text{CO}_2$ is due to abnormally rapid or deep respiration *when the CO_2 transport capacity of the pulmonary alveoli is relatively normal.* The causes are:

- *hysterical overbreathing*, which overrides normal respiratory control;
- *raised intracranial pressure or brain stem lesions*, which may stimulate the respiratory centre;
- *hypoxia*, which may also stimulate the respiratory centre (p. 99);
- *pulmonary oedema*;
- *lobar pneumonia*;
- *pulmonary collapse or fibrosis*;
- *excessive artificial ventilation.*

The fall in $P\text{CO}_2$ slows the carbonate dehydratase mechanism in renal tubular cells and erythrocytes (Figs. 4.1, p. 81; 4.2, p. 83 and 4.3, p. 84). The *compensatory fall in [HCO_3^-]* tends to correct the pH.

It may be difficult to distinguish clinically between the overbreathing due to metabolic acidosis, in which the fall in plasma [HCO_3^-] is the primary biochemical abnormality, and that of respiratory alkalosis in which it is compensatory. In doubtful cases estimation of arterial pH and $P\text{CO}_2$ are indicated.

The arterial blood findings in respiratory alkalosis are:

- $P\text{CO}_2$ always reduced;
- [HCO_3^-] low normal or low;
- pH raised (uncompensated or partly compensated) or normal (fully compensated).

Salicylate overdosage

Salicylates stimulate the respiratory centre directly, and overdosage initially causes respiratory alkalosis. They also uncouple oxidative phosphorylation, and

Table 4.3 Summary of findings in arterial blood in disturbances of hydrogen ion homeostasis

	pH	$P\text{CO}_2$	$[HCO_3^-]$	Plasma $[K^+]$
ACIDOSIS				
Metabolic				
Initial state	↓	N	**↓**	Usually ↑ (↓ in renal tubular acidosis and acetazolamide)
Compensated state	N	↓*	**↓**	
Respiratory				
Acute change	↓	**↑**	N or ↑	↑
Compensation	N	**↑**	↑↑*	
ALKALOSIS				
Metabolic				
Acute state	↑	N	**↑**	↓
Chronic state	↑	N or slightly ↑	**↑↑**	
Respiratory				
Acute change	↑	**↓**	N or ↓	↓
Compensation	N	**↓**	↓↓*	

Bold arrows = Primary change *arrows = compensatory change

Notes 1. Potassium depletion can cause alkalosis, or alkalosis can cause hypokalaemia. *Only the clinical history can differentiate the cause of the combination of hypokalaemia and alkalosis*

2. Overbreathing causes a low $[HCO_3^-]$ in respiratory alkalosis. Metabolic acidosis, with a low $[HCO_3^-]$, causes overbreathing. *Only measurement of blood pH and* $P\text{CO}_2$ *can differentiate these two.*

the consequent impairment of aerobic pathways superimposes a lactic acidosis on the respiratory alkalosis. Both these effects lower plasma $[HCO_3^-]$, but the pH may be high if respiratory alkalosis is predominant, normal if the two 'cancel each other out', or low if metabolic acidosis is predominant. Only measurement of blood pH can reveal the true state of hydrogen ion balance.

The possible findings in disturbances of hydrogen ion homeostasis are summarized in Table 4.3. *The student should read the section on investigation of hydrogen ion homeostasis* *(p. 103)* *to assess how many of these tests are really needed.*

Blood gas levels

In respiratory acidosis it may be important to know the partial pressure of oxygen ($P\text{O}_2$) as well as the pH, $P\text{CO}_2$ and $[HCO_3^-]$.

Normal gaseous exchange across the pulmonary alveoli involves loss of CO_2 and gain of O_2. However, in disease a fall in $P\text{O}_2$ and a rise in $P\text{CO}_2$ do not always coexist. The reasons are as follows.

- *CO_2 is much more soluble than O_2 in water*, and its rate of diffusion is about 20 times as high. For example, in *pulmonary oedema diffusion of O_2 across alveolar walls is hindered by oedema fluid and arterial* $P\text{O}_2$ falls. The hypoxia and alveolar distension stimulate respiration and CO_2 is 'washed out'. However, the rate of oxygen transport through the fluid cannot be increased enough to restore normal arterial PO_2. The result is a *low or normal* PCO_2 and a low PO_2. Only in very severe cases is the PCO_2 raised.

- *Haemoglobin in arterial blood is normally 95 per cent saturated with oxygen*, and the very little O_2 in simple solution in the plasma is in equilibrium with oxyhaemoglobin. Increased respiration at atmospheric PO_2 cannot significantly increase oxygen carriage in blood leaving normal alveoli, but can reduce the PCO_2. Breathing pure oxygen increases arterial PO_2, but not haemoglobin saturation.

In conditions such as *lobar pneumonia, pulmonary collapse and pulmonary fibrosis or infiltration* not all alveoli are affected to the same extent, or in the same way.

- Some alveoli are unaffected. At first the gaseous composition of blood leaving them is normal. Increasing the rate or depth of respiration may later lower the PCO_2 considerably, but does not alter either the PO_2 or the haemoglobin saturation.
- Obstruction of small airways means that air cannot reach those alveoli supplied by them; the composition of blood leaving them will be near that of venous blood (there is a right to left shunt). Only if increased ventilation can overcome the obstruction will it affect the low PO_2 and high PCO_2.
- Some alveoli may have normal air entry, but much reduced blood supply. This is 'dead space'. Increased ventilation will have no effect, because there is no blood to interact with the increased flow of gases.

Blood from unaffected alveoli and from those with obstructed airways mixes in the pulmonary vein before entering the left atrium. The high PCO_2 and low PO_2 stimulate respiration, and if there are enough unaffected alveoli, the very low PCO_2 in blood leaving them may compensate for the high PCO_2 in blood from poorly aerated alveoli; by contrast, neither the PO_2 nor the haemoglobin saturation will be significantly altered by mixing. The systemic arterial blood will then have a *low or normal* PCO_2 with a low PO_2.

If the proportion of affected to normal alveoli is very high, PCO_2 cannot adequately be corrected by hyperventilation and there will be a *high arterial* PCO_2 with a low PO_2.

If there are *mechanical or neurological lesions impairing respiratory movement, or obstruction of large, or most of the small, airways* (as in an acute asthmatic attack), almost all alveoli will have a normal blood supply; the poor aeration will cause a *high* PCO_2 and low PO_2 in the blood draining them, and therefore in systemic arterial blood. Occasionally stimulation of respiration by alveolar stretching can maintain a normal, or even low, PCO_2 early in an acute asthmatic attack.

In the list of examples given below the conditions marked * may fall into either group, depending on the severity of the disease.

Low arterial PO_2 with low or normal PCO_2:

- pulmonary oedema (diffusion defect);
- lobar pneumonia;
- pulmonary collapse*;
- pulmonary fibrosis or infiltration*.

Low arterial PO_2 with a high PCO_2:

- impairment of movement of the chest:
  chest injury;
  gross obesity;
  ankylosing spondylitis.
- neurological lesions affecting the respiratory drive;
- neurological lesions, such as poliomyelitis, affecting innervation of respiratory muscles;
- extensive airway obstruction:
  chronic obstructive airways disease;
  severe asthma;
  laryngeal spasm.
- bronchopneumonia;
- pulmonary collapse*;
- pulmonary fibrosis or infiltration*.

Summary

Hydrogen ion homeostasis

1. CO_2 is of central importance in hydrogen ion homeostasis. Arterial blood PCO_2 is controlled by the respiratory centre at about 5.3 kPa.
2. At a PCO_2 of 5.3 kPa the carbonate dehydratase mechanism in erythrocytes and renal tubular cells maintains the plasma $[HCO_3^-]$ at about 25 mmol/litre.
3. The H^+ produced in erythrocytes is buffered by haemoglobin. It is of physiological importance, but, because of its limited capacity, only plays a minor role in correcting abnormalities in H^+ balance.
4. Renal tubular cells secrete H^+ into the urine in exchange for Na^+. H^+ secretion is essential for HCO_3^- 'reabsorption' and net generation.
5. Normal urine is almost HCO_3^- free. Generation of HCO_3^- to replace its use in buffering depends on the availability of urinary buffers, especially HPO_4^{2-}.
6. Renal correction of either acidosis or alkalosis depends on a normal GFR.
7. A reduction in the ratio $[HCO_3^-]$: PCO_2 causes acidosis. Although the *ratio* is normal in compensated acidosis, both $[HCO_3^-]$ and PCO_2 are abnormal.
8. Acidosis may be due to excessive H^+ production, to dysfunction of the lungs or kidneys, or to excessive loss of HCO_3^-.
9. An increase in the ratio $[HCO_3^-]$: PCO_2 causes alkalosis. In compensated alkalosis the ratio is normal, but the levels of both HCO_3^- and CO_2 are abnormal.

Blood gas levels

Low PO_2 and normal or low PCO_2

1. Carbon dioxide is much more soluble in water than oxygen. Arterial PCO_2 is therefore less affected than PO_2 in pulmonary oedema; it may even be low because of respiratory stimulation.

2. Arterial blood is 95 per cent saturated with oxygen. Increased respiration cannot increase oxygen carriage from normal alveoli, but can reduce the $P\text{CO}_2$. If some alveoli have a normal blood supply, but are poorly ventilated ('ventilation-perfusion mismatch') the mixture of 'shunted' blood with that from normal alveoli results in a low $P\text{O}_2$ and a normal or low $P\text{CO}_2$ in systemic arterial blood.

Low $P\text{O}_2$ and high $P\text{CO}_2$

3. The $P\text{O}_2$ is low and the $P\text{CO}_2$ is high if there is widespread alveolar hypoventilation, because neither gas can be exchanged adequately.

Further reading

Newsholme E A, Leech A R. Metabolism and acid base balance. In: *Biochemistry for the medical sciences*. New York: John Wiley and Sons, 1984; 509–35.

Zilva J F. Disorders involving changes in hydrogen ion and blood gas concentrations. In: Williams D L, Marks V, eds. *Biochemistry in clinical practice*. London: Heinemann, 1983:24–43.

Morris R C, Sebastian A. Renal tubular acidosis and the Fanconi syndrome. In: Stanbury J B, Wyngaarden J B, Fredrickson D S, Goldstein J L, Brown M S. eds. *The metabolic basis of inherited disease*. **5th ed.** New York: McGraw-Hill, 1983:1808–43.

Investigation of hydrogen ion disturbances

Measurements which may be used to assess hydrogen ion balance

Measurement of blood pH indicates only whether there is overt acidosis or alkalosis. If the pH is abnormal the primary abnormality may be in the control of CO_2 by the lungs or respiratory centre, or in the balance between bicarbonate utilization in buffering and its reabsorption and regeneration by renal tubular cells and erythrocytes. A normal pH, however, does not exclude a disturbance of these pathways: compensatory mechanisms may be maintaining it. Chemical assessment of these factors can only be made by measuring components of the bicarbonate buffer system. The concentration of dissolved CO_2 is calculated by multiplying the measured $P\text{CO}_2$ by the solubility constant of the gas (0.23 if $P\text{CO}_2$ is in kPa, or 0.03 if it is in mmHg).

pH and $P\text{CO}_2$. There is a significant arteriovenous difference in pH and $P\text{CO}_2$, and both must be measured in arterial, not venous, blood. Whole, heparinized blood must be used because the result of $P\text{CO}_2$ estimation by some methods may depend on the presence of erythrocytes. The technique of sample collection is important, and details are given below.

Measurement of blood $[HCO_3^-]$. There are two methods which may be used to estimate the circulating bicarbonate concentration.

Plasma total CO_2 ($T\text{CO}_2$): 'plasma bicarbonate concentration' is probably the most commonly measured index of hydrogen ion homeostasis: the specimen tube should be full, to minimize *in vitro* loss of CO_2 into a large dead space of air during centrifugation. If pH and $P\text{CO}_2$ estimations are not needed the assay has the advantage that venous blood can be used, and that it can be performed together with those for urea and electrolytes. It is an estimate of the sum of plasma bicarbonate, carbonic acid and dissolved CO_2. At pH 7.4 the ratio of $[HCO_3^-]$ to the other two components is about 20 to 1 and at pH 7.1 is still 10 to 1. Thus, if the $T\text{CO}_2$ were 21 mmol/litre, $[HCO_3^-]$ would contribute 20 mmol/litre at pH 7.4 and just over 19 mmol/litre at pH 7.1. Only 1 mmol/litre and just under 2 mmol/litre respectively would come from $H_2CO_3 + CO_2$. *Thus $T\text{CO}_2$ is effectively a measure of plasma bicarbonate concentration.*

The bicarbonate concentration may be calculated, usually by a microcomputer in the blood-gas analyser, from the Henderson–Hasselbalch equation, using the measured values of pH and $P\text{CO}_2$ (in kPa) in whole arterial blood.

$$\text{pH} = 6.1 + \log \frac{[HCO_3^-]}{P\text{CO}_2 \times 0.23}$$

It expresses the whole blood $[HCO_3^-]$ and for the reasons discussed above, usually agrees well with plasma $T\text{CO}_2$. It is the estimate of choice if the other two parameters are being measured.

Collection of specimens for blood gas estimations

1. *Arterial* specimens are preferable to capillary ones.
2. The syringe should be moistened with *heparin* and the specimen well mixed.

WARNINGS:

- Excess heparin may dilute the specimen and cause haemolysis.
- If *sodium* heparin is used do *not* estimate sodium on the same specimen. The resultant factitious hypernatraemia is often misinterpreted, and is then a danger to the patient.

3. Gas exchange with the atmosphere should be minimized by *leaving the specimen in the*

syringe and expelling any air bubbles at once. The nozzle should be stoppered.

4. The effect on pH of anaerobic erythrocyte metabolism should be minimized by *performing the assay as soon as possible. The specimen should be kept cool.*

In newborn infants arterial puncture may be technically difficult, but should be used whenever possible. If it is really necessary to perform assays on capillary specimens the following precautions are essential:

1. In order that the composition of the blood is as near arterial as possible the area from which the specimen is taken should be warm and pink. If there is peripheral cyanosis results may be dangerously misleading.
2. The blood should flow freely. Squeezing the skin while sampling may dilute the specimen with interstitial fluid.
3. The capillary tubes must be heparinized, and mixing with the blood must be complete.
4. The tubes must be completely filled with blood. Air bubbles invalidate the results.
5. The ends of the tubes should be sealed immediately.

Suggested diagnostic procedure

We have discussed the mechanisms behind the abnormal findings in disorders of hydrogen ion balance. Understanding is essential for critical assessment of the value of investigation for diagnosis and management. Moreover, without such understanding the results may be dangerously misinterpreted.

Interpretation of plasma $T\text{CO}_2$

As explained above, plasma $T\text{CO}_2$ is effectively a measure of plasma $[HCO_3^-]$. It is safer and less unpleasant for the patient if venous rather than arterial blood is sampled. A careful assessment should be always be made of whether knowledge of the other parameters will really be helpful.

The level of plasma bicarbonate *alone* tells nothing about the state of hydrogen ion balance. For example, a low concentration may be associated with compensated or uncompensated metabolic acidosis, or respiratory alkalosis. The pH is determined by the ratio of $[HCO_3^-]$ to $[CO_2]$, according to the Henderson–Hasselbalch equation. Nevertheless, with a little thought, the $T\text{CO}_2$ usually provides adequate information for clinical purposes.

We suggest the following procedure.

If the $T\text{CO}_2$ is low

(a) *Exclude artefactual causes* due to *in vitro* loss of CO_2 from a small specimen, or one that has been standing for some hours.

(b) *Reassess the clinical picture*, particularly noting the presence of:

- evidence suggesting renal dysfunction (Chapter 1);
- hypotension, volume depletion or other evidence of poor tissue perfusion;
- diarrhoea or intestinal fistulae;
- a history of ureteric transplantation into the ileum, ileal loops or colon;
- a drug history with special reference to biguanides such as phenformin, or acetazolamide.

(c) *Estimate plasma urea and glucose concentrations* and test the urine for ketones.

In the great majority of cases the diagnosis will now be obvious. If it is not, among the few remaining possibilities are:

- respiratory alkalosis due to overbreathing;
- renal tubular acidosis.

In such rare circumstances estimation of arterial blood pH and $P\text{CO}_2$ may help to differentiate respiratory alkalosis from metabolic acidosis. If renal tubular acidosis is suspected, the finding of a high plasma chloride level strengthens the suspicion: an ammonium chloride loading test should be performed.

Ammonium chloride loading test for distal renal tubular acidosis

This test is not necessary if the pH of a urinary specimen collected overnight is already less than 5.5.

The ammonium ion (NH_4^+) is potentially acid because it can dissociate to ammonia and H^+. After ingestion of ammonium chloride the kidneys usually secrete the H^+, and the urinary pH falls.

Procedure. No food or fluid is taken after midnight.

08.00 h. Ammonium chloride is given orally in a dose of 0.1 g/kg body weight.

Urine specimens are collected hourly and the pH of each specimen measured *immediately by the laboratory.*

Interpretation. In normal subjects the urinary pH falls to 5.5 or below at between 2 and 8 hours after the dose. In generalized tubular disease the response of the functioning nephrons may give a normal result. In *distal renal tubular acidosis* this degree of acidification does not occur. Urinary acidification is normal in proximal tubular acidosis.

If the $T\text{CO}_2$ is high

(a) *Reassess the clinical picture*, particularly noting whether:

- there is obstructive airways disease;
- there is a cause for potassium depletion, such as the taking of potassium-losing diuretics;
- bicarbonate has been ingested or infused;
- there is severe vomiting which, especially if there is a history of dyspepsia, might indicate pyloric stenosis.

(b) *Estimate plasma potassium* concentration.

Indications for arterial pH and $P\text{CO}_2$ estimation

In metabolic acidosis the fall of plasma [HCO_3^-] is the primary abnormality of the bicarbonate buffering system, and usually plasma $T\text{CO}_2$ measurement yields adequate information for clinical purposes. Similarly, a patient with chronic obstructive airways disease and a high plasma [HCO_3^-] undoubtedly has respiratory acidosis which may or may not be fully compensated. Unless there is a possibility of improving air entry into the lungs by use of physiotherapy, expectorants and antibiotics, treatment on a respirator is contraindicated. Reducing the $P\text{CO}_2$ will lead to bicarbonate loss by a reversal of the mechanisms described on p. 82; unless the patient continues to be artificially ventilated for life the $P\text{CO}_2$ will return to its initial high level once the respirator is turned off, but there will be a delay of some days before the bicarbonate reaches the previous compensatory concentration. Nothing is to be gained from the therapeutic point of view in such a case by knowing the pH and $P\text{CO}_2$, but occasionally it may be useful to know the $P\text{O}_2$.

Arterial blood estimations may be indicated if:

- there is doubt about the cause of the abnormal $T\text{CO}_2$ (for example, to differentiate metabolic acidosis from respiratory alkalosis);
- respiratory and metabolic disturbances may coexist, such as after cardiac arrest, in renal failure complicated by lung disease, or in salicylate overdosage. In such conditions the estimations are only indicated if the results will influence treatment;
- there is an acute exacerbation of chronic obstructive airways disease, or if there is acute, and potentially reversible lung disease. In such cases vigorous therapy or artificial respiration may tide the patient over until lung function improves; more precise information than the plasma $T\text{CO}_2$ concentration is needed to monitor and control treatment;
- if blood is being taken for estimation of $P\text{O}_2$.

Indications for chloride estimation

Plasma chloride levels are consistently affected in very few conditions. Hyperchloraemia occurs in the metabolic acidosis associated with ureteric transplantation, renal tubular acidosis and acetazolamide administration. Hypochloraemia occurs in the metabolic alkalosis of pyloric stenosis. If the procedure outlined above is followed the level of chloride can usually be predicted, and its estimation adds nothing to diagnostic or therapeutic precision. Chloride estimation may help in two situations:

- If there is a low $T\text{CO}_2$ of obscure origin. The finding of a high plasma $[\text{Cl}^-]$ (and therefore a normal 'anion gap') strengthens the suspicion of renal tubular acidosis. In most other causes of acidosis the $[\text{Cl}^-]$ is normal (and 'anion gap' increased).
- If a patient who is vomiting has a high $T\text{CO}_2$. The finding of an equivalently low chloride concentration favours the diagnosis of pyloric stenosis. Similar findings may be due to compensated respiratory acidosis, but the presence of severe chronic lung disease usually makes this cause obvious.

5

The hypothalamus and pituitary gland

General principles of endocrine diagnosis

Some hormones, such as insulin secreted from the pancreatic islet cells or growth hormone secreted from the anterior pituitary gland, act directly on tissues to influence metabolism. Others, such as the trophic hormones of the pituitary gland, stimulate target endocrine glands. Hormone secretion is always controlled, usually by *negative feedback*, by the circulating target gland hormone or tissue product concentration. For example, a rise in plasma thyroxine concentration inhibits secretion of thyrotrophic hormone (TSH) and hypercalcaemia inhibits that of parathyroid hormone. Trophic hormone secretion may be episodic, and be affected by such factors as circadian variation and stress. These physiological patterns of secretion must be understood if laboratory tests for endocrine disorders are to be correctly interpreted. Some features common to all endocrine glands will be mentioned in this chapter, and the function of individual glands will be discussed in more detail in the relevant chapters.

Endocrine glands may secrete *excessive* or *deficient* amounts of hormone. Abnormalities of target glands may be *primary*, or be *secondary* to dysfunction of the controlling mechanism, usually located in the hypothalamus or anterior pituitary gland. In the latter case the target gland is normal.

Hormone secretion may vary predictably over a 24 hour or longer period, or it may respond predictably to physiological or stressful stimuli. It is logical to take blood samples at a time when hormone concentrations can most readily be distinguished from normal. Accordingly, if *hypofunction* is suspected samples are taken when levels should be high, and *vice versa* when investigating *hyperfunction*.

Simultaneous measurement of both the trophic hormones and their controlling factors (whether hormones or metabolic products) may be more informative than measurement of either alone.

If results of preliminary tests are definitely abnormal the abnormality may be primary, or secondary to a disorder of one of the controlling mechanisms. If they are near the limits of the reference range it is necessary to determine whether they are abnormal. If either of these points is not clear when the results are considered together with the clinical findings, so-called 'dynamic' tests should be carried out. In such tests the response of the gland or the feedback mechanism is assessed after stimulation or suppression by administration of exogenous hormone.

Suppression tests are used mainly for the differential diagnosis of *excessive hormone secretion*. The substance (or an analogue) which normally suppresses secretion by negative feedback is administered and the response measured. *Failure to suppress* implies secretion that is not under normal feedback control *(autonomous secretion)*.

Stimulation tests are used mainly for the differential diagnosis of *deficient hormone secretion*. The trophic hormone that normally stimulates secretion is administered and the response measured. A normal response excludes an abnormality of the target gland whereas *failure* to *respond* confirms it.

In this chapter disorders of the pituitary gland and hypothalamus will be discussed. Diseases of the target endocrine organs, the adrenal cortex, gonads and thyroid gland will be considered in Chapters 6, 7 and 8. The parathyroid glands and endocrine pancreas will be discussed in Chapters 9 and 12.

Hypothalamus and pituitary gland

There is a close relationship between the neural and the endocrine systems, most obvious in the interaction between the hypothalamus and the two lobes of the pituitary gland.

The anterior and posterior lobes of the pituitary gland are developmentally and functionally distinct, but both depend for normal functioning on hormones synthesized in the hypothalamus. The hypothalamus has extensive neural connections with the rest of the brain, and it is not surprising to find that stress and some psychological disorders affect secretion of pituitary hormones, and hence those of many other endocrine glands.

Hypothalamic hormones

The hypothalamus synthesizes two main groups of hormones which are associated with the function of the posterior and anterior lobes of the pituitary gland respectively.

The hypothalamus and the posterior pituitary lobe

Three peptides are synthesized in the hypothalamus, and travel down the nerve fibres of the pituitary stalk to be stored in the posterior lobe of the pituitary gland. Their release into the bloodstream is under hypothalamic control.

These peptides are:

- ***antidiuretic hormone (ADH: arginine vasopressin; AVP)***, which is discussed in Chapter 2;

- ***oxytocin***, a hormone, structurally similar to ADH, which controls ejection of milk from the lactating breast, and which may have some role in the initiation of uterine contractions during labour (although normal labour can proceed in its absence). It can be used in obstetric practice, to induce labour;

- ***neurophysin***, the function of which is doubtful, but which may help transport and storage of the other two hormones in the posterior pituitary gland.

The hypothalamus and the anterior pituitary lobe

The hypothalamus has no direct neural connection with the anterior pituitary gland. It synthesizes small molecules (regulating hormones or factors) which are carried to the cells of the anterior pituitary lobe by the *hypothalamic portal system*. This network of capillary loops in the median eminence forms veins which, after passing down the pituitary stalk, divide into a second capillary network in the anterior pituitary gland. The high local concentration of these hypothalamic hormones stimulates or inhibits hormonal secretion from the cells of the anterior lobe into the systemic circulation.

The cells of the anterior pituitary lobe can be classified simply by their staining reactions as acidophils, basophils or chromophobes. More sophisticated techniques can identify specific hormone-secreting cells.

Acidophils are of two cell types, *somatotrophs* which secrete growth hormone (somatotrophin) and *lactotrophs* which secrete prolactin. These hormones, which are simple polypeptides with some amino acid sequences in common, mainly *affect peripheral tissues directly*. Stimulation and inhibition of secretion via the hypothalamus is markedly influenced by neural stimuli.

Basophils secrete hormones that *affect other endocrine glands*. The hypothalamic control is mainly stimulatory. There are three cell types.

Thyrotrophs secrete thyroid-stimulating hormone (TSH; thyrotrophin) which acts on the thyroid gland. *Gonadotrophs* secrete the gonadotrophins, follicle-stimulating hormone (FSH) and luteinizing hormone (LH), that act on the gonads. These hormones are structurally similar glycoproteins and secretion is influenced more by feedback than by neural mechanisms.

Corticotrophs synthesize a large polypeptide (proopiocortin), which is a precursor for both adrenocorticotrophic hormone (ACTH; corticotrophin) and β-lipotrophin (β-LPH): secretion of these hormones occurs in parallel:

- *ACTH* stimulates secretion of steroids, other than aldosterone, from the adrenal cortex: part of the molecule has melanocyte-stimulating activity and high circulating levels of ACTH are often associated with pigmentation;
- *β-LPH* is probably a precursor of endorphins, which have an opiate-like effect. The endorphins will not be considered further.

Chromophobes, once thought to be inactive, do contain secretory granules. Chromophobe adenomas often secrete hormones, particularly prolactin.

Control of anterior pituitary hormone secretion

Neural and feedback control are the two most important physiological factors influencing secretion of the anterior pituitary hormones (Fig. 5.1).

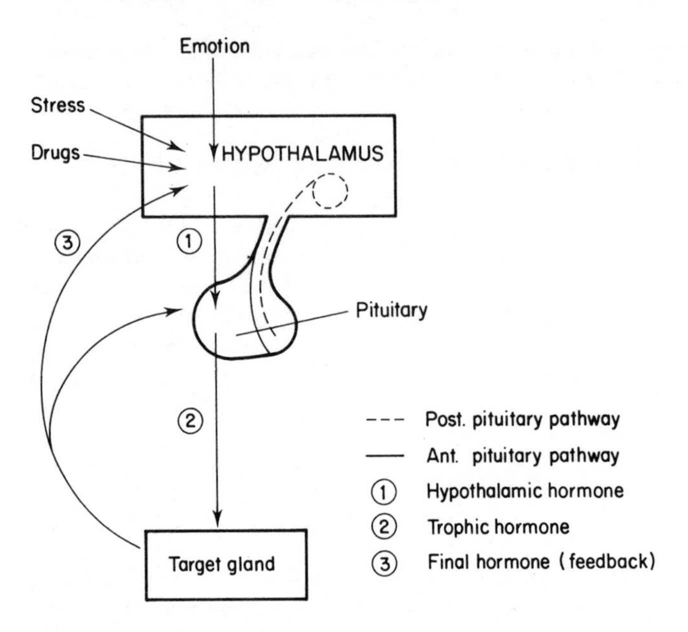

Fig. 5.1 Control of pituitary hormone secretion.

Extrahypothalamic neural stimuli modify and, at times override, other controls. Physical or emotional stress and mental illness may give similar findings to, and even precipitate, endocrine disease. The stress of insulin-induced hypoglycaemia is used to test anterior pituitary function. Stress may also stimulate secretion of the posterior pituitary hormone, ADH.

Feedback control is mediated by the levels of circulating target-cell hormones; a rising level usually suppresses trophic hormone secretion. This negative feedback may directly suppress hypothalamic hormone secretion or may modify its effect on the pituitary cell (long feedback loop). The hypothalamic hormone may also be suppressed by rising levels of pituitary hormone in a short feedback loop.

Inherent rhythms. Release of hypothalamic (and consequently pituitary) hormones occurs intermittently. There is an underlying regular secretory rhythm of some hormone secretion, and disturbances of such rhythms may be of diagnostic value. This subject will be considered further in the relevant sections.

Effect of drugs. The hypothalamus translates neural signals into chemical hormone secretion. It is not surprising therefore that drugs stimulating or blocking the action of neurotransmitters such as catecholamines, acetylcholine and serotonin influence the secretion of hypothalamic, and consequently pituitary, hormones. For example, drugs such as *chlorpromazine* interfere with the action of dopamine. This results in reduced growth hormone secretion (reduced effect of releasing factor) and increased prolactin secretion (reduced inhibition). *Bromocriptine* (2-bromo-α-ergocryptine), which has a dopamine-like action, and *L-dopa*, which is converted to dopamine, have the opposite effect in normal subjects. *Bromocriptine causes a paradoxical suppression of excessive GH secretion in acromegalics*; the reason for this response is unknown.

These effects, which have been used both in the diagnosis and the treatment of hypothalamic-pituitary disorders, will be discussed in later sections.

Assay of pituitary hormones. All the pituitary hormones are peptides, and are usually measured by immunoassay. The clinician should remember that the specificity and precision of peptide hormone assays is not comparable with that of, say, sodium estimation, and should interpret results with caution.

Disorders of pituitary hormone secretion

The main clinical syndromes associated with excessive or deficient anterior pituitary hormone secretion are shown in Table 5.1. Excessive secretion usually involves a single hormone, but deficiencies are often multiple.

Growth hormone (GH)

Growth hormone secretion from acidophil cells of the anterior pituitary gland is mainly controlled by hypothalamic *GH-releasing hormone* (GH-RH). Secretion of GH-RH, and therefore of GH, is pulsatile, and occurs about seven or eight times a day, usually associated with exercise, the onset of deep sleep, or in response to the falling plasma glucose level about an hour after meals. At other times levels may be very low or undetectable, especially in children.

Somatostatin, another hypothalamic hormone, inhibits basal GH release and the GH response to such stimuli as exercise and hypoglycaemia. Somatostatin is found not only in the hypothalamus and elsewhere in the brain, but also in the gastro-intestinal tract and pancreatic islet cells. It inhibits secretion of many gastro-intestinal hormones.

Table 5.1 Syndromes associated with primary abnormalities of anterior pituitary hormone secretion

Hormone	Excess	Deficiency
Growth hormone	Acromegaly Gigantism	Dwarfism
Prolactin	Amenorrhoea Infertility Galactorrhoea	Lactation failure
ACTH (corticotrophin)	Cushing's disease	Secondary adrenal hypofunction
TSH	Hyperthyroidism (very rare cause)	Secondary hypothyroidism
LH/FSH	Precocious puberty	Secondary hypogonadism Infertility

The main function of GH is to *promote growth*. This is illustrated by the fact that, during childhood, deficiency causes dwarfism and excess causes gigantism. GH *promotes protein synthesis* and, in conjunction with insulin, *stimulates amino acid uptake by cells*. Its effect on growth is probably mediated by *somatomedins*, polypeptide growth factors which are synthesized in many tissues, and which may act locally. Plasma levels of one of these, *somatomedin C*, correlate with GH secretion; they are also influenced by other factors, the most important of which is nutritional status. In malnutrition plasma levels are low, whereas those of GH are elevated. This suggests that plasma somatomedin C may influence GH secretion by negative feedback.

Other factors, such as adequate nutrition and thyroxine, are also needed for normal growth. The growth spurt at the onset of puberty is probably due to androgens.

Growth hormone also affects *carbohydrate and fat metabolism*. It antagonizes the insulin-mediated cell uptake of glucose, and excess secretion may produce carbohydrate intolerance. GH stimulates lipolysis and, because the resultant increase in free-fatty acids (FFA) also antagonizes insulin release and its action, it is difficult to distinguish between direct and indirect insulin antagonism.

Growth hormone secretion may be *stimulated* by:

- stress, one cause of which may be hypoglycaemia (or rapidly falling plasma glucose levels);
- glucagon;
- some amino acids, for example arginine;
- drugs such as L-dopa and clonidine.

All these stimuli have been used to assess GH secretory capacity. Oestrogens potentiate the response while it may be impaired in obese patients, in hypothyroidism and hypogonadism, in some cases of Cushing's syndrome and in patients receiving large doses of steroids.

GH secretion is *inhibited* by hyperglycaemia in the normal subject.

In *adults* growth hormone deficiency very rarely causes symptoms.

Growth hormone excess: acromegaly and gigantism

Most patients with GH excess have acidophil adenomas of the anterior pituitary gland, which may be secondary to excessive hypothalamic stimulation. The condition may therefore recur after removal of the pituitary adenoma. The clinical manifestations depend on whether the condition develops before or after fusion of the bony epiphyses.

In childhood GH excess leads to *gigantism*. Epiphyseal fusion may be delayed by accompanying hypogonadism and heights of over 2.1 m (7 feet) may be reached. Acromegalic features may develop after bony fusion, but these giants often die in early adult life from infections, cardiac failure or progressive pituitary tumour growth.

In adults, after epiphyseal fusion, GH excess causes *acromegaly*. Bone and soft tissues increase in bulk and lead to increasing size of the hands and other parts, due to soft tissue thickening. The patient may first complain that rings and shoes are becoming too small. Changes in facial appearance are often marked, due to the increasing size of the jaw and sinuses, but the gradual coarsening of the features may pass unnoticed for many years. There may be excessive hair growth and sebaceous gland secretion, and menstrual disturbances are common. Enlargement of the heart and thyroid gland may be clinically detectable, but acromegalics are usually euthyroid.

A different group of symptoms may occur due to the encroachment of the pituitary tumour on surrounding structures. For example, compression of the optic chiasma produces visual field defects.

If destruction of the gland progresses other anterior pituitary hormones may become deficient. Prolactin levels may, however, be raised (p. 116).

Impaired glucose tolerance may be demonstrated in about 25 per cent of cases. Only about half of these develop symptomatic diabetes mellitus. In most cases the pancreas can secrete enough insulin to overcome the antagonistic effect of GH. Probably only those with a diabetic tendency will develop it under these conditions.

Acromegaly is sometimes one of the manifestations of multiple endocrine adenopathy (p. 417).

Diagnosis. The diagnosis of acromegaly is suggested by the clinical features and radiological findings. Plasma GH levels are usually higher than normal and may reach several hundred mU/litre, but, because of the wide reference range, results from ambulant patients may fail to distinguish acromegalics with only moderately raised levels from normal subjects.

Plasma somatomedin C levels are raised and seem to correlate with the activity of the disease.

The diagnosis is confirmed by the demonstration of a *raised plasma GH level that is not suppressed by a rise in plasma glucose concentration*. During a glucose tolerance test on a normal subject plasma GH falls to very low levels. In acromegalics secretion is autonomous and GH levels may remain constant, may even increase, or may fall only slightly.

Measurement of plasma GH levels may also be used to monitor the efficacy of treatment.

Growth hormone deficiency

A small percentage of normally proportioned dwarfed children have growth hormone deficiency. Isolated GH deficiency is most commonly due to an idiopathic deficiency of hypothalamic GH-releasing hormone. In some cases the secretion of other hormones is also impaired. Sometimes there may be an organic disorder of the anterior pituitary gland or of the hypothalamus, and rare inherited forms have been described. The birth weight may be normal, but the rate of growth is subnormal. It is important to investigate children with reduced growth rate to identify those who may benefit from GH replacement treatment.

Emotional deprivation may be associated with growth hormone deficiency indistinguishable by laboratory tests from that due to organic causes.

Diagnosis. Basal plasma GH levels in normal children are usually low and assays under such conditions rarely exclude the diagnosis. If blood *is taken at a time when physiologically high levels are expected* the need for the more unpleasant stimulation tests may be avoided. Such times are 60 to 90 minutes after the onset of sleep and about 20 minutes after vigorous exercise.

If GH deficiency is not excluded by the above measurements, it is necessary to perform one or more *stimulation tests* (p. 151). An unequivocally normal response to these tests excludes the diagnosis and a clearly impaired one confirms it.

Once GH deficiency is established a cause should be sought by appropriate clinical and radiological means.

Hypopituitarism

Causes of hypopituitarism

- Destruction of, or damage to, the gland or the hypothalamus by a primary or secondary tumour. The most common sources of secondary tumours are the breast and the bronchus.
- Infarction, most commonly postpartum (Sheehan's syndrome) or, rarely, after other vascular catastrophes.
- Pituitary surgery or irradiation.

Rarer causes include:

- head injury;
- infections or granulomas.

Isolated hormone deficiencies may be idiopathic.

Established panhypopituitarism presents with the characteristic clinical picture described below. It is, however, not common and the suspicion of anterior pituitary hypofunction usually arises in patients:

- with clinical and radiological *evidence of a pituitary or localized brain tumour*;
- presenting with *hypogonadism, hypothyroidism or adrenocortical insufficiency* which is shown by preliminary testing to be secondary in origin;
- with *dwarfism* shown to be *due to growth hormone deficiency.*

While isolated hormone deficiency, particularly of growth hormone or gonadotrophins, may sometimes occur, several hormones are usually involved. If a deficiency of one hormone is demonstrated it is therefore important to establish if the secretion of the others is normal or impaired. This evidence is needed, not only to guide replacement therapy, but also to assess the ability of the pituitary gland to respond to stress such as that of infection or surgery.

The anterior pituitary gland has considerable functional reserve. Clinical features of deficiency are usually absent until about 70 per cent of the gland has been destroyed, unless there is associated hyperprolactinaemia (p. 140), when amenorrhoea and infertility may be early symptoms.

Consequences of hormone deficiencies

Progressive pituitary damage usually causes deficiencies of gonadotrophins and of GH first. Plasma ACTH and/or TSH levels may remain normal, or may become deficient months or even years later. The clinical and biochemical consequences, usually those of failure of the target gland, are summarized below.

- **Secondary hypogonadism** due to gonadotrophin deficiency presents as amenorrhoea, infertility, atrophy of secondary sexual characteristics and impotence or loss of libido. Characteristically axillary and pubic hair are lost. Puberty is delayed in children.

- **Retarded growth in children** may be due to growth hormone and thyroid (due to TSH) insufficiency.

- **Secondary hypothyroidism** (TSH deficiency) may be clinically indistinguishable from primary hypothyroidism.

- **Secondary adrenocortical hypofunction** (ACTH deficiency) differs from the primary form (Addison's disease) in several respects. Patients are *not* hyperpigmented because ACTH secretion is deficient, not excessive. The sodium and water deficiency of Addison's disease does not occur because aldosterone secretion, controlled by angiotensin not ACTH, is normal. However, cortisol is needed for normal water excretion, and there may be a marked *dilutional* hyponatraemia due to cortisol deficiency. It is also necessary for maintenance of normal blood pressure, and hypotension is associated with ACTH deficiency. Hyperkalaemia is unusual. Because of cortisol and/or growth hormone deficiency, these patients may be very sensitive to insulin and there may be fasting hypoglycaemia.

- **Lactation failure** due to prolactin deficiency may occur after postpartum pituitary infarction (Sheehan's syndrome). In hypopituitarism due to a tumour, however, prolactin levels are often raised and may cause galactorrhoea (secretion of breast fluid).

The patient with hypopituitarism, like the patient with Addison's disease, may die because of an inability to secrete an adequate amount of cortisol (in this case due to impaired ACTH secretion) in response to the stress caused by, for example, infection or surgery. Other life-threatening complications are hypoglycaemia, water intoxication and hypothermia.

Evaluation of anterior pituitary function

Interpretation of results of basal pituitary hormone assays is often difficult, because low levels are not necessarily abnormal and because 'normal' values do not exclude pituitary disease: the pituitary gland has a large functional reserve. The diagnosis of suspected hypopituitarism is best excluded by *direct measurement of pituitary hormones after stimulation* (p. 151).

Should direct pituitary hormone assays not be available, *indirect* evidence of deficiency should be sought by demonstrating target-gland hyposecretion that responds to administration of the relevant trophic hormone (see Chapters 6, 7, and 8); results are often difficult to interpret.

It must be remembered that *prolonged* hypopituitarism may lead to secondary atrophy of the target gland with a consequent diminished response to stimulation.

Laboratory tests only establish the presence or absence of hypopituitarism. The *cause* must be sought by clinical and radiological means.

Hypothalamus or pituitary?

In the above discussion we have not distinguished between hypothalamic and pituitary causes of pituitary hormone deficiency or, more correctly, between deficient releasing factor and deficient pituitary hormone secretion. Isolated hormone deficiencies are more likely to be of hypothalamic than of pituitary origin. The coexistence of diabetes insipidus suggests a hypothalamic disorder, but symptoms may be masked initially by ACTH, and therefore cortisol, deficiency.

Some tests evaluate both hypothalamic and pituitary function, and some only the latter. It is sometimes possible to distinguish the anatomical level of the lesion. The TSH response to TRH (p. 169), for example, differs in hypothalamic and pituitary causes of secondary hypothyroidism. In cases of hypogonadism due to gonadotrophin deficiency the differentiation on the basis of the response to gonadotrophin-releasing hormone (Gn-RH) is less clear cut. The role in diagnosis of the recently available releasing hormones for GH and ACTH is currently under investigation.

Pituitary tumours

The clinical presentation of pituitary tumours depends on the types of cell involved and on the size of the tumour.

Tumours of secretory cells may produce the clinical effects of excessive hormone secretion. Excessive

- prolactin causes infertility, amenorrhoea and varying degrees of galactorrhoea;
- GH causes acromegaly or gigantism;
- ACTH causes Cushing's syndrome.

Large tumours may present with:

- such symptoms as visual disturbances and headache due to pressure on the optic chiasma or to raised intracranial pressure;
- deficiency of some or all of the pituitary hormones.

Non-secreting tumours are difficult to diagnose by biochemical tests, although the combined pituitary stimulation test may indicate subclinical impairment of function. *Hyperprolactinaemia*, which may be asymptomatic, is a valuable biochemical marker for the presence of a pituitary tumour. Prolactin may be secreted by the tumour cells, or it may be secreted by unaffected lactotrophs if tumour growth interferes with the normal inhibition of prolactin secretion.

Summary

1. The anterior pituitary gland secretes growth hormone (GH), prolactin, adrenocorticotrophic hormone (ACTH; corticotrophin), β-lipotrophin, thyroid-stimulating hormone (TSH) and the two pituitary gonadotrophins, follicle-stimulating hormone (FSH) and luteinizing-hormone (LH). The posterior lobe secretes antidiuretic hormone (ADH; arginine vasopressin; AVP) and oxytocin.

2. The secretion of anterior pituitary hormones is controlled by regulating hormones secreted by the hypothalamus. These, in turn, are controlled by circulating levels of hormones or metabolic products (feedback), or respond to stimuli from higher cerebral centres.

3. GH controls growth and has a number of effects on intermediary metabolism. *Excessive GH* secretion causes gigantism or acromegaly. Laboratory evidence of autonomous GH secretion is obtained by failure of suppression of plasma levels of GH during a glucose tolerance test. *GH deficiency* in childhood leads to dwarfism. Diagnosis of such deficiency is made by demonstrating a subnormal GH response to appropriate stimuli.

4. Anterior pituitary hormone deficiency usually involves several hormones. Less commonly isolated deficiency occurs. The clinical features of hypopituitarism are those of gonadotrophin and sex hormone deficiencies and of secondary hypofunction of the adrenal cortex and thyroid gland. Diagnosis is made by demonstrating reduced anterior pituitary reserve after stimulation.

5. Pituitary tumours may produce excess of GH, ACTH or prolactin, or may cause hypopituitarism. In patients without obvious endocrine disturbance hyperprolactinaemia may be found.

Further reading

Ho K Y, Evans W S, Thorner M O. Disorders of prolactin and growth hormone secretion. *Clin Endocrinol Metab* 1985; **14**: 1–32.

Jadresic A, Banks L M, Child D F, Diamont L, Doyle F H, Fraser T R, Joplin G F. The acromegaly syndrome. Relation between clinical features, growth hormone values and radiological characteristics of the pituitary tumours. *Quart J Med* 1982; **51**: 189–204.

Abboud C F. Laboratory diagnosis of hypopituitarism. *Mayo Clin Proc* 1986; **61**: 35–48.

Schaff-Blass E, Burstein S, Rosenfield R L. Advances in diagnosis and treatment of short stature, with special reference to the role of growth hormone. *J Pediatr* 1984; **104**: 801–13.

6

Adrenal cortex: ACTH

The adrenal glands are divided into the functionally distinct cortex and medulla. The *cortex* is part of the hypothalamic-pituitary-adrenal endocrine system. Histologically the adult adrenal cortex is divided into three layers. The outer thin layer (*zona glomerulosa*) secretes only aldosterone. The inner two layers, the *zona fasciculata* and the *zona reticularis*, form a functional unit and secrete most of the adrenocortical hormones. The wider fourth layer present in the fetal adrenal gland disappears after birth: during fetal life one of its most important functions, in association with the placenta, is to synthesize oestriol (p. 139). Although the *medulla* is part of the sympathetic nervous system, glucocorticoids are probably needed for the final stage of adrenaline (epinephrine) synthesis.

Chemistry and biosynthesis of steroids

Steroid hormones, bile salts and the D vitamins are all derived from cholesterol, and steroid-producing tissues are rich in cholesterol. If the molecule contains 21 carbon atoms it is referred to as a C_{21} steroid. The carbon atom at position 21 of the molecule is written as C-21.

Fig. 6.1 shows the internationally agreed numbering of the 27 carbon atoms, and the lettering of the four rings of the cholesterol molecule: the products of cholesterol are also indicated. The bile acids and D vitamins are dealt with elsewhere in the book (pp. 300 and 174).

As shown in Fig. 6.1, the side chain on C-17 is the main determinant of the type of hormonal activity, but substitutions in other positions modify activity within a group.

The first hormonal product of cholesterol is pregnenolone. Several important synthetic pathways diverge from it (Fig. 6.2). The final product depends on the tissue and the enzymes it contains.

Adrenal cortex

Zonae fasciculata and reticularis.

- *Cortisol* (the most important C_{21} steroid) is formed by progressive addition of hydroxyl groups at C-17, C-21 and C-11.

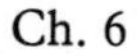

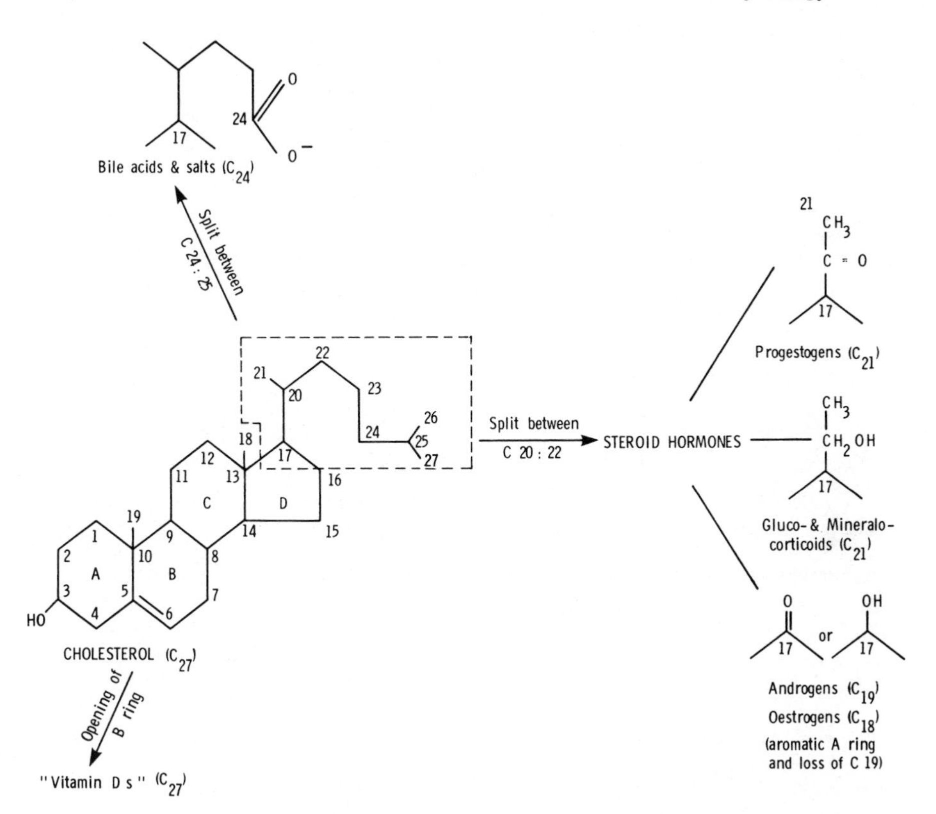

Fig. 6.1 Numbering of the steroid carbon atoms and products of cholesterol.

- *Androgens* (for example, androstenedione) are formed after removal of the side chain to produce C_{19} steroids.

Synthesis of these two groups is stimulated by ACTH secreted by the anterior pituitary gland (p. 109). ACTH secretion is influenced by negative feedback from plasma cortisol levels. Impaired cortisol synthesis, due, for example, to an inherited deficiency of 21- or 11-hydroxylase (congenital adrenal hyperplasia) results in increased ACTH stimulation with increased activity of *both* pathways. The resultant excessive androgen production causes virilization (p. 130).

Zona glomerulosa. A different series of hydroxylations produces *aldosterone*. Synthesis of this steroid is controlled by the renin-angiotensin system and not normally by ACTH (p. 27). In ACTH deficiency, therefore, aldosterone secretion is not significantly affected.

Physiology

The adrenocortical hormones can be classified into three groups depending on their predominant physiological effects.

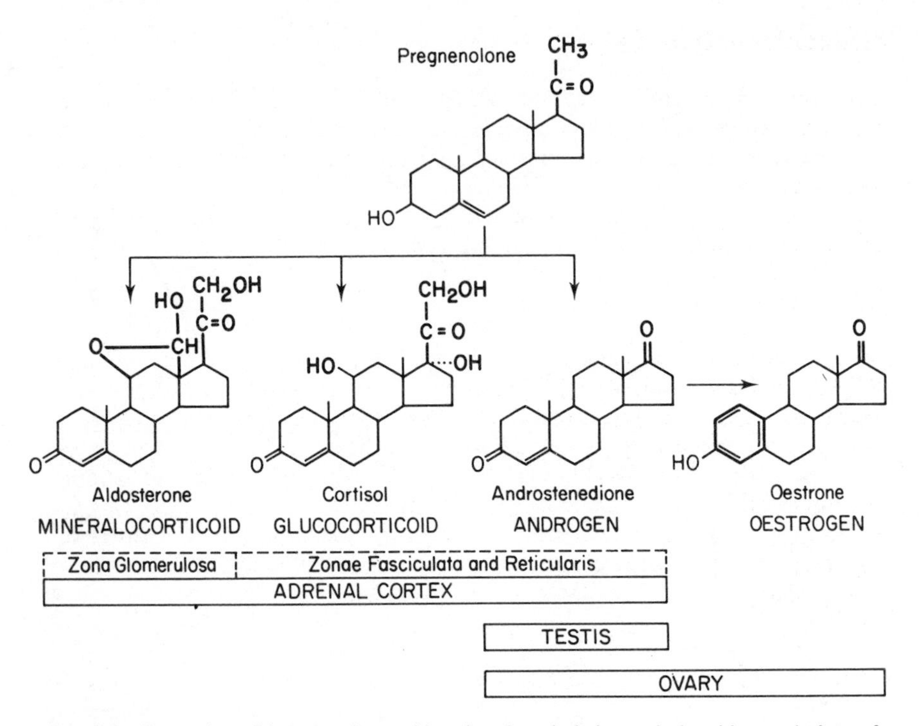

Fig. 6.2 Examples of hormonal steroids, showing their interrelationships and sites of synthesis. (*Note*: intermediate steps have been omitted.)

Glucocorticoids

Naturally occurring glucocorticoids are *cortisol* and *corticosterone*. Glucocorticoids stimulate gluconeogenesis and the breakdown of protein and fat and so oppose some of the actions of insulin. In excess they impair glucose tolerance and alter the distribution of adipose tissue.

Cortisol helps to maintain the extracellular fluid volume and normal blood pressure. Most is bound to cortisol-binding globulin (CBG; transcortin) and to albumin. Only the unbound free fraction (about 5 per cent of the total at normal levels) is physiologically active (compare with thyroxine, p. 159). Increased CBG levels, such as those that occur during normal pregnancy, produce high bound, and therefore high total, cortisol levels with normal, or only slightly raised, free fractions: low levels, such as occur in the nephrotic syndrome, have the opposite effect (again compare with thyroxine, p. 162).

The liver metabolizes glucocorticoids to inactive conjugates: because these conjugates are more water soluble than the mainly protein-bound parent hormones, they can be excreted in the urine (compare bilirubin, p. 289).

Cortisone is not secreted in significant amounts by the adrenal cortex, but may be used therapeutically. It is biologically inactive until it has been converted *in vivo* to cortisol (hydrocortisone).

Mineralocorticoids

Aldosterone is discussed on p. 27. It stimulates sodium exchange for potassium and hydrogen ions across cell membranes, and its renal action is especially important for homeostasis. Like the glucocorticoids it is excreted in the urine after hepatic conjugation.

There is some overlap in the actions of the C_{21} steroids. Cortisol in particular may, at high levels, have a significant mineralocorticoid effect.

Adrenal androgens

The main adrenal androgens are dehydroepiandrosterone (DHA) and its sulphate (DHAS), and androstenedione. They promote protein synthesis and are only mildly androgenic at physiological concentrations. The powerful androgen, testosterone, comes from the testis or ovary, not the adrenal cortex. Most circulating androgens, like cortisol, are protein bound, mainly to sex-hormone-binding globulin (SHBG) and to albumin.

There is extensive peripheral interconversion of adrenal and gonadal androgens. The end-products, *androsterone* and *aetiocholanolone*, together with DHA, are conjugated in the liver and excreted as glucuronides and sulphates.

Control of adrenal steroid secretion

The hypothalamus, anterior pituitary gland and adrenal cortex form a functional unit (the hypothalamic-pituitary-adrenal axis, Fig. 6.3).

Cortisol is secreted in response to *adrenocorticotrophic hormone* (*ACTH*; corticotrophin) from the anterior pituitary gland. ACTH secretion is, in turn, dependent on a hypothalamic *releasing hormone* (corticotrophin-releasing factor; CRF). At least three mechanisms influence CRF secretion.

Feedback mechanism. High free cortisol levels suppress and low ones stimulate CRF and ACTH secretion. This negative feedback probably acts on both the hypothalamus and on the anterior pituitary gland. The melanocyte-stimulating effect of ACTH (or of related peptides that are released at the same time, p. 109) accounts for the pigmentation of Addison's disease, in which cortisol levels are persistently low and ACTH levels therefore persistently high. Marked pigmentation may also occur in patients after bilateral adrenalectomy for Cushing's disease: in this case removal of cortisol feedback causes a further rise in ACTH levels (Nelson's syndrome).

Inherent rhythms. Sampling at frequent intervals has shown that ACTH is secreted *episodically*, each burst of secretion being followed 5 to 10 minutes later by cortisol secretion. These episodes are most frequent in the early morning (between the 5th and 8th hour of sleep) and least frequent in the few hours before sleep; plasma cortisol levels are usually *highest* between about 07.00 and 09.00 h and *lowest* between about 23.00 and 04.00 h.

ACTH and cortisol secretion usually vary inversely and the almost parallel

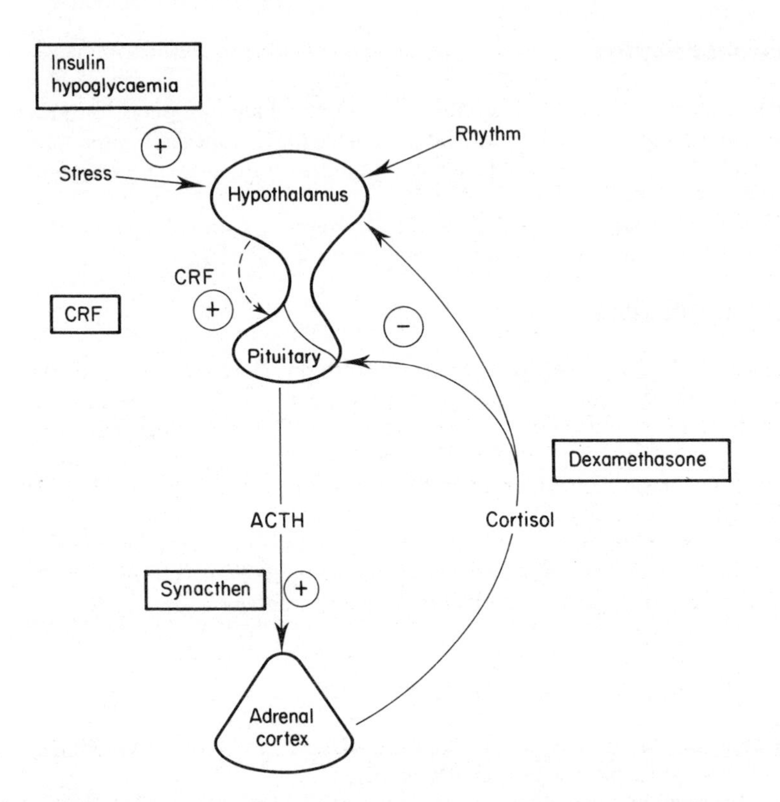

Fig. 6.3 Control of cortisol secretion (+ = stimulates; − = inhibits). Dynamic tests shown in boxes.

circadian rhythm may be due to a cyclical change in the sensitivity of the hypothalamic feedback centre to cortisol levels. Inappropriately high cortisol levels at any time of day suppress ACTH secretion and inappropriately low levels stimulate it. The first effect can be tested by the *dexamethasone suppression test*.

Loss of circadian rhythm is one of the earliest features of Cushing's syndrome.

Stress. Physical or mental stress may override the first two mechanisms and cause sustained ACTH secretion. The consequent diagnostic difficulties are discussed on p. 126. An inadequate stress response may cause acute adrenal insufficiency. The stress caused by insulin-induced hypoglycaemia can be used to test the axis.

ACTH (Corticotrophin)

ACTH is a single-chain polypeptide made up of 39 amino acids. Biological activity depends on the 24 N-terminal amino acids. A peptide consisting of this sequence has been synthesized (*tetracosactrin*; 'Synacthen') and can be used for diagnosis and treatment in place of ACTH. ACTH *stimulates cortisol* synthesis and secretion by the adrenal cortex. It has much less effect on adrenal androgen production and, at physiological levels, virtually no effect on aldosterone production.

Estimations used in the diagnosis of adrenocortical disorders

Cortisol is usually measured in plasma or in urine by immunoassay or by competitive protein binding. It is important to note that cortisol, like thyroxine, is largely protein bound. Primary changes in cortisol-binding globulin levels will alter total cortisol values (p. 120) and should not be misinterpreted as indicating adrenocortical dysfunction.

Other plasma corticosteroids and androgens may be measured by immunoassay. Their use is discussed in the relevant sections.

Plasma ACTH assay is used in the differential diagnosis of Cushing's syndrome (p. 153).

The selection of tests is discussed in the following pages. Blood is easy to collect but *plasma* levels reflect adrenocorticotrophic activity *at that moment*. Because of the episodic nature of cortisol secretion, isolated values may be misleading. The measurement of steroids in a 24-hour *urine* sample reflects the *overall daily secretion*, but it is difficult to ensure the accuracy of timed collections.

Disorders of the adrenal cortex

The main disorders of adrenocortical function are:

Hyperfunction

Excessive secretion of:

- cortisol — Cushing's syndrome;
- aldosterone — primary aldosteronism (Conn's syndrome), p. 50;
- androgens — adrenocortical carcinoma; congenital adrenal hyperplasia.

Hypofunction

- primary adrenal disorders, causing deficiency of both cortisol and aldosterone:
  Addison's disease;
  congenital adrenal hyperplasia.
- secondary to ACTH deficiency, causing deficiency of cortisol only.

Cushing's syndrome

Cushing's syndrome is caused by an excess of circulating cortisol.

Clinical and metabolic features

Mild to moderate *obesity*, typically involving the trunk and face, is the most common presenting symptom of Cushing's syndrome. Like the round, red 'moon' face it is characteristic, but not diagnostic, of cortisol excess. Although the reason for these features is not understood, many of the other clinical and metabolic disturbances can, at least partly, be explained by the known actions of cortisol.

The most prominent metabolic abnormalities reflect the glucocorticoid action. About two-thirds of the patients have impaired glucose tolerance and of these many have *hyperglycaemia* and therefore *glycosuria*. As cortisol has an opposite action to insulin, the resultant picture in some ways resembles diabetes mellitus. Protein breakdown is accelerated and the carbon chains of the liberated amino acids may be converted to glucose (gluconeogenesis); the nitrogen is lost in the urine leading to a negative nitrogen balance. This catabolic effect not only causes *osteoporosis* and *muscle wasting* and therefore *weakness*, but also thinning of the skin. The tendency to *bruising*, and the *purple striae*, most obvious on the abdominal wall, are probably due to this thinning.

The mineralocorticoid effect of cortisol leads to urinary retention of sodium, and therefore of water, with *hypertension*. Increased urinary potassium loss may cause hypokalaemia and aggravate muscle weakness.

High circulating adrenal androgen concentrations may account for the common findings of a greasy skin with acne vulgaris, and for the hirsutism and menstrual disturbances in women. Psychiatric disturbance, particularly agitated depression, may be a prominent feature.

The typical picture outlined above may vary depending on the cause of the cortisol excess.

Causes of Cushing's syndrome

The adrenal gland may be overactive because it is stimulated by ACTH, or because its hormone secretion is autonomous.

ACTH is secreted in excessive amounts in two conditions.

- *Cushing's disease* is the commonest form of Cushing's syndrome, and usually occurs in women of reproductive age. It is associated with bilateral adrenal hyperplasia, often secondary to a basophil adenoma of the anterior pituitary gland. It is not known whether the pituitary tumour is the primary lesion, or is due to excessive CRF secretion from the hypothalamus.
- Tumours other than those of the anterior pituitary gland may secrete ACTH (*ectopic ACTH secretion*) and may also produce bilateral adrenal hyperplasia.

  In *overt* ectopic ACTH secretion, usually from a small-cell carcinoma of the bronchus, the malignancy is obvious, and ACTH levels may be high enough to cause pigmentation. The most striking feature is often hypokalaemic alkalosis, and the patient may not have the clinical features of Cushing's syndrome.
  In *occult* ectopic ACTH secretion, often from pulmonary or mediastinal carcinoid tumours, the clinical features may be indistinguishable from those of Cushing's disease.

Primary tumours of the adrenal cortex secrete cortisol *despite suppression of*

ACTH secretion by the consequent high plasma cortisol concentration. The tumour may be benign or malignant. Benign adenomas occur in the same sex and age group as Cushing's disease and produce a similar clinical picture. Carcinomas secrete a variety of steroids which may cause virilization. This is the commonest form of Cushing's syndrome in children.

Iatrogenic Cushing's syndrome may be produced by steroid or ACTH treatment.

Basis of investigation of suspected Cushing's syndrome

Because untreated Cushing's syndrome has a high morbidity and mortality rate, any suspected case must be adequately investigated. Obese, red-faced, hairy, hypertensive women usually do *not* have Cushing's syndrome, and the problem is more often one of exclusion than of confirmation. However, as the tests available are not specific, it is *essential* to apply them in a logical sequence.

What we need to know is:

(a) is there abnormal cortisol secretion?
(b) if so, does the patient have any other condition that may cause it?
(c) if it is Cushing's syndrome, what is the cause?

(a) Is there abnormal cortisol secretion?

Plasma cortisol levels in the morning may be normal or only moderately raised. However, because one of the earliest features of Cushing's syndrome is a reduction in the amplitude, or loss, of the circadian rhythm, *high levels in the late evening*, when secretion is normally at a minimum, are suggestive but not diagnostic of the disorder.

As discussed on p. 120, most plasma cortisol is protein-bound. Only the small free fraction is filtered at the glomerulus and excreted in the urine *(urinary 'free' cortisol)*. Normally, however, CBG is almost fully saturated, and increased cortisol secretion produces a disproportionate rise in the free fraction, with a corresponding increase in urinary free cortisol. Because of the loss of circadian rhythm, the raised plasma values are present for longer than normal, and daily urinary cortisol excretion is further increased. The *urinary free cortisol excretion* will be more markedly raised than plasma levels. Plasma and urinary cortisol levels are much higher when Cushing's syndrome is due to adrenocortical carcinoma or overt ectopic ACTH secretion.

The *dexamethasone suppression test* (p. 153) can be used to test the ability to respond to negative feedback, because a small dose of this synthetic steroid inhibits ACTH, and therefore cortisol, secretion. Suppression does not occur in patients with Cushing's disease, in whom the feedback centre is less sensitive than normal, nor in patients with ectopic ACTH secretion, or adrenal tumours in whom ACTH secretion is already suppressed. The dexamethasone suppression test is thus a sensitive, but not completely specific test in evaluating such patients. A normal fall in cortisol levels makes the diagnosis of Cushing's syndrome very unlikely.

(b) Is there another cause for the abnormal cortisol secretion?

Abnormal cortisol secretion may be due to causes other than Cushing's syndrome.

The following are of particular importance.

- *Stress* overrides the other controlling mechanisms of ACTH secretion and leads to a loss of the normal circadian variation of plasma cortisol and to a reduced feedback response. Urinary free cortisol excretion may be markedly increased. This may occur in even relatively minor physical illness or mental stress.
- *Endogenous depression* is often associated with sustained high plasma cortisol and ACTH levels which are not suppressed by dexamethasone. These patients, however, usually have a normal cortisol response to insulin-induced hypoglycaemia, while those with Cushing's syndrome do not.
- *Severe alcohol abuse* can produce hypersecretion of cortisol that mimics Cushing's syndrome clinically and biochemically. The abnormal findings revert to normal when alcohol is stopped.

(c) What is the cause of the Cushing's syndrome?
Only when Cushing's syndrome has been confirmed should tests to establish the cause be carried out.

The test of choice is *estimation of plasma ACTH*. Levels are moderately raised in patients with Cushing's disease or occult ectopic ACTH production, markedly raised in cases of overt ACTH production and very low in patients with adrenocortical tumours.

If ACTH assay is not available, a *high-dose dexamethasone test* may be helpful. The principle of the test is as described earlier, except that a high enough dose of dexamethasone is given to suppress the relatively insensitive feedback centre of pituitary-dependent disease. In the other two categories, when pituitary ACTH is already suppressed, even this high dose will usually have no effect, although suppression sometimes occurs with occult ectopic ACTH secretion.

It must be emphasized that these tests can only be interpreted if the diagnosis of Cushing's syndrome has been established. For example, in an ill patient with high (but appropriate) cortisol levels, plasma ACTH levels are raised but may be suppressed only by a high dose of dexamethasone. This may lead to an erroneous diagnosis of pituitary-dependent Cushing's syndrome.

Table 6.1 summarizes the results of tests for Cushing's syndrome.

Adrenocortical hypofunction

Primary adrenocortical hypofunction (Addison's disease)

Addison's disease is caused by destruction of all zones of the adrenal cortex, usually as the result of an autoimmune process. Tuberculosis is an important cause in countries where this disease is common. Rare causes of destruction (which must be bilateral to cause adrenal failure) include amyloidosis, mycotic infections, and secondary deposits often originating from a bronchial carcinoma.

Clinical features. The presentation of Addison's disease depends on the degree of adrenal destruction. The patient may be shocked and volume depleted, and such an adrenal crisis should be treated as a matter of urgency. It may be difficult to make a diagnosis if destruction is less extensive; the prolonged period of

Table 6.1 Usual test results in Cushing's syndrome

		Adrenocortical hyperplasia			Adrenocortical tumour	
		Pituitary-dependent (Cushing's disease)	Ectopic ACTH: Occult	Ectopic ACTH: Overt	Carcinoma	Adenoma
DIAGNOSIS	Plasma cortisol morning	Raised, may be N		Raised		Raised, may be N
	evening	← Raised →				
	after 2 mg dexamethasone	← No suppression →				
	Urinary cortisol	Raised		Usually very high		Raised
AETIOLOGY	Plasma ACTH	High normal or moderately raised		Raised	Very low	
	Plasma cortisol after 8 mg dexamethasone	Suppression	← No suppression →			

vaguė ill-health, tiredness, weight loss, mild hypotension, and pigmentation of the skin and buccal mucosa are common in many severe chronic diseases. The cause of these vague symptoms may only become evident if an Addisonian crisis is precipitated by the stress of some other, perhaps mild, illness or of surgery.

The most serious consequences are due to the *mineralocorticoid (aldosterone) deficiency, with sodium and water depletion*. While sodium and water loss parallel each other the plasma sodium concentration remains within the reference range (see Fig. 2.1c, p. 42). The volume depletion and consequent haemoconcentration are nevertheless evident from the clinical state and usually from the raised haematocrit and total protein concentration. During a crisis, however, stress and acute contraction of the circulating volume may stimulate ADH secretion: water is then reabsorbed in excess of sodium and hyponatraemia develops. Because the loss is due to inability of the tubules to reabsorb sodium adequately in the absence of aldosterone, the urine will contain an inappropriately high sodium concentration for the state of volume depletion (p. 51). (Estimation of urinary sodium is, however, not a useful diagnostic test). The Addisonian crisis is therefore the result of massive sodium depletion. Other abnormalities usually include hyperkalaemia and metabolic acidosis (compare the reverse that occurs with the mineralocorticoid excess of hyperaldosteronism, p. 50). The fluid depletion leads to a reduced circulating volume and renal circulatory insufficiency with a reduced glomerular filtration rate and a moderately raised plasma urea concentration. In the more chronic form of Addison's disease only some of these features may be present.

The *glucocorticoid deficiency* contributes to the hypotension and causes marked sensitivity to insulin: hypoglycaemia may be a presenting feature.

Androgen deficiency is not clinically evident because testosterone production by the testes is unimpaired, and because androgen deficiency does not produce obvious effects in women.

The pigmentation that develops in Addison's disease is probably due to the high circulating levels of ACTH or related peptides resulting from the lack of cortisol suppression of the feedback mechanism.

Secondary adrenal hypofunction (ACTH deficiency)

ACTH release may be impaired by disease of the hypothalamus or of the anterior pituitary gland, most commonly due to tumour or infarction. Corticosteroids suppress ACTH release and after taking such drugs, especially for a long time, the ACTH-releasing mechanism may be slow to recover. There may be temporary adrenal atrophy after prolonged lack of stimulation.

If there is extensive destruction of the anterior pituitary gland the picture is that of panhypopituitarism (p. 115), but if it is only partial there may be sufficient ACTH for basal requirements. As in partial adrenal cortical destruction the deficiency may only become evident under conditions of stress, which in these patients, as in those on corticosteroid treatment, may precipitate acute adrenal insufficiency. The most usual causes of stress are infections and surgery.

Patients with acute cortisol deficiency may present with *nausea, vomiting* and *hypotension. Hyponatraemia,* which may contribute to mental disturbance, is due to dilution, not sodium loss as in Addison's disease. Cortisol is needed for normal free water excretion by the kidney, and the biochemical picture therefore resembles that of inappropriate ADH secretion. Because aldosterone secretion is normal, hyperkalaemia does not occur.

In the absence of stress the patient presents with non-specific symptoms such as weight loss and tiredness. There is marked insulin sensitivity and hypoglycaemia may occur. The pigmentation of Addison's disease is absent because ACTH levels are low.

Basis of investigation of suspected adrenocortical hypofunction

If the patient presents with a probable diagnosis of acute adrenal insufficiency, blood should be taken so that the plasma cortisol concentration can be measured later, but *treatment must be started at once.*

The performance of further tests can await recovery from the crisis: the underlying abnormality will not be affected by treatment. Results of plasma cortisol assay on the initial sample should distinguish between adrenocortical insufficiency (inappropriately low values) and clinically similar crises when levels are markedly raised due to stress. It will not, however, distinguish primary from secondary adrenal failure.

In the less acutely ill patient, a morning plasma cortisol level should be measured, and, if unequivocally high, no further tests are necessary. In most cases, however, levels will be within the reference range. *The essential abnormality in adrenocortical hypofunction is that the adrenal gland cannot adequately increase cortisol secretion in response to stress.* Because this may be due to adrenal (primary) or hypothalamic-pituitary (secondary) pathology, it may be necessary to test the whole axis.

Plasma ACTH levels are technically difficult to measure, but ACTH secretion from the anterior pituitary gland in response, for example, to insulin-induced hypoglycaemia, can be assessed by demonstrating a rise in plasma cortisol levels; such a rise indicates adrenocortical stimulation. An impaired response only indicates pituitary dysfunction if the adrenal cortices have already been shown to be able to increase cortisol secretion in response to exogenous ACTH. Tetracosactrin (for example, 'Synacthen') has the same biological action as ACTH but, because it lacks the antigenic part of the molecule, there is much less danger of an allergic reaction. If the patient cannot be taken off steroids, then a steroid such as dexamethasone, which does not interfere with the assay, should be prescribed.

Plasma ACTH assay is of value in some cases. When inappropriately low cortisol values have been found, a raised plasma ACTH concentration indicates primary, and a low level secondary, insufficiency. This will distinguish between true adrenocortical disease (when mineralocorticoid replacement is needed) and reversible atrophy due to prolonged ACTH deficiency.

Urinary free cortisol levels are low, but this estimation is of little help in diagnosis.

Corticosteroid therapy

There is a risk of adrenocortical hypofunction when long-term corticosteroid treatment is stopped suddenly. This may be due either to secondary adrenal atrophy or to impairment of ACTH-releasing mechanisms.

A simple means of testing the feedback centre and the responsiveness of the pituitary-adrenal axis is to estimate the morning plasma cortisol level two or three days after stopping steroid treatment. A level within the reference range indicates a functioning adrenal, pituitary and feedback centre. It must be emphasized, however, that this does not test the all-important stress pathway (p. 122).

A suggested sequence of testing is described on p. 153.

Congenital adrenal hyperplasia

Rarely there may be an inherited deficiency of one of the enzymes involved in the biosynthesis of cortisol (Fig. 6.4). As a result, plasma cortisol levels tend to fall, leading to increased secretion of ACTH from the anterior pituitary gland. This in turn leads to hyperplasia of the adrenal cortex, with increased synthesis of cortisol precursors before the enzyme block. The precursors may then be metabolized by alternative pathways, especially those of androgen synthesis.

C-21-hydroxylase, which is involved in the synthesis of both cortisol and aldosterone, is the enzyme most commonly affected.

Increased androgen production may cause:

- *female pseudohermaphroditism* by affecting the development of the female external genitalia *in utero*;

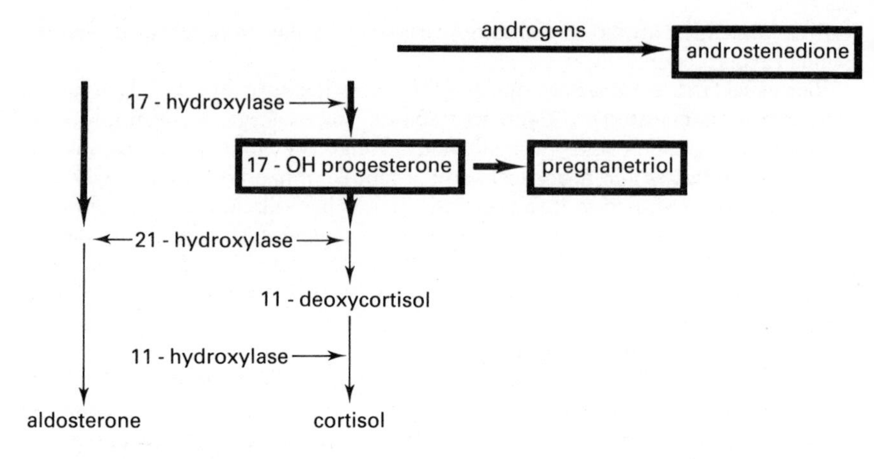

Fig. 6.4 The abnormalities occurring in congenital adrenal hyperplasia due to 21-hydroxylase deficiency. Substances of diagnostic importance are shown in boxes.

- *virilization in childhood* with phallic enlargement in either sex, development of pubic hair and a rapid growth rate;
- *milder virilization in females at or after puberty, with amenorrhoea.*

Aldosterone synthesis may be markedly reduced and may cause an Addisonian-like picture during the first few weeks of life. Vomiting is severe and there is hyponatraemia and hyperkalaemia with marked volume depletion. This occurs in more than half the patients with 21-hydroxylase deficiency, but even in those with normal plasma sodium levels, demonstrably increased plasma renin activity may suggest lesser degrees of sodium and water depletion.

All female babies with ambiguous genitalia should have plasma electrolytes estimated. In male babies with no obvious physical abnormalities the diagnosis may not be suspected.

Diagnosis

The investigation of suspected congenital adrenal hyperplasia is best undertaken in special centres. Only the principles of the diagnosis of 21-hydroxylase deficiency will be outlined here (see Fig. 6.4).

17-hydroxyprogesterone can only be metabolized by the cortisol pathway in the presence of 21-hydroxylase. Plasma levels of this steroid are raised and may be measured directly, or its metabolite, *pregnanetriol*, may be measured in the urine.

Excessive androgen synthesis is best shown by demonstrating *high plasma androstenedione* concentrations.

Detailed evaluation of the pattern of urinary steroid secretion on a random sample will indicate which enzyme is deficient.

Congenital adrenal hyperplasia is *treated* by giving cortisol or one of the glucocorticoid analogues with, if necessary, a mineralocorticoid. This treatment not only replaces the deficient hormones, but, by feedback, suppresses excessive ACTH secretion and therefore excessive androgen production.

The efficacy of treatment is monitored by measuring 17-hydroxyprogesterone and androstenedione concentrations in plasma or saliva. Salivary steroid concentrations correlate well with those in plasma and collection of saliva may be more acceptable to the patient than repeated venepuncture. The adequacy of mineralocorticoid replacement may be assessed by measuring the plasma renin activity.

Summary

1. The steroids of the adrenal cortex may be classified into three groups:

- *glucocorticoids*, for example cortisol, which influence intermediary metabolism, fluid balance and blood pressure. Secretion is controlled by ACTH from the anterior pituitary gland;
- *mineralocorticoids*, for example aldosterone, which influence the balance of sodium, and secondarily of water, and potassium. Secretion is controlled by the renin-angiotensin system;
- *androgens*, for example androstenedione, which probably have an anabolic action on protein synthesis.

2. ACTH (and therefore cortisol) secretion is stimulated by CRF from the hypothalamus. The important controlling factors are:

- an *inherent rhythm* which produces an overall circadian variation in plasma cortisol levels. These are lowest at around midnight and highest in the morning;
- *negative feedback* by circulating *free cortisol* levels;
- *stress* which may override both the above controls.

3. Estimations of plasma and urinary cortisol are the most useful tests. Plasma ACTH measurements are of value in the differential diagnosis of Cushing's syndrome and, more rarely, to distinguish primary from secondary adrenal failure.

4. *Cushing's syndrome* is due to inappropriately high circulating cortisol levels. The causes are hyperplasia or tumour of the adrenal cortex. Hyperplasia is due to excess ACTH, either from the pituitary (pituitary-dependent) or from a non-endocrine tumour.

There are three stages in investigation.

(a) Is there abnormal cortisol secretion?
(b) If so, can it be accounted for by a condition other than Cushing's syndrome?
(c) If not, what is the cause of the Cushing's syndrome?

5. Primary adrenocortical hypofunction *(Addison's disease)* is due to destruction of the adrenal cortex with deficiency of all its hormones. Aldosterone deficiency causes sodium and consequent water depletion.

Diagnosis is made by demonstrating that the adrenal cortex cannot respond to stimulation by exogenous ACTH (or its analogue).

6. *Secondary adrenal insufficiency* is caused by diminished ACTH secretion by the anterior pituitary gland. This may be due to disease of the hypothalamus or of the pituitary, or may be the result of corticosteroid treatment.

It is usually diagnosed by demonstrating a definite but impaired response to tetracosactrin (synthetic ACTH) stimulation. In cases without adrenal atrophy the hypothalamic-pituitary-adrenal axis may be assessed by the insulin stress test.

7. *Congenital adrenal hyperplasia* is due to an inherited enzyme deficiency in the biosynthesis of cortisol. Symptoms are due to a deficiency of cortisol and to excessive androgen secretion. Diagnosis of the commonest form is made by estimation of plasma 17-hydroxyprogesterone. Plasma androgen levels are raised.

Further reading

Cushing H. The basophil adenomas of the pituitary body and their clinical manifestations (pituitary basophilism). *Bull Johns Hopkins Hosp* 1932; **50**: 137–95.

Urbanic R C, George J M. Cushing's disease – 18 years' experience. *Medicine (Baltimore)* 1981; **60**: 14–24.

Besser G M, Rees L H eds. The pituitary-adrenocortical axis. *Clin Endocrinol Metab* 1985; **14**: 765–1038.

(This issue covers all aspects of normal and abnormal adrenocortical function).

Crapo L. Cushing's syndrome: a review of diagnostic tests. *Metabolism* 1979; **28**: 955–77.

Howlett T A, Rees L H. Is it possible to diagnose pituitary-dependent Cushing's disease? *Ann Clin Biochem* 1985; **22**: 550–8.

7

The reproductive system: pregnancy

The study of the endocrinology of the male and female reproductive systems is a specialized field. Nevertheless many disorders can be assessed using the general principles of endocrine diagnosis, based on a knowledge of normal function. This chapter will deal only with such principles and for more details the reader should consult the references given at the end of the chapter.

The hypothalamic-pituitary-gonadal axis

The reproductive system is responsible, not only for the production of hormones, but also for maturation of the germ cells in the gonads. It is essential to understand the relations between hormones with respect to these two functions if the results of tests are to be correctly interpreted.

Hypothalamic hormones

The *hypothalamus* secretes *gonadotrophin-releasing hormone* (Gn-RH) which regulates the secretion of the pituitary gonadotrophins, and the neurotransmitter *dopamine* which controls prolactin secretion.

Pituitary hormones

The pituitary gonadotrophins, *luteinizing hormone* (LH) and *follicle-stimulating hormone* (FSH) secreted by basophil cells, control the function of, and secretion of hormones by, the ovary and testis. The secretion of Gn-RH, and consequently of LH and FSH, is pulsatile. Although there is only one releasing hormone, secretion of LH and FSH does not always occur in parallel and may also be modified by feedback from the circulating levels of gonadal oestrogens or androgens. The actions of these two gonadotrophins overlap, but LH is the more important in stimulating the production of hormones by the gonads, while FSH stimulates the development of the germ cells.

Prolactin, secreted by the anterior pituitary acidophil cells, is important during pregnancy and the post-partum period (p. 139). Abnormally high levels at other times may affect gonadal function. It differs from all other pituitary hormones because its secretion is normally *inhibited by dopamine: impairment of hypothalamic control* therefore causes *hyperprolactinaemia*. Plasma levels of prolactin and GH are affected by similar factors; for example, both rise during sleep and in response to physical or psychological stress. Prolactin secretion is regulated by a short feedback loop involving only the anterior pituitary gland and hypothalamus. Oestrogens, however, enhance prolactin secretion, and plasma concentrations are normally higher in women. Although thyrotrophin-releasing hormone (TRH) stimulates both TSH and prolactin secretion, it does not seem to be of physiological importance in the control of prolactin secretion.

Ovarian hormones

The ovary secretes oestrogens, androgens and progesterone. They are produced by ovarian follicles, which consist of germ cells (ova) surrounded by granulosa and theca cells. Androgens (C_{19} steroids), synthesized by theca cells are converted into oestrogens (C_{18} steroids) in the granulosa cells, a process which involves aromatization of the A ring and loss of the C-19 methyl group (Figs. 6.1 and 6.2, pp. 119 and 120). Oestrogens are essential for the development of female secondary sex characteristics and for normal menstruation: levels are usually undetectable in children.

Oestradiol, the most important ovarian oestrogen, is derived from locally synthesized androgens. The liver and subcutaneous fat convert ovarian and adrenal androgens to *oestrone*. Both oestradiol and oestrone are metabolized to the relatively inactive *oestriol*.

The main androgen secreted by the ovaries is *androstenedione*, which is converted in extraovarian tissues to oestrone, and to the more active *testosterone*. A small amount of testosterone is secreted directly by the ovaries. Plasma testosterone levels in women are about a tenth of those in men.

Progesterone is secreted by the corpus luteum and is chemically similar to the adrenocortical progestogens (p. 130). It prepares the endometrium to receive a fertilized ovum, and is necessary for the maintenance of early pregnancy.

Testicular hormones

Testosterone is secreted by the Leydig cells, which lie in the interstitial tissue of the testis between the seminiferous tubules. Its production is stimulated by LH, and testosterone, in turn, inhibits LH secretion. The Sertoli cells in the seminiferous tubules produce *inhibin* which inhibits FSH secretion.

Testosterone stimulates anabolism in either sex. In the male it is essential for the development of secondary sexual characteristics, an effect that depends on intracellular conversion to the even more potent androgen, dihydrotestosterone.

Testosterone and, to a lesser extent, oestradiol, are transported in the plasma bound to a carrier protein, sex-hormone-binding globulin (SHBG) as well as to albumin. Only the free fraction is metabolically active.

Sexual development from conception to puberty

The complex series of events leading to the development of sexual competence depends on each step occurring at the correct time. The following simplified account aims to provide a background for the discussion of abnormalities of the system.

Chromosomal sex is determined at fertilization by the chromosomes present in the ovum and sperm, each of which contributes 22 autosomes and one sex chromosome, X or Y (p. 362). Normal females have a 46XX karyotype, and normal males 46XY. Abnormalities occurring at this stage may result in defective gonadal development, such as Turner's syndrome (45X) or Klinefelter's syndrome (47XXY).

The sex chromosomes determine whether the primitive gonad becomes a testis or an ovary. We will consider the development and disorders of gonadal function in the male and female separately.

The male

Development of male characteristics

In the presence of the Y chromosome the fetal gonads develop into testes at about 7 weeks gestation. The testes secrete:

- a factor which causes *degeneration of the potential female genitalia*;
- *testosterone* which stimulates development of the *potential male internal genitalia*.

However, intracellular conversion of testosterone to dihydrotestosterone by the enzyme 5α-reductase is needed for normal development of male external genitalia. If this enzyme is deficient, varying degrees of feminization may occur, and produce the condition known as *male pseudohermaphroditism*. If the tissues are insensitive to the action of testosterone, female sex characteristics develop despite normal testosterone secretion *(testicular feminization)*. The very rare *true hermaphrodite* has both testicular and ovarian tissue.

Puberty. During childhood the secretion of anterior pituitary gonadotrophins is low. As puberty approaches the pulses of gonadotrophin secretion, and of LH in particular, increase in amplitude and frequency. Initially they occur during sleep, but later continue throughout the day.

Gonadotrophin secretion stimulates the germ cells in the seminiferous tubules, which have remained dormant during childhood, to undergo meiosis and so to produce sperm: this process, which is not self-limiting like the production of ova (p. 137), may continue throughout life. The rate of testosterone secretion increases, and this stimulates the development of secondary male characteristics.

- *Deficient secretion of either pituitary or of gonadal hormones* may cause *delayed puberty*.
- *Premature secretion of gonadotrophins* may cause *early (true precocious) puberty*.
- *Abnormally high androgen secretion*, such as from gonadal tumours, leads to the development of secondary sex characteristics before maturation of the testis *(pseudoprecocious puberty)*.

The adult male

By contrast with the finite number of ova in the female, the male produces spermatozoa continuously after puberty. Spermatogenesis occurs in the seminiferous tubules which contain the germ cells and Sertoli cells, and is dependent on both normal Sertoli cell function and on testosterone secretion by the Leydig cells. FSH stimulates spermatogenesis, and FSH secretion is inhibited by negative feedback by inhibin from the Sertoli cells. As discussed earlier, LH stimulates the Leydig cells to secrete testosterone and its secretion is inhibited as circulating testosterone levels rise. Therefore, both LH and FSH are needed for normal spermatogenesis, but testosterone secretion can occur in the absence of normal seminiferous tubules.

Disorders of gonadal function in the male

Men with gonadal dysfunction may present with symptoms of *androgen deficiency*, with *infertility* or with both. Hormone measurement has a limited but important role in the assessment of these patients. Only the principles will be discussed here.

Androgen deficiency is the result of impaired testosterone secretion by the Leydig cells. The patient may present with *delayed puberty* or with regression of previously established male characteristics that are dependent on testosterone (hair distribution, potency and libido). There may be primary testicular dysfunction, in which case the low plasma testosterone concentration is accompanied by raised plasma LH levels (hypergonadotrophic hypogonadism) or it may be secondary to pituitary or hypothalamic disease, with low plasma LH levels (hypogonadotrophic hypogonadism).

Hyperprolactinaemia is much less common in males than in females, but its presence may indicate a pituitary tumour. Clinical features may suggest a chromosomal abnormality such as Klinefelter's syndrome.

Although androgen deficiency always causes *infertility*, most infertile males have normal plasma androgen concentrations. Sertoli cell function is dependent on both FSH secretion from the anterior pituitary gland and on locally produced testosterone. Semen analysis is the most important investigation for male infertility. If there is evidence of a failure of spermatogenesis the plasma FSH concentration should be measured.

- If the cause is *primary testicular failure the plasma testosterone level is low*, and reduced inhibin production by Sertoli cells causes a *rise in FSH levels*.

- If the cause is secondary to *anterior pituitary failure, both testosterone and FSH levels are low.*
- If Leydig cell function is normal neither testosterone nor LH secretion is affected.

The investigation of male gonadal dysfunction is considered on p. 157.

The female

Development of female characteristics

If there is *no Y chromosome* the fetus starts to develop female characteristics at about 12 weeks of gestation. If androgens are produced at this stage as, for example, in congenital adrenal hyperplasia (p. 129), they may cause masculinization of the external genitalia *(female pseudohermaphroditism).*

Proliferation of fetal germ cells produces several million oocytes. By late fetal life all the germ cells have degenerated and no more oocytes can be produced. Those present enter the first stage of meiosis and their numbers decline throughout the rest of the intrauterine period and childhood; the inability to replenish them explains the limit to the span of reproductive life in women in contrast to the continued ability to produce sperm in men. If the rate of decline is abnormally high there is a premature menopause.

Puberty. At the onset of puberty gonadotrophin secretion increases, as it does in the male; ovarian oestrogen secretion rises and stimulates the development of female secondary sex characteristics and the onset of menstruation (menarche).

As in the male, abnormal hormone secretion may cause *delayed puberty, true precocious puberty, or pseudoprecocious puberty* (p. 136).

The adult female

Normal gonadal function

At puberty the ovaries contain 100 000 to 200 000 primordial follicles. During each menstrual cycle a small number develop, but only one reaches the stage at which the ovum is extruded from the ovary (ovulation) and is ready for fertilization. The events of the menstrual cycle are regulated by changing hormone levels (Fig. 7.1) and by changing sensitivity of ovarian tissue.

Follicular (pre-ovulatory) phase

At the beginning of the menstrual cycle ovarian follicles are undeveloped, and oestrogen levels are low. The diminished negative feedback allows FSH and LH secretion to increase.

- *FSH and LH together cause growth of a group of follicles and maturation of follicular cells. LH also stimulates secretion of hormones* from these cells and

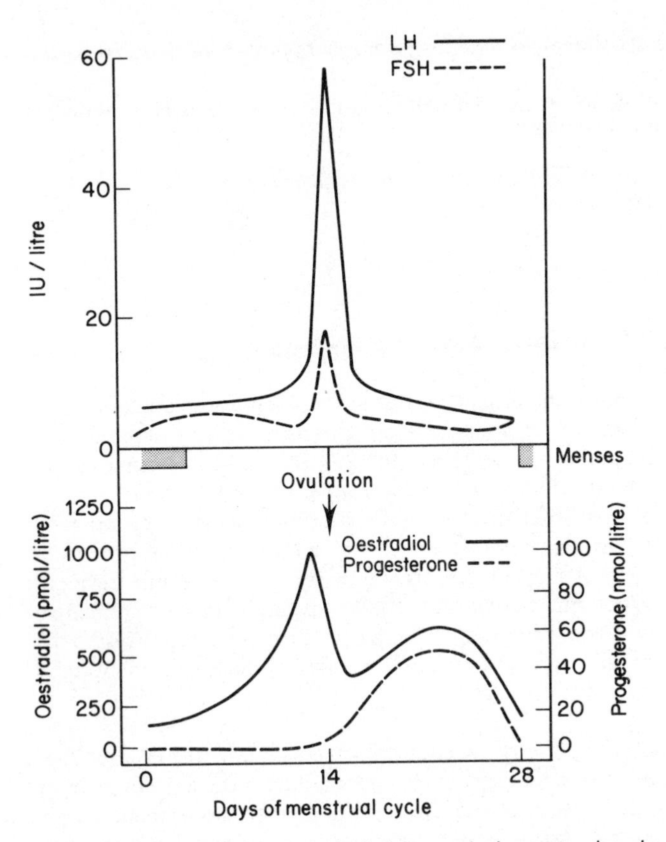

Fig. 7.1 Plasma hormone levels during a typical menstrual cycle.

circulating oestrogen levels rise steadily. This stimulates regeneration of the previously shed endometrium.

• The rising oestrogen levels cause a slight fall in FSH secretion by negative feedback to the anterior pituitary gland.

By about the seventh day of the cycle one of the group of follicles has become especially sensitive to pituitary FSH and continues to grow while the rest atrophy.

Ovulation

The dominant follicle develops rapidly and secretes a large amount of oestradiol; this triggers a surge of LH release from the anterior pituitary gland by *positive* feedback. Ovulation occurs at about 16 hours after this. The very high LH level inhibits oestrogen and stimulates progesterone secretion. The follicle gradually develops into the corpus luteum.

Luteal (postovulatory or secretory) phase

This phase is characterized by the rise and fall of the corpus luteum, which takes over ovarian hormone production.

- *LH stimulates the development of the corpus luteum*, and the secretion of *progesterone and oestrogens* from it.
- *Progesterone prepares the endometrium* to receive a fertilized ovum.

The subsequent events depend on whether the released ovum is fertilized. If it is not, the menstrual cycle takes its course; if it is, pregnancy may supervene.

- Falling levels of ovarian hormones, as the corpus luteum regresses, cause endometrial sloughing and menstrual bleeding.
- As ovarian hormone levels fall, FSH and LH levels begin to rise. The cycle recommences.

Prolactin levels do not change cyclically during the menstrual cycle.

Interpretation of sex-hormone levels must be made in relation to the stage of the cycle. For example, it may be important to establish whether a patient complaining of infertility has ovulated, either spontaneously or as a result of treatment to induce ovulation. The plasma progesterone concentration should be measured on a blood sample taken during the second half of the menstrual cycle. A value within the reference range for the time of the cycle is good presumptive evidence of ovulation, while a value in the range expected in the follicular phase (Fig. 7.1) indicates the absence of a corpus luteum, and therefore of ovulation.

Progesterone secretion is associated with a rise in body temperature, which may be monitored serially to determine the time of ovulation.

Pregnancy and lactation

If the ovum is fertilized it may implant in the endometrium which has been prepared by progesterone (see 'Luteal phase'). Gonadotrophin secretion is soon taken over by the chorion and developing placenta. This *chorionic gonadotrophin (HCG) is similar in structure and action to LH*, and prevents the involution of the corpus luteum as pituitary gonadotrophin levels fall. *Oestrogen and progesterone levels therefore continue to rise*, and endometrial sloughing is prevented. After the first trimester these hormones are produced by the placenta.

Probably because of the high oestrogen (mostly oestriol) levels, prolactin secretion increases progressively after the eighth week of pregnancy and at term may be 10 to 20 times that of non-pregnant women.

Prolactin, oestrogens, progesterone and placental lactogen (HPL, p. 143) stimulate breast development for lactation. High levels of oestrogen inhibit milk secretion and lactation can only start when levels fall after delivery of the placenta.

Early lactation depends on prolactin and in its absence (resulting from, for example, pituitary infarction or administration of bromocriptine, which inhibits prolactin secretion) milk production ceases. Suckling stimulates prolactin secretion; nevertheless, despite continuing successful lactation, plasma concentrations fall progressively post-partum and reach non-pregnant levels after 2 or 3 months. Apart from the effects on the breast, the high concentration of prolactin interferes with gonadotrophin-gonadal function, producing a period of relative infertility.

The menopause

The menopause occurs when all the ovarian follicles have atrophied and plasma oestrogen levels fall. Removal of normal premenopausal feedback to the pituitary causes increased FSH and, to a lesser extent, LH secretion. The findings are identical with those of primary gonadal failure (p. 155).

Disorders of gonadal function in the female

Women with possible gonadal dysfunction may present with any or all of the following complaints:

- delayed puberty and primary amenorrhoea;
- infertility with or without amenorrhoea;
- hirsutism;
- virilism.

The investigation of these disturbances cannot be covered comprehensively here and the attention of the reader is drawn to the references given at the end of the chapter for further details. The following is a guide to the more obvious points, and stresses the underlying biochemical abnormalities.

Amenorrhoea

Amenorrhoea is defined as the absence of menstruation. It is called *primary* when the patient has never menstruated and *secondary* when previously established menstrual cycles have stopped. Note that 'primary' and 'secondary' used here have different meanings from 'primary' and 'secondary' endocrine disorders.

Amenorrhoea may be due to hormonal abnormalities. If there is ovarian failure pituitary gonadotrophin levels are high (hypergonadotrophic hypogonadism); if the cause is in the hypothalamus or anterior pituitary gland gonadotrophin secretion is reduced (hypogonadotrophic hypogonadism).

Primary amenorrhoea is most commonly associated with delayed puberty. The age of the menarche is very variable; unless there are other signs of endocrine disturbances (such as hirsutism and virilization) or of chromosomal abnormalities, extensive investigation should probably be postponed until the age of 18. Turner's syndrome and testicular feminization (p. 135) usually present with primary amenorrhoea.

Secondary amenorrhoea is most commonly due to such physiological factors as pregnancy and the menopause. Other causes include severe illness, weight loss for any reason, or stopping oral contraceptives. These causes should be considered before extensive and potentially dangerous investigations are started. Other endocrine disorders, such as hyperthyroidism, Cushing's syndrome and acromegaly, may present with amenorrhoea.

Hyperprolactinaemia is an important cause of amenorrhoea and of infertility. High plasma prolactin concentrations inhibit the normal pulsatile release of Gn-RH and inhibit gonadal steroid hormone synthesis directly. This causes low gonadotrophin and oestrogen levels and the symptoms of oestrogen deficiency. About a third of patients with hyperprolactinaemia have galactorrhoea.

The most important cause of hyperprolactinaemia is a prolactin-secreting microadenoma of the pituitary gland. Patients with no apparent cause for hyperprolactinaemia should be followed clinically in case they develop an obvious pituitary adenoma. The higher the prolactin level the greater the likelihood that a tumour is the cause.

The finding of hyperprolactinaemia should be interpreted with caution. It is particularly important to take samples for prolactin estimation at least 2 to 3 hours after waking to eliminate the misleading elevated levels found during sleep, and to remember that even the minor stress of venepuncture may cause prolactin secretion.

Other causes of hyperprolactinaemia include:

- drugs:
  - phenothiazines;
  - cimetidine;
  - methyldopa;
  - reserpine;
  - oestrogens;
  - metoclopramide;
  - pimozide;
  - haloperidol.
- failure of hypothalamic inhibitory factors to reach the anterior pituitary gland:
  - damage to the pituitary stalk by non-prolactin-secreting tumours of the pituitary gland or hypothalamus;
  - surgical section of the pituitary stalk.
- possibly reduced clearance of prolactin from the plasma:
  - chronic renal failure;
  - severe primary hypothyroidism (also possibly due to anterior pituitary stimulation by high TRH levels).

The investigation of amenorrhoea is considered on p. 155.

Hirsutism and virilism

Increased androgen levels, or *increased tissue sensitivity to androgens*, produce effects ranging from increased hair growth to marked masculinization. Testosterone is the most important of these hormones, although a marked increase in the other androgens may have the same effect. In normal women about half the plasma testosterone comes from the ovary, both by direct secretion and by peripheral conversion of androstenedione. The rest is derived from peripheral conversion of adrenal androstenedione and dehydroepiandrosterone (DHA). Because of the extensive interconversion of androgens, the source of a slightly raised plasma testosterone concentration is difficult to establish, but this is of little clinical relevance. In general, markedly raised plasma levels of DHA or DHAS indicate an adrenocortical, and of testosterone an ovarian, origin.

Hirsutism is defined as an excessive growth of hair in a male distribution. It is a common complaint, and extensive investigations should not be started without considering the possibility that it is of familial or racial origin. Plasma testosterone levels may be slightly raised, but are often within the female reference range. However, the active free hormone level may be increased significantly if the plasma SHBG concentration is low.

A common cause of hirsutism is the *polycystic ovary syndrome*. Presenting symptoms may include infertility, menstrual disturbances, hirsutism and obesity. The plasma LH concentration is increased, and there is an abnormally high ratio of LH to FSH, often greater than 3:1. Plasma levels of testosterone are usually, and of prolactin sometimes, high. Multiple small subcapsular ovarian cysts may be demonstrated using ultrasound.

Virilism is uncommon but much more serious than hirsutism and is always associated with increased androgen levels. It is characterized by *other evidence of excessive androgen secretion* such as enlargement of the clitoris, increased hair growth of male distribution, receding temporal hair, deepening of the voice and breast atrophy. The main causes are:

- *ovarian tumours* (such as arrhenoblastomas and hilus-cell tumours) which secrete androgens, mainly testosterone;
- *adrenocortical pathology*:
    - tumours, usually carcinomas;
    - hyperplasia:
        - pituitary-dependent Cushing's syndrome (rarely);
        - congenital adrenal hyperplasia.

Plasma DHA and DHAS levels are increased.

Effects of some drugs on the female hypothalamic-pituitary-adrenal axis

Oral contraceptives contain synthetic oestrogens and/or progestogens. By suppressing pituitary gonadotrophin secretion they *prevent ovulation*. Withdrawal mimics involution of the corpus luteum and results in menstrual bleeding.

Clomiphene, by blocking oestrogen receptors in the hypothalamus, prevents negative feedback and may stimulate gonadotrophin release even when circulating oestrogen levels are high. It may be used therapeutically to *induce ovulation* in patients with amenorrhoea or with infertility.

Gonadotrophin treatment may be used if clomiphene fails to induce ovulation. Human menopausal gonadotrophin (FSH and LH) is given to mimic the follicular phase. This treatment may cause dangerous follicular enlargement due to hyperstimulation, or may stimulate many follicles and so cause multiple pregnancy: it *must therefore be monitored* by frequent plasma or urinary oestrogen estimations, or by ultrasound examination. Human chorionic gonadotrophin (HCG) is then given to induce ovulation by mimicking the LH peak. Ovulation may be assessed by demonstrating rising plasma progesterone levels.

Bromocriptine may rapidly *reduce high plasma prolactin levels to normal*, whatever the cause, after which menstruation restarts and fertility is restored.

Monitoring pregnancy

The change from the hormonal pattern of the luteal phase of the menstrual cycle into that of pregnancy has been outlined on p. 139. Here we will consider some biochemical changes that occur during pregnancy. These fall into two types:

- substances produced by the fetus or placenta which may be measured to detect abnormalities or to monitor the progress of the pregnancy;
- changes in the mother that result from pregnancy and that need to be recognized as such to avoid misdiagnosis. There may be similar changes in the newborn infant, who has been exposed *in utero* to the mother's hormonal background.

Fetoplacental products

Substances secreted by the fetoplacental unit may be detected throughout pregnancy and are usually measured in those patients known to be at risk of miscarriage. Changes in the composition of maternal urine or plasma often reflect changes in fetal and placental metabolism, and sampling these is safe and simple. Occasionally there are indications for testing amniotic fluid obtained by amniocentesis, but the ability to visualize the fetus using ultrasound has reduced the need for this invasive procedure.

Monitoring early pregnancy (placental function)

Chorionic gonadotrophin secretion by the placenta reaches a peak at about 13 weeks of pregnancy and then falls. The fetoplacental unit then takes over hormone production and secretion of both oestrogen and progesterone rises rapidly.

Relatively crude tests for plasma or urinary HCG, which give positive results at one or two weeks after the first missed menstrual period, are most commonly used to diagnose pregnancy. However, by using more sensitive immunoassay techniques plasma HCG may be detected soon after implantation of the ovum and before the first missed period; such early diagnosis may be of value if an ectopic pregnancy is suspected, or if the patient is being treated for infertility.

In selected cases serial HCG measurements may be used to assess the progress of early pregnancy; single values are difficult to interpret because of the wide reference range. As a rough guide levels should double in two days.

Human placental lactogen (HPL) is a peptide hormone synthesized by the placenta. It is detectable by rapid immunoassay methods after about the eighth week of gestation, and has been used in the assessment of threatened abortion, and to monitor late pregnancy.

Monitoring late pregnancy (fetoplacental products)

Assessment of the condition of the fetus and of placental function may indicate the need for early obstetric intervention. Biochemical tests, which included estimation of placental total oestrogen ('oestriol') and HPL output, have largely been superseded by, for example, ultrasound examination.

Detection of fetal abnormalities

Some fetal abnormalities may be diagnosed by tests carried out on maternal plasma, or on amniotic fluid obtained through a needle inserted into the uterus through the maternal abdominal wall at any time after about 14 weeks gestation (amniocentesis). The procedure carries a very small risk to the fetus, even in the best hands, and *should only be performed for very strong clinical indications, and if the diagnosis cannot be made by non-invasive procedures*. Analytical results may be *dangerously misleading* if, for example, the specimen is *contaminated with maternal or fetal blood or maternal urine, or is not fresh and properly preserved*. Both the safety and the reliability of the procedure are improved if it is performed by someone with experience, using ultrasound to indicate the position of the fetus, placenta and maternal bladder. The laboratory analysing the sample should also have experience and, as always, close liaison between the clinician and the laboratory staff helps to ensure the suitability of the specimen and the speed of the assay.

Amniotic fluid is probably derived from both maternal and fetal sources, but its value in reflecting fetal abnormalities arises from its intimate contact with the fetus, and from the increasing contribution of fetal urine in later pregnancy.

Neural tube defects

α-fetoprotein (AFP) is a low-molecular-weight glycoprotein synthesized mainly in the fetal yolk sac and liver. Its production is almost completely repressed in the normal adult. Because of its relatively low molecular weight it can diffuse slowly through the capillary membranes, and appears in the fetal urine, and hence in amniotic fluid, and in maternal plasma. Severe fetal neural-tube defects, such as open spina bifida and anencephaly, are associated with abnormally high concentrations in these fluids: the reason for this is not clear, but the protein may leak from the exposed neural-tube vessels. Many other causes of a raised AFP concentration in amniotic fluid and maternal plasma have been reported; one of these is multiple pregnancy, but almost all the others are associated with serious fetal abnormalities, such as exomphalos.

Maternal AFP levels increase during normal pregnancy, but may be abnormally high in those carrying a fetus with a neural tube defect. In some countries all pregnant women attending for antenatal care have plasma AFP measured between 16 and 18 weeks of gestation. The gestational age should be confirmed by ultrasound, which will also exclude multiple pregnancy as a cause of high levels. Positive results must be confirmed on a fresh sample and, if still high, and if the diagnosis has not been confirmed by ultrasound, should be followed by AFP estimation on *amniotic fluid*. This is a more precise diagnostic tool and yields fewer false positive results than plasma assay if sampling is properly performed. It should be reserved for subjects known to be at risk either because of a family history of neural-tube defects, or because of the finding of a high level in maternal plasma with a normal or equivocal ultrasound scan. Amniocentesis should not usually be performed for this purpose unless the parents are willing to consider termination of pregnancy if the result is positive. It is important to remember that the AFP concentration in normal fetal plasma is high at 16 to 18 weeks, and that *bloodstained amniotic fluid can therefore yield dangerously misleading results.*

Amniotic fluid acetylcholinesterase assay may be used to detect many serious fetal malformations, including neural tube defects and exomphalos, and gives reliable results up to 22 or 23 weeks of gestation. The interpretation of the result is less dependent on fetal age than that of AFP, but is equally invalidated by contamination with fetal or maternal blood. The assay is less widely available than that for AFP.

Other fetal abnormalities

Chromosomal abnormalities and some inborn errors of metabolism may be detected by cytogenetic, chemical or enzymatic assays on cells cultured from amniotic fluid or after biopsy of chorionic villi. These tests are performed only in special centres and only on subjects with a genetic history of the condition.

Assessment of fetomaternal blood group incompatibility

Amniotic fluid levels of *bilirubin* are used, in conjunction with maternal antibody titres, to assess the effects on the fetus of rhesus or other blood group incompatibility. Normally amniotic fluid bilirubin levels decrease during the last half of pregnancy. The level at any stage may be correlated with the severity of haemolysis: used with tests of maturity, the optimum time for induction of labour or the need for intrauterine transfusion may be assessed.

Assessment of fetal lung maturity

Examination of amniotic fluid may be useful to assess pulmonary maturity. The lungs will not expand normally at birth if they are immature and the infant may then suffer from the *respiratory distress syndrome* (hyaline membrane disease): respiratory support may then be needed. It is therefore important to have evidence of pulmonary maturity before labour is induced.

In late pregnancy (after 32 to 34 weeks of gestation) the cells lining the fetal alveolar walls start to synthesize a surface-tension-lowering complex ('surfactant'), 90 per cent of which is lipid in nature; most of this lipid is *lecithin,* which contains *palmitic acid.* Surfactant is probably washed from, or is perhaps secreted by, the alveolar walls into the surrounding amniotic fluid, in which both lecithin and palmitic acid concentrations increase steadily. The concentration of another lipid, sphingomyelin, remains constant, and the *lecithin/sphingomyelin (L/S) ratio rises;* measurement of this ratio is the most commonly used estimate of pulmonary maturity. Other estimates have included total lecithin and total palmitic acid concentrations. The predictive value of all these parameters is similar.

Maternal biochemical changes in pregnancy

The level of many plasma constituents is affected by steroid hormones: for example, the reference ranges of plasma urate and iron differ in males and females after puberty. It is therefore not surprising to find that pregnancy, which is associated with very high circulating levels of oestrogens and progesterone, affects plasma concentrations of many substances.

The plasma concentration of many specific *carrier proteins* is increased during pregnancy. If this fact is not recognized an erroneous diagnosis may be made. In most cases the rise in the level of the carrier protein is accompanied by a proportional increase of the substance bound to it, without any change in the unbound fraction. Because the protein-bound fraction is a transport form and because, in most cases, it is the free substance that is physiologically active, this rise in concentration is only of importance in the interpretation of the results of such assays as those of thyroxine and cortisol.

Many of these changes may also be found in subjects taking oral contraceptives, particularly those with a high oestrogen content and in newborn infants.

Other changes in maternal plasma are due to progressive *haemodilution* due to the fluid retention that occurs during pregnancy. This is maximal at about the 30th week and the effects are most evident in the reduced concentration of albumin, and of calcium, which is bound to albumin. These changes are more marked in preeclamptic toxaemia, in which fluid retention may be greater.

Renal glycosuria is common both in pregnancy and in subjects taking oral contraceptives. The glomerular filtration rate increases by about 50 per cent during pregnancy and glycosuria may partly be due to an increased glucose load in normal tubules.

The positive protein and purine balance during growth of the fetus, and the increase in glomerular filtration rate that occurs during pregnancy, result in lowered maternal plasma urea and urate levels. These findings are of little clinical significance, but a raised plasma urate may be found in pre-eclamptic toxaemia.

Plasma alkaline phosphatase activity rises during the last three months of pregnancy due to the presence of the placental isoenzyme (p. 316). This physiological cause should be remembered if a high alkaline phosphatase activity is found during pregnancy.

Some of the changes which may occur during pregnancy and in patients taking high oestrogen oral contraceptives are summarized in Table 7.1.

Detection and follow-up of trophoblastic tumours

Trophoblastic tumours (hydatidiform mole, choriocarcinoma) which may follow abnormal pregnancy (miscarriage), and some teratomas, secrete HCG. For diagnostic purposes the simple pregnancy tests are usually adequate but they are not sensitive enough for follow-up studies. HCG estimation in plasma or urine by sensitive tests allows early detection and treatment of recurrence. It will *not* differ-

Table 7.1 Some metabolic effects of pregnancy which may confuse diagnosis

Test	Effect	Comment
Plasma		
total T_4*	Increased	Due to increased thyroxine-binding globulin. *Free T_4 probably normal*
cortisol*	Increased	Due to increased cortisol-binding globulin. *Free cortisol probably normal*
transferrin or TIBC*	Increased	See Chapter 20
iron*	Increased	See Chapter 20
alkaline phosphatase	Increased	Due to placental isoenzyme
total protein and albumin	Decreased	Due to dilution by fluid retention
urea and urate	Decreased	Anabolism due to fetal growth and probably raised GFR
Urinary glucose	May be renal glycosuria	Probably due to raised GFR

*These changes may also occur in women taking high oestrogen oral contraceptives

entiate pregnancy from recurrence of a tumour, because in both cases HCG levels rise.

Summary

Reproductive system

1. The secretion of gonadal hormones and the development of the germ cells is controlled by the pituitary gonadotrophins LH and FSH. These, in turn, are influenced by hypothalamic-releasing factors and, by negative feedback, by circulating gonadal hormone levels.

2. The main gonadal hormones are testosterone, secreted by the Leydig cells of the testis, and oestradiol, secreted by the cells of the ovarian follicle. There is extensive interconversion between these hormones and androgens secreted from the adrenal cortex.

3. In the male LH stimulates testosterone secretion from the testicular Leydig cells, and FSH stimulates spermatogenesis. Testicular failure may be confined to failure of spermatogenesis (raised FSH only) or may also involve the Leydig cells when levels of testosterone are low and those of both LH and FSH are raised.

4. In the female cyclical hormonal changes prepare the endometrium for implantation of a fertilized ovum. In the first half of the menstrual cycle LH and FSH stimulate ovarian follicle development and oestrogen secretion. A mid-cycle LH surge induces ovulation from one follicle, and converts it into a corpus luteum which secretes oestrogens and progesterone. In the absence of successful implantation the corpus luteum involutes and, as hormone levels fall, endometrial breakdown and shedding occur (menstruation). If implantation occurs the developing

placenta produces HCG which maintains the corpus luteum and prevents menstruation.

5. Gonadal hormone levels must be interpreted in relation to the stage of the menstrual cycle.

6. In primary ovarian failure, and at the menopause, plasma oestrogen levels are low and gonadotrophin levels are high (negative feedback). In secondary ovarian failure both oestrogen and gonadotrophin levels are low.

7. Hyperprolactinaemia is commonly due to a pituitary adenoma. In females it produces amenorrhoea and infertility, sometimes accompanied by galactorrhoea. Reducing prolactin secretion by bromocriptine restores fertility.

8. Gonadal hormones and gonadotrophin estimations are of value in:

- detection of ovulation;
- monitoring of treatment designed to induce ovulation;
- assessment of amenorrhoea;
- evaluation of hypogonadism and infertility;
- evaluation of delayed puberty.

Pregnancy

1. The measurement of HCG is of value in diagnosing pregnancy and in monitoring the progress during the early stages in selected cases.

2. Tests on amniotic fluid may be used to assess fetal development. These include:

- α-fetoprotein (AFP), which is raised in cases of severe neural tube defect. Maternal plasma AFP levels may also be abnormally high;
- bilirubin, to assess the severity of materno-fetal blood group incompatibility;
- the lecithin/sphingomyelin (L/S) ratio, or similar assay, to assess fetal lung maturity.

3. Several biochemical parameters alter in maternal plasma and urine during pregnancy, and in the newborn infant. The most obvious are the increases in carrier proteins such as thyroxine- and cortisol-binding globulins, which cause increases in plasma total thyroxine and cortisol concentrations. It is important to recognize these, and other changes, as physiological rather than pathological.

Further reading

Fritz M A, Speroff L. The endocrinology of the menstrual cycle: interaction of folliculogenesis and neuroendocrine mechanisms. *Fertil Steril* 1982; **38**: 509–29.

Morris D V, Adams J, Jacobs H S. The investigation of female gonadal dysfunction. *Clin Endocrinol Metab* 1985; **14**: 125–43.

Grossman A, Besser G M. Prolactinomas. *Br Med J* 1985; **290**: 182–4.

Handelsman D J, Swerdloff R S. Male gonadal dysfunction. *Clin Endocrinol Metab* 1985; **14**: 89–124.

Ismail A A A, Astley P, Burr W A, Cawood M, Short F, Wakelin K, Wheeler M J. The role of testosterone measurement in the investigation of androgen disorders. *Ann Clin Biochem* 1986; **23**: 113–34.

Investigation of pituitary, adrenal and gonadal disorders

These investigation sections are considered together because pituitary disorders, particularly hypofunction, may present with features of adrenal or gonadal deficiency.

Specimen collection

Your laboratory should always be consulted *before* starting any test, both to ensure the most efficient and speedy analysis, and to check local details of specimen collection and handling and variations in test protocols.

Many hormones are stable in blood for several hours or even days. Specimens for others, such as insulin, renin and ACTH, need to be kept cool and sent to the laboratory without delay.

The *time* of sampling should be recorded on the request form.

Factors affecting results of cortisol assays

Plasma cortisol is usually measured by immunoassay, or a similar method. Some laboratories use fluorescence assays for 11-hydroxycorticosteroids. Some drugs may affect results. Examples are:

- *hydrocortisone (cortisol) and cortisone (converted to cortisol by metabolism), which are measured by both methods*;
- *oestrogens and some oral contraceptives* which increase the CBG, and therefore protein-bound cortisol concentration, and give *falsely high results by both methods*;
- *prednisolone*, which will contribute to the 'cortisol' value using *most immunoassay, but not fluorescence, methods*;
- *spironolactone*, which produces falsely *high plasma and urinary results using fluorescence methods*;
- *dexamethasone*, which does not affect plasma assay by either method, but metabolites of which may interfere with *some immunoassay methods in urine.*

Procedure for repeated sampling

Many 'dynamic' tests of endocrine function need several blood samples to be taken over a short period of time. Repeated venepuncture is unpleasant for the patient and is also undesirable because it may cause stress, and so interfere with the results. Insertion of an indwelling needle or cannula helps to minimize these problems and enables intravenous therapy to be given without delay if there should be an untoward reaction such as hypoglycaemia.

We suggest the following procedure:

(a) A needle or cannula at least as large as 19G is inserted and is secured in position with adhesive strapping.

(b) Isotonic saline is infused *slowly* to keep the needle open. Heparin should not be used because it interferes with some assays.

(c) The first specimen should not be taken for *at least* 30 minutes, so that any stress-induced elevation of hormones may return to basal levels.

All samples may be taken through this needle as follows:

(a) Disconnect the saline infusion, or use a three-way tap between the infusion set and the needle.

(b) Aspirate and *discard* 1 to 2 ml of the saline-blood mixture.

(c) *Using a different syringe* aspirate the specimen for assay.
(d) Reconnect and restart the saline flow.

Investigation of suspected hypopituitarism

Deficiency of pituitary hormones causes hypofunction of the target endocrine glands. Investigation aims to confirm such deficiency, to exclude disease of the target gland and then to test pituitary hormone secretion after maximal stimulation of the gland.

1. Measure plasma concentrations of:

- LH, FSH and oestradiol (female) or testosterone (male);
- thyroxine and TSH;
- prolactin (to test for hypothalamic or pituitary-stalk involvement);
- cortisol at 09.00 h, to assess the risk of adrenocortical insufficiency during later testing.

2. If low target-gland and raised trophic hormone levels are found, test the target gland affected.
3. If both target gland and trophic hormone levels are low or low normal, proceed to a combined pituitary stimulation test.
4. Investigate the pituitary region using radiological techniques.

Combined pituitary stimulation test

This test is potentially dangerous and must be done under direct medical supervision. It is contraindicated in patients with ischaemic heart disease or epilepsy. *Glucose* for intravenous administration should be *immediately available* in case severe hypoglycaemia develops. After the test the patient should be given something to eat.

The plasma levels of anterior pituitary hormones are measured after stimulation by stress, TRH and Gn-RH. Plasma cortisol is usually measured as an index of ACTH secretion: the entire hypothalamic-pituitary-target gland axis is therefore tested. It will be referred to again when discussing investigation of individual target glands. If glucose needs to be given, *continue with the sampling*. The stress has certainly been adequate to stimulate hormone secretion.

After an overnight fast:

1. Insert an indwelling intravenous cannula.
2. *After at least 30 minutes take basal samples.*
3. *Inject soluble insulin* in a high enough dose to lower plasma glucose levels to less than 2.5 mmol/litre (45 mg/dl) *and to cause symptomatic hypoglycaemia*. The usual dose is 0.15 U/kg body weight. If pituitary or adrenocortical hypofunction are probable, or if a low fasting glucose concentration has been found before, reduce the dose to 0.1 or 0.05 U/kg. Conversely, if there is resistance to the action of insulin because of Cushing's syndrome, acromegaly or obesity, 0.2 or 0.4 U/kg may be needed.
4. *Immediately inject 200 μg of TRH and 100 μg of Gn-RH.*
5. *Take blood samples at 30, 45, 60, 90 and 120 minutes* after the injections.

Interpretation. Methods of hormone assay vary and results should not be compared with reference values issued by other laboratories. The following is intended as a guide only.

If hypoglycaemia has been adequate plasma cortisol levels should rise by more than 200 nmol/litre (7 μg/dl) and exceed 550 nmol/litre (20 μg/dl). GH values should exceed 20 mU/litre.

TSH values should increase by at least 2 mU/litre and exceed the upper limit of the reference range. Prolactin and gonadotrophin levels should rise significantly for the reference range for the method.

If the plasma cortisol level does not rise a tetracosactrin test (p. 154) should be performed to exclude primary adrenocortical hypofunction.

Neither cortisol nor GH respond in Cushing's syndrome.

Investigation of suspected growth hormone deficiency

1. A single high plasma GH value probably excludes deficiency. The initial sample should be taken when the highest physiological levels occur – immediately after exercise or during sleep.

2. If GH levels do not exclude deficiency, stimulate the pituitary. A combined stimulation test (see above) can be used, or only insulin need be given.

Investigation of suspected acromegaly or gigantism

A raised plasma GH level which fails to suppress normally in response to a rising plasma glucose concentration suggests autonomous hormone secretion. Basal levels of GH may be, but are not always, high enough to confirm the diagnosis. It saves time to start with a glucose suppression test.

Glucose suppression test

After an *overnight fast*:

1. Insert an indwelling intravenous cannula.
2. *After at least 30 minutes* take basal samples for glucose and GH estimation.
3. The patient drinks 75 g of glucose, or glucose equivalent (p. 226), dissolved in about 300 ml of water.
4. Samples are taken for glucose and GH assay at 30, 60, 90 and 120 minutes after the glucose has been taken.

Interpretation. In normal subjects plasma GH will fall to less than 4 mU/litre. Although failure to suppress suggests acromegaly or gigantism, it may be found in severe liver or renal disease, in heroin addicts or in patients taking levodopa.

Investigation of suspected Cushing's syndrome

Because of the serious nature of Cushing's syndrome, and because it may be treatable, the diagnosis must be excluded even when clinical features are only suggestive. Initial tests may exclude the diagnosis, but may yield some 'false positive' results.

A. Has the patient got Cushing's syndrome?

The initial tests can be carried out without admitting the patient.

1. Perform an overnight dexamethasone suppression test (see next page).
2. Estimate urinary free cortisol in several 24 hour collections. If these tests are normal, it is unlikely that the patient has Cushing's syndrome. If either is abnormal, or if there is a strong clinical suspicion, further tests should be carried out in hospital.
3. Assess the circadian rhythm of plasma cortisol by collecting blood samples at 09.00 h and 23.00 h. This may be followed by repeating the overnight dexamethasone test.

If all these tests are normal, follow up the patient in the outpatient clinic. The manifestations of Cushing's syndrome may be intermittent, and tests may have to be repeated later.

If the results show abnormal cortisol secretion consider alternative causes such as stress, alcoholism or endogenous depression. If the latter is likely, perform an insulin stress test (p. 151). A normal cortisol response to this suggests depression rather than Cushing's syndrome.

Once the diagnosis of Cushing's syndrome has been made, proceed to the next step.

B. What is the cause of the Cushing's syndrome?

1. *Extremely high cortisol levels are in favour of adrenocortical carcinoma (especially if the patient is virilized), or of ectopic ACTH production.* Investigate clinically and using other investigations, especially bearing in mind the possibility of carcinoma of the adrenal gland or bronchus. Severe hypokalaemic alkalosis suggests ectopic ACTH secretion.

2. It is highly desirable to estimate plasma *ACTH* levels, even if the specimen has to be sent to a special centre. Moderately raised levels suggest Cushing's disease or occult ectopic ACTH secretion. Very high levels are found with overt ectopic ACTH secreting tumours, while in patients with adrenocortical tumours the plasma ACTH concentration is low. Remember that stress increases the ACTH and cortisol levels.

3. Alternatively perform a *high dose dexamethasone test. Suppression suggests either pituitary-dependent Cushing's syndrome or an occult ACTH-secreting tumour. Failure to suppress* suggests either an adrenocortical tumour (low ACTH levels) or overt ectopic ACTH production (very high ACTH levels).

4. If there is virilization estimate plasma androgens, especially DHAS. High levels suggest adrenocortical carcinoma.

The distinction between occult ectopic ACTH-secreting tumours and Cushing's disease may be difficult. CT scans of the chest and abdomen may detect the tumour.

Dexamethasone suppression tests

Anticonvulsant drugs, particularly *phenytoin*, may interfere with dexamethasone suppression tests. They induce liver enzymes which increase the rate of metabolism of dexamethasone, and plasma levels may therefore be too low to suppress the feedback centre.

1. Overnight test

Dexamethasone (2 mg) is given as a single oral dose at 23.00 h. Plasma cortisol levels are measured on a specimen taken at 09.00 h the next morning. Suppression is defined as a plasma cortisol of less than 190 nmol/litre (7mg/dl).

2. High dose test

Dexamethasone (2 mg) is given orally every 6 hours for two days, starting at 09.00 h. Plasma cortisol is measured in a specimen taken at 09.00 h on the third day. Suppression is defined as levels less than 50 per cent of previously measured values.

Investigation of suspected adrenal hypofunction

A. Suspected Addisonian crisis

1. Take blood *first* for immediate electrolyte, and later cortisol, estimation.
2. Start steroid treatment at once. Do not wait for laboratory results.
3. Hyponatraemia, hyperkalaemia and uraemia, although compatible with an Addisonian crisis, are common in many clinically similar acute conditions. Treat appropriately (see Chapters 2 and 3).

4. *Plasma cortisol* may be estimated later:

- if it is *very high an Addisonian crisis is excluded*;
- if it is *very low or undetectable*, and if there is no reason to suspect CBG deficiency, an *Addisonian crisis is confirmed.*

5. *Plasma cortisol levels which would be 'normal' under basal conditions may be inappropriately low for a stressed patient. Perform a short tetracosactrin test* (see below).

B. Suspected chronic adrenal hypofunction

1. Measure plasma cortisol levels. A high level *at any time of day* excludes Addison's disease.
2. If the plasma cortisol results are equivocal, perform a *short tetracosactrin test. A normal result excludes Addison's disease* and makes long-standing secondary adrenal insufficiency unlikely. Prolonged ACTH deficiency causes reversible adrenal insensitivity to trophic stimulation.
3. If the results of the short tetracosactrin test are equivocal, send a blood specimen to the laboratory before proceeding further, in case ACTH assay seems indicated later (see 7).
4. Admit the patient and perform a *5-hour tetracosactrin test*. A *normal result* excludes primary adrenal hypofunction: *if there was a subnormal response to the short tetracosactrin test it suggests secondary adrenal hypofunction.*
5. If doubt remains perform a *3-day tetracosactrin test*. The same impaired response to the short, 5-hour and 3-day tetracosactrin test would confirm *primary adrenal hypofunction.*
6. An *increasing response* to the short, 5-hour and 3-day tetracosactrin tests indicates gradual recovery of the adrenal cortex following prolonged lack of ACTH, and suggests *hypothalamic or pituitary hypofunction*. Perform a combined pituitary function test (see p. 151).
7. If doubt remains, and if facilities are available, assay ACTH on the plasma specimen taken earlier. *If the cortisol level is low or low normal:*

- *a high ACTH level* confirms *primary adrenal hypofunction*;
- *a low ACTH level* suggests *secondary adrenal hypofunction*, and a combined pituitary function test should be performed.

Tetracosactrin tests of adrenal function

Tetracosactrin is marketed as 'Synacthen' (Ciba) or 'Cortrosyn' (Organon).

1. 30-minute stimulation test

The patient should be resting quietly.

(a) Blood is taken for basal cortisol assay.

(b) 250 μg of tetracosactrin, dissolved in about 1 ml of sterile water or isotonic saline, is given by intramuscular injection.

(c) 30 minutes later blood is taken for cortisol assay.

Normally plasma cortisol increases by at least 200 nmol/litre (7 μg/dl), to a level of at least 550 nmol/litre (20 μg/dl).

2. 5-hour stimulation test

(a) Blood is taken for basal cortisol assay.

(b) 1 mg of depot tetracosactrin is injected intramuscularly.

(c) Blood is taken 1 and 5 hours later.

Normally plasma cortisol rises to between 600 and 1300 nmol/litre (22 and 46 μg/dl) at 1 hour, and to between 1000 and 1800 nmol/litre (37 and 66 μg/dl) at 5 hours.

3. 3-day stimulation test

Repeated injections of depot tetracosactrin are painful and may lead to sodium and water retention: *this test is therefore contraindicated in patients, such as those with congestive cardiac failure, in whom such retention may be dangerous.*

(a) 1 mg of depot tetracosactrin is given daily by intramuscular injection.
(b) Plasma cortisol is estimated in blood withdrawn 5 hours after each injection.

C. Patients on steroid treatment with suspected adrenal hypofunction

It is not uncommon for patients to be treated before adrenal hypofunction has been proved. Steroids such as cortisone or hydrocortisone may interfere with the assays and should be stopped. If there is a danger that this may precipitate an adrenal crisis, stimulation tests may be performed while the biologically active dexamethasone, which does not interfere with the assays, is being given. Consult your laboratory to find out which steroids interfere with their assays.

Investigation of amenorrhoea

A full clinical assessment should be made, and the patient should be questioned carefully to determine whether the amenorrhoea is primary or secondary. Amenorrhoea may accompany any severe disease.

The aim of laboratory investigation is to detect hypogonadism and to distinguish between pituitary and ovarian causes. *Before embarking on such investigation it is essential to test for pregnancy.*

If the pregnancy test is negative proceed as follows.

1. Measure plasma LH, FSH and oestradiol concentrations.

Raised gonadotrophin (especially FSH) with *low oestradiol* levels indicate *ovarian failure.* If the amenorrhoea is primary, chromosome studies are indicated.

Normal gonadotrophin with *low oestradiol* levels suggest the possibility of *testicular feminization. Plasma testosterone* should be measured. A value *within the male reference range supports the diagnosis,* which may be confirmed by chromosomal analysis.

Low oestradiol with *low or low normal gonadotrophin* levels suggest a hypothalamic or pituitary cause.

2. *Measure the plasma prolactin* level at least three hours after waking (p. 141). A high level supports a hypothalamic or pituitary cause. Because prolactin levels respond variably to stress, a raised value should be confirmed on a specimen taken on a different occasion. Exclude drugs as a cause (p. 141).

3. Perform a gonadotrophin-releasing hormone (Gn-RH) test.

- *A normal gonadotrophin response* suggests a probable *hypothalamic* cause, perhaps involving only gonadotrophin-releasing hormone.
- *A subnormal response* suggests a *pituitary lesion.*

4. Proceed to a combined pituitary stimulation test (p. 151).

Gonadotrophin-releasing hormone test

Synthetic Gn-RH stimulates the release of gonadotrophins from the normal anterior pituitary gland. 100 μg of Gn-RH is given by rapid intravenous injection *Plasma LH and FSH levels are measured in blood drawn before and at 20 and 60 minutes after* the injection. In normal subjects plasma LH rises by at least 5 U/litre, but this rise fails to occur in patients with pituitary hypofunction.

Other endocrine disorders such as hyperthyroidism and Cushing's syndrome must be considered as causes of amenorrhoea and, if necessary, excluded by appropriate tests. If there is hirsutism or virilism, proceed as below.

Investigation of hirsutism and virilism

The aim of investigation, after full clinical assessment, is to detect cases with significant elevation of plasma androgen levels and to identify the source as the ovary or the adrenal cortex.

Depending on the clinical presentation, some of the following plasma assays may help:

- testosterone (total level and estimate of free fraction by measuring SHBG);
- DHA or DHAS;
- 17-hydroxyprogesterone if congenital adrenal hyperplasia is suspected;
- gonadotrophins.

Plasma findings					
Testosterone	DHAS	LH	FSH	17-OH-progesterone	
N or slightly ↑	N or slightly ↑	N	N	N	Simple hirsutism
↑	N	↑	N or ↓	N	Polycystic ovaries
↑↑	N	N or ↓	N or ↓	N	Ovarian tumour
N or slightly ↑	↑↑	N or ↓	N or ↓	N	Adrenocortical tumour
↑	↑	N or ↓	N or ↓	↑	Congenital adrenal hyperplasia

If there is evidence of other endocrine disease such as Cushing's syndrome, investigate accordingly.

Investigation of infertility

It is important to make a full clinical assessment of, and to test, both partners if a woman complains of infertility.

Female infertility

A woman may be infertile despite clinically normal menstrual cycles, or may have amenorrhoea or oligomenorrhoea. Even if the cycle seems regular it is important to try to determine whether ovulation is occurring and if luteal development is normal.

1. If the patient is menstruating regularly measure plasma progesterone during the luteal phase.

- A *normal* level is strong evidence that the patient has *ovulated.*
- A *low* level suggests either *ovulatory failure* or *impaired luteal function.*

2. Follicular development and ovulation may be monitored by ultrasound examination.
3. Histological examination of an endometrial biopsy specimen should indicate whether luteal function is normal.

The subject of detailed investigation of female infertility is beyond the scope of this book.

Male infertility

Laboratory investigations may detect early hormonal deficiency or distinguish between testicular and pituitary causes.

Measure plasma testosterone, LH and FSH and perform semen analysis.

1. A *raised plasma LH* with a *low testosterone* level indicate Leydig cell failure. A *raised LH with normal testosterone* suggests a lesser degree of damage with a compensatory increase in LH secretion (compare TSH in primary hypothyroidism).

2. *Low LH and testosterone* levels suggest pituitary or hypothalamic disease.

   (a) Measure plasma *prolactin*. If *markedly raised* this is suggestive of a *pituitary tumour*;
   (b) Perform a combined pituitary stimulation test if clinically indicated.

3. A *raised FSH* indicates *seminiferous tubular failure, whatever the testosterone levels.* There is usually oligospermia.

4. *Oligospermia with low FSH* levels suggest *pituitary or hypothalamic disease* and the patient should be investigated for the condition.

8

Thyroid function: TSH

Thyroxine (T_4), triiodothyronine (T_3) and calcitonin are secreted by the thyroid gland. T_4 and T_3 are products of the follicular cells, and influence general metabolism. Calcitonin is produced by the specialized C-cells, and influences calcium metabolism. This is an anatomical rather than a functional relationship and in some lower animals the calcitonin-secreting cells are completely separated from the thyroid gland. Calcitonin is considered briefly on p. 173.

Physiology

The thyroid hormones are synthesized in the thyroid gland by iodination and coupling of two molecules of the amino acid tyrosine, a process which is dependent on an adequate supply of iodide.

Iodide in the diet is absorbed rapidly from the small intestine. Most natural foods contain adequate amounts of iodide except in regions where the content of the soil is very low. In these areas there used to be a high incidence of enlargement of the thyroid gland (*goitre*), but general use of artificially iodized salt has made this a less common occurrence. Sea foods have a high iodide content and fish and iodized salt are the main dietary sources of the element.

Normally about a third of absorbed iodide is taken up by the thyroid gland and the other two thirds is excreted by the kidneys.

Synthesis of thyroid hormones

The biosynthesis and secretion of the thyroid hormones are outlined below. A knowledge of this pathway is essential for the understanding of the mode of action of drugs, whether used to diagnose or to treat thyroid disorders, and of the effects of congenital enzyme deficiencies (p. 166).

- *Iodide is actively taken up* by the thyroid gland. The concentration in the gland is normally about twenty times that in plasma but may exceed it by a hundred times or more. The salivary glands, gastric mucosa and mammary glands are also capable of concentrating iodide. *Uptake is blocked by thiocyanate and perchlorate.*
- Trapped *iodide is rapidly converted to iodine.*

- *Tyrosine residues* in a large glycoprotein, thyroglobulin, are *iodinated* to form mono- and diiodotyrosine (MIT and DIT). This step is *inhibited, for example, by carbimazole and propylthiouracil.*
- *The iodotyrosines are coupled to form thyroxine (T_4) (DIT and DIT) and triiodothyronine (T_3) (DIT and MIT).* (Fig. 8.1). Normally much more T_4 than T_3 is synthesized but if there is an inadequate supply of iodine the ratio of T_3 to T_4 in the gland increases. The thyroid hormones, still incorporated in thyroglobulin, are stored in the colloid of the thyroid follicle.

Before secretion of the thyroid hormones, thyroglobulin is taken up by the follicular cells. T_4 and T_3 are released from it by proteolytic enzymes into the bloodstream where they are immediately bound to plasma proteins. Mono- and diiodotyrosine released at the same time are deiodinated and the iodine is reutilized.

Each step is controlled by specific enzymes and congenital deficiency of any of these enzymes can lead to goitre and, if severe, hypothyroidism. The uptake of iodide as well as synthesis and secretion of thyroid hormones is regulated by thyroid-stimulating hormone (TSH) secreted from the anterior pituitary gland.

Protein binding of thyroid hormones in plasma

More than 99 per cent of plasma T_4 and T_3 is protein-bound, mainly to an α-globulin, *thyroxine-binding globulin (TBG)*, and to a lesser extent, to albumin and thyroxine-binding pre-albumin. It is the small free fractions which are physiologically active, and which regulate pituitary TSH secretion (compare cortisol, p. 120, and calcium, p. 173).

Changes in the levels of the binding proteins, particularly that of TBG, alter total T_4 and T_3, but not free hormone concentrations.

Peripheral conversion of thyroid hormones

Some of the circulating T_4 is deiodinated by enzymes in peripheral tissues, especially in the liver and kidneys. Removal of an iodine atom from the outer (β) ring produces about 80 per cent of the circulating active hormone T_3, the other 20 per cent being secreted by the thyroid gland. Deiodination of the inner (α) ring

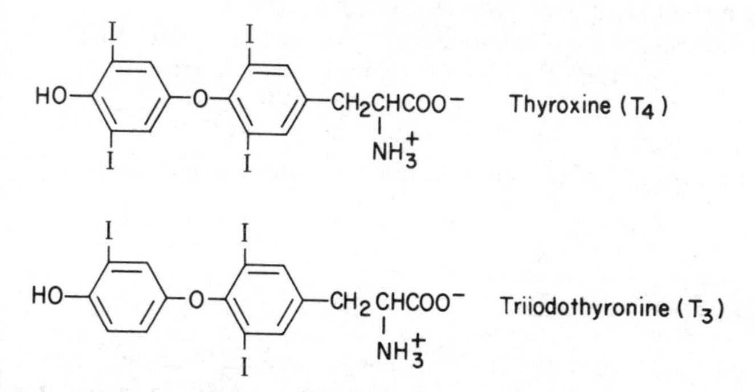

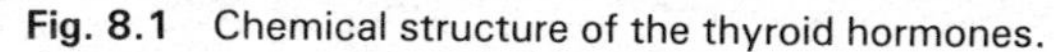
Fig. 8.1 Chemical structure of the thyroid hormones.

produces the probably inactive reverse T_3.

The conversion of T_4 to T_3 may be:

- *reduced* by many factors, of which the most important are:
  - systemic illness;
  - prolonged fasting;
  - drugs such as:
    - propranolol;
    - amiodarone;
    - some radiocontrast media (transient effect).
- *increased* by drugs such as phenytoin, which *induce hepatic enzymes*.

The plasma T_3 concentration is a poor indicator of thyroid secretion because it is influenced by many non-thyroidal factors. Its measurement is rarely indicated.

Actions of thyroid hormones

Thyroid hormones influence and speed up many metabolic processes in the body. They are essential for normal growth, mental development and sexual maturation. They also increase the sensitivity of the cardiovascular and central nervous systems to catecholamines and so influence cardiac output and heart rate. Many of these actions are mediated by T_3 which, by binding to specific receptors in cell nuclei, alters the expression of some genes.

Control of TSH secretion

TSH secretion is controlled by:

- circulating concentrations of thyroid hormones;
- thyrotrophin-releasing hormone (TRH).

Effect of thyroid hormones. These hormones reduce TSH secretion by negative feedback. T_3 binds to pituitary nuclear receptors, as it does in other cells. In the pituitary most of the intracellular T_3 is derived from circulating free T_4 and therefore the gland is more sensitive to plasma T_4 than to T_3 concentrations. For example, in early hypothyroidism plasma T_3 levels are often normal; their effect on peripheral tissues minimizes the clinical effects of low T_4 levels, but plasma TSH secretion may be increased in response to falling T_4 levels.

Effect of TRH. TSH secretion is also stimulated by the hypothalamic TRH. Its stimulatory effect can be overridden by abnormally high circulating free T_4 levels. This is why exogenous TRH has little or no effect on TSH secretion in hyperthyroidism, and can be used to test for this condition (p. 169).

The metabolism and control of thyroid hormones are summarized in Fig. 8.2.

Thyroid function tests

Treatment of thyroid disease may need to be prolonged and, in the case of hypothyroidism, life-long, and before starting treatment it is essential to confirm the

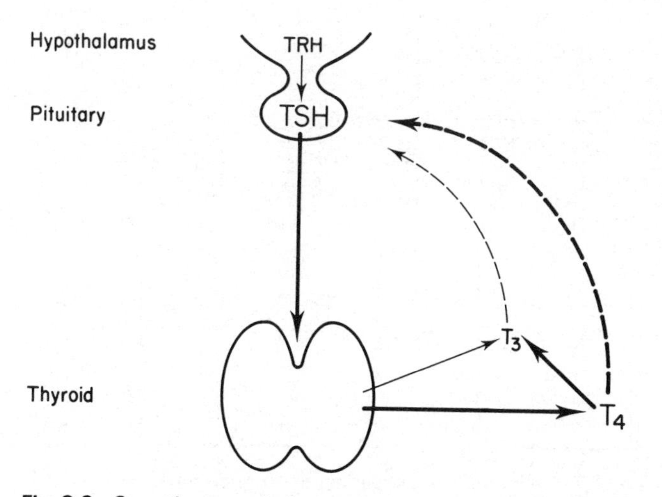

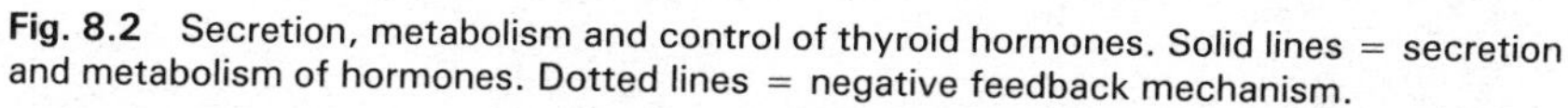

Fig. 8.2 Secretion, metabolism and control of thyroid hormones. Solid lines = secretion and metabolism of hormones. Dotted lines = negative feedback mechanism.

diagnosis by laboratory tests. During treatment the original clinical picture disappears, and it is not uncommon to be faced with a patient who has taken thyroxine for many years for probable 'thyroid trouble'. Without an adequately established diagnosis it may be difficult to assess the past history.

Assessment, by either direct or indirect measurement, of the plasma free T_4 concentration is the best single index of thyroid hormone secretion and is the first step in laboratory diagnosis of thyroid function. Until recently the free T_4 level could only be inferred by adjusting the total T_4 result for changes in binding-protein concentration. Newer 'direct' assays of 'free T_4' are now available, but also have their methodological limitations. The results of either type of assay may indicate the need for further investigations.

Total thyroxine and free thyroxine index

Plasma T_4 is more than 99 per cent protein bound; results of total T_4 assays therefore reflect the protein-bound rather than the free fraction. Changes in thyroid secretion produce parallel changes in total and free plasma T_4 levels. Total T_4 reflects free T_4 concentrations *unless there are abnormalities of binding proteins.*

- In the *euthyroid* state, whatever the level of the binding protein, about a third of the TBG binding sites are occupied by T_4, and the remaining two-thirds are unoccupied (Fig. 8.3).
- In *hyperthyroidism* both total and free T_4 levels are increased and the number of unoccupied binding sites on TBG decreased (Fig. 8.3a).
- In *hypothyroidism* the opposite occurs (Fig. 8.3b).

An increase in TBG (Fig. 8.3c) causes an increase in both bound T_4 and unoccupied binding sites. Such an increase may occur:

- due to high oestrogen levels in pregnancy or in the newborn infant;

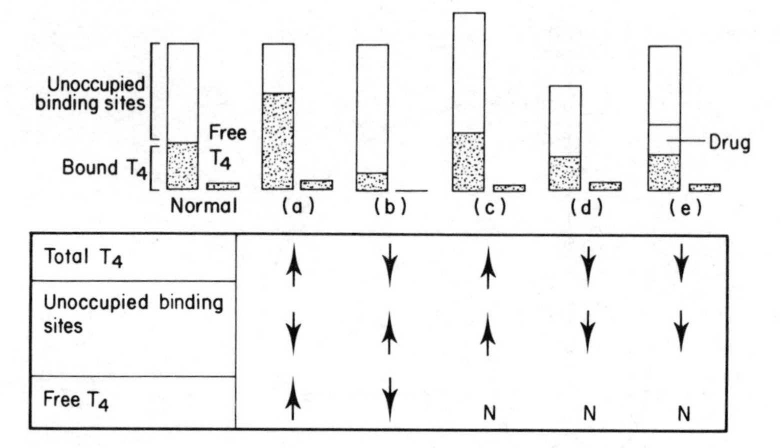

Fig. 8.3 Interpretation of results of tests of circulating thyroid hormones (see text).

- during oestrogen therapy, or while taking oral contraceptives;
- rarely, with inherited TBG excess.

A decrease in TBG (Fig. 8.3d) decreases both bound T_4 levels and unoccupied binding sites. Such changes may occur:

- temporarily in severely ill patients;
- due to loss of low molecular-weight proteins, usually in the urine (nephrotic syndrome);
- in patients taking androgens or danazol;
- rarely, in inherited TBG deficiency.

These changes might be misinterpreted as diagnostic of hyperthyroidism and hypothyroidism respectively if only total T_4 were measured. Usually, if there is doubt, results of the tests discussed later in the chapter, together with the clinical picture, will elucidate the true thyroid status.

- *Displacement of T_4 by drugs* (Fig. 8.3e). Some drugs such as salicylates and danazol, are bound by TBG and displace T_4. The change in unoccupied binding sites is variable and TBG levels are unaffected. Measures to correct the total T_4 result for protein abnormalities may not indicate the problem. Low T_4 values can only be interpreted with a knowledge of what drugs the patient is taking.

Correction for altered TBG levels

Assessment of unoccupied binding sites enables some correction for changes in TBG levels to be made. This may be done by adding a known amount of radioactive thyroid hormone (T_3) to a sample of the patient's plasma and then assessing how much is bound (T_3 uptake). With the results of this and of total T_4 assay an assessment of the free hormone may be made (for example, as the *free-thyroxine index; FTI*).

- *If there is a change in the level of binding protein* the measured T_4 and unoccupied binding sites change in the same direction. The FTI 'corrects' the abnormal total T_4 levels.
- In disorders of thyroid function, if TBG levels are unchanged, the measured T_4 and unoccupied binding sites vary inversely. The FTI exaggerates the abnormal total T_4 level.

Measurement of TBG concentration may be made directly and a T_4: TBG ratio calculated.

The FTI is simpler, cheaper and more generally available of these two assessments of plasma free T_4. It should be noted that whichever method is used, 'correction' is often incomplete when the concentrations or binding properties of TBG, or of other binding proteins, are very abnormal. Comparison of the total T_4 and FTI values will indicate if there is abnormal binding.

Free thyroxine assays

True free thyroxine assays are difficult to perform and are not readily available. A number of newer assays use analogues of T_4 which do not bind to TBG and so assess only the unbound fraction. In uncomplicated situations, when total T_4 assay is often adequate, they perform well but they are not always reliable when abnormal binding proteins or abnormal albumin levels are present. The user must be aware of which 'free T_4' assay is being used in his laboratory and what its limitations are.

Plasma TSH

TSH may be measured by immunoassay. Some assays are unable to distinguish between low normal and subnormal levels and are only useful to diagnose primary hypothyroidism, in which TSH secretion is increased. Newer sensitive assays, however, seem able to distinguish low from low normal levels, and so can be used to diagnose hyperthyroidism.

Plasma total or free triiodothyronine(T_3)

Plasma T_3 immunoassay may occasionally be helpful as a test for hyperthyroidism, but should not be used to diagnose hypothyroidism. T_3, like T_4, is bound to protein, and if total T_3 is measured it may be necessary to assess binding sites to interpret the results.

TRH test

The normal anterior pituitary gland responds to administration of TRH by increasing the secretion, and therefore the plasma levels, of TSH.

The rise in plasma TSH concentration in response to TRH will be reduced in two conditions.

- In *hyperthyroidism* negative feedback due to increased plasma T_4 levels overrides the stimulatory effect of TRH. The impaired response to TRH may be demonstrable when *plasma T_4 levels are still within the upper reference range.* A normal response to TRH excludes the diagnosis of hyperthyroidism.
- In *hypopituitarism with secondary hypothyroidism* the anterior pituitary gland is unable to respond normally to TRH. In these cases the *plasma T_4 levels will tend to be low.*

The response will be exaggerated:

- in *hypothyroidism*, when negative feedback is reduced because plasma T_4 levels are low.

Disorders of the thyroid gland

The commonest presenting features in patients with thyroid disease are:

- hyperthyroidism, due to excessive thyroid hormone secretion;
- hypothyroidism, due to deficient thyroid hormone secretion;
- goitre (enlargement of the thyroid gland), either diffuse, or due to one or more nodules within the gland. There may or may not be abnormal hormone secretion.

Excess or deficiency of circulating thyroid hormone produces characteristic clinical changes. Disease of the thyroid gland may, however, be present without hyper- or hypofunction.

Hyperthyroidism (thyrotoxicosis)

The syndrome produced by a sustained high plasma level of thyroid hormone may easily be recognized or may remain unsuspected for a long time. The key feature is a speeding up of metabolism. Clinical features include tremor, tachycardia (and sometimes arrhythmias), weight loss, tiredness, sweating and diarrhoea. There may be severe anxiety and emotional symptoms. In some cases a single feature predominates, such as weight loss, diarrhoea, tachycardia or atrial fibrillation.

The *common causes* of hyperthyroidism are:

- autonomous secretion:
  Graves' disease;
  toxic multinodular goitre or a single functioning nodule (occasionally an adenoma);
- ingestion of thyroid hormones.

Rare causes are:

- secretion of thyroid-stimulating hormone by tumours of trophoblastic origin;
- struma ovarii (thyroid tissue in an ovarian teratoma);
- administration of iodine to a subject with iodine-deficiency goitre;
- TSH-secreting tumour of the pituitary (exceedingly rare).

Graves' disease occurs at any age and is commoner in females. It is a systemic condition characterized by:

- autonomous secretion from a diffuse goitre causing hyperthyroidism;
- exophthalmos, due to swelling and infiltration of retroorbital tissues;
- sometimes localized thickening of the subcutaneous tissue over the shin ('pretibial myxoedema').

Graves' disease is one of the autoimmune thyroid diseases. Thyroid antibodies are detectable in some cases, in addition to *thyroid-stimulating immunoglobulins* which probably cause hyperfunction.

Toxic nodules, single or multiple, in a nodular goitre may secrete thyroid hormones *autonomously. TSH is suppressed* by negative feedback (as in Graves' disease) and the thyroid tissue not in the nodules is inactive. These 'hot' nodules may be detected by their uptake of radioactive iodine or technetium. Toxic nodules are found more commonly in the older age groups and the patient may present with only one of the features of hyperactivity, most commonly cardiovascular symptoms.

Pathophysiology

Plasma concentrations of T_3 and T_4 are both usually abnormally high in hyperthyroidism. Much of the T_3 is secreted directly by the thyroid and the increase in plasma levels is greater, and usually evident earlier, than that of T_4. Rarely only T_3 levels are elevated (T_3 toxicosis). The abnormally high plasma T_4 concentration suppresses TSH secretion even if TRH is given. This failure of plasma TSH to rise normally after TRH administration may be used to test for mild hyperthyroidism.

The diagnostic approach to hyperthyroidism is considered on p. 169.

Hypothyroidism

Hypothyroidism is due to suboptimal circulating levels of one or both thyroid hormones. The presenting signs and symptoms depend to some extent on the aetiology of the thyroid disorder.

The condition develops insidiously and in its early stages the cause of the vague symptoms may not be recognized. Many of the features of severe hypothyroidism, which is less common than it was, are the opposite of those of hyperthyroidism. There is generalized slowing down of metabolism, with mental dullness and physical slowness.

If the hormone deficiency is due to primary disease of the thyroid gland the patient may present with weight gain and menstrual disturbances. The skin is dry, hair falls out and the voice becomes hoarse. The face is puffy and the subcutaneous tissues thickened; this pseudooedema, with a histologically myxoid appearance, accounts for the term myxoedema which is used to describe severe hypothyroidism.

In the most severe cases myxoedema coma, with profound hypothermia, may develop. Congenital thyroid hormone deficiency leads to cretinism, and in children growth may be impaired.

Primary hypothyroidism. The common causes are:

- atrophic autoimmune thyroiditis;
- Hashimoto's disease.

In these there is progressive destruction of thyroid tissue. Circulating thyroid antibodies are present, often in high concentration.

- other inflammatory diseases of the gland;
- following treatment of hyperthyroidism:
  - post-thyroidectomy;
  - post radioiodine treatment;
  - drugs:
    - propylthiouracil;
    - carbimazole;
    - lithium carbonate.

Rare causes of *primary* hypothyroidism are:

- exogenous goitrogens;
- dyshormonogenesis. The term 'dyshormonogenesis' includes inherited deficiencies of any of the enzymes involved in thyroxine synthesis. Although the biochemical and clinical features differ, the end result is hypothyroidism due to reduce thyroxine synthesis. In most cases prolonged TSH stimulation causes goitre. The commonest form is due to failure to incorporate iodine into tyrosine.

Secondary hypothyroidism is due to anterior pituitary or hypothalamic deficiency and is much less common than the primary form. In long-standing secondary hypothyroidism the thyroid gland may atrophy irreversibly.

The essential difference in laboratory findings between the primary and secondary forms is in the level of plasma TSH; this is high in primary and inappropriately low in secondary hypothyroidism.

Pathophysiology

As primary hypothyroidism develops the falling plasma T_4 level stimulates TSH secretion from the anterior pituitary gland; the plasma T_4 level may be within the population reference range, although abnormally low for the individual, and for this reason plasma TSH is the most sensitive index of early disease. Plasma T_3 levels are even more often normal, and its assay is of no help in making the diagnosis.

Secondary hypothyroidism may be due to impaired TSH secretion, either due to a disorder of the pituitary gland itself, or to reduced hypothalamic secretion. In the former case there is a subnormal rise in plasma TSH levels after administration of exogenous TRH.

The diagnosis of hypothyroidism is considered on p. 170.

Special problems in the diagnosis of hypothyroidism. Two situations warrant special consideration.

Neonatal hypothyroidism. The incidence of neonatal hypothyroidism in most populations is about 1/5000. It is therefore commoner than many of the other inborn errors for which routine screening is advocated (Chapter 18). The clinical signs of hypothyroidism in the newborn are minimal but if treatment is not started within the first few months of life permanent brain damage results. Plasma T_4 values are variably high during the first month of life (due to the high maternal oestrogen levels in late pregnancy, which increase TBG levels in both mother and infant), and their assay is therefore of little diagnostic value. Plasma TSH levels are usually raised in hypothyroidism, and in some countries all newborn infants are screened by plasma TSH assay at about a week after delivery: the test should, in any case, be done on any infant thought on clinical evidence to be hypothyroid.

Interpretation of results in very ill patients. Any severe illness may cause low plasma total T_4 concentrations ('sick euthyroid'). Unless there is primary hypothyroidism plasma TSH levels are usually normal or slightly high. T_3 levels may be low and there may be an impaired response to TRH.

Because of these problems, the assessment of thyroid function is best deferred until the patient has recovered from the illness. If hypo- or hyperthyroidism is suspected as a part of the illness, the results must be interpreted in the context of the changes outlined above.

Euthyroid goitre

Thyroxine synthesis may be impaired by iodine deficiency, by drugs such as para-aminosalicylic acid, or possibly by minor degrees of enzyme deficiency. The tendency of the plasma T_4 concentration to fall increases TSH secretion. The thyroid cells synthesize and secrete T_4 and T_3 and eventually become hyperplastic. Thus plasma thyroid hormones levels are maintained at the expense of the development of a goitre. In areas with a low soil iodide content iodine deficiency goitre was endemic before prophylactic iodinization of salt was introduced.

Inflammation of the thyroid gland (thyroiditis), whether acute or subacute, may produce marked but temporary aberrations of thyroid function tests. These conditions are relatively uncommon.

Other biochemical findings in thyroid disease

Cholesterol level. In *hypothyroidism* the synthesis of cholesterol is impaired but its catabolism is even more impaired and plasma cholesterol levels are high.

Plasma creatine kinase (CK) activity is often raised in hypothyroidism but the estimation does not help diagnosis. CK activity may also be raised in thyrotoxic myopathy.

Hypercalcaemia is very rarely found with severe thyrotoxicosis. There is increased turnover of bone, probably due to direct action of thyroid hormone (p. 176).

Summary

1. There are two thyroid hormones (apart from calcitonin), thyroxine and triiodothyronine. Their synthesis depends on an adequate supply of iodine, and is controlled by circulating thyroid hormone levels *via* TSH secreted from the anterior pituitary gland.

2. Circulating thyroid hormone levels may be assessed by estimating the plasma total T_4 concentration.

3. An excess of circulating thyroid hormones produces the syndrome of hyperthyroidism. This may be due to the generalized hyperplasia of Graves' disease, or hyperfunctioning nodule(s).

Basal TSH levels are below normal and fail to rise after TRH stimulation.

4. A decreased circulating thyroid hormone level produces the syndrome of hypothyroidism. This may be primary, due to disease of the thyroid gland, or be secondary to pituitary or hypothalamic disease. TSH levels are raised in the first group and low in the second.

The diagnosis of primary hypothyroidism is made by the finding of a high plasma TSH, and usually a low T_4 concentration. If both T_4 and TSH levels are low TRH may be administered to stimulate TSH secretion. This will confirm secondary hypothyroidism and may distinguish between pituitary and hypothalamic causes.

5. Euthyroid goitre represents compensated thyroid disease. Thyroid function tests may be normal.

Further reading

Kaplan M M, Larsen P R eds. Symposium on thyroid disease. *Med Clin North Am* 1985; **69**: No 5.

This contains a series of articles covering the physiological and clinical aspects of thyroid disease. The following are of special relevance to the student:

Kaplan M M. Clinical and laboratory assessment of thyroid abnormalities. pp. 863–80. (this includes a discussion of drug effects).

Spaulding S W, Lippes H. Hyperthyroidism. Causes, clinical features, and diagnosis. pp. 937–51.

Sawin C T. Hypothyroidism. pp. 989–1004.

Tibaldi J M, Surks M I. Effects of non-thyroidal illness on thyroid function. pp. 899–911.

Ladenson PW. Diseases of the thyroid gland. *Clin Endocrinol Metab* 1985; **14**: 145–73.

This is a particularly clear and comprehensive article.

Wensel K W. Pharmacological interference with *in vitro* tests of thyroid function. *Metabolism* 1981; **30**: 717–32.

Investigation of thyroid function (Table 8.1)

Table 8.1 Results of thyroid function tests

	True T_4 abnormalities			TBG abnormalities	
	Hyperthyroidism	Hypothyroidism			
		Primary	Secondary	Raised	Lowered
Plasma					
Total T_4	↑	↓	↓	↑	↓
FTI	↑	↓	↓	N†	N†
Free T_4	↑	↓	↓	N*	N*
TSH	↓	↑	↓	N	N
TRH response	Absent	Increased	See text	Normal	Normal

†Correction may not be complete if TBG levels very abnormal
*Free T_4 assay may be affected if albumin levels are very abnormal

A total T_4 assay should usually be performed first.

Investigation of suspected hyperthyroidism

1. Measure plasma total T_4. If the value is clearly high in a clinically thyrotoxic patient, no further tests are necessary. If it is normal or slightly elevated:
2. Measure plasma TSH if a sensitive assay is available. A value below the reference range may confirm the diagnosis of hyperthyroidism.
3. If a sensitive TSH assay is not available, measure plasma total T_3. Any binding protein abnormality will also affect the total T_3 value and must be allowed for. A clearly elevated T_3 confirms the diagnosis of hyperthyroidism.

Rarely both T_4 and T_3 results are equivocal. If hyperthyroidism is still considered to be likely on clinical grounds it is possible that the levels are abnormally high for that particular patient. If so TSH secretion from the anterior pituitary gland will be suppressed. If a sensitive TSH assay is not available, or if such an assay gives an equivocal result, the TSH response to exogenous TRH should be assessed.

TRH test

It is best to contact your laboratory *before* starting the test.

(a) A basal blood sample is taken.
(b) 200 μg of TRH in 2 ml saline is injected intravenously over about a minute.
(c) Further blood samples are taken 20 and 60 minutes after the TRH injection.
TSH is measured on all samples.

In normal subjects plasma TSH levels increase by at least 2 mU/litre and exceed the upper limit of the reference range. The maximum response occurs at 20 minutes.

A normal rise in plasma TSH levels after administration of TRH rules out the diagnosis of hyperthyroidism. A subnormal response is suggestive but not diagnostic, and if there is clinical doubt it is best to observe and reassess the patient at a later date.

Monitoring treatment of hyperthyroidism

The progress of a patient being treated for hyperthyroidism is monitored by estimating plasma total T_4 levels, or those of T_3 if the latter is the predominant abnormality. Over-treatment may induce hypothyroidism: plasma TSH levels should therefore be monitored after the plasma T_4 concentration has fallen to normal.

Investigation of suspected hypothyroidism

Measure plasma TSH levels and total T_4 levels

The lower limit of 'normal' for T_4 is poorly defined. Levels within the reference range may be suboptimal for that individual patient. Because of the T_4-pituitary feedback, however, this will be established by the resultant increased TSH secretion.

A *very low plasma T_4* value indicates hypothyroidism unless there is gross TBG deficiency. In most cases the TSH level will be high (primary hypothyroidism). If it is low, proceed to a TRH test (p. 169). If the patient is very ill investigations should be deferred ('sick euthyroid').

A *borderline low T_4* value with:

(a) a *slightly elevated TSH* may indicate early hypothyroidism. Measure circulating thyroid antibodies to assess the presence of autoimmune thyroiditis. Note, however, that their presence does not confirm hypothyroidism. As replacement treatment is probably not indicated at this stage, repeat the tests after a period of 3 to 6 months.

(b) a *normal TSH* value suggests either that there is competition by a drug for binding sites on TBG, or that hypothyroidism is *secondary to pituitary or hypothalamic disease* (p. 166). A slightly low plasma T_4 with a normal TSH level may also be due to severe non-thyroidal disease. Reassess the drug history. If secondary hypothyroidism is suspected, proceed to a *TRH test* (p. 169).

(a) A subnormal rise of TSH confirms the diagnosis of *secondary hypothyroidism of pituitary origin.*

(b) A normal, or exaggerated but delayed rise, with TSH levels higher at 60 minutes than at 20 minutes suggests *secondary hypothyroidism due to hypothalamic dysfunction.*

If clinically indicated, pituitary and hypothalamic function should be investigated (p. 151).

Monitoring treatment of hypothyroidism

The progress of a patient with primary hypothyroidism on thyroid hormone replacement should be monitored by estimating plasma TSH and T_4 levels.

TSH assays are of no value in monitoring secondary hypothyroidism.

Investigation of a patient already on treatment

No patient should be given thyroid replacement therapy until the clinical diagnosis has been confirmed and documented by laboratory tests. If it is necessary to confirm the diagnosis in a patient already on treatment, but without adequate laboratory confirmation of the disease, plasma T_4 and TSH should be measured and, unless hypothyroidism is confirmed by a low T_4 and raised TSH, treatment should be stopped. The tests should then be repeated at intervals, but the diagnosis cannot be excluded until at least 6 weeks after stopping treatment.

Table 8.2 Drug effects on thyroid function tests

Drug	Total T_4	F-T_4	T_3	Remarks
Oestrogens	↑	N	↑	↑ TBG
Oral contraceptives	↑	N	↑	↑ TBG
Some radiocontrast media (eg ioponoate)	↑	N	↓	Blocking $T_4 \rightarrow T_3$ (transient effect)
Amiodarone	↑	N or ↑	N	Blocking $T_4 \rightarrow T_3$
Propranolol	N	N	↓	Blocking $T_4 \rightarrow T_3$
Carbimazole	↓	↓	↓	Therapeutic effect
Propylthiouracil	↓	↓	↓	Therapeutic effect
Androgens	↓	N	↓	↓ TBG
Danazol	↓	N		Reduced TBG binding
Salicylates	↓	N		Reduced TBG binding
Phenytoin	↓	↓	N	Increased $T_4 \rightarrow T_3$
Carbamazepine	↓	↓	N	Increased $T_4 \rightarrow T_3$

Drug effects on thyroid function tests

Drugs may alter plasma total T_4 and total T_3 values as measured by immunoassay or competitive protein-binding techniques. The commoner effects are summarized in Table 8.2. If the primary change is in binding protein levels, free T_4 (F-T_4) values are usually normal but correction of measured T_4 by, for example, the FTI, may not be complete.

A more complete listing is provided in the references quoted on p. 168.

9

Calcium, phosphate and magnesium metabolism

Most calcium is in bone, and prolonged deficiency causes bone disease. However, the extraosseous fraction, although amounting to only 1 per cent of the whole, is of great importance because of its effect on neuromuscular excitability and on cardiac muscle.

Phosphate is the most important anion associated with calcium *in vivo* and a knowledge of its plasma level is needed to interpret disturbances of calcium metabolism. Net renal tubular secretion of hydrogen ion and formation of bicarbonate depend on the presence of phosphate and sodium in the glomerular filtrate (p. 85). Cellular 'high energy' phosphate compounds are of great biological importance.

More than half the body magnesium is in bone: most of the remainder is intracellular but about 1 per cent is in the extracellular compartment and is important because of its effect on neuromuscular activity. Because of these similarities, calcium and magnesium will be discussed in one chapter.

Calcium metabolism

Total body calcium

The total body calcium depends on the balance between intake and loss.

Factors affecting intake

The amount of calcium entering the body depends on the amount *absorbed* from the *diet*. About 25 mmol (1 g) is ingested per day of which between about 6 and 12 mmol (0.25 to 0.5 g) is absorbed. *Vitamin D,* in the form of 1,25-dihydroxycholecalciferol (p. 176), is needed for adequate calcium absorption.

Factors affecting loss

Calcium is lost in urine and in faeces.

Much more calcium enters the gut lumen in the large volume of intestinal secretions than in the diet. Most *faecal calcium* consists of that which has not been reabsorbed from these secretions and is therefore lost from the body. Calcium in the intestine, whether endogenous or exogenous in origin, may be rendered insoluble by complexing with large amounts of *phosphate* or *fatty acids* and so cannot be absorbed. Oral phosphate may be used therapeutically to reduce calcium absorption and reabsorption (p. 188): an excess of fatty acids in the intestinal lumen in steatorrhoea contributes to calcium malabsorption.

Urinary calcium excretion depends on the *amount of calcium reaching the glomeruli* in plasma, on *renal function* (p. 187), on *parathyroid hormone* and *1,25-dihydroxycholecalciferol levels* and to a lesser extent on *urinary phosphate excretion*. The amount of calcium reaching the glomeruli increases if there is hypercalcaemia, after calcium ingestion, or during bony decalcification not due to calcium deficiency, such as that of osteoporosis or acidosis. The latter rarely cause hypercalcaemia because calcium is rapidly cleared through the glomeruli. Conversely, hypercalcaemia, whatever the aetiology, causes hypercalciuria if glomerular function is normal. Glomerular impairment causes hypocalciuria even if there is hypercalcaemia.

Plasma calcium

The normal plasma calcium concentration is about 2.5 mmol/litre (10 mg/dl). The estimation should be precise to about 0.05 mmol/litre (0.2 mg/dl) and a change of less than this in two consecutive samples should not be considered to be significant.

Calcium circulates in the plasma in two main forms. That bound to albumin accounts for a little less than half the total calcium as measured by routine analytical methods: it is physiologically inactive and is a transport form comparable with iron bound to transferrin (p. 391). Most of the rest of the calcium is free and ionized (Ca^{2+}), and is the physiologically important fraction (compare with thyroxine, p. 159, and cortisol, p. 120). Techniques for the measurement of free-ionized, rather than total, calcium are still not suitable for assaying large batches of specimens. Whenever a plasma total calcium result is interpreted, an assessment of the level of the free-ionized fraction should be made by noting the albumin concentration in the same specimen. The *total* (but not free-ionized) calcium concentration is lower in the supine than in the erect position, because of the effect of posture on plasma protein concentration (p. 459).

The proportion of free-ionized calcium liberated from its salts in aqueous solution falls as the pH rises: an increase in pH also enhances binding to plasma proteins. In alkalosis, therefore, tetany may occur despite a normal total calcium level. Acidosis increases the release of calcium from bones into the extracellular fluid and increases the proportion of plasma calcium in the free-ionized form: hence calcium loss in the urine is increased and prolonged acidosis may cause osteomalacia despite a normal supply of calcium to bones.

Control of plasma calcium

The most important factor controlling the level of circulating free-ionized calcium within narrow limits is parathyroid hormone (PTH), secreted by the parathyroid glands. *Calcitonin*, produced in the C-cells of the thyroid gland, has the opposite effect to that of PTH on calcium levels by decreasing osteoclastic activity. Its importance in normal calcium homeostasis is less certain than that of PTH. Circulating levels may be very high in cases of medullary carcinoma of the thyroid, although hypocalcaemia has not been reported in this condition. It has been used to treat hypercalcaemia and Paget's disease of bone. The bone provides a reservoir from which both PTH and calcitonin control plasma free-ionized calcium levels. 1,25-dihydroxycholecalciferol increases calcium absorption from the intestine, and at physiological levels seems to be necessary for the action of PTH.

Action and control of parathyroid hormone secretion. Parathyroid hormone increases the plasma free-ionized calcium concentration. It has two direct actions on plasma calcium and phosphate levels.

- It stimulates osteoclasts directly, and so releases calcium and phosphate into the extracellular fluid. This action increases the plasma concentrations of *both* calcium and phosphate.
- It decreases phosphate reabsorption from the renal tubular fluid, causing phosphaturia. This action tends to *decrease* the plasma phosphate concentration.

PTH also increases renal tubular reabsorption of calcium.

PTH secretion is not controlled by any other endocrine gland. It depends on the concentration of free-ionized calcium ions circulating through the parathyroid glands. If the concentration is reduced the rate of hormone release is increased, and continues until the free-ionized calcium level returns to normal.

The consequences of most disturbances of calcium metabolism on plasma calcium and phosphate levels can be predicted from a knowledge of the actions of PTH on bone and on the renal tubules, and the control of its secretion by free-ionized calcium. A low free-ionized calcium concentration, unless due to primary hypoparathyroidism, stimulates the parathyroid glands and causes phosphaturia; the loss of phosphate overrides the tendency to hyperphosphataemia due to the action of PTH on bone, and consequently plasma phosphate levels are low normal or low. Conversely, a high calcium concentration, unless due to an inappropriate excess of PTH, inhibits hormone secretion and causes a high phosphate level. Therefore, plasma calcium and phosphate levels vary in the same direction unless:

- there is an inappropriate excess or deficiency of PTH, due to primary diseases of the parathyroid gland or to ectopic secretion of the hormone: in such cases calcium and phosphate vary in opposite directions;
- renal glomerular dysfunction impairs the phosphaturic, and therefore hypophosphataemic, effect of PTH.

Metabolism and action of vitamin D (Fig. 9.1). Most cholecalciferol (vitamin D_3) is formed in the skin by the action of ultraviolet light on 7-dehydrocholesterol; this is the form found in animal tissues, especially the liver. Ergocalciferol (D_2) can be obtained from plants after artificial ultraviolet irradiation. These D vitamins are transported in the plasma bound to specific carrier proteins: they are inactive until metabolized.

In normal adults much more cholecalciferol is derived from the skin than from food: dietary sources are important at times when requirements are high, such as

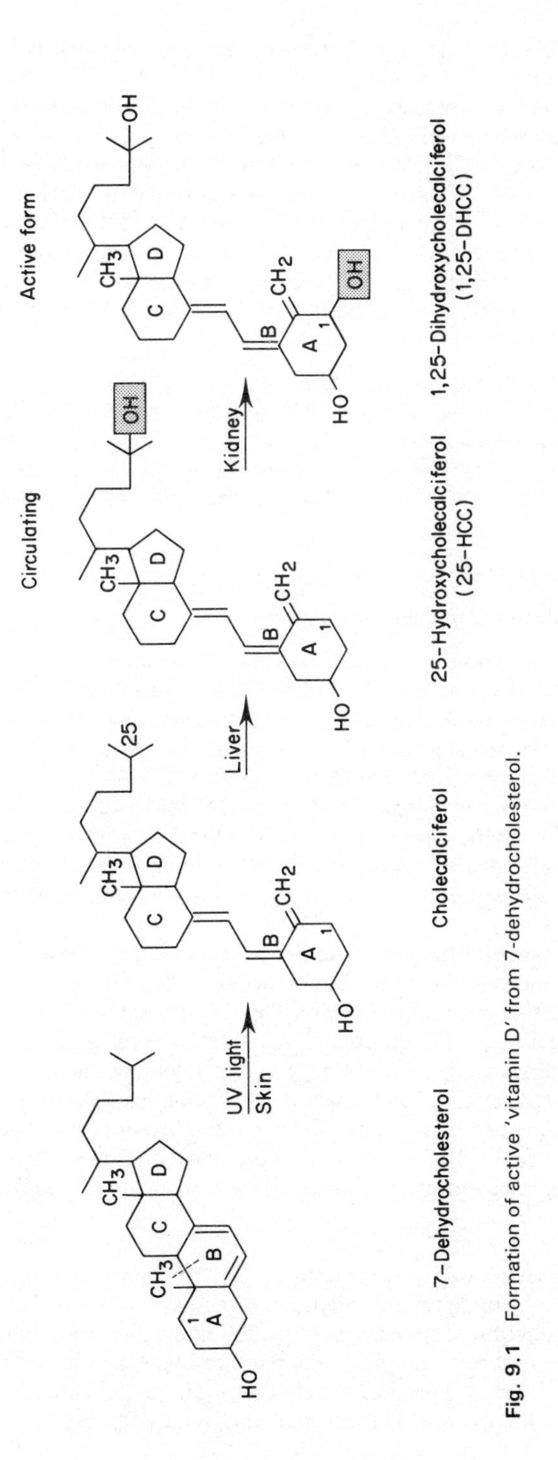

Fig. 9.1 Formation of active 'vitamin D' from 7-dehydrocholesterol.

during growth or pregnancy, or in subjects such as those elderly people who are confined indoors.

In the *liver* 25-hydroxylase catalyses the hydroxylation of cholecalciferol to *25-hydroxycholecalciferol (25-HCC)*, the main circulating form of the vitamin, and the rate of formation of 25-HCC is affected by the supply of calciferol derived from the skin and intestine. Some is excreted in the bile after inactivation, but most circulates bound to protein and forms the largest store of 'vitamin D' in the body.

Further hydroxylation of 25-HCC to *1,25-dihydroxycholecalciferol* (1,25-DHCC) in the *kidney*, catalysed by *1α-hydroxylase* confers biological activity on the vitamin. The activity of the enzyme, and hence the production of 1,25-DHCC may be *stimulated* by:

- *a low plasma phosphate concentration*;
- *an increase in the plasma PTH concentration*, possibly because of its phosphate-lowering effect;
- *oestrogens, prolactin and growth hormone*, which increase 1,25-DHCC production, and therefore increase calcium absorption, during pregnancy, lactation and growth.

It may be *inhibited* by:

- *a high plasma phosphate concentration.*

1,25-DHCC acts on the intestinal mucosal cells, increasing calcium absorption and reabsorption; in conjunction with PTH it releases calcium from bone. The kidney is thus an endocrine organ, manufacturing and releasing the hormone 1,25-DHCC; impairment of the final hydroxylation probably explains the hypocalcaemia so common in renal disease.

The lag between administration of vitamin D for therapeutic purposes and its action (p. 198) may be due to the time taken for the two-stage hydroxylation.

Other hydroxylated metabolities of vitamin D are formed in the kidney, but their functions have not been elucidated and they will not be discussed further.

Interrelationship of parathyroid hormone and 1,25-DHCC. The action of PTH on bone is impaired in the absence of 1,25-DHCC.

A fall in plasma free-ionized calcium levels stimulates PTH secretion by the parathyroid glands. PTH stimulates 1,25-DHCC synthesis, and the two hormones act synergistically on the bone reservoir, releasing calcium into the circulation; 1,25-DHCC also increases calcium absorption from the intestinal lumen. In short term homeostasis the bone effect is the more important: after prolonged hypocalcaemia more efficient absorption becomes important. As soon as the free-ionized calcium concentration is corrected both PTH and 1,25-DHCC secretion is reduced.

Action of thyroid hormone on bone. Thyroid hormone excess may be associated with the histological appearance of osteoporosis, and with increased faecal and urinary excretion of calcium, probably following its release from bone. Hypercalcaemia is a very rare complication of severe hyperthyroidism; unless there is gross excess of thyroid hormone the effects on plasma calcium are overridden by homeostatic reduction of PTH secretion and by urinary loss.

Calcium homeostasis follows the general rule that extracellular concentrations are controlled, rather than total body content. The effectiveness of this control depends mainly on:

- normal functioning of the
  - parathyroid glands;
  - kidneys;
  - intestine.
- an adequate supply of
  - calcium;
  - vitamin D.

If any of these factors is impaired plasma concentrations may be controlled at the expense of bone calcification.

Disorders of calcium metabolism

The plasma phosphate concentration, *interpreted with the plasma urea level* (see Fig. 9.2, p. 195), usually indicates whether either hyper- or hypocalcaemia is due to parathyroid hormone-like activity, or to some other cause.

Hypercalcaemia

Clinical effects of an increased free-ionized calcium concentration

Effects on the kidneys. *Renal damage* is the most serious clinical consequence of prolonged mild hypercalcaemia. Because of the high plasma free-ionized calcium concentration the solubility product of calcium phosphate may be exceeded, and the salt precipitates in extraosseous sites, of which the kidney is the most important. Renal tubular calcification may impair the ability to concentrate the urine and thus contribute to the *polyuria* so characteristic of chronic hypercalcaemia: acute hypercalcaemia may cause reversible inhibition of the tubular response to ADH rather than tubular damage.

Every attempt should be made to diagnose the cause of even mild hypercalcaemia and treat it at an early stage because of the danger of renal damage (but see p. 188).

Hypokalaemia, often with alkalosis, is a common finding when there is a high plasma free-ionized calcium level. Calcium may inhibit potassium reabsorption from the tubular lumen.

The high free-ionized calcium concentration in the glomerular filtrate may, if circumstances favour it, cause precipitation of calcium salts in the urine: patients may present with *renal calculi* without significant renal parenchymal damage. Any patient with calcium-containing renal stones should have plasma calcium estimated on more than one occasion.

Effects on neuromuscular excitability. High extracellular free-ionized calcium levels depress neuromuscular excitability in both voluntary and involuntary muscle. The patient may complain of *constipation* and *abdominal pain*. There may also be muscular hypotonia.

Effects on the central nervous system. *Depression* is a common complaint even in patients with only mild hypercalcaemia. This, and the *anorexia, nausea* and *vomiting* associated with higher calcium levels are probably due to an effect on the central nervous system.

Effects on the stomach. Calcium stimulates gastrin, and therefore gastric acid, secretion (p. 253): there is an association between peptic ulceration and chronic free-ionized hypercalcaemia.

Effects on blood pressure. Some patients with hypercalcaemia have hypertension, which may sometimes respond to reducing the plasma calcium level.

Effects on the heart. Severe hypercalcaemia causes changes in the electrocardiogram. If levels exceed about 3.75 mmol/litre (15 mg/dl) there is a risk of sudden *cardiac arrest*, and for this reason such marked hypercalcaemia should be treated as a matter of urgency.

Causes of hypercalcaemia

Hypercalcaemia with hypophosphataemia relative to the GFR

True hypercalcaemia with relative hypophosphataemia is due to inappropriate secretion of parathyroid hormone, or of a substance with a PTH-like effect. The term 'inappropriate secretion' is used here to mean the release of the hormone into the circulation when the free-ionized calcium concentration is high.

Such inappropriate secretion occurs in three conditions.

- The first two result from the production of the hormone by the parathyroid gland. These are:
  - primary hyperparathyroidism;
  - tertiary hyperparathyroidism.
- The third results from the ectopic production of PTH, or a PTH-like substance, by non-parathyroid tissue (other terms which have been used to describe this condition are 'pseudohyperparathyroidism' and 'quarternary' hyperparathyroidism).

If renal glomerular function is normal, the abnormally high circulating PTH levels cause not only hypercalcaemia, but also a low normal or low plasma phosphate concentration. If glomerular damage develops, often due to hypercalcaemia, both plasma calcium and phosphate concentrations tend to return towards normal; the kidney cannot then respond normally to the phosphaturic effect of PTH, and, because of impaired hydroxylation of 25-HCC, plasma calcium levels may be lowered in renal disease (p. 184). Diagnosis at this stage may be difficult.

The *clinical features* in these cases are due to:

- excess of circulating free-ionized calcium;
- the effect of PTH on bone.

Effects on bone of a high parathyroid hormone concentration with a normal supply of vitamin D and calcium. Parathyroid hormone stimulates osteoclastic activity. If the supply of vitamin D and calcium is normal and if PTH has been circulating in inappropriately high concentrations *over a long period of time* the effects are as follows:

- *clinical*
  bone pain;
  bony swellings.
- *radiological*
  generalized decalcification;
  subperiosteal erosions;
  cysts in bones.
- *histological*
  increased number of osteoclasts.

Prolonged decalcification of bone causes a secondary osteoblastic reaction. Osteoblasts are rich in alkaline phosphatase and if they multiply more of this enzyme than normal is released into the extracellular fluid and the *plasma alkaline phosphatase activity rises (p. 315). All these effects on bone only become evident in some long-standing cases.* The differences between the syndromes associated with inappropriately high PTH concentrations depend mainly on the duration of the disease.

Primary hyperparathyroidism. Primary hypersecretion of PTH by the parathyroid glands is usually due to an adenoma, occasionally in an ectopic parathyroid gland. Parathyroid tumours are almost always benign, although parathyroid carcinomas have been reported. Occasionally the syndrome is due to diffuse hyperplasia of all four glands.

Although the incidence of parathyroid adenoma increases with age, it may be found in young people, and occasionally in children. The apparent preponderance in elderly women may be due to their longer life expectancy: in lower age groups the sex incidence seems to be equal.

The most common presenting signs and symptoms of primary hyperparathyroidism are depression, nausea, anorexia and abdominal pain, or, less commonly, confusion, renal calculi, polyuria or even renal failure; all these are due to a high free-ionized calcium concentration. Many cases may be diagnosed after the chance finding of high plasma calcium with low plasma phosphate levels. *In most there are no bone changes, and the plasma alkaline phosphatase activity is normal or only slightly elevated.*

Occasionally the patient is admitted as an emergency with abdominal pain, vomiting and constipation. Severe hypercalcaemia should be considered as one cause of an 'acute abdomen'.

Tertiary hyperparathyroidism. If the parathyroid glands have been subjected to long-standing and sustained positive feedback by low free-ionized calcium concentrations, they hypertrophy and hormone secretion becomes partly autonomous. The diagnosis is usually made when the cause of the original hypocalcaemia is removed, for example by renal transplantation or correction of long-standing calcium or vitamin D deficiency. This condition is known as tertiary hyperparathyroidism. Only the history of previous hypocalcaemia, and the finding of a very *high plasma alkaline phosphatase activity* due to the prolonged osteomalacia of calcium and vitamin D deficiency, distinguish it from primary hyperparathyroidism. In some cases, after the hypocalcaemia has been corrected, the glandular hypertrophy gradually regresses and the plasma calcium concentration returns to normal.

Ectopic production of parathyroid hormone or of a parathyroid hormone-like substance. Some malignant tumours of non-endocrine tissues synthesize peptides normally foreign to that tissue: PTH may be one of these peptides. The production of hormones at these sites is not subject to normal feedback control, and initially the findings are identical with those of primary hyperparathyroidism. Because the underlying disease is usually either fatal, or successfully treated in a relatively short time, the bony lesions due to excess of circulating PTH are not evident: however, the plasma alkaline phosphatase activity may be raised because of secondary deposits in the bone, the liver, or both. In malignant disease the plasma calcium may rise from normal to a dangerously high concentration within a day.

The subject of ectopic hormone production is discussed more fully in Chapter 22.

Familial hypocalciuric hypercalcaemia (familial benign hypercalcaemia). Hypercalcaemia has been reported in blood relations, in none of whom was a parathyroid adenoma found at operation. High PTH concentrations in the presence of hypercalcaemia, and hypocalciuria, have been reported. The condition is probably inherited as an autosomal dominant trait. The combination of reported findings is very difficult to explain, and is extremely rare.

Hypercalcaemia with hyperphosphataemia relative to the GFR

Hypercalcaemia not due to inappropriate PTH secretion suppresses release of the hormone, and the plasma phosphate concentration tends to rise from its basal level.

The causes of hypercalcaemia with relative hyperphosphataemia are:

- vitamin D excess;
- sarcoidosis;
- idiopathic hypercalcaemia of infancy;
- possibly some cases of extensive bony secondary malignant deposits and myelomatosis;
- very rarely, severe hyperthyroidism.

The clinical picture in these cases is due to the high free-ionized calcium concentration (p. 177). Generalized decalcification of bone is rare in this group, and is only associated with thyrotoxicosis, sarcoidosis, and sometimes with myelomatosis or bony deposits. The plasma alkaline phosphatase activity is usually normal in myelomatosis, and in cases which do not involve bone.

Vitamin D excess. Vitamin D overdose, which can be caused by overvigorous treatment of hypocalcaemia, increases calcium absorption and reabsorption and may cause dangerous hypercalcaemia. Vitamin D therapy should always be monitored by frequent estimation of plasma calcium levels and, if there is osteomalacia, of alkaline phosphatase activity. If the cause of hypercalcaemia is obscure a careful drug history should be taken. Occasionally patients overdose themselves with vitamin D.

PTH secretion is suppressed, and phosphate concentrations are therefore usually normal or high.

Hypersensitivity to the action of vitamin D may cause hypercalcaemia and is associated with:

- sarcoidosis;
- idiopathic hypercalcaemia of infancy.

Hypercalcaemia of sarcoidosis. Hypercalcaemia is a rare complication of sarcoidosis. *Chronic beryllium* poisoning produces a granulomatous reaction very similar to that of sarcoidosis, and may also be associated with hypercalcaemia: beryllium is used in several industrial processes, including the manufacture of fluorescent lamps.

Idiopathic hypercalcaemia of infancy. In addition to hypersensitivity to vitamin D as a cause of hypercalcaemia, a rare and more severe genetic form is associated with various stigmata, especially mental deficiency and 'elfin' facies.

Malignant disease of bone. The ectopic production of PTH, or a PTH-like hormone, by malignant tumours was discussed on p. 180. A different syndrome has been said to occur in patients with multiple bony metastases or with myelomatosis: in these the parallel rise of plasma phosphate has been said to indicate that the hypercalcaemia is caused by direct bone breakdown due to the total action of malignant deposits. Critical examination of the findings in most of these cases shows an inappropriately low plasma phosphate concentration for that of the urea; such findings are compatible with the action of excess PTH (or PTH-like substance) when there is renal glomerular dysfunction. Moreover, there is little correlation between the degree of hypercalcaemia and the extent of bony lesions. Probably almost all such cases are due to ectopic hormone, or hormone-like, production (see last reference at the end of this chapter). The paraproteins of myelomatosis very rarely bind calcium to any significant extent, and are unlikely to account for hypercalcaemia in this condition.

Malignant deposits in bone stimulate a local osteoblastic reaction, and hence a rise in plasma alkaline phosphatase activity (p. 179). This osteoblastic reaction does not occur in the bone erosion due to the marrow expansion of myelomatosis, in which the plasma level of alkaline phosphatase of bony origin is therefore normal. This fact may be a useful pointer to the diagnosis of myelomatosis if there are extensive bone deposits of unknown origin.

Hypercalcaemia in hyperthyroidism. Thyroid hormones probably have a direct calcium-releasing effect on bone. Rarely severe hyperthyroidism causes hypercalcaemia. If the plasma calcium concentration fails to fall when the hyperthyroidism is controlled, and especially if there is hypophosphataemia, the possibility of coexistent hyperparathyroidism should be considered; these may be part of the syndrome of multiple endocrine adenopathy (p. 417).

Hypocalcaemia

Clinical effects of a reduced free-ionized calcium concentration

Low free-ionized calcium levels (including those associated with a normal total calcium concentration in alkalosis, p. 173) cause increased neuromuscular activity leading to *tetany*.

Prolonged hypocalcaemia, even when mild, interferes with the metabolism of the lens in the eye and causes *cataracts*. Because of this, asymptomatic hypocalcaemia should be sought when there has been a known risk of parathyroid damage, such as after partial or total thyroidectomy. Hypocalcaemia may also cause *depression* and other *psychiatric symptoms*.

Hypocalcaemia with hypophosphataemia relative to the GFR

High PTH concentrations cause phosphaturia with hypophosphataemia if glomerular function is normal.

Secondary hyperparathyroidism ('appropriate' secretion of parathyroid hormone). In secondary hyperparathyroidism the parathyroid glands respond appropriately to low plasma free-ionized calcium concentrations. If the response is effective, the plasma calcium concentration returns to normal, the stimulus to secretion is removed, and hormone production is inhibited by negative feedback: if the hormone action is inadequate to correct the abnormality the plasma calcium concentration is low. It is *never high*. Hypercalcaemia, or a plasma calcium concentration in the upper part of the reference range, in renal failure suggests either primary hyperparathyroidism (perhaps causing the renal disease), or that prolonged calcium deficiency has led to the development of tertiary hyperparathyroidism.

Unless the secondary hyperparathyroidism is due to renal disease, *both the plasma calcium and phosphate levels tend to be low.*

Effects on bone of a high PTH concentration in vitamin D and calcium deficiency. PTH cannot act effectively on bone in the absence of 1,25-dihydroxycholecalciferol. In cases of vitamin D deficiency PTH levels may be very high, but the effects on bone differ from those of inappropriate secretion. Unless there is an adequate supply of calcium and phosphate, osteoid cannot be calcified despite marked osteoblastic proliferation; uncalcified osteoid (*osteomalacia* in adults, or *rickets* in children) is the characteristic histological finding in advanced cases. Plasma alkaline phosphatase activity increases because of osteoblastic proliferation.

The effects in advanced cases are therefore as follows:

- *clinical*
  bone pain;
  rarely bony swelling.
- *radiological*
  generalized decalcification;
  pseudofractures (cortical fractures; Looser's zones);
  rarely subperiosteal erosions or cysts in bone.
- *histological*
  uncalcified osteoid and wide osteoid seams;
  osteoblastic proliferation;
  occasionally an increased number of osteoclasts.
- *in the plasma*
  low calcium concentration;
  low phosphate concentration;
  raised alkaline phosphatase activity.

Secondary hyperparathyroidism with osteomalacia or, before fusion of the epiphyses, with rickets is due to long-standing deficiency of calcium, phosphate and vitamin D. Predisposing factors include:

- reduced dietary intake of vitamin D, calcium and phosphate in *malnutrition*;
- impaired absorption of vitamin D in *steatorrhoea* or after *gastrectomy*;
- impaired metabolism of vitamin D to 1,25-DHCC due to *renal disease*;
- increased inactivation of vitamin D due to *anticonvulsant therapy* (p. 184).

Secondary hyperparathyroidism without osteomalacia or rickets occurs in:

- early calcium and vitamin D deficiency;
- most cases of the very rare syndrome of pseudohypoparathyroidism (p. 185).

Osteomalacia or rickets without secondary hyperparathyroidism may occur in phospate depletion due to the very rare renal tubular disorders of phosphate reabsorption (p. 186).

Reduced intake and absorption of calcium and vitamin D. 1,25-DHCC is essential for normal absorption of calcium from the intestinal tract: calcium deficiency is more commonly due to a poor intake of vitamin D than of calcium itself. In relatively affluent countries malabsorption is the commonest precipitating cause of calcium and vitamin D deficiency: in the world as a whole dietary deficiency is of more importance. *Children and pregnant women*, in whom the increased needs may not be met by the normal supply from the skin, and *subjects not exposed to sunlight* because they are confined indoors, are at greatest risk of osteomalacia or rickets.

Vitamin D can only be absorbed from the intestine if it is dissolved in fat. In *steatorrhoea* there is impaired fat, and therefore vitamin D, absorption: absorption of calcium is also impaired by combination with the excess of fatty acids to form insoluble soaps in the intestinal lumen.

In Britain there is a relatively high incidence of osteomalacia in the Asian community: rickets may occur in children, but osteomalacia is especially common during puberty and pregnancy. The reasons for this are not clear. The condition is most common in those Asian Muslims not integrated into the Western community; dietary habits, lack of sunlight and even genetic factors may all play a part.

Malnutrition, whatever the cause, is rarely selective, and protein deficiency may reduce the concentration of protein-bound calcium. In this group the free-ionized level may not, therefore, be as low as that of the total calcium would suggest.

Impaired metabolism of vitamin D. Patients with *renal disease* are relatively resistant to calciferol. This is partly due to the effect of the disease on 1α-hydroxylation of 25-HCC, but is primarily due to inhibition of the enzyme by the hyperphosphataemia associated with a low GFR. Hypocalcaemia may develop within a few days of the onset of renal damage. Low plasma protein levels often contribute to the reduction in the total calcium concentration. Tetany is rare, perhaps because the accompanying acidosis maintains the free-ionized calcium above tetanic levels. Treatment of the acidosis of renal disease with bicarbonate is contraindicated because the rise in pH may lead to precipitation of calcium in extraosseous sites; if this occurs in the kidney it may cause further deterioration of renal function.

Osteomalacia, usually mild, is common in *chronic liver disease*, especially if there is cholestasis. 25-HCC levels are low, but can be corrected by vitamin D supplements. It is therefore unlikely that there is sufficient impairment of 25-hydroxylation to account for the osteomalacia. Factors such as reduced vitamin D intake and impaired absorption due to bile-salt deficiency are probably more important.

Effect of anticonvulsants. Patients on prolonged anticonvulsant therapy (especially if both barbiturates and phenytoin are being taken) sometimes develop hypocalcaemia and even osteomalacia; these drugs probably induce the hepatic synthesis of enzymes catalysing inactivation of vitamin D.

In all these conditions the low free-ionized calcium concentration stimulates PTH secretion, and the plasma phosphate concentration tends to be low. Phosphate is retained if there is renal glomerular dysfunction, but plasma phosphate levels are relatively lower than in cases with the same level of plasma urea, but normal PTH concentrations. In chronic cases a rising plasma alkaline phosphatase activity heralds the onset of the bone changes of osteomalacia or rickets.

Hypocalcaemia with hyperphosphataemia relative to the GFR

A low level of circulating PTH tends to cause hyperphosphataemia.

Primary hypoparathyroidism

Surgical causes. Hypoparathyroidism is most commonly due to surgical damage to the parathyroid glands, either directly or by impairment of their blood supply, during partial *thyroidectomy*. Total thyroidectomy or *laryngectomy* is almost inevitably associated with removal of the parathyroid glands. Post-thyroidectomy hypocalcaemia is not always due to damage to the glands: a fall in plasma total calcium concentration is a common accompaniment of the temporary hypoalbuminaemia which follows most operations.

Parathyroidectomy carried out to treat primary or tertiary hyperparathyroidism also carries a slight risk of damage to the remaining parathyroid tissue. However, there are several causes of temporary hypocalcaemia, which usually need no treatment:

- apparent hypocalcaemia may be due to post-operative hypoalbuminaemia;
- in those rare cases of primary hyperparathyroidism with overt bone disease, rapid entry of calcium into bone when PTH concentrations fall may cause true, but temporary, postoperative hypocalcaemia;
- if the antecedent hypercalcaemia has been severe and prolonged the remaining parathyroid tissue may have been suppressed by feedback for so long that there may be true hypoparathyroidism with a rising plasma phosphate concentration, which recovers within a few days.

Evidence for asymptomatic hypocalcaemia should always be sought after partial thyroidectomy or parathyroidectomy. Although early postoperative parathyroid insufficiency may recover, a low calcium level persisting for more than a few weeks should be treated because of the danger of cataract formation (p. 182). Even with highly skilled surgery there is danger of parathyroid damage.

Autoimmune hypoparathyroidism. Hypoparathyroidism is occasionally due to autoimmune damage to the parathyroid glands.

Pseudohypoparathyroidism. A very rare inborn error is associated with an impaired response of both kidney and bone to PTH. Plasma calcium levels therefore fall and stimulate PTH secretion. The high plasma PTH concentration is ineffective, and the biochemical changes of hypoparathyroidism, not hyperparathyroidism, are present. The stimulatory effect of PTH on the renal 1α-hydroxylase is also impaired.

The hypocalcaemia of 'shock'

After any episode causing 'shock', such as a surgical operation or acute *pancreatitis*, there is temporary hypoalbuminaemia, with a consequent fall in plasma *total* calcium concentration. Although free-ionized calcium levels very occasionally fall in acute pancreatitis, they are almost always normal. Calcium should only be given if there is tetany.

Diseases of bone not affecting the plasma calcium concentration

Some diseases producing biochemical abnormalities other than those of plasma calcium and phosphate, or important in the differential diagnosis from those already discussed, are:

- *osteoporosis*, in which the primary abnormality is a reduction in the mass of bone matrix, with secondary loss of calcium. There is no osteoblastic reaction. Clinically and radiologically it may resemble osteomalacia, but *normal plasma concentrations of calcium, phosphate and especially alkaline phosphatase activity* are more in favour of the diagnosis of osteoporosis;
- *Paget's disease of bone*. Plasma calcium and phosphate concentrations are rarely affected in Paget's disease. The *plasma alkaline phosphatase* activity is typically *very high*;
- *renal tubular disorders of phosphate reabsorption*. In a group of inborn errors of renal tubular function *less phosphate than normal is reabsorbed* from the glomerular filtrate. The consequent *rickets* or *osteomalacia*, unlike the usual form, responds poorly to vitamin D therapy: the syndrome has therefore been called *'resistant rickets'*. *Familial hypophosphataemia* is an X-linked dominant trait; the syndrome may also be part of a more generalized reabsorption defect in the *Fanconi syndrome* (p. 16).

In these syndromes failure to calcify bone is probably due to phosphate deficiency. Plasma concentrations of phosphate are usually very low and fail to rise when vitamin D is given alone: there is phosphaturia inappropriate to the plasma level. Osteomalacia is reflected by the *high alkaline phosphatase activity*. The plasma calcium concentration is usually normal, and this differs from the finding in classical osteomalacia; because of the normocalcaemia the parathyroid glands are not overstimulated, and evidence in the bone of hyperparathyroidism is rare. The conditions respond to large doses of oral phosphate, perhaps with a small dose of calciferol.

Tests for disorders of calcium metabolism

If careful attention is paid to the clinical picture, concentrations of plasma calcium, phosphate *in relation to that of urea*, activity of alkaline phosphatase, and to haematological and radiological findings, other biochemical tests are usually unnecessary and occasionally misleading. The non-invasive technique of isotope subtraction scanning of the neck may help to localize the parathyroid adenoma before operation.

Plasma parathyroid hormone assay

In our experience parathyroid hormone assay is rarely necessary, and results can sometimes be misleading. It may occasionally help if, after considering all the factors mentioned at the beginning of this section, the cause of *hypercalcaemia* is still in doubt.

If the feedback mechanism is normally responsive, true (free-ionized) hypercalcaemia should reduce the plasma PTH concentration to very low levels. However raised values may be found, not only in the presence of a parathyroid adenoma, but also in those tumours of non-parathyroid origin which can secrete immunoreactive PTH inappropriate to the calcium level; by contrast, some proven parathyroid adenomas secrete biologically active hormone which fails to react *in vitro*, even with different antibodies to more than one peptide sequence. Thus 'negative' results of PTH assay do not always exclude the diagnosis of primary hyperparathyroidism, nor high ones necessarily confirm it.

The assay may help if the exceedingly rare pseudohypoparathyroidism (p. 185) is suspected as a cause of *hypocalcaemia*. Very high PTH levels despite hypocalcaemia *and hyperphosphataemia* suggest that the feedback mechanism is intact, but that there is end-organ unresponsiveness.

Urinary calcium

Estimation of urinary calcium is almost never of diagnostic value in the differential diagnosis of hypercalcaemia. If glomerular function is normal, hypercalcaemia, whatever the cause, increases the calcium load on the glomerulus and causes hypercalciuria: although excess PTH, by increasing tubular reabsorption, might be expected to reduce excretion of calcium below that appropriate for the plasma level, the 'normal' response is too variable for the test to be helpful. If renal glomerular function is impaired calcium excretion is low, even if there is hypercalcaemia.

Hypercalciuria occurs without hypercalcaemia in so-called *idiopathic hypercalciuria*, some cases of *osteoporosis* in which calcium cannot be deposited in normal amounts because the matrix is reduced, and in cases, such as acidosis, in which release of free-ionized calcium is increased.

Steroid suppression test

This test is rarely necessary in the differential diagnosis of hypercalcaemia if the factors mentioned earlier are taken into account, and results are difficult to interpret if the plasma calcium concentration is less than about 3 mmol/litre. It can be used if the cause of severe hypercalcaemia remains obscure.

It usually differentiates primary and tertiary hyperparathyroidism from any other cause. Large doses of hydrocortisone, or cortisone, reduce high plasma calcium concentrations to within the reference range *except in primary or tertiary hyperparathyroidism*. Most of the exceptions to this rule are associated with very extensive malignant deposits in bone, in which case the clinical diagnosis is already clear and the test unnecessary. Even in proven cases of ectopic PTH secretion, the plasma calcium concentration is usually suppressed by steroids, but is not when the hormone is produced by the parathyroid glands. The reason for this difference in response to steroids is unknown.

The investigation of disorders of calcium metabolism is considered in more detail on p. 194.

Biochemical basis of treatment

Hypercalcaemia

Mild to moderate hypercalcaemia. If the plasma calcium concentration is below about 3.75 mmol/litre (15 mg/dl), and if there are no significant electrocardiographic changes attributable to hypercalcaemia, there is no *immediate* need for therapy. However, treatment should be instituted as soon as the abnormality is found because of the danger of renal damage.

If possible, the primary cause should be treated.

The patient should be volume repleted. The plasma total calcium concentration will almost certainly fall as albumin is diluted, but the free-ionized calcium concentration is probably little affected; however, correcting haemoconcentration enables a more realistic assessment of the degree of true hypercalcaemia to be made.

Medical measures used to treat mild hypercalcaemia include giving oral sodium phosphate or steroids, or calcitonin or mithramycin by injection.

- *Oral sodium phosphate*, by forming calcium phosphate, prevents the reabsorption of calcium entering the gut lumen in intestinal secretions: it therefore removes calcium from the body. This treatment is effective, safe and cheap. The tendency to develop osmotic diarrhoea is minimized if the phosphate is dissolved in enough water to make it isosmolar. It may be used in conjunction with steroids to treat intractable hypercalcaemia due to extensive malignancy.
- *Steroids* may lower the plasma calcium concentration in some cases, but are more likely to have side-effects than phosphate. They are most useful if indicated for other reasons, such as for the treatment of sarcoidosis.
- *Calcitonin* is very expensive and should only be used if oral phosphate treatment is impossible, for example because the patient is vomiting.
- *Mithramycin* may be used to treat the hypercalcaemia of malignancy. It inhibits calcium mobilization from bone and the effect lasts for several days. It is a bone marrow suppressant, and should only be used at low dosage and for a short time.

Not everyone agrees that patients with apparently asymptomatic mild hypercalcaemia due to primary hyperparathyroidism need parathyroidectomy. It is argued that the prolonged hypercalcaemia does not always cause obvious renal dysfunction, and that the risk of surgery is greater than that of mild hypercalcaemia. Nevertheless, almost all patients volunteer the information that a successful operation has improved their feeling of well-being, and has made them aware of previous vague ill-health. In the hands of an experienced surgeon the morbidity of the procedure is minimal and short-lived. The decision as to whether to operate must be made on clinical grounds: of particular importance are the fitness of the patient for operation if other disease is present, and age. Symptomatic hypercalcaemia in a patient unfit for operation must be treated medically.

Severe hypercalcaemia. The patient should be volume repleted. Frusemide may be given in an attempt to increase urinary calcium clearance.

Oral phosphate, steroids and calcitonin have no significant effect for about 24 hours. Although the exact level at which urgent treatment is indicated varies in different subjects, it is usually necessary if the plasma calcium concentration exceeds about 3.75 mmol/litre (15 mg/dl) because of the danger of cardiac arrest. If there is any doubt about the degree of urgency the electrocardiogram should be inspected for abnormalities associated with hypercalcaemia.

The intravenous administration of sodium, sometimes with potassium, *phosphate* to lower the plasma calcium concentration within a few hours depends, at least in part, on precipitation of insoluble calcium salts. Because some of this precipitation may occur in the kidney, the treatment carries a slight theoretical risk of initiating or aggravating renal dysfunction. This is not a problem in practice, probably because it only needs very little calcium precipitation to reduce the plasma calcium concentration acutely. Calcitonin is often used to treat severe hypercalcaemia. In our experience intravenous phosphate is a very safe and effective method of reducing hypercalcaemia acutely.

As with most other extracellular constituents, rapid changes in calcium concentration may be dangerous because time is not allowed for equilibration across cell membranes. The aim of emergency treatment should be temporarily to lower the level to one which is not immediately dangerous, while initiating treatment for mild hypercalcaemia. Too rapid a reduction, even to only normal or slightly high concentrations, may cause tetany or, more seriously, hypotension; there may also be tetany despite normocalcaemia during the rapid fall of plasma calcium concentration after removal of a parathyroid adenoma.

Hypocalcaemia

Postoperative hypocalcaemia. Hypocalcaemia during the first week after thyroidectomy or parathyroidectomy should only be treated if there is severe tetany, and then only with calcium supplements, which, unlike calciferol, have a rapid effect and a short half-life: mild hypocalcaemia is of no immediate danger, and helps to stimulate recovery of suppressed parathyroid cells. Persistent hypocalcaemia may indicate that the glands are permanently damaged and that long-standing, or even life-long calciferol treatment is necessary.

Asymptomatic hypocalcaemia. Whatever the cause, hypocalcaemia which is asymptomatic or accompanied by only mild clinical symptoms, can usually be treated effectively with large oral doses of vitamin D. It is difficult to give enough oral calcium by itself to make a lasting and significant difference to plasma calcium concentrations, and, if a normal diet is being taken, vitamin D, by increasing absorption of calcium from the intestine, is usually adequate without calcium supplements.

The hypocalcaemia of renal disease with hyperphosphataemia should be treated cautiously, unless there are such signs of osteomalacia as a raised plasma alkaline phosphatase activity, because of the danger of ectopic calcification. The plasma phosphate should first be lowered by giving oral aluminium hydroxide to bind phosphate in the intestinal lumen. 1,25-DHCC or 1α-hydroxycholecalciferol have been used. However, the hypocalcaemia of renal disease responds to very high doses of the much cheaper calciferol. The risk of ectopic calcification is the same whichever compound is used.

Apparent hypocalcaemia due to low albumin concentrations should not be treated.

Hypocalcaemia with severe tetany. In the presence of severe tetany due to hypocalcaemia, intravenous calcium, usually as the gluconate, should be given at once.

Abnormalities of plasma phosphate concentration

Hypophosphataemia

Hypophosphataemia associated with disturbances of calcium metabolism is usually due to *high circulating PTH concentrations*. In these condition, and in *renal disorders of phosphate reabsorption* (p. 186), phosphate is lost from the body.

Phosphate, like potassium, enters cells from the extracellular fluid if the rate of glucose metabolism is increased; this occurs during treatment of diabetic coma with *insulin*. This redistribution of phosphate is a common cause of hypophosphataemia in patients receiving parenteral nutrition with insulin and glucose. *Long-term parenteral feeding* without phosphate supplementation may cause true phosphate depletion. Hypophosphataemia in such circumstances, whether due to deficiency or redistribution, has been said to cause convulsions. These patients are very ill: if the hypophosphataemia alone is the cause of the symptoms it is difficult to explain why those with high plasma PTH concentrations, or with hypophosphataemic rickets, and with equally low phosphate levels, do not have the same symptoms.

Hypophosphataemia may be caused by severe and prolonged dietary deficiency.

Hyperphosphataemia

The commonest cause of hyperphosphataemia is *renal glomerular dysfunction*; it is important not to correct hypocalcaemia until this abnormality has been corrected (p. 189).

Less common causes of hyperphosphataemia are *hypoparathyroidism* and *acromegaly*.

The normal plasma phosphate concentration is higher in children than in adults.

Magnesium metabolism

Bone salts contain magnesium as well as calcium, and the two ions tend to move in and out of bone together. Cells contain magnesium at much higher concentrations than the extracellular fluid, and magnesium, potassium and phosphate tend to enter and leave cells under the same conditions.

Large amounts of magnesium can be lost in faeces in diarrhoea, or in fluid lost through intestinal fistulae.

Plasma magnesium and its control

About 35 per cent of the plasma magnesium is protein-bound. Less is known about the importance of this binding than in the case of calcium.

The mechanism by which magnesium is controlled is poorly understood. Aldosterone is known to increase its renal excretion: PTH may affect its absorption and excretion similarly to that of calcium.

Clinical effects of abnormal magnesium concentrations

Hypomagnesaemia

The symptoms of hypomagnesaemia are very similar to those of hypocalcaemia. If the plasma calcium and protein concentrations and blood pH are normal in a patient with tetany, plasma magnesium should be assayed. If the deficiency is severe, magnesium may be given intravenously, and this may be used as a therapeutic test if the assay cannot be performed as an emergency: in less severe cases oral treatment is adequate (p. 199).

Hypermagnesaemia

Hypermagnesaemia causes muscular hypotonia, but, as it is rarely found without other abnormalities such as hypercalcaemia or those due to renal failure, it is not always easy to distinguish those signs and symptoms specifically due to hypermagnesaemia.

Causes of abnormal plasma magnesium concentrations

Hypomagnesaemia

Excessive loss of magnesium. By far the most important causes of clinical disturbances of magnesium metabolism severe enough to warrant treatment are extensive losses in *severe, prolonged* diarrhoea, or through intestinal fistulae. Loss in urine in renal tubular dysfunction is usually less severe. Cytotoxic drugs, such as cisplatin, have been reported to cause hypomagnesaemia by impairing renal tubular reabsorption.

Hypomagnesaemia accompanied by hypocalcaemia. Hypocalcaemia and hypomagnesaemia often coexist. This may occur in hypoparathyroidism, or during the fall of plasma calcium concentration after removal of a parathyroid adenoma which is associated with severe bone disease; the finding is uncommon and rarely causes clinical problems. Hypomagnesaemia can cause temporary hypoparathyroidism, which is corrected by giving magnesium.

Hypomagnesaemia accompanied by hypokalaemia. Since magnesium moves in and out of cells with potassium, hypomagnesaemia tends to occur with hypokalaemia, for example during diuretic therapy, or in primary hyperaldosteronism. Such hypomagnesaemia is rarely of clinical importance. The low plasma magnesium concentration due to renal tubular dysfunction may also be associated with hypokalaemia.

Hypomagnesaemia can also occur in alcoholic patients and may sometimes cause symptoms.

Hypermagnesaemia

The commonest cause of hypermagnesaemia is probably renal glomerular dysfunction. It rarely needs specific treatment, and responds to measures to treat the underlying condition. Magnesium salts should never be given in cases of renal glomerular failure.

Summary

Calcium metabolism

1. About half the plasma calcium is protein-bound, and half in the free-ionized form.
2. The free-ionized calcium is the physiologically active fraction and total calcium concentrations should be interpreted together with albumin levels.
3. The plasma calcium concentration is controlled by parathyroid hormone and vitamin D. Parathyroid hormone secretion is increased if the free-ionized calcium concentration is reduced.
4. Parathyroid hormone acts on:

   - bone, stimulating osteoclastic activity and so releasing calcium and phosphate into the plasma: if prolonged, this action increases the number of osteoblasts and so plasma alkaline phosphatase activity;
   - kidneys, stimulating calcium and inhibiting phosphate reabsorption, and so lowers the plasma phosphate concentration. It may also increase the rate of 1α-hydroxylation of 25-HCC.

5. Symptoms and findings in disturbances of calcium metabolism can be related to the concentrations of free-ionized calcium and parathyroid hormone, and to renal function.
6. Inappropriately high levels of parathyroid hormone are present in primary and tertiary hyperparathyroidism, or if the hormone is secreted from ectopic sites. In these circumstances the plasma calcium concentration is high.
7. Appropriately high levels of parathyroid hormone are present whenever the free-ionized calcium concentration falls below normal.
8. Parathyroid hormone levels are low in hypoparathyroidism (associated with a low plasma calcium concentration), or in the presence of hypercalcaemia other than that due to inappropriate hormone secretion.

Magnesium metabolism

1. Hypomagnesaemia may cause tetany in the absence of hypocalcaemia.
2. The commonest cause of significant hypomagnesaemia is severe diarrhoea.
3. Magnesium tends to move in and out of bone with calcium, and in and out of cells with potassium.

Further reading

Kanis JA, Paterson AD, Russell RGG. Disorders of calcium and skeletal metabolism. In: Williams DL, Marks V, eds. *Biochemistry in clinical practice.* London: Heinemann, 1983: 299–325.

Palmer M, Adami H-O, Bergström R, Jakobsson S, Åkerström G, Ljunghall S. Survival and renal function in untreated hypercalcaemia: population-based cohort study with 14 years of follow-up. *Lancet* 1987; **1**: 59–62.

Suva LJ, Winslow GA, Wettenhall REH et al. A parathyroid hormone-related protein implicated in malignant hypercalcaemia: cloning and expression. *Science* 1987; **237**: 893–7.

Investigation of disorders of calcium metabolism

Plasma calcium should never be interpreted without taking into account albumin and phosphate levels, nor plasma phosphate without that of urea.

Differential diagnosis of hypercalcaemia

Three groups of causes must be differentiated:

- raised protein-bound, with normal free-ionized calcium;
- raised free-ionized calcium due to inappropriately high PTH;
- raised free-ionized calcium due to other causes, and associated with low PTH levels.

The following procedure is simple and very reliable. The diagnosis is usually obvious before all the steps have been followed.

A. Is the rise in total calcium concentration due only to a high protein-bound fraction?

1. What is the plasma calcium level?

2. *Is the plasma albumin high?* If so, and if the total calcium concentration is less than about 3 mmol/litre (12 mg/dl), *in vivo* or artefactual haemoconcentration may be the cause. If the plasma calcium is much higher than this true hypercalcaemia is likely, even if the free-ionized calcium is lower than the total level would suggest.

   (a) If clinically indicated, rehydrate the patient: this will correct *in vivo* haemoconcentration.

   (b) Take a specimen without venous stasis to eliminate artefactual haemoconcentration (p. 447).

   (c) Repeat the plasma calcium and albumin assays. If hypercalcaemia is confirmed a cause for a high free-ionized calcium must be sought.

   (d) If the calcium is now in the high reference range repeat the assay at three monthly intervals, to exclude developing hypercalcaemia.

3. *Is the plasma albumin level low?* If so, plasma calcium concentrations within the high reference range may indicate a high free-ionized fraction.

Many formulae have been proposed in an attempt to 'correct' total calcium for abnormal protein levels. Unfortunately there is evidence that the relationship between albumin concentration and calcium-binding is not simple, and results of such calculations may be misleading.

B. What is the cause of a raised free-ionized calcium concentration?

Hypercalcaemia found in a specimen with a normal albumin concentration, or a plasma calcium in the high reference range in a specimen with hypoalbuminaemia, should be assumed to be due to an abnormally high concentration of the physiologically active free-ionized fraction. The commonest causes are primary hyperparathyroidism and malignancy; the latter is usually obvious on other evidence.

1. Take a careful history, with special reference to the taking of vitamin D-containing preparations.

2. *Is the plasma phosphate low?*

   (a) If it is at the *lower end of the reference range or is low* there is likely to be *inappropriate PTH secretion.*

   (b) If it is normal or high, *is the urea concentration in the same specimen high?* Both phosphate and urea are retained if there is glomerular dysfunction. *If there is hyper-*

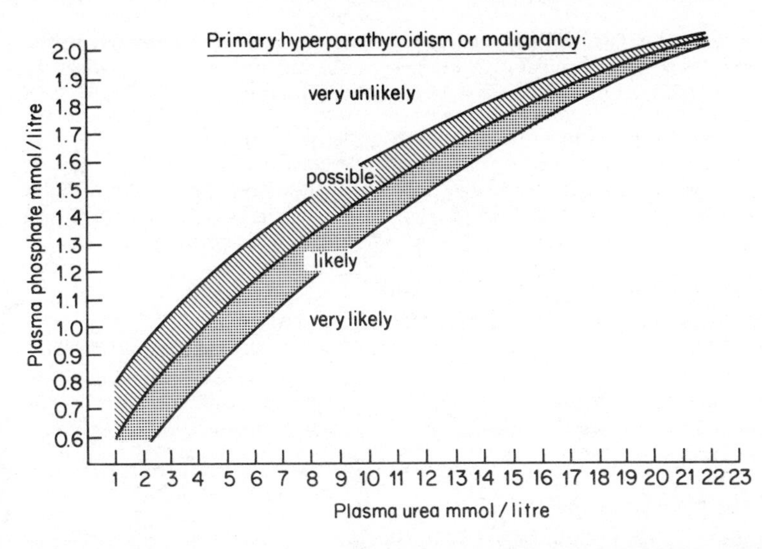

Fig. 9.2 Plasma phosphate in relation to urea in the differential diagnosis of PROVEN hypercalcaemia. The lower the phosphate the higher, and the higher the phosphate the lower, the probability of primary hyperparathyroidism or malignancy. The differentiation between the two diagnoses can usually be made on non-chemical evidence (see text) or, if doubt remains, by PTH assay. *Use this graph ONLY in cases with PROVEN hypercalcaemia.*

calcaemia Fig. 9.2 gives a rough guide to the significance of the plasma phosphate concentration.

3. *If the concentration of phosphate is low in relation to that of urea* seek a cause of a PTH-like effect. The most important differential diagnosis is now between primary hyperparathyroidism and malignancy.

(a) A history of *renal calculi or peptic ulceration suggests that the hypercalcaemia has been chronic* and is therefore due to primary hyperparathyroidism. However, either of these conditions may occur coincidentally.

(b) Repeat the clinical examination, paying special attention to palpation of the breasts and pelvic examination. Seek evidence of carcinoma of the bronchus.

(c) If myeloma is suspected perform serum electrophoresis and test the urine for Bence Jones protein.

(d) A very high plasma alkaline phosphatase activity is unlikely to be due to uncomplicated primary hyperparathyroidism: a normal activity is the rule, although it may be slightly raised if there is radiological evidence of bone involvement. If it is very high it suggests malignancy, or some other intercurrent disease. Many elderly subjects have a high plasma alkaline phosphatase activity, perhaps because of Paget's disease of bone.

(e) If all the findings in (b), (c) and (d) are negative, if the haemoglobin concentration and ESR are normal, and especially if the history suggests chronic hypercalcaemia, primary hyperparathyroidism is by far the most likely cause, and other investigations are rarely necessary. Isotope subtraction scanning of the neck may help to localize the adenoma.

(f) In rare cases in which doubt remains about the diagnosis, and if the plasma calcium concentration is 3 mmol/litre (12 mg/dl) or above, a steroid suppression test may help. Failure of suppression in a case without obvious malignancy is very suggestive of primary hyperparathyroidism. Results are difficult to interpret at lower calcium levels.

(g) PTH assay rarely helps to make a diagnosis. Remember that:

- the results can only be interpreted if there is concomitant hypercalcaemia; a high level may be appropriate in its absence;

- failure to detect a high PTH concentration does not exclude primary hyperparathyroidism, because the circulating PTH may not be immunoreactive. Nor does the finding of a high level prove primary hyperparathyroidism.

4. *Is the plasma phosphate high in relation to the urea?*

(a) By far the commonest cause is *ingestion of vitamin D*, alone or in multivitamin preparations, in which case the patient may not know that he has taken vitamin D. Some people seem to be sensitive to diets high in calcium and vitamin D. A fall in both plasma calcium and phosphate during a supervised stay in hospital suggests vitamin D as a cause.

(b) Look for evidence of *sarcoidosis.*

(c) Is there severe thyrotoxicosis?

(d) In infants consider 'idiopathic' hypercalcaemia (p. 181).

In all these conditions the calcium concentration falls to normal during a steroid suppression test.

It is unwise to embark upon parathyroidectomy in a hypercalcaemic patient, whatever the results of plasma PTH assay, unless the plasma phosphate concentration is low in relation to that of urea.

Steroid suppression test

All specimens should be taken without venous stasis.

Procedure

1. At least two specimens for calcium and albumin assay are taken on two different days before starting the test.
2. The patient takes 40 mg of oral hydrocortisone 8 hourly for 10 days.
3. Blood is taken for plasma calcium and albumin estimation on the 5th, 8th and 10th days after starting the hydrocortisone.
4. After the 10th day the hydrocortisone is stopped.

Interpretation

If the plasma calcium concentration does *not* fall to within the reference range by the end of the test the diagnosis of *primary hyperparathyroidism* is very probable.

Significant fluid retention during high-dose steroid administration may dilute the protein-bound calcium. This effect must be allowed for when interpreting the results.

Differential diagnosis of hypocalcaemia

As in the case of hypercalcaemia, the causes of hypocalcaemia fall into three groups:

- reduced protein-bound, with normal free-ionized calcium;
- reduced free-ionized calcium due to primary PTH deficiency;
- reduced free-ionized calcium due to other causes and associated with appropriately high PTH concentrations.

A. Is the fall in total plasma calcium concentration due only to a low protein-bound fraction?

Cases with a low albumin concentration should not be given calcium or vitamin D unless there is clinical evidence of an associated low free-ionized calcium concentration.

B. What is the cause of a low free-ionized calcium concentration?

1. *Is the plasma phosphate low?* If so, calcium deficiency with normal secretion of PTH in response to feedback is likely.
   (a) *Is the plasma alkaline phosphatase activity high?* This finding would suggest prolonged secondary hyperparathyroidism due to calcium deficiency.
   (b) *Do bone X-rays show signs of osteomalacia?* This confirms very prolonged calcium deficiency.
   (c) Look for causes of malnutrition, especially malabsorption.
2. *Is the plasma phosphate high?*
   (a) *If the urea level is high* glomerular dysfunction is the likely cause.
   (b) If the urea level is normal hypoparathyroidism is the most likely cause. Is there a history of an operation on the neck?
   (c) Hypoparathyroidism may be of autoimmune origin. This may be distinguished from the even rarer 'pseudohypoparathyroidism' by PTH assay. Concentrations will be low in true hypoparathyroidism but very high if there is the end-organ unresponsiveness of pseudohypoparathyroidism.

Treatment of disorders of plasma calcium

Hypercalcaemia

Emergency treatment of hypercalcaemia

The solution for intravenous infusion should contain a mixture of mono- and dihydrogen phosphate such that the pH is 7.4.

$Na_2HPO_4.12H_2O$	29.00 g/litre
KH_2PO_4 (anhydrous)	2.59 g/litre

500 ml is infused over 4 to 6 hours. This 500 ml will contain a total of 81 mmol of sodium, 9.5 mmol of potassium and 50 mmol of phosphorus.

Long term treatment of hypercalcaemia with oral phosphate

Oral phosphate is given as the disodium or dipotassium salt. The choice depends on the plasma potassium concentration.

1. The *solution* should contain 32 mmol of phosphorus in 100 ml.
   This is 114 g/litre of $Na_2HPO_4.12H_2O$
   or 56 g/litre of K_2HPO_4 (anhydrous)

100 to 300 ml a day is given in divided doses. 100 ml contains approximately 65 mmol of sodium or potassium respectively.

2. *Phosphate Sandoz Effervescent tablets* each contain:

Phosphorus	16 mmol
Sodium	20 mmol
Potassium	3 mmol

The dose is 1 to 6 tablets a day. *These MUST be dissolved in water*, as directed, to reduce the concentration and therefore the likelihood of osmotic diarrhoea.

3. As an alternative the phosphate can be given as *sodium cellulose phosphate*. One sachet contains 5g, given three times a day, *dissolved in water or sprinkled on food.*

Note that hypokalaemia is a common accompaniment of hypercalcaemia. When this is present, the phosphate preparation of choice is K_2HPO_4.

Hypocalcaemia

Emergency treatment of hypocalcaemia with severe tetany

Calcium gluconate injection (BP). 0.22 mmol (9 mg) calcium/ml.
Dose. 10 ml intravenously in the first instance.

Long-term treatment of hypocalcaemia

Vitamin D therapy

WARNING. There may be several days lag in response to either stopping or starting therapy with calciferol: plasma calcium concentratons may continue to rise for weeks after stopping treatment. Caution must be exercised, using intelligent anticipation based on laboratory assessment of plasma calcium and, in osteomalacia, alkaline phosphatase levels.

Patients on maintenance doses should be seen at regular intervals, because requirements may change, with the danger of the development of hypercalcaemia. Plasma calcium concentrations must be assessed with those of albumin.

Osteomalacia with raised alkaline phosphatase activity.

1. *Initially.* 12.5 mg (500 000 IU) of calciferol daily orally or intramuscularly. Very rarely doses as high as 18.75 to 25 mg (750 000 to 1 000 000 IU) may be needed.
2. When the plasma calcium concentration reaches about 1.75 mmol/litre (7.0 mg/dl) and the plasma alkaline phosphatase activity is near normal, reduce to 2.50 to 6.25 mg (100 000 to 250 000 IU) daily. At this stage bone calcification is nearing normal, and hypercalcaemia may develop rapidly if high dosage is continued.
3. *Maintenance doses* vary between 0.63 and 2.5 mg (25 000 and 100 000 IU) daily, and are determined by trial. The aim should be to keep plasma calcium, phosphate and alkaline phosphatase levels normal.

Hypocalcaemia of hypoparathyroidism with normal alkaline phosphatase activity

The lower dosage of 2.50 mg (100 000 IU) daily should be used from the outset, monitoring being based on the plasma calcium concentrations; again, the exact dosage is found by trial.

Oral calcium tablets

Calcium gluconate (BP) (600 mg) = 1.35 mmol (54 mg) of calcium per tablet.
Calcium gluconate effervescent (BP) (1000 mg) = 2.25 mmol (90 mg) of calcium per tablet.
'Sandocal' effervescent (4.5 g calcium gluconate) = 10 mmol (400 mg) of calcium per tablet.
Calcium lactate.$5H_2O$ (BP) (300 mg) = 1 mmol (40 mg) of calcium per tablet.
Calcium lactate.$5H_2O$ (BP) (600 mg) = 2 mmol (80 mg) of calcium per tablet.

Treatment of disorders of plasma magnesium

Emergency treatment of hypomagnesaemia

Magnesium chloride ($MgCl_2.6H_2O$) 40 g/100 ml.
or Magnesium sulphate ($MgSO_4.7H_2O$) 50 g/100 ml.

These solutions contain 2 mmol of magnesium per ml and may be added to other intravenous fluids. If renal function is normal, up to 40 mmol (20 ml) may be infused in 24 hours.

Oral magnesium treatment

Magnesium is poorly absorbed from the intestinal lumen and so acts as an osmotic purgative. It should not be given orally to a patient with diarrhoea, and intravenous supplementation is always more effective.

Magnesium chloride ($MgCl_2.6H_2O$) 20 g/100 ml
Dose. 5 to 10 ml (5 to 10 mmol) four times a day.

10

Carbohydrate metabolism and its interrelationships

In most parts of the world carbohydrate is the major source of energy intake. Under normal circumstances starch is the main dietary carbohydrate, disaccharides contribute significantly and monosaccharides are a minor component of the diet.

Chemistry

The main important *monosaccharide hexoses* are all reducing sugars; they therefore reduce copper compounds, such as those incorporated in Clinitest tablets (Ames), to produce a colour change. Reducing sugars include:

- glucose;
- fructose;
- galactose.

The common *disaccharides* are:

- sucrose (fructose + glucose);
- lactose (galactose + glucose);
- maltose (glucose + glucose).

Lactose and maltose are reducing sugars, but sucrose is not.

Naturally occurring *polysaccharides* are long-chain carbohydrates composed of glucose subunits:

- starch, found in plants, is a mixture of amylose (straight chains) and amylopectin (branched chains);
- glycogen, found in animal tissue, is a highly branched polysaccharide.

Physiology

The importance of extracellular glucose concentrations

The brain cells are most vulnerable to hypoglycaemia. They derive their energy from aerobic metabolism of glucose and they cannot:

- *store glucose in significant amounts;*
- *synthesize glucose;*
- *metabolize substrates other than glucose and ketones.* Plasma ketone concentrations are normally very low; in physiological conditions they are of little importance as an energy source;
- *extract enough glucose for their needs from the extracellular fluid at low concentrations, because entry of glucose into the brain is not facilitated by insulin.*

The brain is very dependent on extracellular concentrations of glucose for its energy supply and hypoglycaemia is therefore likely to impair cerebral function. Hyperglycaemia, especially of rapid onset, can also cause cerebral dysfunction by its effect on extracellular osmolality (p. 33). In normal subjects the plasma (extracellular) glucose concentration usually remains between about 4.5 and 11 mmol/litre (about 80 and 200 mg/dl) despite the intermittent load entering the body from the gastrointestinal tract.

The maintenance of plasma glucose levels below about 11 mmol/litre also minimizes loss of this energy source from the body. The renal tubular cells reabsorb almost all glucose from the glomerular filtrate so that normal urine is nearly glucose-free, even after a carbohydrate meal. Glycosuria detected by routine testing will normally be present when the plasma glucose concentration exceeds about 11 mmol/litre (the 'renal threshold'). The retained glucose can be stored until needed.

Maintenance of extracellular concentrations

The plasma glucose concentration depends on the balance between glucose entering and glucose leaving the extracellular compartment. Because little is normally lost unchanged from the body, maintenance of the relatively narrow range of 4.5 to 11 mmol/litre in the face of widely varying input from the gastrointestinal tract is most likely to depend on exchange with cells. If we understand the interaction between tissues which effects this control we can also explain pathological disturbances of carbohydrate metabolism, including ketosis and lactic acidosis.

The liver

The liver is the most important single organ ensuring a constant energy supply for other tissues, including the brain, under a wide variety of conditions. It is also of importance in helping to control the plasma glucose concentration postprandially. It is well adapted to these roles for many reasons.

- Portal venous blood leaving the absorptive area of the intestinal wall reaches the liver first. The hepatic cells are in a key position to buffer the hyperglycaemic effect of a high carbohydrate meal (Fig. 10.1).
- The liver cells can store some of the excess glucose as glycogen. The rate of glycogen synthesis (*glycogenesis*) from glucose-6-phosphate (G-6-P) may be increased by insulin (Fig. 10.1) which is secreted by the β-cells of the pancreas in response to systemic hyperglycaemia.
- The liver can convert some of the excess glucose to fatty acids, which are

ultimately transported as triglyceride in very low density lipoprotein (VLDL) and stored in adipose tissue (Fig. 10.1 and p. 235).

- The entry of glucose into liver and cerebral cells is not affected directly by insulin, but depends on the extracellular glucose concentration. The conversion of glucose to G-6-P, the first step in glucose metabolism in all cells, is catalysed in the liver by the enzyme glucokinase, which has a low affinity for glucose compared with that of hexokinase found in most other tissues. Glucokinase activity is induced by insulin. Therefore proportionally less glucose is extracted by hepatic cells during fasting, when levels in portal venous plasma are low, than after carbohydrate ingestion. This helps to maintain a fasting supply of glucose to vulnerable tissues such as the brain.
- Under aerobic conditions the liver can synthesize glucose (*gluconeogenesis*) using the metabolic products from other tissues, such as glycerol, lactate or the carbon chains resulting from deamination of most amino acids (mainly alanine).
- The liver contains the enzyme, glucose-6-phosphatase, which, by hydrolysing the G-6-P derived from glycogen breakdown (*glycogenolysis*) or by gluconeogenesis, releases glucose and helps to maintain extracellular fasting levels. Hepatic glycogenolysis is stimulated by the hormone glucagon, secreted by the α-cells of the pancreas, and by catecholamines such as adrenaline (epinephrine) or noradrenaline (norepinephrine).
- During fasting the liver can convert fatty acids, released from adipose tissue as a consequence of low insulin activity, to ketones. These can be used by other tissues, including the brain, as an energy source when glucose is in short supply.

This combination of properties is unique to the liver. The renal cortex is the only other tissue capable of gluconeogenesis, and of converting G-6-P to glucose. The gluconeogenic capacity of the kidney is probably mainly of importance in hydrogen ion homeostasis (p. 86) and during prolonged fasting.

Other tissues can store glycogen to a greater or lesser extent, but because they do not contain glucose-6-phosphatase they can only use it locally; this glycogen plays no part in maintaining the plasma glucose concentration.

Systemic effects of a glucose load (Fig. 10.1)

We have seen that the liver modifies the potential hyperglycaemic effect of a high carbohydrate meal by extracting relatively more glucose from the portal plasma than in the fasting state. Some glucose, however, passes through the liver unchanged and the rise in systemic concentration stimulates the β-cells of the pancreas to secrete insulin, which may further stimulate hepatic and muscle glycogenesis. More importantly, *entry of glucose into adipose tissue and muscle cells*, unlike that into liver and brain, *is stimulated by insulin*, and plasma glucose falls rapidly to near fasting levels. The relatively high insulin activity also *inhibits* the breakdown of triglycerides (*lipolysis*) and of protein (*proteolysis*). If there is relative or absolute insulin deficiency (diabetes mellitus) these actions are impaired. Conversion of intracellular glucose to G-6-P in adipose and muscle cells is catalysed by the enzyme *hexokinase which, because its affinity for glucose is greater than*

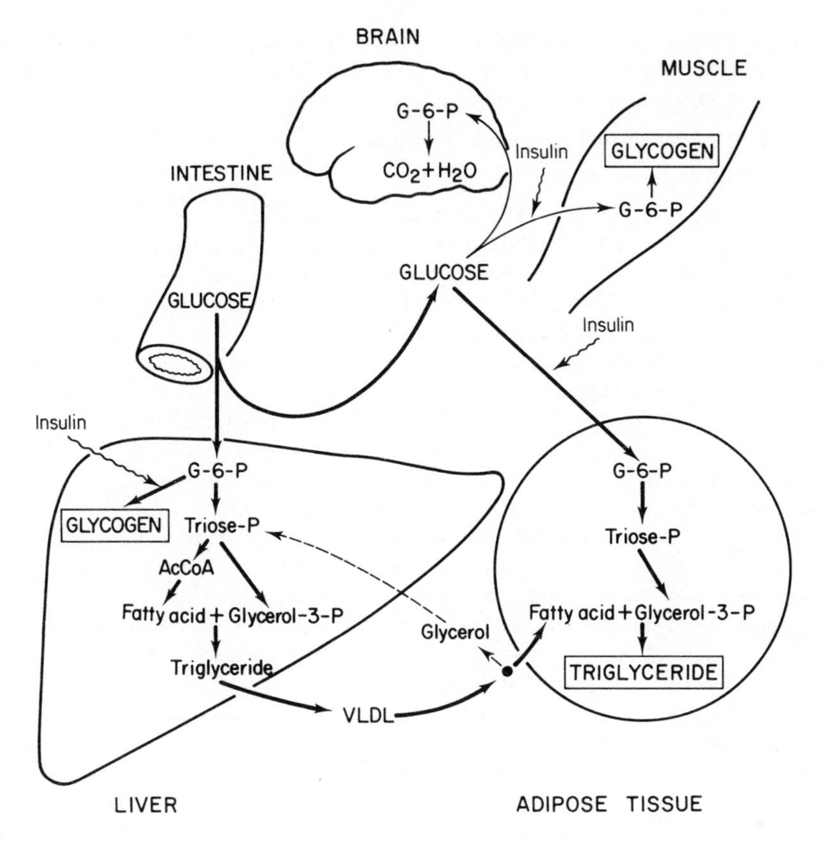

Fig. 10.1 Postprandial storage of glucose.

that of hepatic glucokinase, ensures that glucose enters the metabolic pathways in these tissues at lower extracellular concentrations than in the liver.

Both muscle and adipose tissue store the excess postprandial glucose, but the mode of storage and the function of the two types of cell is very different: by examining the interrelationships of each of these tissues with the liver, many of the disturbances of carbohydrate metabolism can be explained.

Ketosis

Adipose tissue and the liver

Adipose tissue triglyceride is the most important long-term energy store in the body. Greatly increased utilization of fat stores is associated with ketosis.

Adipose tissue, in conjunction with the liver, converts excess glucose to triglyceride and stores it in this form rather than as glycogen. The component fatty acids are derived from glucose entering the liver and the component glycerol from glucose entering adipose tissue cells.

In the *liver* triglycerides are formed from:

- glycerol-3-phosphate (from triose phosphate);
- fatty acids (from acetyl CoA).

This triglyceride is transported to adipose tissue in VLDL, where it is hydrolysed by lipoprotein lipase (p. 236). The released fatty acids (of hepatic origin) condense with glycerol-3-phosphate derived from glucose entering *adipose tissue* under the influence of insulin, and the resultant triglyceride is stored. Far more energy can be stored as triglyceride than as glycogen.

During *fasting*, when exogenous glucose is unavailable and the plasma insulin concentration is low, endogenous triglycerides are reconverted to free fatty acids (FFA) and glycerol by *lipolysis* (Fig. 10.2). Both are transported in the blood stream to the liver, the plasma FFA being protein-bound, predominantly to albumin. Glycerol enters the gluconeogenic pathway at the triose phosphate stage; the glucose synthesized can be released into the blood stream at a time when the plasma glucose concentration would otherwise tend to fall. Most tissues, other than the brain, can convert fatty acids to acetyl CoA by oxidation, and use the resultant acetyl CoA as an energy source. When the rate of production of acetyl CoA exceeds the rate of its utilization, the liver can form *acetoacetic acid* by enzymatic condensation of two moles of acetyl CoA; acetoacetic acid can be reduced to *3-hydroxybutyric acid* and decarboxylated to *acetone*. These 'ketones' can be used as an energy source by brain and other tissues at a time when glucose is in relatively short supply.

Ketosis therefore occurs when fat stores are the main energy source and may

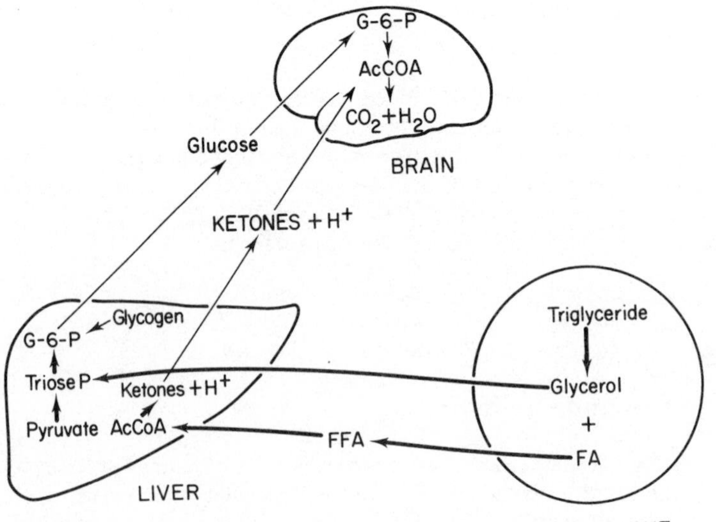

Fig. 10.2 Fasting pathways: ketosis.

result from *fasting*, or from reduced nutrient absorption due to vomiting. Mild ketosis may occur after as little as 12 hours fasting (and should not be misinterpreted as diabetic ketosis). After short fasts acidosis is not usually detectable, but after longer periods more hydrogen ions may be produced than can be dealt with by homeostatic mechanisms, and the plasma bicarbonate levels fall. For many weeks the plasma glucose concentration is maintained, principally by hepatic gluconeogenesis, but in prolonged starvation (as in anorexia nervosa) or in childhood (p. 222) ketotic hypoglycaemia may occur. The brain may suffer less from ketotic hypoglycaemia than from the same degree of insulin-induced hypoglycaemia; in the former the brain adapts to ketone metabolism, while in the latter ketone concentrations are low, thus depriving the brain of its only non-glucose energy source.

Diabetic ketoacidosis is more severe than the ketosis of fasting, from which it is differentiated by hyperglycaemia. The mechanism of ketone production is, however, the same; it is due to intracellular glucose deficiency in the presence of low insulin activity.

During *starvation* the reduced supply of glucose to cells, and therefore reduced adipose tissue glycolysis and lipogenesis, is the result of low extracellular concentrations; this tendency to *hypoglycaemia* also *inhibits the normal physiological secretion of insulin by feedback*, and the rate of lipolysis, and therefore of fatty acid and ketone production increases (Table 10.1, p. 209).

In *diabetes mellitus low insulin activity is the primary abnormality*; intracellular glucose deficiency occurs *despite hyperglycaemia* because of impaired entry into adipose tissue cells in the absence of the hormone; the intracellular consequences are identical to those of starvation.

Thus ketosis always reflects excessive utilization of fat as an energy source due to:

- intracellular glucose deficiency;
- low insulin activity.

The low insulin activity increases the rate of production of gluconeogenic substrates by glycolysis and proteolysis, and the rate of hepatic gluconeogenesis. The resultant increased rate of glucose release into the ECF is appropriate in starvation, but aggravates the hyperglycaemia in diabetes mellitus.

Lactate production and lactic acidosis

Striated muscle and the liver

Glucose enters muscle postprandially under the influence of insulin and is stored as glycogen. Because of the absence of glucose-6-phosphatase, this glycogen cannot be reconverted to glucose and can only supply local needs. Quantitatively, muscle glycogen stores are second only to those in the liver.

Muscular contraction (Fig. 10.3). During muscular activity glycogenolysis is stimulated by adrenaline (epinephrine), and the resultant G-6-P is drawn upon by rapid glycolysis and by oxidation in the tricarboxylic acid (TCA) cycle to supply

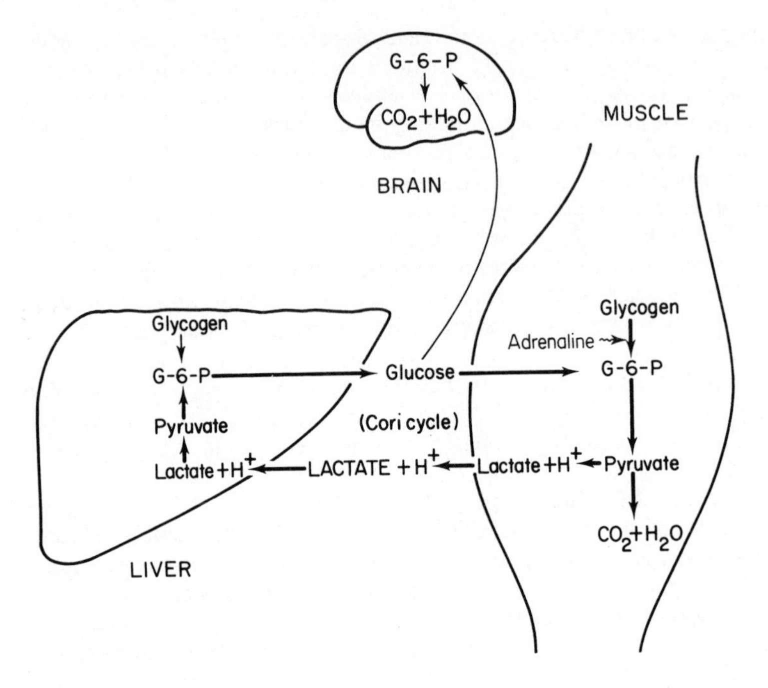

Fig. 10.3 Pathways during muscular contraction.

necessary energy. Under these conditions the rate of glycolysis may outstrip the availabililty of oxygen, and glycolytic products then exceed the immediate aerobic capacity to oxidize them.

The overall reaction for anaerobic glycolysis is

$$\text{Glucose} \rightarrow 2\ \text{Lactate}^- + 2\ \text{H}^+$$

The lactate is transported in the blood stream to the liver where it can be used for gluconeogenesis, providing further glucose for the muscle (Cori cycle). During gluconeogenesis H^+ is also reutilized. Under aerobic conditions the liver consumes much more lactate than it produces.

This physiological accumulation of lactic acid during muscular contraction is a temporary phenomenon, and rapidly disappears at rest, when slowing of glycolysis allows aerobic processes to 'catch up'.

Pathological lactic acidosis

Lactic acid produced by anaerobic glycolysis may be oxidized to CO_2 and water in the TCA cycle, or be reconverted to glucose by gluconeogenesis in the liver. Both the TCA cycle and gluconeogenesis require oxygen; *anaerobic glycolysis is the only pathway that does not need oxygen.*

Pathological accumulation of lactate might be due to increased production or to decreased utilization.

- Production may be increased by:
  an increased rate of anaerobic glycolysis.
- Utilization may be decreased by:
  impairment of the TCA cycle;
  impairment of gluconeogenesis.

The *clinical syndromes* associated with lactic acidosis are usually associated with more than one of these factors.

Tissue hypoxia, due to the poor tissue perfusion of the 'shock' syndrome, *is the commonest and most important cause of lactic acidosis* (Fig. 10.4). Under these circumstances tissue hypoxia increases plasma lactate levels because:

- the TCA cycle cannot function anaerobically and oxidation of pyruvate and lactate to CO_2 and water is impaired;
- hepatic and renal gluconeogenesis from lactate cannot occur anaerobically;
- anaerobic glycolysis is stimulated, because the falling ATP levels cannot be regenerated by the TCA cycle as they are in aerobic conditions.

The combination of impaired gluconeogenesis and increased anaerobic glycolysis converts the liver from a lactate and H^+ consuming organ to one generating large amounts of lactic acid.

Severe hypoxia, such as may follow cardiac arrest, causes marked acidosis. If diabetic ketoacidosis is associated with significant volume depletion this hypoxic syndrome may aggravate the acidosis.

Other causes of lactic acidosis are:

- *severe illnesses*, such as leukaemias and lymphomas. A variety of factors, including poor tissue perfusion and increased anaerobic glycolysis by malignant tissue, may contribute to the lactic acidosis;
- *metformin or phenformin*; these drugs are rarely used now to treat diabetes mellitus because they can cause severe lactic acidosis. They inhibit both the TCA cycle and gluconeogenesis;

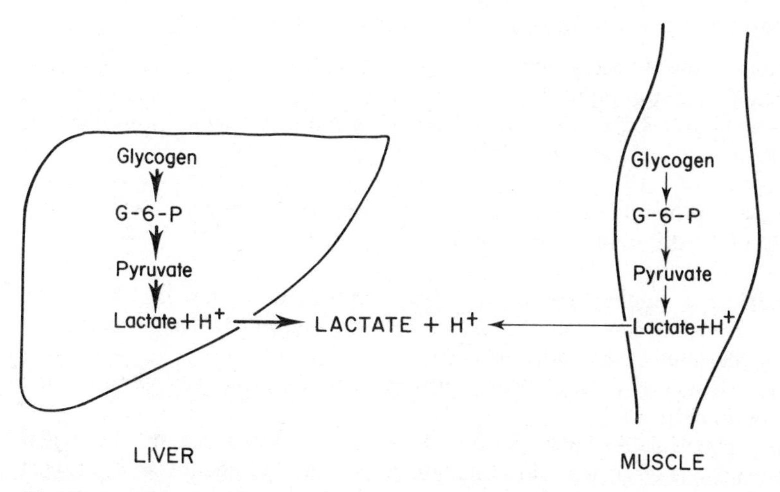

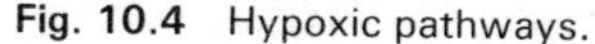
Fig. 10.4 Hypoxic pathways.

- *glucose-6-phosphatase deficiency* (von Gierke's disease, p. 220); the rate of glycolysis increases if G-6-P cannot be converted to glucose by the liver and kidney.

The treatment of lactic acidosis is that of the cause, and of the acidosis. *Measurement of plasma lactate levels is rarely necessary, because it is the acidosis that is dangerous: lactate itself is harmless.* Lactic acidosis may be suspected if the plasma T_{CO_2} concentration is very low in the absence of another obvious cause of metabolic acidosis.

Hormones concerned with glucose homeostasis

Some of the important effects of these hormones have already been described. The more detailed actions are summarized in Table 10.1.

Insulin is the most important hormone controlling the metabolic pathways described earlier. The β-cells of the pancreatic islets produce proinsulin, which incorporates the 51-amino acid polypeptide insulin, and a 33-amino acid linking peptide which joins one end of the A chain to the other end of the B chain of insulin. Proteolysis of proinsulin releases insulin, two amino acids (one from each end of the connecting peptide), and the rest of the connecting peptide (called *C-peptide*). Both insulin and C-peptide are stored in islet cells and are released into plasma in equimolar amounts, mainly in response to rising plasma glucose levels.

Insulin binds to specific receptors on the surface of insulin-sensitive cells of adipose tissue and muscle: the most important effect is stimulation of glucose entry into these cells with a resultant fall in plasma levels. As we have already discussed, it also promotes glycogen synthesis in the liver and in muscle, and inhibits gluconeogenesis, lipolysis and proteolysis.

The normal response to hyperglycaemia therefore depends on:

- adequate insulin secretion;
- normal insulin receptors;
- normal intracellular responses to receptor binding of insulin ('post-receptor events').

C-peptide is probably of little physiological importance, but its measurement may help in the differential diagnosis of hypoglycaemia (p. 218).

Glucagon is a single-chain polypeptide synthesized by the α-cells of the pancreatic islets and secretion is *stimulated by hypoglycaemia*. Glucagon stimulates hepatic glycogenolysis and gluconeogenesis.

A brief discussion of some of these principles as applied to parenteral feeding will be found on p. 270.

When plasma insulin levels are low, for example during fasting, the hyperglycaemic actions of growth hormone, glucocorticoids, adrenaline and glucagon become apparent, even if there is no increase in secretion rates. Secretion of these

Table 10.1 Actions of hormones on intermediary metabolism

		Insulin	Glucagon	Growth Hormone	Glucocorticoids	Adrenaline
Carbohydrate metabolism (a) in liver	• glycolysis	+				
	• glycogenesis	+				
	• glycogenolysis		+			+
	• gluconeogenesis	−	+		+	
(b) in muscle	• glucose uptake	+		−	−	
	• glycogenesis	+				
	• glycogenolysis					+
Protein	• synthesis	+		+		
	• breakdown	−			+	
Fat	• synthesis	+				
	• lipolysis	−		+	+	+
Secretion	• stimulated by	Hyperglycaemia Amino acids Glucagon Gut hormones	Hypoglycaemia Amino acids Fasting	Hypoglycaemia Stress Sleep	Hypoglycaemia Stress	Stress
	• inhibited by	Adrenaline Fasting	Insulin			
Results		Uses and stores available glucose	Provides glucose	Spares glucose	Provides glucose	
				Provide FFA as alternative fuel		
Plasma FFA levels		Fall		Rise		
Plasma glucose levels		Fall		Rise		

+ stimulates − inhibits

hormones may increase during stress and in patients with acromegaly (GH) (p. 113), Cushing's syndrome (glucocorticoids) (p. 124) or phaeochromocytoma (adrenaline and noradrenaline) (p. 414).

Urinary glucose

Glycosuria is defined as a concentration of urinary glucose which is detectable using relatively insensitive, but specific, screening tests which depend on the action of the enzyme, glucose oxidase, incorporated into a diagnostic strip such as Diastix (Ames) (p. 229). Usually most filtered glucose is reabsorbed by the proximal tubular cells. Although very low urinary concentrations may be detectable, even in normal subjects, by more sensitive methods, glycosuria as defined above occurs only when the plasma, and therefore glomerular filtrate levels, greatly exceed the tubular reabsorptive capacity. This may be because:

- the plasma and glomerular filtrate concentrations are more than about 11 mmol/litre, and therefore the normal tubular reabsorptive capacity is significantly exceeded;
- the tubular reabsorptive capacity is reduced, so that glycosuria occurs at a lower filtrate concentration (*'renal glycosuria'*). This is usually a harmless condition.

Very occasionally, if the GFR is much reduced, there may be no glycosuria despite plasma glucose levels above 11 mmol/litre; if the volume of glomerular filtrate is very low the *total amount* of glucose delivered to tubular cells may be less than normal, even if the concentration is high. In such rare cases urine testing cannot be used to assess the dose of antidiabetic drug treatment.

Glycosuria should be sought on a specimen secreted *by the kidney* at about an hour after a meal, when peak plasma concentrations are reached. The *double void technique* of collection should be used to ensure that the specimen being tested has not been stored in the bladder after secretion by the kidney at an earlier time. The patient should empty his bladder and discard the specimen; a further specimen passed 10 to 15 minutes later is tested. Specimens collected after a period of fasting will yield positive results only when fasting plasma glucose levels are above about 11 mmol/litre (severe diabetes mellitus, or during infusion of glucose), or if there is gross renal glycosuria.

Reducing substances in the urine (including glucose) can be detected using copper-containing reagents such as those incorporated in Clinitest tablets. It is important to use this test in the neonatal period, when the finding of reducing substances other than glucose may suggest an inborn error of metabolism, such as galactosaemia. It may also be used as a semiquantitative test to assess diabetic control (p. 213).

Hyperglycaemia and diabetes mellitus

Hyperglycaemia occurs:

- in the syndrome of diabetes mellitus;
- in patients receiving intravenous glucose-containing fluids;
- temporarily in severe stress;
- sometimes after cerebrovascular accidents.

Diabetes mellitus

Diabetes mellitus is due to absolute or relative insulin deficiency. It has been defined by the World Health Organization (WHO), on the basis of laboratory findings, as a fasting plasma venous glucose concentration greater than 7.8 mmol/litre (140 mg/dl) or a concentration of 11.1 mmol/litre (200 mg/dl) or more two hours after a carbohydrate meal or two hours after oral ingestion of the equivalent of 75 g of glucose, even if the fasting concentration is normal (p. 225). Severe cases have persistent hyperglycaemia. The WHO classification divides diabetes mellitus into the following categories.

Insulin-dependent diabetes mellitus (IDDM, Type 1) is the term used to describe patients for whom, because they are prone to develop ketoacidosis, insulin therapy is essential. The onset is most commonly during childhood, and subjects with HLA tissue types DR3 and DR4 are most at risk. It has been suggested that many cases follow a viral infection which has destroyed the β-cells of the pancreatic islets.

Non-insulin-dependent diabetes mellitus (NIDDM, Type 2) is the commonest variety. Patients are much less likely to develop ketoacidosis and although insulin may sometimes be needed it is not essential for survival. Onset is most common during adult life. It is further subdivided into *obese and non-obese* NIDDM. A variety of inherited disorders may be responsible for the syndrome, either by reducing insulin secretion, or by causing relative insulin deficiency despite high plasma hormone levels, because of resistance to its action or because of post-receptor defects.

Diabetes associated with other conditions includes:

- *absolute insulin deficiency*, due to pancreatic disease (chronic pancreatitis, haemochromatosis, cystic fibrosis);
- *relative insulin deficiency*, due to excessive growth hormone (acromegaly) or glucocorticoid secretion (Cushing's syndrome), or increased glucocorticoid levels due to administration of steroids;
- *drugs*, such as thiazide diuretics.

Impaired glucose tolerance

The WHO definition of impaired glucose tolerance includes patients with a fasting plasma venous glucose concentration between 5.5 and 7.8 mmol/litre (100 and 140 mg/dl) and/or with a plasma glucose concentration between 7.8 and 11.1 mmol/litre (140 and 200 mg/dl) at two hours after taking a standard glucose load. Only a small proportion of such cases may develop diabetes mellitus later, and because of the serious psychological, social and economic implications, patients must not be diagnosed as diabetic merely on the basis of impaired glucose tolerance. It is not possible to predict the outcome at the time of discovery of such impaired tolerance.

Subjects at risk of developing diabetes mellitus

A strong family history of diabetes mellitus or the birth of an abnormally large baby may suggest that, despite a normal result of a glucose tolerance test, an individual is at risk of developing diabetes mellitus. Such a subject should be given dietary advice and told to lose weight if necessary.

The terms prediabetic, latent diabetic or potential diabetic should not be used.

The clinical and metabolic features of insulin-dependent diabetes mellitus

Most of the metabolic changes of diabetes mellitus are a consequence of insulin deficiency.

Hyperglycaemia at some time is, by definition, an invariable finding. If plasma glucose levels exceed about 11 mmol/litre, and renal function is normal, *glycosuria* will be present. High urinary glucose concentrations produce an osmotic diuresis and therefore *polyuria*. Polyuria and the increased plasma osmolality due to hyperglycaemia causes *thirst*, (polydipsia) (p. 28). A prolonged osmotic diuresis may cause excessive urinary electrolyte loss. *These 'classical' symptoms of diabetes are only present in advanced cases.*

Abnormalities in *lipid* metabolism may be secondary to insulin deficiency. Lipolysis is enhanced and plasma FFA levels rise. In the liver FFA are converted to acetyl CoA and ketones, or are reesterified to form endogenous triglycerides and incorporated into VLDL. If insulin deficiency is very severe chylomicrons may accumulate in the blood, because lipoprotein lipase needs insulin for optimal activity. The rate of cholesterol synthesis is also increased, with an associated increase in LDL.

Increased breakdown of *protein* may cause muscle wasting.

Long-term effects

Vascular disease is a common complication of diabetes mellitus and may present as cerebrovascular or peripheral vascular insufficiency. Abnormalities of small blood vessels particularly affect the retina (*diabetic retinopathy*) and the kidney. *Kidney disease* is associated with a wide variety of clinical disorders, including proteinuria and progressive renal failure. Diffuse nodular glomerulosclerosis (Kimmelstiel-Wilson lesions) may cause the nephrotic syndrome. The presence of small amounts of albumin in the urine (microalbuminuria) is associated with an increased risk of developing progressive renal disease in the future (p. 341); it is possible that this may be prevented by more stringent control of plasma glucose levels and of blood pressure. The renal complications may partly be due to the increase in glycosylation of structural proteins within the arterial wall of the glomerular basement membrane: similar vascular changes in the retina may account for the high incidence of diabetic retinopathy, and glycosylation of protein in the lens may cause cataracts.

Haemoglobin and plasma proteins may also be glycosylated, and may be assayed to assess long-term diabetic control. Glycosylated haemoglobin (HbA_{1c}) is most commonly measured for this purpose, but glucose bound to plasma proteins undergoes a rearrangement to form fructosamine, and this sugar amine may also be estimated to assess control.

Infections are common and may aggravate renal and peripheral vascular disease. There is an increased incidence of fetal abnormalities in babies born to poorly controlled diabetics, and they tend to be large at birth.

Principles of management of diabetes mellitus

The management of diabetes mellitus will be considered only briefly.

Outpatients may assess their plasma glucose concentrations at predetermined times by semiquantitative measurement of urinary reducing substances using Clinitest tablets. The renal threshold for glucose in that individual must first be established and the patient be instructed to collect urine by the *double void technique* (p. 210). Patients may be able to measure their own blood glucose levels using glucose oxidase-containing reagent strips: the colour change of the strip may be quantitated visually or by using a portable reflectance meter. Although this procedure involves the discomfort of several skin punctures many patients are able to adjust their insulin dose more accurately than by testing their urine. The patient should still remain under regular medical supervision. It is *essential* that both reagent strips and meters are regularly checked by the laboratory staff.

Measurement of the percentage of total haemoglobin which is glycosylated (HbA_{1c}) gives a retrospective assessment of the mean plasma glucose concentration during the preceding 6 to 8 weeks: the higher the percentage the poorer the mean control. It will not detect brief hypoglycaemic episodes. Since the result depends not only on the glucose levels, but on the life-span of the red cell, falsely low values may be found in patients with haemolytic anaemia. *Measurement of HbA_{1c} is an adjunct to, not a replacement for, serial plasma glucose estimations,* which are necessary to reveal potentially dangerous short-term swings. The measurement of plasma fructosamine levels may be used to assess glucose control over a shorter time course than that of HbA_{1c} (about two to four weeks), but the assay is not yet widely available.

Insulin requirements may vary in a patient with IDDM. For example, the dose may need to be increased during any illness or during pregnancy. In patients with NIDDM, plasma glucose concentrations can often be controlled by *diet* with *weight reduction*, but insulin may be needed during periods of stress. In this group insulin secretion may be stimulated by the *sulphonylurea* drugs, such as tolbutamide or glibenclamide. *Biguanides*, such as metformin or phenformin, have been used to lower plasma glucose levels; the mechanism of their action has not been completely elucidated, but they inhibit gluconeogenesis either directly or indirectly. The danger of lactic acidosis is mentioned on p. 207. They are now rarely prescribed.

Acute metabolic complications of diabetes

The diabetic patient may develop one of several metabolic complications needing emergency treatment. The main ones are:

- diabetic ketoacidosis;
- hyperosmolal non-ketotic coma;
- hypoglycaemia due to taking insulin in excess of needs (p. 218).

Diabetic ketoacidosis

Diabetic ketoacidosis is a more severe form of the metabolic events outlined in p. 212. It may be precipitated by infections or by vomiting. Insulin may be mistakenly withheld by the patient, who reasons 'no food, therefore no insulin'. The consequences are due to three main factors: *glycosuria, plasma hyperosmolality and acidosis*.

Plasma glucose levels are usually in the range of 20 to 40 mmol/litre (about 350 to 700 mg/dl) but may be considerably higher. This causes severe *plasma hyperosmolality*, with *glycosuria* which causes an *osmotic diuresis*. To the renal loss of water and electrolytes may be added loss due to the vomiting which is common in this syndrome. Extracellular loss may cause haemoconcentration and, by reducing renal blood flow and therefore the GFR, may cause prerenal uraemia. The extracellular hyperosmolality causes severe cellular dehydration, and loss of water from cerebral cells is probably the reason for the confusion and coma. Thus, there is both cellular and extracellular volume depletion.

Decreased insulin activity with intracellular glucose deficiency accelerates *lipolysis*. More FFA are produced than can be metabolized by peripheral tissues and they are either converted to ketones by the liver, or are incorporated into VLDL as endogenous triglycerides, sometimes causing *hyperlipidaemia*. The hydrogen ions produced with ketones other than acetone are buffered by plasma bicarbonate. H^+ secretion causes a fall in urinary pH. Despite equimolar generation of bicarbonate, the rate of H^+ production exceeds that of bicarbonate generation causing a fall in plasma $T\text{CO}_2$ concentration. The deep, sighing respiration (*Kussmaul respiration*) and the odour of acetone on the breath are classical features of diabetic ketoacidosis.

Plasma *potassium levels may be raised* before treatment is started, due to acidosis, due to reduced entry into cells because of impaired glucose metabolism, and due to the low GFR. Despite hyperkalaemia, there is a total body deficit due to increased urinary potassium loss in the presence of an osmotic diuresis. This deficit will be revealed as potassium reenters cells during treatment, with resultant and possibly severe hypokalaemia.

The plasma *sodium* concentration at presentation may be low or low normal, partly because of the osmotic effect of the high extracellular glucose concentrations which draws water from the cells. If there is severe hyperlipidaemia the possibility of pseudohyponatraemia must be considered (p. 36). When insulin is given gluconeogenesis is inhibited, glucose enters cells, and sodium-free water follows along the osmotic gradient. If plasma sodium levels rise rapidly the patient may remain confused or even comatose due to persistence of plasma hyperosmolality despite a satisfactory fall in plasma glucose levels. This is especially common if isosmolar or stronger saline solutions are given.

The most usual findings are:

- *clinical*
  confusion and later coma (hyperosmolality);
  hyperventilation (acidosis);
  volume depletion (osmotic diuresis).
- *plasma*
  hyperglycaemia;
  acidosis with a low total CO_2 (bicarbonate) concentration;
  high normal or high plasma potassium concentration;
  haemoconcentration and mild uraemia.
- urine
  glycosuria } if the glomerular filtration rate is adequate;
  ketonuria }
  low pH (unless there is renal impairment).

Associated findings. Changes in plasma phosphate levels parallel those of potassium and may remain low for several days after recovery from diabetic coma. Amylase activities may be markedly elevated in both plasma and urine, and, even in the presence of abdominal pain mimicking an 'acute abdomen', do not necessarily indicate acute pancreatitis. In many patients the amylase is of salivary rather than of pancreatic origin.

These incidental findings are mentioned only to avoid the dangers of misinterpretation.

Hyperosmolal non-ketotic coma

In diabetic ketoacidosis there is always hyperosmolality due to the hyperglycaemia, and many of the symptoms, including those of confusion and coma, are probably related to it. The term 'hyperosmolal' coma (or 'precoma'), however, is usually confined to a condition in which there is marked hyperglycaemia, but no detectable ketoacidosis. The reason for these different presentations is not clear. It has been suggested that insulin activity is sufficient to suppress lipolysis, but not to maintain normal plasma glucose concentrations, but this does not explain why the plasma glucose is often higher in non-ketotic coma than in ketoacidosis. Hyperosmolal non-ketotic coma is more common in older subjects. Plasma glucose levels are very high and may exceed 50 mmol/litre (900 mg/dl). The resulting glycosuria produces an osmotic diuresis, with severe water and electrolyte depletion. There is often uraemia due to volume depletion and hypernatraemia due to predominant water loss, and both these aggravate plasma hyperosmolality. The coma is probably due to cerebral cellular dehydration, which may also cause hyperventilation: the consequent respiratory alkalosis may cause a slight fall in the plasma $T\text{CO}_2$ concentration which should not be confused with that of metabolic acidosis.

Other causes of coma in a patient with diabetes mellitus

In addition to the metabolic complications described above a known diabetic may present in coma, or in a confused state (precoma), due to hypoglycaemia, a cerebrovascular haemorrhage or to an unrelated cause. The onset of coma due to these causes is often more sudden than that due to hyperglycaemia.

- *Hypoglycaemia* is most commonly due to accidental overadministration of insulin. It may result from a failure to eat normally, or from participation in excessive exercise after the usual dose of insulin or oral antidiabetic drugs.
- *Cerebrovascular accidents* are relatively common in diabetics because of the increased incidence of vascular disease. Such an event may be associated with hyperglycaemia and glycosuria ('piqûre diabetes'). Stimulation of the respiratory centre may cause hyperventilation and consequent respiratory alkalosis, with a low plasma $T\text{CO}_2$ level. This should not be confused with metabolic acidosis. Diagnosis depends on clinical findings.

The assessment of a diabetic patient presenting in coma or precoma is considered on p. 227.

Principles of treatment of diabetic coma

Only the outline will be discussed here. For details of management the reader should consult the references given at the end of the chapter. The treatment of hypoglycaemia is considered on p. 222.

Ketoacidosis. *Repletion of fluid and electrolytes* should be vigorous. If the plasma sodium concentration is low normal or low, isosmolal saline should be given initially and continued unless hypernatraemia develops. A careful watch should be kept for this finding, and if it develops hypoosmolar saline solutions should be used. If the metabolic acidosis is very severe (pH below 7.0) bicarbonate may be infused, but only until the blood pH rises to between about 7.15 and 7.20. It is unnecessary and often dangerous to correct the plasma $T\text{CO}_2$ level completely; it rapidly returns to normal following adequate fluid and insulin therapy. 8.4 per cent sodium bicarbonate is grossly hyperosmolar (Table 2.3, p. 54) and so may cause hypernatraemia and aggravate hyperosmolality, and a rapid rise in pH may aggravate the hypokalaemia associated with treatment. *Potassium* should be given as soon as plasma levels start to fall. Urinary volume should be monitored; if it fails to rise despite adequate rehydration further fluid and potassium should only be given if required, and then with care.

Insulin should be given immediately by continuous intravenous infusion, or by intermittent intramuscular injections, as soon as the plasma glucose and potassium concentrations are known. The insulin dose should be 'titrated' against the plasma glucose concentration.

The factor which precipitated the coma, such as infection, should be sought and treated.

Frequent monitoring of plasma glucose, potassium and sodium levels is necessary to assess progress and to detect developing hypoglycaemia, hypokalaemia or hypernatraemia.

Hyperosmolal non-ketotic coma. Treatment of hyperosmolal coma is similar to that of ketoacidosis; a sudden reduction of extracellular osmolality may do more harm than good (p. 33), and it is especially important to give *small doses of insulin* to reduce plasma glucose levels slowly. These patients are often very sensitive to the action of insulin. Hypoosmolal solutions should be used to correct volume depletion, but these too should be given slowly.

Hypoglycaemia

By definition hypoglycaemia is present if the plasma glucose level is less than 2.5 mmol/litre (45 mg/dl) in a specimen collected into a *tube which contains an inhibitor of glycolysis*: red cells continue to metabolize glucose *in vitro* and low values found in a specimen collected without such an inhibitor can be dangerously misleading. Symptoms of hypoglycaemia may develop at higher, or even normal, levels if there has been a rapid fall from a previously elevated value; by contrast, some people may have no symptoms at levels below 2.5 mmol/litre, especially if the concentration has fallen gradually. As discussed earlier, cerebral metabolism depends on an adequate supply of glucose from the plasma and the symptoms of hypoglycaemia resemble those of cerebral hypoxia. Faintness, dizziness or lethargy may progress rapidly to coma and, if untreated, permanent cerebral damage or death may result. If the plasma glucose concentration has fallen rapidly adrenaline secretion may be stimulated and cause sweating, tachycardia and agitation, but these symptoms may not occur if the fall is gradual, or if the autonomic nervous system is unresponsive to stress because the patient is taking β-blocking agents or has severe peripheral neuropathy. Existing cerebral or cerebrovascular disease may aggravate the condition. It is essential to restore the plasma glucose concentration quickly to prevent permanent cerebral damage.

There is no completely satisfactory classification of the causes of hypoglycaemia, as overlap between different groups occurs. A practical approach to diagnosis is based on the history. Particular attention should be paid to drugs being taken and to the relation of symptoms to meals, or to ingestion of a particular food. The patient can usually provisionally be allocated to one of two main groups.

Those with fasting hypoglycaemia. Symptoms typically occur at night or in the early morning, or may be precipitated by a prolonged fast or strenuous exercise. This pattern suggests excessive utilization of glucose or an abnormality of the glucose-sparing or glucose-forming mechanisms (p. 201).

The main causes are:

- inappropriately high insulin levels due to a tumour, or to hyperplasia of the pancreatic islet cells;
- deficiency of glucocorticoids;
- severe liver disease;
- some non-pancreatic tumours.

Those with non-fasting hypoglycaemia. Symptoms typically occur within five or six hours after a meal and may be related to ingestion of a particular type of food, or may be associated with medication. Substances that may provoke hypoglycaemia include:

- drugs (especially insulin);
- alcohol;
- glucose (reactive hypoglycaemia);
- galactose (in milk);
- fructose (in sucrose-containing foods);
- leucine (an amino acid especially abundant in casein, a protein found in milk).

Galactose, fructose or leucine are important causes of hypoglycaemia in infants rather than in adults.

It is not always possible to distinguish between provoked and early fasting hypoglycaemia on the basis of the history alone.

Hypoglycaemia, particularly in adults

Symptoms can only be attributed to hypoglycaemia if hypoglycaemia has been demonstrated. This statement may seem superfluous, but it is not unknown for a patient to be submitted to multiple tests for the differential diagnosis of hypoglycaemia that does not exist, or worse, to undergo treatment for it.

Once demonstrated the following main causes must be considered.

Insulin- or other drug-induced hypoglycaemia. These are probably the commonest causes and unless ingestion is deliberately concealed by the patient the offending drug should be easily identifiable. Hypoglycaemia in a diabetic may follow accidental *insulin* overdosage, be due to changing requirements, or to failure to eat after insulin has been given. Self-administration for suicidal purposes or to gain attention is not unknown, and homicidal use is a remote possibility. *Sulphonylureas* may also induce hypoglycaemia, especially in the elderly, and *salicylate poisoning* may be complicated by hypoglycaemia. Other drugs, such as antihistamines, have been suspected in some cases and it is always important to take a careful drug history.

Hypoglycaemia due to exogenous insulin suppresses insulin and C-peptide secretion by feedback. Measurement of C-peptide may help to differentiate exogenous insulin administration from endogenous insulin secretion, whether the latter is from an insulinoma, follows pancreatic stimulation by sulphonylurea drugs, or is derived from non-pancreatic insulin-secreting tumours.

Insulinoma. An insulinoma is a primary tumour of the islet cells of the pancreas. As with other functioning endocrine tumours, hormone secretion is inappropriate and usually excessive. It may occur at any age. It is usually single and benign, but may rarely be malignant. Multiple tumours may occur. It may be part of the syndrome of multiple endocrine adenopathy (p. 417). *C-peptide is released in parallel with insulin and plasma levels are therefore inappropriately high.*

Attacks of hypoglycaemia may be sporadic, with symptom-free intervals. They typically occur at night and before breakfast, and may be precipitated by strenuous exercise. Personality or behavioural changes may be the first feature and many patients present to psychiatrists.

Alcohol-induced hypoglycaemia. Hypoglycaemia may develop between two and 10 hours after ingestion of large amounts of alcohol. It is found most frequently in chronic alcoholics in whom there is also malnutrition, but it may occur in young subjects when they first drink alcohol. The hypoglycaemia is probably due to reduced hepatic output of glucose because of suppression of gluconeogenesis during metabolism of alcohol.

During prolonged fasting, when glycogen stores are depleted, gluconeogenesis is the main source of plasma glucose, and hypoglycaemia is potentially severe. Differentiation from alcoholic stupor may be impossible unless the plasma glucose level is estimated. It may be necessary to infuse glucose frequently during treatment, until glycogen stores are repleted.

Non-pancreatic tumours. Although carcinomas, especially of the liver, and sarcomas have been reported to cause hypoglycaemia, it occurs most commonly in association with retroperitoneal tumours resembling fibrosarcomas. The tumours are slow-growing and may become very large. Hypoglycaemia may be the presenting feature. The mechanism is not always clear, but may sometimes be due to secretion of insulin or of an insulin-like substance, or to excessive glucose utilization by the tumour (p. 422).

Functional (reactive) hypoglycaemia (sensitivity to glucose). Some people develop symptomatic hypoglycaemia between two and four hours after a meal or a glucose load. Loss of consciousness is very rare. A similar 'reactive' hypoglycaemia may be found after *gastrectomy*, when rapid passage of glucose into the intestine and rapid absorption may stimulate excessive insulin secretion (one type of the 'dumping syndrome', p. 255).

Except after gastrectomy, reactive hypoglycaemia is probably rare and is diagnosed too often.

Endocrine causes. Hypoglycaemia may occur in pituitary or adrenal insufficiency (Chapters 5 and 6). It is rarely the presenting manifestation of these conditions.

Impaired liver function. The functional reserve of the liver is so great that, despite its central role in the maintenance of plasma glucose levels, liver disease is a rare cause of hypoglycaemia. It may complicate very severe *hepatitis* or *liver necrosis* in which the whole liver is affected. This is less a problem of the differential diagnosis of hypoglycaemia than of the awareness that it may occur.

Hypoglycaemia in children

Hypoglycaemia is not uncommon in infancy and is important because it may cause permanent brain damage, especially in the first few months of life. Only the main causes will be outlined.

Neonatal period

There may be no obvious clinical signs attributable to hypoglycaemia even if the plasma glucose level is as low as 1.7 mmol/litre (about 30 mg/dl) in the first 72 hours of life, or 2.2 mmol/litre (about 40 mg/dl) in the later neonatal period. Clinical signs may be absent in very premature infants weighing less than 2.5 kg (5.5 lb) even at levels below 1.1 mmol/litre (about 20 mg/dl). Signs of hypoglycaemia at this age include convulsions, tremors and attacks of apnoea with cyanosis; because of the danger of brain damage the need for treatment is urgent. Symptomatic hypoglycaemia may last for up to a week. Neonatal hypoglycaemia occurs particularly:

- *in babies of diabetic mothers*. If the fetus is exposed to hyperglycaemia during pregnancy fetal islet-cell hyperplasia may occur, and the consequent hyperinsulinism may cause hypoglycaemia when the high glucose supply from the mother is removed after parturition. The incidence of asymptomatic hypoglycaemia in babies of diabetic mothers is about 50 per cent;
- *in erythroblastosis fetalis*, which may also be associated with islet-cell hyperplasia and neonatal hypoglycaemia;
- *in babies suffering from intrauterine malnutrition* (for example, infants of mothers with toxaemia of pregnancy, or the smaller of twins). These babies are usually 'small for dates' and may develop hypoglycaemia during the first week of life. Prematurity is an aggravating factor, because most of the liver glycogen is laid down after 36 weeks of gestation. Low fat stores also limit the availability of ketones as an alternative energy source for the brain.

Early infancy

Soon after birth, or after the introduction of milk to the diet, hypoglycaemia may be due to one of the following causes.

Glycogenoses. A deficiency of one of the enzymes involved in glycogenesis or glycogenolysis results in the accumulation of normal or abnormal glycogen. In von Gierke's disease, the least rare glycogen storage disorder, there is a *deficiency of glucose-6-phosphatase*. Because this enzyme is essential for the conversion of glucose-6-phosphate to glucose there is *fasting hypoglycaemia*. There may also be *ketosis* and *endogenous hypertriglyceridaemia* due to the excessive lipolysis caused by low insulin activity and intracellular glucose deficiency, *lactic acidosis*, due to excessive anaerobic glycolysis, and *hyperuricaemia* (p. 384).

Accumulation of glycogen causes *hepatomegaly*.

The *diagnosis* is made directly by demonstrating the absence of the enzyme in a liver biopsy specimen, or indirectly by demonstrating the failure of plasma glucose levels to rise after giving glucagon to stimulate glycogenolysis. Infusion of galactose or fructose, which are normally converted to glucose *via* G-6-P, also fails to increase plasma glucose levels because the G-6-P cannot be converted to glucose.

Treatment. Frequent meals should be given to maintain normal plasma glucose levels and so to prevent cerebral damage.

Galactosaemia. Galactose is necessary for the formation of cerebrosides, of some glycoproteins and, during lactation, of milk. Any excess is rapidly converted to glucose.

The commonest form of galactosaemia is due to deficiency of *hexose-1-phosphate uridylyltransferase* (Galactose-1-phosphate uridyltransferase).

The condition only becomes apparent after milk has been added to the infant's diet. The main features are:

- hypoglycaemia;
- vomiting and diarrhoea, with failure to thrive;
- hepatomegaly with jaundice and cirrhosis;
- cataract formation;
- mental retardation;
- renal tubular damage due to deposition of galactose-1-phosphate in the cells ('Fanconi syndrome').

Galactose is a reducing substance and the urine may give a positive reaction with Clinitest tablets. This feature may be absent if the subject is not receiving milk and therefore galactose.

Tubular damage may cause a generalized aminoaciduria.

The *diagnosis* is made by identifying the urinary sugar as galactose by thin-layer chromatography and by demonstrating a deficiency of the enzyme in erythrocytes. Cord blood of all newborn infants with affected siblings should be tested in this way.

Treatment. Galactose in milk and milk-products should be eliminated from the diet. Sufficient galactose for the body's needs can be synthesized endogenously as UDP-galactose.

Hereditary fructose intolerance. Hereditary fructose intolerance is a rare cause of hypoglycaemia. It is due to deficiency of *fructose-1-phosphate aldolase*: the accumulation of *fructose-1-phosphate* in several tissues causes many of the clinical features and the hypoglycaemia may be due to inhibition of glycogenolysis and gluconeogenesis. Symptoms only start after sucrose (fructose + glucose) and fructose-containing fruit or fruit drinks have been introduced into the diet. Hypoglycaemia, with nausea and vomiting and abdominal pain, and fructosuria follow about 30 minutes after fructose ingestion or intravenous infusion: this can be used as a diagnostic test. The infant fails to thrive, and liver deposits of fructose-1-phosphate cause hepatomegaly with jaundice, and sometimes cirrhosis and ascites due to progressive liver damage.

Later infancy

Idiopathic hypoglycaemia of infancy. The diagnosis of this condition is made by excluding other causes. Symptoms usually develop after fasting or after a febrile illness. There is a high incidence of brain damage.

In some cases there is excessive insulin secretion and it may not be possible to differentiate this condition from an insulinoma. Islet cell hyperplasia (nesidioblastosis) is a difficult diagnosis to make and is an uncommon cause of hypoglycaemia occurring before the age of the three years.

Leucine sensitivity There is often a familial incidence of leucine sensitivity. During the first six months of life *casein* may precipitate severe hypoglycaemia. This is due to its high leucine content. Leucine sensitivity is probably due to stimulation of insulin secretion by the amino acid. The condition appears to be self-limiting and does not usually persist beyond the age of six years. The *diagnosis* is confirmed by demonstrating hypoglycaemia within 30 minutes after an oral dose of leucine or casein. Normal subjects do not respond to leucine with a significant fall in plasma glucose level, but a number of patients with insulinoma are leucine sensitive.

The *treatment* is to give a low leucine diet.

Ketotic hypoglycaemia is the commonest cause of hypoglycaemia in the second year of life and develops after fasting or a febrile illness. These children were usually 'small for dates' babies. As in starvation, ketonuria precedes the hypoglycaemia and the diagnosis can be established by feeding a high fat, low calorie ('ketogenic') diet for 48 hours, during which time clinical hypoglycaemia occurs.

'Adult' causes of hypoglycaemia, including insulinoma, must always be considered.

Treatment of hypoglycaemia

Hypoglycaemia should be treated by urgent intravenous administration of 10 to 20 ml of at least 10 per cent, and in adults 50 per cent, glucose solution *after withdrawal of a blood sample for glucose and insulin assays* (p. 228). Some cases may need to be maintained on a glucose infusion until the cause has been established and treated.

An *insulinoma* should be removed surgically. If this is contraindicated a combination of diazoxide and chlorothiazide may maintain normoglycaemia.

Summary

1. Glucose is the main product of dietary carbohydrate metabolism.
2. The brain is almost entirely dependent on extracellular glucose as an energy source and maintenance of plasma glucose levels is important for normal cerebral function.
3. After a carbohydrate-containing meal excess glucose is:

- stored as glycogen in liver and muscle;
- converted to fat and stored in adipose tissue.

Insulin stimulates these processes.

4. During fasting:

- glycogen breakdown in the liver (and kidney) releases glucose into the plasma;

- triglyceride breakdown in adipose tissue releases glycerol, which can be converted to glucose, and fatty acids which can be metabolized by most tissues *other than the brain.*

5. The liver converts excess fatty acids to ketones which can be used as an energy source by the brain and other tissues.
6. If ketoacid formation exceeds the capacity of homeostatic mechanisms ketoacidosis may develop.
7. Anaerobic glycolysis produces lactic acid:

- lactic acid production occurs temporarily in contracting muscles;
- lactic acid is produced by hypoxic tissues. The hypoxic liver becomes a major lactic acid-producing rather than a lactic acid-consuming organ, and lactic acidosis results.

Other factors increasing glycolysis or reducing utilization of lactic acid may also cause lactic acidosis.

8. Diabetes mellitus is the result of relative or absolute insulin deficiency. It is characterized by hyperglycaemia, which may be intermittent. In severe diabetes mellitus excessive lipolysis may cause ketosis and later acidosis. Confusion and coma are probably due to hyperosmolality, and in both ketoacidosis and non-ketotic hyperosmolal coma there is severe water and electrolyte depletion.
9. Hypoglycaemia may occur during fasting, or be provoked by drugs or by some foods.
10. *In adults insulinoma is the most important cause of fasting hypoglycaemia; it is diagnosed by demonstrating high plasma insulin and C-peptide levels despite hypoglycaemia.* In diabetics *insulin administration* is the *commonest* cause of hypoglycaemia.
11. Factors causing hypoglycaemia in childhood vary with age. Many inborn errors of carbohydrate metabolism may cause hypoglycaemia.

Further reading

Diabetes mellitus. Report of a WHO study group. WHO Tech Rep Ser No. **727**. Geneva, World Health Organization, 1985.

Cryer PE, Gerich JE. Glucose counterregulation, hypoglycaemia, and intensive insulin therapy in diabetes mellitus. *N Engl J Med* 1985; **313**: 232–41.

Raskin P, Rosenstock J. Blood glucose control and diabetic complications. *Ann Intern Med* 1986; **106**: 254–63.

Turner RC, Williamson DH. Control of metabolism and the alterations in diabetes. In: O'Riordan JLH ed. *Recent Advances in Endocrinology and Metabolism* **No. 2.** Edinburgh, Churchill Livingstone, 1982: 73–97.

Wilkin T, Armitage M. Markers for insulin dependent diabetes: towards early detection. *Br Med J* 1986: **293**: 1323–26.

Bunn HF. Nonenzymatic glycosylation of protein: relevance to diabetes. *Am J Med* 1981; **70**: 325–30.

Marks V, Rose FC eds. *Hypoglycaemia.* **2nd ed.** Oxford: Blackwell Scientific, 1981.

Historical interest

Banting FG, Best CH. The internal secretion of the pancreas. *J Lab Clin Med* 1922; **7**: 251–66.
Banting FG, Best CH. Pancreatic extracts. *J Lab Clin Med* 1922; **7**: 464–72.

Investigation of disorders of carbohydrate metabolism

Estimation of plasma or blood glucose

Glucose levels are measured by enzymatic methods specific for glucose.

The supply of glucose to cells depends on extracellular concentrations and these are reflected in *plasma* levels. Since intracellular concentrations are kept low by metabolism, inclusion of erythrocytes in the sample 'dilutes' the plasma, giving results 10 to 15 per cent lower than those of plasma or serum, the actual figure depending on the haematocrit. Plasma assay is therefore desirable, but whole blood is sometimes used.

Either whole blood or plasma must be assayed immediately, or mixed with an inhibitor of glycolysis. *In vitro* metabolism of glucose by the cellular elements of the blood will produce falsely low levels; tubes containing fluoride or iodoacetate (both of which inhibit glycolysis) mixed with an anticoagulant are used.

Capillary blood from a finger prick is used for home glucose monitoring. Results on such samples usually fall between those of venous whole blood and venous plasma.

Investigation of suspected diabetes mellitus

The diagnosis of diabetes mellitus should not be made unless unequivocally high plasma glucose concentrations have been found in *two* specimens taken on different occasions. Falsely diagnosing a patient as diabetic may have serious social and economic consequences (for example, on life insurance premiums or mortgages). If the plasma glucose concentrations are not within the reference range and are not above that defined as diabetic, the patient is said to have impaired glucose tolerance.

Initial tests

If the patient presents with the symptoms of diabetes mellitus, with glycosuria, or if it is desirable to exclude the diagnosis because, for example, of a strong family history, take blood for plasma glucose estimation.

The finding of one random plasma (venous) glucose level (at least two hours after a meal) of more than 11.1 mmol/litre (200 mg/dl) is very suggestive of diabetes mellitus; two such levels are diagnostic of the disease. However, it is preferable to take blood after a fast of at least 10 hours, especially if one abnormal random level has been found.

	Interpretation	
	Venous plasma glucose mmol/litre (mg/dl)	
	Fasting	Random
Diabetes unlikely		5.5 (100) or less
Diabetic	7.8 (140) or more	11.1 (200) or more

(a) The diagnosis of diabetes mellitus is confirmed if:

(i) the *fasting* plasma level is 7.8 mmol/litre or more *on two occasions*;

or (ii) the *random* plasma level is 11.1 mmol/litre or more *on two occasions*;

or (iii) *both* a fasting level of more than 7.8 mmol/litre and a random level of more than 11.1 mmol/litre are found.

(b) The diagnosis of diabetes mellitus is usually excluded if the fasting plasma glucose level is less than 5.5 mmol/litre (100 mg/dl) on two occasions. Samples taken at random times are less reliable for excluding than for confirming the diagnosis.

(c) If the fasting values lie between 5.5 and 7.8 mmol/litre, the levels on random sampling are between 7.8 and 11.1 mmol/litre or if there is a high index of clinical suspicion, proceed to an oral glucose tolerance test. If the glucose concentration is measured in whole blood the levels will be approximately 1 mmol/litre (18 mg/dl) lower (see above). The two hour glucose level is about 1 mmol/litre higher in capillary whole blood than in the corresponding venous sample.

A patient *without symptoms* may be suspected of having diabetes mellitus because of the chance finding of glycosuria or hyperglycaemia. If the plasma glucose level is more than 11.1 mmol/litre in such a patient two hours after a standard glucose load the finding should be confirmed before a definitive diagnosis is made.

Oral glucose tolerance test

Contact your laboratory *before* starting this test, because local details may vary.

The patient should be resting and may not smoke during the test.

1. The patient fasts overnight (for at least 10, but not more than 16 hours). Water, but no other beverage, is allowed.

2. A venous sample is withdrawn for plasma glucose estimation and a double voided urine specimen (p. 210) collected.

3. The equivalent of 75 g of anhydrous glucose (for children 1.75 g/kg body weight up to a maximum of 75 g) is given. 75 g of glucose dissolved in 300 ml of water is hyperosmolar, and may not only cause nausea and occasionally vomiting and diarrhoea, but because of delayed absorption, may affect the results of the test. It is therefore more usual to give a solution of a mixture of glucose and its oligosaccharides, which, although all converted to glucose at the brush border, have less osmotic effect in the lumen when present as larger molecules (for example 105 ml of 'Hycal' in an equal volume of water, or 353 ml of 'Lucozade' contain the equivalent of 75 g of hydrated glucose). The patient must drink this within about five minutes.

4. Further blood and urine samples are taken at two hours after the dose.

The plasma glucose concentrations are measured and the urine samples tested for glucose.

Factors influencing the result of the glucose tolerance test

Previous diet. No special restrictions are necessary if the patient has been on a normal diet for three to four days. If, however, the test is performed after a period of carbohydrate restriction, perhaps as part of a reducing diet, there may be abnormal glucose tolerance. This is probably because metabolism is set in the 'fasted' state and so favours gluconeogenesis. Slightly abnormal tests should be repeated after appropriate dietary preparation.

Time of day. Most glucose tolerance tests are performed in the morning and the reference values quoted are for this time of day. There is evidence that tests performed in the afternoon yield higher plasma glucose levels and that the accepted 'normal values' may not be applicable. This may be due to a circadian variation in islet cell responsiveness.

Drugs. Drugs such as steroids, oral contraceptives and thiazide and loop diuretics may impair glucose tolerance.

	Interpretation	
	Venous plasma glucose mmol/litre (mg/dl)	
	Fasting	2 hours
Diabetes unlikely	5.5 (100) or less	7.8 (140) or less
Impaired glucose tolerance	5.5 to 7.8	7.8 to 11.1
Diabetic	7.8 (140) or more	11.1 (200) or more

The interpretation of the oral glucose tolerance test is identical in the pregnant and non-pregnant patient.

Initial investigation of a diabetic presenting in coma

A diabetic patient may be in coma due to hyperglycaemia, hypoglycaemia or any of the causes shown in Table 10.2.

After a thorough clinical assessment proceed as follows.

1. *Notify the laboratory that specimens are being taken and ensure that they are delivered promptly.* In this way delays are minimized.
2. Take blood *immediately* for estimation of:

plasma glucose
potassium
sodium
$T\text{CO}_2$ and/or arterial pH and $P\text{CO}_2$.

Repeated arterial puncture is undesirable and often unnecessary. If arterial pH is measured initially, plasma $T\text{CO}_2$ may be estimated if monitoring of acid-base balance is needed.

3. Test a urine sample for glucose and ketones. *It is unwise to rely solely on urine testing to diagnose hyperglycaemia.* The urine may have been in the bladder for some time and reflect earlier and very different plasma glucose levels.
4. If it is really necessary to obtain a rapid assessment of blood glucose using a reagent strip, for example because the laboratory is at such a distance that there may be an unacceptable delay, *it is essential to remember that improperly stored reagent strips, or failure to follow the manufacturers' instructions for use, may give completely and dangerously wrong results.*

Table 10.2 Presenting clinical and biochemical features of a diabetic presenting in coma

		Plasma		Urine	
Diagnosis	Clinical features	glucose	$[HCO_3^-]$	glucose	ketones
Ketoacidosis	Dehydration Hyperventilating	High	Low	+ + +	+ + +
Hyperosmolal coma	Dehydration May be hyperventilating	Very high	N or slightly reduced	+ + +	Neg.
Hypoglycaemia	Non-specific	Low	N	Neg.	Neg.
Cerebrovascular accident	Neurological May be hyperventilating	May be raised	May be low	May be +	Usually neg.

5. If hypoglycaemia is suspected on clinical grounds, or because of the results obtained using reagent strips, glucose should be given immediately while waiting for the laboratory results. It is less dangerous to give glucose to a hyperglycaemic patient than to give insulin to a hypoglycaemic one.

Blood samples must be sent to the laboratory immediately, but treatment should never be delayed until the results are available. *Results of side-room tests must be interpreted with caution.*

Investigation of hypoglycaemia

The most important test in a patient with proven hypoglycaemia is measurement of the plasma insulin level *when that of glucose is low.* This should differentiate exogenous insulin administration or endogenous insulin production (the most important cause of the latter being an insulinoma) from other causes of hypoglycaemia. *If plasma insulin levels are inappropriately high*, and if doubt remains about the cause of this finding, C-peptide assay may help. If it is high it suggests endogenous insulin secretion (including that due to pancreatic stimulation by sulphonylureas): an undetectable level suggests exogenous insulin administration.

1. *If a patient is seen during an episode of hypoglycaemia, take blood for glucose, insulin and C-peptide assay before giving glucose.* Plasma for the last two assays should be separated from cells *immediately* and the plasma stored frozen until hypoglycaemia has been proven. This step may considerably shorten the time needed for further investigation.

2. More commonly the patient has been referred for investigation of previously documented hypoglycaemia or with a history strongly suggestive of hypoglycaemic attacks.

A full assessment should be made, paying special attention to:

(a) the time of attacks in relation to meals (reactive hypoglycaemia);
(b) drug (especially antidiabetic) or alcohol ingestion. This information may not be freely given and a negative history does not exclude it;
(c) possible hypopituitarism or adrenocortical hypofunction. If either of these is considered likely, investigate as outlined on p. 151;
(d) a possible non-pancreatic tumour (p. 219).

3. If no cause is identified an attempt should be made to induce hypoglycaemia by fasting, for up to 72 hours if necessary, and even accompanied by exercise under close supervision. Blood should be taken every six hours for glucose and insulin estimations and, when symptoms occur, should be assayed for glucose immediately. The test can be stopped if hypoglycaemia is demonstrated and only this specimen assayed for insulin.

If hypoglycaemia is not induced by prolonged fasting, endogenous hyperinsulinism is unlikely to be the cause of the symptoms.

4. *Insulin should be measured in a sample taken at the time of proven hypoglycaemia.* If insulin administration is suspected, C-peptide should also be assayed.

The interpretation of results is summarized in Table 10.3.

Table 10.3 Results of plasma insulin and C-peptide estimations *during hypoglycaemia* (spontaneous or after a prolonged fast)

Hypoglycaemia due to:	Plasma insulin	Plasma C-peptide
Insulin administration	Inappropriately high	Low
Insulinoma or ectopic insulin secretion	Inappropriately high	High
Sulphonylurea administration	Inappropriately high	High
Alcohol	Appropriately low	
Non-pancreatic non-insulin-secreting tumour	Appropriately low	
Pituitary or adrenal failure	Appropriately low	

5. If results of fasting glucose and insulin assays are equivocal, an *insulin suppression test* can be performed. Hypoglycaemia induced by intravenous injection of insulin should suppress endogenous insulin and C-peptide secretion. Failure of C-peptide levels to fall confirms autonomous insulin secretion, usually due to an insulinoma.

Insulin suppression test

The precautions outlined on p. 151 must be observed.

1. The patient must fast overnight for 10 to 16 hours, to ensure an adequate response to insulin, and is allowed only small amounts of water.
2. Insert an indwelling intravenous cannula (p. 150) and keep it patent with sodium citrate solution or heparinized saline.
3. Take fasting samples for glucose and C-peptide estimation.
4. Give soluble insulin (0.15 U/kg body weight) intravenously.
5. Take further samples for glucose and C-peptide assay at 30, 60, 90, 120 and 150 minutes after the insulin injection. If clinically indicated, cortisol may also be measured to assess hypothalamic-pituitary-adrenal function (p. 151).

Interpretation

In normal subjects plasma C-peptide levels fall by more than 50 per cent of the initial value. In most cases of insulinoma this degree of fall does not occur.

Glycosuria

Glycosuria is best detected by enzyme reagent strips such as Diastix (Ames). Directions for use are supplied with the strips. Glycosuria may be due to:

- diabetes mellitus;
- glucose infusion;
- renal glycosuria, which may be inherited as an autosomal dominant trait;
- pregnancy.

False negative results may occur if the urine contains large amounts of ascorbic acid after ingestion of therapeutic doses, or after injection of tetracyclines which contain ascorbic acid as a preservative.

False positive results may occur if the urine container is contaminated with detergent.

Reducing substances

The reagents in Clinitest tablets (Ames) react with any reducing substance. These tablets should be used to screen the urine in newborn infants or children because of the diagnostic importance of non-glucose reducing substances, such as galactose, in these age groups.

Clinitest tablets must be used as directed. They are caustic and should be handled carefully. The test is less sensitive for glucose than Diastix, but more accurate for quantitation.

Causes of a positive result with Clinitest

Substances that give a positive reaction with Clinitest are:

glucose; } common
glucuronates; }
lactose; common in pregnancy

galactose;
fructose;
pentoses;
homogentisic acid.
} rare

A weakly positive reaction may be due to high concentrations of urate or creatinine.

In adults a positive reaction with Clinitest is usually due to glucose. Testing with Diastix will confirm its presence or absence. Non-glucose reducing substances are identified by chromatography and specific tests.

The significance of a positive result varies with the substance.

Glucose (see p. 210).

Glucuronates. A large number of drugs, such as salicylates and their metabolites, are excreted in the urine after conjugation with glucuronate in the liver (p. 288). Glucuronates are relatively common urinary reducing substances.

Galactose. Galactose is found in the urine in galactosaemia (p. 221).

Fructose. Fructose may appear in the urine after very large oral doses of sucrose, or after excessive fruit ingestion, but usually fructosuria is due to two rare inborn errors of metabolism, both transmitted by autosomal recessive genes.

- *Essential fructosuria* is a harmless condition.
- *Hereditary fructose intolerance* (p. 221) is a serious disease characterized by hypoglycaemia that may lead to death in infancy.

Lactose. Lactosuria may occur in:

- late pregnancy and during lactation;
- lactase deficiency (p. 261).

Pentoses. Pentosuria is very rare. It may occur in:

- *alimentary pentosuria* after excessive ingestion of fruits such as cherries and grapes. The pentoses are arabinose and xylose;
- *essential pentosuria*, a rare recessive disorder due to a block in glucuronate metabolism characterized by the excretion of xylose. It is harmless.

Homogentisic acid. Homogentisic acid appears in the urine in the rare inborn error alkaptonuria (p. 371). It is usually recognizable because it forms a blackish precipitate.

Ketonuria

Most simple urine tests for ketones are more sensitive for detecting acetoacetate than acetone; acetoacetic acid is also present at a higher concentration. 3-hydroxybutyrate does not react in these tests.

Ketostix and Acetest (Ames) are strips and tablets respectively, impregnated with ammonium sulphate and sodium nitroprusside. Instructions are provided with them.

Ketonuria may be detected by these methods after the patient has fasted for several hours. Occasional colour reactions resembling, but not identical with, that of acetoacetate may be given by phthalein compounds such as phenolphthalein.

11

Plasma lipids and lipoproteins

Current understanding of the physiology and pathology of the plasma lipids is based on the concept of lipoproteins, the form in which lipids circulate in the plasma. The first part of this chapter will describe the classifications in current usage and show how terminologies are related.

Terminology and classification

Plasma lipids

The chemical structures of the four forms of lipid present in the plasma are illustrated in Fig. 11.1.

Fatty acids are straight-chain compounds of varying lengths. They may be *saturated* (containing no double bonds), or *unsaturated* (with one or more double

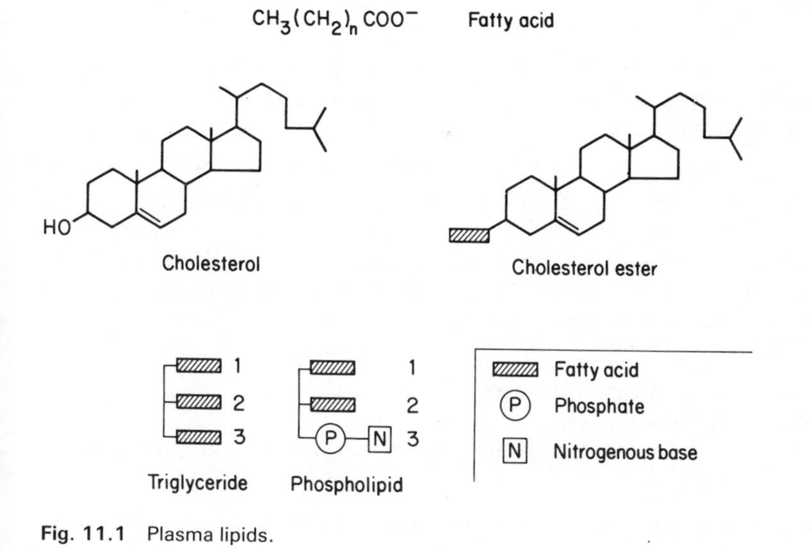

Fig. 11.1 Plasma lipids.

bonds). The main saturated plasma fatty acids are *palmitic* (16 carbon atoms) and *stearic* (18 carbon atoms). Fatty acids may be *esterified* with glycerol to form glycerides, or they may be free, when they are called *free fatty acids* (FFA) or *non-esterified fatty acids* (NEFA). In the plasma FFA are carried mainly bound to albumin. Free fatty acids are an immediately available energy source and provide a significant proportion of the energy requirements of the body. This aspect of fat metabolism is considered further on p. 204.

Triglycerides consist of glycerol, each molecule of which is esterified with three fatty acids.

Phospholipids are complex lipids, resembling triglycerides, but containing phosphate and a nitrogenous base.

Cholesterol is a steroid from which other steroids are derived. About two-thirds of the plasma cholesterol is esterified with fatty acids to form *cholesterol esters.* Assays in routine use measure the total cholesterol level, but do not distinguish between the unesterified and esterified forms.

Lipoproteins

Plasma lipids are derived from food (exogenous) or are synthesized in the body (endogenous). They are relatively insoluble in water and are carried in body fluids as soluble protein complexes known as lipoproteins: a core of insoluble (non-polar) cholesterol esters and triglycerides is surrounded by proteins, phospholipids and free cholesterol with their water-soluble (polar) groups facing outwards.

Lipoproteins are classified by their density which in turn reflects their size. The more lipid a complex contains, the larger it is and the lower its density. There are four main classes.

Two are *large, triglyceride-rich* complexes:

- chylomicrons, which transport exogenous lipid from the intestine to all cells;
- VLDL (Very Low Density Lipoproteins), which transport endogenous lipid from the liver to cells.

Because of their large size these complexes reflect light and plasma with increased levels appears turbid or milky (lipaemia). If a turbid plasma sample is left standing for 18 hours at 4°C the large chylomicrons, because of their low density, rise to form a creamy layer on the surface. The smaller, denser VLDL particles do not rise and the sample remains diffusely turbid. If hyperlipidaemia is gross the distribution may be less easy to distinguish.

Two *smaller* lipoproteins contain mostly *cholesterol*:

- LDL (Low Density Lipoproteins), which are formed from VLDL and transport cholesterol to cells;
- HDL (High Density Lipoproteins) are involved in the transport of cholesterol from the cells to the liver.

These small lipoproteins do not scatter light and even very high levels in plasma do not produce lipaemia.

A fifth class, IDL (Intermediate Density Lipoproteins), is usually a transient intermediate lipoprotein formed during the conversion of VLDL to LDL: it contains both cholesterol and endogenous triglycerides. IDL is undetectable in normal plasma.

Although the classification of lipoproteins is based on their density, determined by ultracentrifugation, the lipoprotein composition of plasma can usually be inferred from simple lipid assays. Plasma taken from a fasting subject contains only HDL, LDL and VLDL, both in normal individuals and in many cases of hyperlipidaemia (but see p. 242). The measured plasma cholesterol concentration reflects LDL levels and the plasma triglyceride concentration those of VLDL. If necessary the cholesterol in HDL and LDL may be separated and quantitated by simple precipitation techniques.

In some cases of hyperlipidaemia it may be necessary to define the lipoprotein pattern with greater precision. Electrophoresis is the simplest technique. It separates the complexes by electrical charge into four principal bands, named according to their relative positions as α, pre-β and β (which correspond to HDL, VLDL, and LDL respectively) and chylomicron fractions. IDL excess may produce a broad β band.

The nomenclature and composition of the main lipoprotein classes are summarized in Table 11.1.

Metabolism of lipoproteins

General

Lipoproteins are *synthesized* in the *liver* or *intestine*. After secretion they are *modified by enzymes* and their remnants are *taken up by receptors* on cells. These processes are regulated by the protein component of the complex, the *apolipoproteins*.

There are several groups of apolipoproteins, such as apo-A, apo-B etc: members of these groups, such as A-I and C-II, have specific functions. Some apolipoproteins, especially apo-B, are incorporated into the lipoprotein, but others interchange freely between lipoproteins. Normal lipid secretion by cells, activation of

Table 11.1 The composition and electrophoretic mobility of the main lipoproteins

		Composition (% mass)					
Lipoprotein	Source	Prot.	Chol.	TG.	PL.	Apolipoproteins	Electrophoretic mobility
Chylomicrons	Intestine	1	4	90	5	A, B, (C), (E)	Origin
VLDL	Liver	8	25	55	12	B, C, E	pre-β
LDL	VLDL (via IDL)	20	55	5	20	B	β
HDL	Liver, intestine	50	20	5	25	A, C, E	α

Prot. = apolipoprotein. Chol. = cholesterol TG. = triglyceride PL. = phospholipid
The apolipoproteins in brackets are acquired after secretion

Table 11.2 The main apolipoproteins

Apolipoprotein	Occurrence	Known functions
A	Chylomicrons, HDL	Cofactor for LCAT (A-I)
B	Chylomicrons, VLDL, IDL, LDL	Secretion of chylomicrons Secretion of VLDL Binding of LDL to receptors
C	HDL, VLDL, IDL, chylomicrons (from HDL)	Cofactor for lipoprotein lipase (C-II)
E	HDL, VLDL, IDL, chylomicrons (from HDL)	Binding of IDL and remnant particles to receptors

enzymes concerned with lipid metabolism and receptor recognition of lipoproteins all depend on apolipoproteins. These functions are summarized in Table 11.2, and described in the text.

Exogenous lipid pathways (Fig. 11.2)

Fatty acids and cholesterol which have been released by digestion of dietary fat, together with cholesterol from the bile, are absorbed into the intestinal mucosal cells, where they are re-esterified to form triglycerides and cholesterol esters. These, together with phospholipids, apo-A and apo-B, are secreted from the cell into the lymphatic system as chylomicrons, and enter the systemic circulation by the thoracic duct: secretion from cells depends on the presence of apo-B. Apo-C and apo-E, both derived from HDL, are added to the chylomicrons in lymph and in plasma.

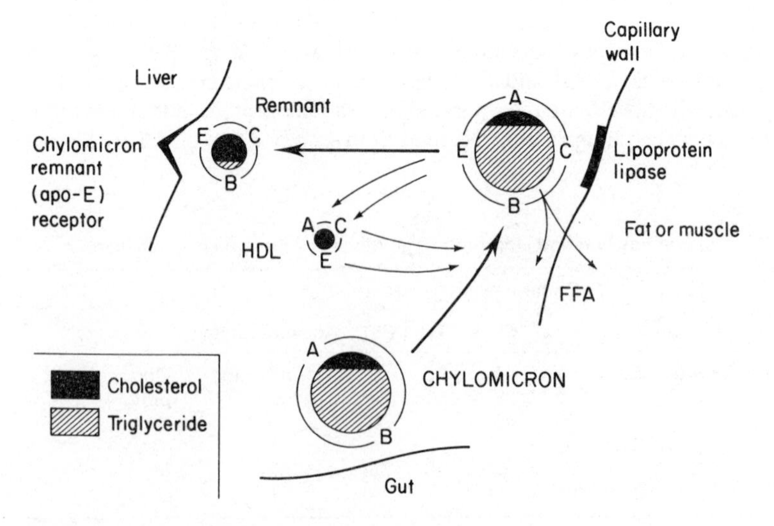

Fig. 11.2 Exogenous lipid pathways.

Chylomicron metabolism

Most chylomicrons are metabolized in adipose and muscle tissue. The enzyme *lipoprotein lipase*, located on capillary walls, is activated by apoprotein C-II, and triglycerides in the chylomicrons are hydrolysed to glycerol and fatty acids. The fatty acids are taken up by adipose or muscle cells, or are bound to albumin in the plasma, and glycerol enters the hepatic glycolytic pathway. As the chylomicron shrinks surface material containing apo-A, and some apo-C and phospholipid, is released and incorporated into HDL.

The small *chylomicron remnants* are composed mainly of cholesterol and apo-B and apo-E. They rapidly bind to *hepatic chylomicron-remnant receptors*, which recognize the constituent apo-E; the remnants then enter the liver cells, where the protein is catabolized and the cholesterol is released into the cells. The uptake of chylomicron remnants, unlike that of LDL (p. 236) is *not* influenced by the amount of cholesterol in hepatic cells.

At the end of this pathway dietary triglycerides have been delivered to adipose tissue and muscle, and cholesterol to the liver.

Endogenous lipid pathways (Fig. 11.3)

The liver is the main source of endogenous lipids. Triglycerides are synthesized from glycerol and fatty acids, which may reach the liver from the fat stores or be synthesized from glucose. Hepatic cholesterol may be synthesized locally or be derived from lipoproteins, such as chylomicron remnants, after they have been taken up by the cell. These lipids are transported from the liver in VLDL.

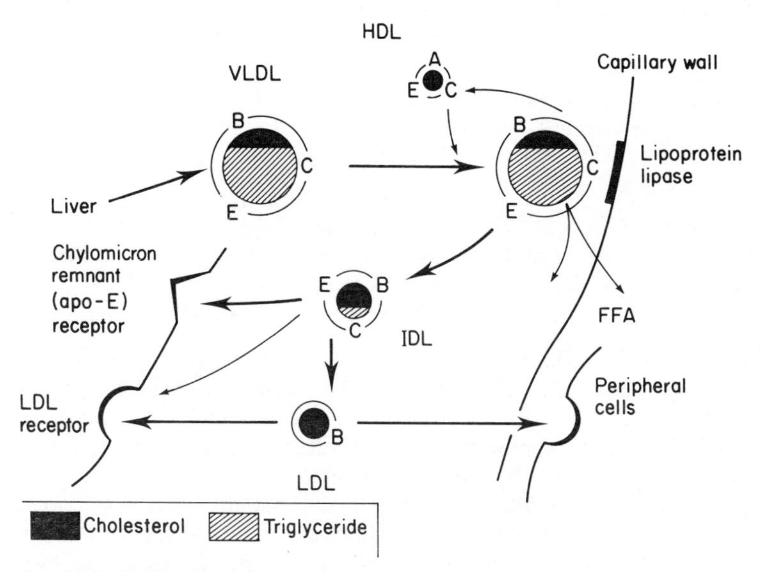

Fig. 11.3 Endogenous lipid pathways.

VLDL metabolism

VLDL is a large triglyceride-rich complex incorporating apoproteins B, C and E. After secretion it gains more apo-C from HDL. In peripheral tissues, triglycerides are removed after hydrolysis by lipoprotein lipase. Up to this stage the metabolism of VLDL is analogous to that of chylomicrons, although it occurs more slowly. The disposal of the resulting remnant particle, however, differs.

The 'VLDL remnant', *IDL* (intermediate density lipoprotein), contains both triglycerides and cholesterol, with apo-B and apo-E, and may follow one of two pathways. Some is rapidly taken up by the liver and some loses the remaining triglycerides and apo-E to become LDL. The mechanism and site of this conversion is unknown.

LDL metabolism

LDL is a small cholesterol-rich lipoprotein containing only apo-B. It has a longer life than its precursors and accounts for most of the measured cholesterol in plasma. Because of its small size it can infiltrate tissues such as the arterial wall. LDL is taken up by LDL receptors on cells. In view of its great importance we will consider this in some detail.

LDL receptors, although present on all cells, are most abundant in the liver. They recognize apo-B and apo-E and so can take up either LDL or IDL. After entering the cell the lipoprotein is broken down and the cholesterol released, and much of it contributes to membrane formation or, in the adrenal cortex and gonads, to steroid synthesis. Most cells can also synthesize cholesterol, but several feedback mechanisms prevent its intracellular accumulation. Cholesterol taken up by receptors *inhibits intracellular cholesterol synthesis and, most importantly, by reducing the synthesis of LDL receptors*, prevents further uptake.

Most of the plasma LDL is removed by LDL receptors. Especially if plasma levels are high, some may also enter cells by a passive, unregulated route.

Factors influencing plasma levels of LDL

The plasma concentration of LDL, and so of cholesterol, is determined mainly by the rate of uptake by LDL receptors. The *liver* has a central role in cholesterol metabolism because it:

- contains most of the LDL receptors in the body;
- synthesizes most of the endogenous cholesterol;
- receives cholesterol from the diet and from lipoproteins;
- is the only organ that can excrete cholesterol from the body.

The concentration of LDL receptors on hepatic cell surfaces depends on the amount of cholesterol in the cells. As cholesterol accumulates the number of receptors is reduced (p. 239). Factors which lead to cholesterol accumulation in the liver will, by reducing receptor numbers, increase plasma LDL levels. One of these factors is the amount of cholesterol reaching the liver from the gut.

Cholesterol enters the intestinal lumen from the diet and in bile. About 30 to 60 per cent of the total is absorbed or reabsorbed; absorption increases if the diet is rich in saturated fat, possibly because of more efficient micelle formation (p. 250).

In affluent societies dietary cholesterol is derived mainly from egg yolk (the richest source), dairy products and meat, and the daily intake is about 1.5 to 2.0 mmol (600 to 800 mg). Absorbed cholesterol reaches the liver *via* chylomicron remnants which are taken up by the chylomicron remnant receptors (*not* affected by hepatic cholesterol concentration).

Control of hepatic cholesterol synthesis by negative feedback may not be able to prevent accumulation if dietary intake is excessive. As in all other cells, cholesterol accumulation leads to a *reduction in LDL receptors*. LDL entry into cells falls and plasma levels rise.

Cholesterol may be secreted into plasma from hepatic cells in VLDL, with no net loss to the body, or may be excreted from the body in bile as cholesterol or after conversion to bile acids. Some bile acids are reabsorbed from the intestinal lumen and return to the liver in the enterohepatic circulation (p. 250); they then control further bile acid synthesis. Any interruption to this enterohepatic circulation by, for example, bile acid sequestrants (p. 243) will lead to increased conversion of cholesterol to bile acids, reduction in hepatic cholesterol stores and an *increase* in LDL receptors.

Hepatic LDL receptors are also increased by oestrogens and thyroid hormones.

The role of high density lipoprotein (HDL)

If cholesterol synthesized in cells could not be removed it would accumulate progressively, but the only excretory route is in the bile. The transport of cholesterol from cells to liver involves HDL.

HDL is synthesized in hepatic and intestinal cells and secreted from them as small complexes of phospholipid and apo-A and apo-E. It may also be formed from the surface coat of the large triglyceride-rich lipoproteins, VLDL and chylomicrons (p. 236).

Free cholesterol derived from cells and from other lipoproteins is esterified in HDL: esterification is catalysed by the enzyme *lecithin-cholesterol acyltransferase* (LCAT) which is part of the HDL complex and which requires apo-A-I. Most of the esterified cholesterol is transferred to LDL, VLDL and chylomicron remnants and so reaches the liver. A small amount is stored in the core of HDL and some cholesterol-rich HDL may be taken up directly by the liver.

The role of apoprotein C-II in activating lipoprotein lipase has been mentioned. When VLDL or chylomicron levels are low most of the plasma apo-C is carried on HDL. As the levels of the triglyceride-rich lipoproteins rise, these particles take up apo-C-II from HDL. After the triglyceride has been hydrolysed, the apoprotein returns to HDL.

Disorders of lipid metabolism

Most common disorders of lipid metabolism are associated with hyperlipidaemia. Very rare inherited disorders may be associated with accumulation of lipid in tissues and not in plasma.

Clinical manifestation of hyperlipidaemia

Accumulation of lipid in tissues is usually, but not always, the result of severe and prolonged hyperlipidaemia, and causes cell damage. Lipid accumulations, for example under skin or mucous membranes, may be visible.

Lipid may accumulate in:

- ***arterial walls.*** *(This is by far the commonest and most important manifestation of lipid disorders).* Cholesterol accumulation and associated cellular proliferation and fibrous-tissue formation produce the *atheromatous plaque. Atherosclerosis* is due to the distortion and obstruction of the artery which may result from calcification and ulceration of the plaque. The small lipoproteins LDL and IDL are atherogenic;

- ***subcutaneous tissue***, causing *xanthomatosis*. The nature of the lipid fraction most affected usually seems to determine the clinical appearance:

  eruptive xanthomata are crops of small, itchy yellow nodules. They are associated with very high *VLDL* or *chylomicron (triglyceride)* levels, and soon resolve if plasma lipid concentrations fall to normal;
  tuberous xanthomata are yellow plaques found mainly over the elbows and knees, and can be large and disfiguring. These, and raised linear lipid deposits in the palmar creases, are associated with high levels of *IDL* (which contain both triglyceride and cholesterol);
  xanthelasma is the name given to lipid deposits under periorbital skin, and may be associated with high *LDL cholesterol* levels.

- ***tendons.*** Tendinous xanthomata;

- ***cornea.*** A *corneal arcus* is due to corneal deposits.
  Both tendinous xanthomata and corneal arcus in relatively young subjects (under the age of 40) are associated with high *LDL cholesterol* levels, and have a bad prognosis.

Hypertriglyceridaemia, whether due to chylomicrons, VLDL, or both, causes *turbid plasma*. Sustained and very high concentrations of *chylomicrons* are associated with *abdominal pain* and even *pancreatitis*, as well as eruptive xanthomata, but many cases of hypertriglyceridaemia are symptom free. These large lipoproteins are probably unlikely to cause atheroma. However, many patients with increased levels of VLDL triglyceride have reduced HDL (necessary for transport of cholesterol from tissues) and some also have increased LDL or IDL (which contain cholesterol); the effects on cholesterol metabolism may explain the slightly increased risk of atheroma which has been attributed to hypertriglyceridaemia.

Lipids and cardiovascular disease

Evidence suggests that there is a positive correlation between the risk of developing ischaemic heart disease and plasma LDL cholesterol levels, and a negative one with those of HDL cholesterol. Lowering high LDL cholesterol concentrations does reduce the risk of cardiovascular disease, but it has not been proved that low

HDL levels increase this risk; high plasma HDL levels may have no direct effect, but may be associated with other beneficial factors.

Most of our knowledge is based on plasma total, or on LDL or HDL, cholesterol levels. As methods for estimation of apo-A and apo-B become more generally available they may prove better indicators of cardiovascular risk.

In practice lipoprotein disorders are recognized by measuring lipids, rather than lipoproteins themselves. We will therefore consider them under three headings:

- predominant hypercholesterolaemia;
- predominant hypertriglyceridaemia;
- mixed hyperlipidaemia.

Predominant hypercholesterolaemia

What is hypercholesterolaemia?

Plasma cholesterol at birth (cord blood) is usually below 2.5 mmol/litre (about 100 mg/dl). Levels increase slowly, mostly during the first year of life, but do not usually exceed 4.0 mmol/litre (about 160 mg/dl) in children. In most affluent populations plasma concentrations again increase progressively after the second decade, more in men than in women during the reproductive years. The upper 95 per cent limit of the 'reference range' in many societies is as high as 8.5 mmol/litre (about 330 mg/dl) in the fifth and sixth decades. This rise does not occur in less affluent communities, in which the incidence of ischaemic heart disease is much lower. It is likely that the progressive rise in plasma cholesterol reflects a decreasing concentration of LDL receptors in the liver.

Because of these factors, the term 'reference range' for plasma cholesterol has little significance. It has been shown that the risk of cardiovascular disease increases as plasma cholesterol levels rise above 5.5 mmol/litre (about 215 mg/dl).

Causes of hypercholesterolaemia

Hypercholesterolaemia, with little or no elevation of triglycerides, is almost always due to raised *LDL* levels. The coexistence of an underlying genetic defect or the development of a disorder that affects LDL levels will cause a greater increase in plasma cholesterol with age than the factors already discussed.

The main disorders that may produce a *secondary* increase in plasma LDL and total cholesterol are:

- hypothyroidism;
- diabetes mellitus;
- nephrotic syndrome;
- cholestasis.

The occurrence of hypercholesterolaemia in families, often associated with an increased risk of ischaemic heart disease, suggests an *inherited disorder*, the exact nature of which can only be established by extensive family studies.

In most families with a moderate hypercholesterolaemia, a graph of the plasma cholesterol values of all individual members shows a continuous distribution: this contrasts with the clear trimodal distribution found in the monogenic pattern of familial hypercholesterolaemia (see below), in which values in homozygotes for the normal gene, heterozygotes, and homozygotes for the abnormal gene form three distinct peaks. The continuous distribution, with a high mean value within a family, is thought to be due to several gene abnormalities affecting LDL or cholesterol synthesis and disposal. It is therefore called *polygenic* hypercholesterolaemia. Environmental and dietary factors may determine the expression of the defect. Xanthomata are relatively rare but there is an increased risk of cardiovascular disease.

In the following two disorders with moderate to severe hypercholesterolaemia, the pattern of inheritance is autosomal dominant.

Familial combined hyperlipidaemia is the commoner and is associated with excessive hepatic production of apolipoprotein B and therefore of LDL.

Plasma VLDL levels are high in about a third of affected family members, because of an associated primary or secondary increase in triglyceride synthesis. There is no increase in triglyceride synthesis in another third, who only have increased LDL levels. Increases in the concentrations of both fractions are found in the final third. The lipid abnormalities only become apparent after the third decade, and the risk of ischaemic heart disease in all cases is higher than that for an age and sex matched population. High triglyceride levels may cause eruptive xanthomata.

Familial hypercholesterolaemia is due to deficiency of LDL receptors. There are several variants which are inherited as autosomal dominant traits. The reduced uptake of LDL by cells, particularly in the liver, results in increased plasma LDL and cholesterol levels. Plasma triglyceride levels are normal, or only slightly raised. It is the most lethal of the inherited disorders.

In homozygotes LDL receptors are virtually absent and plasma LDL cholesterol levels are three or four times higher than those in normal subjects; patients rarely survive beyond the age of 20, because of ischaemic heart disease. In heterozygotes the number of LDL receptors is reduced by about 50 per cent and the plasma levels are about twice those in normal subjects, and there is a 10- to 20-fold higher risk of ischaemic heart disease. Tendinous xanthomata and xanthelasma develop in early childhood in homozygotes, but only after the second decade in heterozygotes.

In families with this *monogenic* mode of inheritance there is a clear distinction between unaffected, homozygous and heterozygous subjects; this differs from the commoner polygenic disease. In most countries monogenic hypercholesterolaemia accounts for less than 5 per cent of all cases of primary hypercholesterolaemia.

Predominant hypertriglyceridaemia

Elevated plasma triglyceride levels may be due to increased VLDL or chylomicrons or both. Hypertriglyceridaemia is usually *secondary* to another disorder, of which the most common are:

- obesity, and excessive carbohydrate intake;
- excessive alcohol intake;
- diabetes mellitus; } with hypercholesterolaemia
- hypothyroidism; } with hypercholesterolaemia
- nephrotic syndrome; } with hypercholesterolaemia
- renal failure;
- pancreatitis;
- oestrogen therapy or oral contraceptives;
- some β-blocking drugs.

Primary hypertriglyceridaemia is rarer than primary hypercholesterolaemia.

Familial combined hyperlipidaemia has been discussed above. About a third of affected individuals have raised VLDL levels.

Familial endogenous hypertriglyceridaemia is due to hepatic triglyceride overproduction with increased VLDL secretion from the liver. The condition seems to be transmitted as an autosomal dominant trait, and usually becomes apparent only after the fourth decade of life. It may be associated with:

- obesity;
- glucose intolerance;
- hyperuricaemia.

Superimposed secondary factors, such as diabetes mellitus or alcoholism, may cause very high VLDL levels and often, in addition, chylomicronaemia.

Hyperchylomicronaemia is usually due to an acquired or inherited deficiency of the enzyme lipoprotein lipase that catalyses the hydrolysis of triglycerides in the large lipoproteins. Insulin is needed for optimal action of the enzyme, and hyperchylomicronaemia may occur in those with poorly controlled diabetes mellitus. It may also be found in cases of pancreatitis.

Inherited *lipoprotein lipase deficiency* may be due to:

- true deficiency of the enzyme;
- failure to activate the enzyme because of apoprotein C-II deficiency.

The plasma is very turbid because of the accumulation of chylomicrons. True lipoprotein lipase deficiency usually presents during childhood, with signs and symptoms due to excess of fat at various sites, such as:

- skin (eruptive xanthomata);
- liver (hepatomegaly);
- retinal vessels (lipaemia retinalis);
- abdomen (abdominal pain).

Hyperchylomicronaemia due to apoprotein C-II deficiency is more likely to present in adults.

Mixed hyperlipidaemia

Raised plasma levels of both cholesterol and triglycerides are most common in patients with poorly controlled diabetes mellitus or severe hypothyroidism or nephrotic syndrome.

The commonest primary cause is *familial combined hyperlipidaemia* with elevated LDL and VLDL levels (p. 240).

Less commonly, increased plasma cholesterol and triglyceride may be due to accumulation of *intermediate density lipoprotein (IDL)* which contains about equal amounts of these lipids: the subjects are homozygous for an uncommon apolipoprotein-E variant. The variant does not in itself produce hyperlipidaemia but, if there is another cause for primary or secondary hyperlipidaemia the impaired recognition of this apo-E by hepatic receptors leads to accumulation of IDL (dysbetalipoproteinaemia, broad-β disease, Type III hyperlipidaemia). These patients have:

- a high incidence of vascular disease;
- tuberous xanthomata;
- lipid deposition in the palmar creases.

Rare disorders of lipid metabolism

A few very rare disorders will be briefly mentioned as they illustrate aspects of the normal pathway. They are associated with reduced plasma lipid levels and *tissue lipid accumulation despite low plasma levels.*

Several inherited *deficiencies of HDL* associated with premature coronary heart disease have been described. In *Tangier disease,* an abnormal apo-A results in excessive catabolism of HDL. Plasma levels are low and cholesterol esters accumulate in the reticuloendothelial system. Patients characteristically have large yellow tonsils and enlargement of the liver and lymph nodes.

Apoprotein B deficiency (abetalipoproteinaemia; LDL deficiency) results in impaired synthesis of chylomicrons and VLDL (and therefore of LDL). Lipids cannot be transported from the intestine or liver. The clinical syndrome consists of steatorrhoea, progressive ataxia, retinitis pigmentosa and acanthocytosis ('thorny' red cells).

Deficiency of LCAT, the enzyme necessary for esterification of free cholesterol (p. 237) results in accumulation of free, mostly unesterified, cholesterol in tissues. This produces premature atherosclerosis, corneal opacities, renal damage and haemolytic anaemia which may be due to a cell-membrane defect.

Principles of treatment of hyperlipidaemia

The decision as to whether to treat a patient with hyperlipidaemia must be based on clinical considerations as well as plasma lipid levels. Some forms of treatment carry risks which must be weighed against possible benefits.

General measures

Secondary causes of hyperlipidaemia and aggravating factors, such as obesity and excessive alcohol consumption, should be sought and treated.

The diet should be controlled whatever the cause of hyperlipidaemia. The type of diet depends on the nature of the abnormality.

Hypercholesterolaemia

It is now established that the higher the plasma LDL level, the greater the risk of ischaemic heart disease and furthermore, that lowering the plasma cholesterol concentration reduces this risk. The vigour with which hypercholesterolaemia is treated depends on the clinical circumstances. It is important to understand the nature of the lipid disorder to ensure rational therapy.

Restriction of dietary animal fats, eggs and dairy products reduces the intake of both cholesterol and saturated fatty acids. The intake of polyunsaturated fatty acids should be increased. These dietary measures are not always completely successful because endogenous cholesterol synthesis may increase.

Bile-salt sequestrants such as *cholestyramine* and *colestipol* are resins which bind bile salts derived from cholesterol, and so prevent their reabsorption and reutilization. More cholesterol is diverted to bile acid synthesis and, despite a compensatory increase in cholesterol synthesis, liver cell cholesterol is decreased. This causes an increase in the number of hepatic LDL receptors (p. 237) and plasma LDL levels fall.

The bile-salt sequestrants may cause constipation or other gastrointestinal symptoms and some patients cannot tolerate them.

Nicotinic acid may reduce VLDL secretion and therefore formation of LDL, but may cause uncomfortable side-effects such as flushing. It may be given together with bile-salt sequestrants, a combination that is the most effective treatment currently available for heterozygous familial hypercholesterolaemia.

New drugs such as *mevinolin* and *compactin*, which inhibit the synthesis of cholesterol are being tested in clinical trials, with promising results.

Hypertriglyceridaemia

Dietary restriction may be the only treatment needed for hypertriglyceridaemia:

- fat restriction may be effective in lowering plasma chylomicron levels;
- carbohydrate restriction reduces endogenous triglyceride synthesis and can be used to treat high VLDL levels.

Clofibrate is a drug which may activate lipoprotein lipase, and so increase the rate of clearance from the plasma. It is used to treat familial dysbetalipoproteinaemia when dietary measures are ineffective and may be used in cases of endogenous hypertriglyceridaemia (increased VLDL levels), if clinically indicated. It may also be used together with nicotinic acid to lower high VLDL levels in combined hyperlipidaemia.

There is an increased incidence of gall stones in patients taking clofibrate: other side-effects include muscle cramps and rarely, impotence. It also potentiates the action of warfarin. As always, the risks of treatment must be weighed against its possible benefits.

Summary

1. The main plasma lipids are cholesterol, triglycerides, phospholipids and free fatty acids. Of these, cholesterol and triglycerides are the most frequently measured.

2. Lipids are transported in plasma incorporated in lipoproteins.
   - Exogenous (dietary) lipid is carried in chylomicrons.
   - Endogenous lipid from the liver is carried in VLDL (very low density lipoprotein) which is metabolized to LDL (low density lipoprotein).
   - HDL (high density lipoprotein) is important in the removal of cholesterol from the cells.

3. Lipoproteins are modified by enzymes and their remnants are taken up by receptors on cells, mainly in the liver. The metabolism of lipoproteins is controlled by the protein components, the apolipoproteins.

4. Plasma LDL levels are regulated mainly by hepatic LDL receptor concentrations. The higher the plasma LDL, the greater the risk of ischaemic heart disease.

5. Hyperlipidaemia may be primary or secondary to other disease. The nature of the lipoprotein abnormality can usually be inferred from the plasma cholesterol and triglyceride levels. In primary hyperlipidaemia it may be necessary to define the lipoprotein abnormality more fully for the purpose of treatment.

6. Different genetic defects may produce similar lipoprotein abnormalities. Extensive family studies are required to differentiate them. This is rarely practicable, because of the difficulty of tracing family members.

7. Initially primary hyperlipidaemia should be treated by dietary control. If necessary, drugs may be added.

Further reading

Shepherd J, Packard CJ. Lipoprotein receptors and atherosclerosis. *Clin Sci Mol Med* 1986; **70**: 1–6.

Brown MS, Goldstein JL. How LDL receptors influence cholesterol and atherosclerosis. *Sci Am* 1984; **251** No 5: 52–60.

Levy RI. Primary prevention of coronary heart disease by lowering lipids: results and implications. *Am Heart J* 1985; **110**: 1116–22.

Schaeffer EJ, Levy RI. Pathogenesis and management of lipoprotein disorders. *New Engl J Med* 1985; **312**: 1300–10.

Pocock SJ, Shaper AG, Phillips AN, Walker M, Whitehead TP. High density lipoprotein cholesterol is not a major risk factor for ischaemic heart disease in British men. *Br Med J* 1986; **292**: 515–9.

Sampling blood for plasma lipid investigation

Plasma lipid levels and lipoprotein patterns are labile and affected by eating, smoking, alcohol intake, changes in posture and stress. For initial evaluation and follow-up studies it is *essential* that the sample be taken under *standard conditions*. It may, in some cases, be necessary to repeat the analysis to establish the most consistent pattern. The following points are important.

1. The patient must have fasted for 12 hours. Obviously, he should not be receiving a lipid infusion.
2. The patient should have been on his 'normal' diet and his weight should have remained constant for two weeks before the test.
3. The patient must not be on any treatment designed to lower plasma lipid levels, unless treatment is being monitored.
4. Like all large particles, lipoprotein concentrations are affected by venous stasis and posture (pp. 447 and 446). Plasma cholesterol concentration may be up to 10 per cent higher in the upright than in the recumbent position; triglyceride concentration may change slightly more than this. A standardized collection procedure is important in serial estimations to assess the effect of treatment. For example, the patient should remain seated for 30 minutes before blood is taken.
5. Investigation for, and typing of, hyperlipidaemia is preferably deferred for three months after a myocardial infarction, a major operation, or any serious illness because stress may alter lipid levels. However, samples taken within 12 hours of myocardial infarction seem to reflect 'true' values.
6. The blood sample should *not* be heparinized and plasma or serum must be separated from cells as soon as possible.

Investigation of suspected hyperlipidaemia

There should be a logical progression in the evaluation of a patient suspected of having an abnormality of plasma lipids.

1. Is there true hyperlipidaemia?

Hyperlipidaemia may be diagnosed because of the lipaemic appearance of the plasma, or by measuring cholesterol and triglyceride levels in plasma. While cholesterol levels do not vary greatly after fat intake, triglyceride levels do, so that specimens for analysis of both should be taken after the patient has fasted for 12 hours.

Lipid infusion is a common cause of grossly lipaemic plasma taken from hospital patients and no lipid should be infused for several hours before taking blood for *any* such investigation (p. 271).

2. Is the cause primary or secondary?

Many cases of hyperlipidaemia are *secondary* to other factors and may be corrected by modification of the diet or treatment of the underlying disease. *Diabetes mellitus, hypothyroidism* and the *nephrotic syndrome* should be considered in all cases of hyperlipidaemia. Abuse of *alcohol* is a common cause of secondary hypertriglyceridaemia.

3. What is the nature of the abnormality?

If no secondary cause is found, the hyperlipidaemia must be considered a primary abnormality.

Selection of appropriate treatment sometimes depends on an accurate definition of the basic lipoprotein disturbance. Valuable information may be obtained by visual inspection of the plasma. On the basis of inspection and the results of simple chemical tests we can distinguish three main groups:

A predominant increase in LDL cholesterol is associated with *clear plasma*. It may be due to:

- familial hypercholesterolaemia;
- polygenic hypercholesterolaemia;
- familial combined hyperlipidaemia.

A predominant increase in triglycerides is associated with *turbid* or even *milky plasma* due to large lipoproteins, which scatter light. It may be due to:

- familial combined hyperlipidaemia (VLDL);
- familial endogenous hypertriglyceridaemia (VLDL);
- familial hyperchylomicronaemia.

The treatment of these conditions differs and it is important to determine whether VLDL or chylomicrons are causing the turbidity.

The turbid plasma sample is left standing for 18 hours at 4°C. During this time the large, low-density chylomicrons form a creamy layer on the surface. The smaller, denser VLDL particles do not rise and the sample remains diffusely turbid. If hyperlipidaemia is gross the distribution may be less easy to distinguish.

An increase in both cholesterol and triglycerides in the same proportion may be due to:

- familial combined hyperlipidaemia (LDL and VLDL);
- familial dysbetalipoproteinaemia (IDL).

In such cases lipoprotein electrophoresis or ultracentrifugation are often necessary to distinguish the relative contributions from LDL, IDL and VLDL.

4. Family studies

These are necessary both to identify the disorder and to detect other affected, possibly asymptomatic, individuals.

The Fredrickson (WHO) classification

Fredrickson introduced a classification of hyperlipidaemia based on electrophoretic patterns. However, even the same individuals with primary hyperlipidaemia may have different patterns at different times and the same pattern may be found in different disorders. As the terms still appear in the medical literature the student may use Table 11.3 for comparison.

Table 11.3 Fredrickson classification

Fredrickson Type	Electrophoretic picture	Lipoprotein increased
I	Increased chylomicrons	Chylomicrons
II	Increased β-lipoprotein	LDL
IIb	Increased pre-β and β lipoprotein	VLDL and LDL
III	'Broad-β' band	IDL
IV	Increased pre-β lipoprotein	VLDL
V	Increased chylomicrons and pre-β lipoprotein	Chylomicrons and VLDL

12

Intestinal absorption: gastric and pancreatic function

The most important functions of the gastrointestinal tract are digestion and absorption of nutrient. A small amount of nutrient is lost into the lumen during desquamation of intestinal cells; most is reabsorbed, but, for example, almost all the fat in normal faeces is of endogenous origin. This must be remembered when interpreting faecal fat values.

Digestion and absorption can only proceed effectively if large volumes of fluid, mostly filtered from the extracellular fluid through the 'tight junctions' between epithelial cells (p. 26), dilute the food and the products of its digestion. This ultrafiltration occurs mainly in the duodenum. As in the kidney, energy is expended in reclaiming the bulk of filtered water and electrolytes. Some is reabsorbed in the proximal jejunum, along the osmotic gradient created by reabsorption of the products of digestion, but a large amount remains in the lumen to be reclaimed more distally.

A much smaller volume of fluid enters the intestinal lumen by active secretion. As a result the pH and the electrolyte and enzyme concentrations change during passage of fluid through the tract in such a way that enzyme activities are near optimal for digestion. Adjustment of pH and electrolyte concentrations occurs in the distal ileum and colon (Fig. 4.5, p. 88). In this way the extracellular fluid changes produced by intestinal secretion are corrected.

Disturbances of water, electrolyte and hydrogen ion homeostasis are common in diarrhoea due to extensive small intestinal or colonic disease; they also occur when there is loss of a large amount of fluid and electrolyte from the upper intestinal tract because of vomiting or because the function of intestinal cells is so grossly impaired that the amount of fluid and electrolyte entering the distal parts exceeds the reabsorptive capacity.

This chapter is primarily concerned with disturbance of digestion and absorption. Water, electrolyte and hydrogen ion disturbances may be marked in diarrhoea but are relatively unimportant in malabsorption syndromes unless gross malabsorption causes severe intestinal hurry.

Normal digestion and conversion of nutrient to an absorbable form

Complex molecules such as protein, polysaccharides and fat are usually broken down by digestive enzymes. This process starts in the mouth, where food is mechanically broken down by chewing and is mixed with saliva which contains *α-amylase*. In the stomach further fluid is added and the low pH initiates protein digestion by *pepsin*. The stomach also secretes intrinsic factor essential for vitamin B_{12} absorption from the terminal ileum. However, most *digestion* takes place in the duodenum and upper jejunum, where alkaline fluid is added to the now liquid food. Pancreatic enzymes in this fluid digest proteins to amino acids and small peptides, polysaccharides to monosaccharides, disaccharides and oligosaccharides (consisting of a small number of monosaccharide units), and fat to monoglycerides and fatty acids respectively.

Severe generalized malabsorption due to failure of digestion is rare, but if present is most commonly due to pancreatic insufficiency.

Normal absorption

Absorption depends on:

- the integrity and large surface area of absorptive cells;
- the presence of nutrient in an absorbable form (and therefore on normal digestion);
- a normal ratio of the rate of absorption to the rate of passage of contents through the intestinal tract.

The *absorptive area* of the small intestine is normally very large. Macroscopically the mucosa forms *folds*, increasing the area considerably. Microscopically, these folds are covered with *villi* lined with absorptive cells; this further increases the area about eight-fold.

Each intestinal absorptive cell (enterocyte) has on its surface a large number of minute projections (*microvilli*), detectable with the electron microscope, which increase the absorptive area by about a further 20-fold.

Minute spaces exist between the microvilli (*microvillous spaces*).

If the villi are flattened, as they are, for example, in gluten-sensitive enteropathy, the absorptive area is much reduced.

To be absorbed molecules must be relatively small, such as those that result from normal digestion. The method of absorption depends on whether a molecule is lipid-soluble or water-soluble. Lipid-soluble nutrients are absorbed dissolved in fat and can diffuse across the lipid membrane of the enterocyte; water-soluble nutrients are absorbed either by *active* transport against a physicochemical gradient, or by *passive* diffusion along physicochemical gradients.

Gut peptides

The release of intestinal secretions and the control of gut motility is, in part, controlled by a series of peptide hormones produced in the mucosa of the gastrointestinal tract. Like other hormones, release is under feedback control by the product, or by the physiological response of the target tissue. However, unlike most other hormones, gut peptides are synthesized by cells dispersed throughout the length of the gastrointestinal tract rather than by discrete glands, although the secretory cells are often concentrated in one segment of the tract. Many of these peptides are present in other organs, particularly the brain, where they probably act as neurotransmitters.

Some of these peptides have well-recognized physiological functions.

Gastrin is released from the G-cells in the gastric antrum in response to distension and to protein. It stimulates the contraction of the stomach muscles and secretion of gastric acid. Acid inhibits gastrin release by negative feedback.

Cholecystokinin stimulates contraction of the gall bladder and stimulates release of pancreatic digestive enzymes.

Secretin stimulates the release of pancreatic fluid rich in bicarbonate.

The physiological response of other gut peptides is less well understood. Abnormal secretion of some causes well-recognized clinical and pathological syndromes, examples of which are mentioned briefly in Chapter 22.

A more detailed account of gut peptides will be found in the references listed at the end of this chapter.

Lipid absorption

Digestion of fats

Triglycerides are the main form of dietary fat. They are esters of glycerol with three, usually different, fatty acids and are insoluble in water. Cholesterol in the intestinal lumen is derived from bile salts and from the diet; only about 30 per cent of the dietary cholesterol is absorbed.

Primary bile acids (p. 300) are synthesized in the liver from cholesterol, conjugated with the amino acids glycine or taurine, and enter the intestinal lumen in the bile. In the alkaline duodenal fluid their sodium salts act as detergents, facilitating the digestion and absorption of fats. They are actively reabsorbed in the *distal ileum* after bacterial conversion to secondary bile salts. They are recirculated to the liver and resecreted into the bile (the '*enterohepatic circulation*'). Some bacteria within the gut lumen contain enzymes which catalyse the deconjugation of bile salts: unconjugated bile salts emulsify fat less effectively than conjugated ones.

Triglycerides are emulsified by bile salts within the duodenum. They are hydrolysed by pancreatic *lipase* at the glycerol/fatty acid bond, primarily in positions 1 and 3 (Fig. 11.1, p. 231). The end products are mainly 2-monoglycerides, with some diglycerides, and free fatty acids. *Colipase*, a peptide coenzyme secreted by the pancreas, is essential for lipase activity. It anchors lipase at the fat/water interface, prevents the inhibition of lipase by bile salts and reduces the pH optimum of lipase from about 8.5 to about 6.5, the luminal pH in the upper intestinal lumen.

Micelle formation. The monoglycerides and free fatty acids aggregate with bile salts to form water-miscible *micelles*: the micelles also contain free *cholesterol* (liberated by hydrolysis of cholesterol esters in the lumen) and *phospholipids*, as well as the *fat-soluble vitamins* (A, D and K). The diameter of the negatively charged micelle is between 100 and 1000 times smaller than that of the emulsion particle. Both its small size and its charge allow it to pass through the microvillous spaces.

Lipids in the intestinal cell. In the enterocyte triglycerides are resynthesized from monoglycerides and fatty acids, and cholesterol is reesterified. The triglycerides, cholesterol esters and phospholipids, together with fat-soluble vitamins, combine with apolipoproteins manufactured in the enterocyte to form *chylomicrons* (p. 235). These are suspended in water and pass into the lymphatic circulation.

Some short- and medium-chain free fatty acids pass through the intestinal cell directly into the portal blood stream.

The absorption of *lipids and fat-soluble vitamins* depends on:

- the presence of *bile salts*;
- digestion of triglyceride by *lipase* and therefore on normal pancreatic function;
- an adequate *absorptive area of the intestinal mucosa* for chylomicron formation.

Carbohydrate absorption

Polysaccharides such as *starch* consist either of straight chains of glucose molecules joined by 1:4 linkages, called *amylose*, or of branch chains, in which the branches are joined by 1:6 linkages, called *amylopectin*. Salivary and pancreatic α-amylase hydrolyses starch to 1:4 disaccharides such as maltose (glucose + glucose). Pancreatic amylase is quantitatively the more important. A few larger branch-chain saccharides (limit dextrans) remain undigested.

Disaccharides (maltose, sucrose (glucose + fructose) and lactose (glucose + galactose)) are hydrolysed to their constituent monosaccharides by the appropriate disaccharidases (maltase, sucrase or lactase), which are located on the intestinal cell surface ('brush border'), especially in the proximal jejunum. Sucrase also hydrolyses the 1:4, and isomaltase the 1:6, linkages of limit dextrans.

Monosaccharides are absorbed in the duodenum. Glucose and galactose are probably absorbed by a common active process, while fructose is absorbed by a different mechanism.

Thus carbohydrate absorption depends on:

- the presence of *amylase, and therefore on normal pancreatic function (polysaccharides only)*;
- *the presence of disaccharidases* on the luminal membrane of the intestinal cell (disaccharides);
- normal *intestinal mucosal cells* with normal active transport mechanisms (monosaccharides).

Polysaccharides can only be absorbed if all three mechanisms are functioning.

Protein absorption

The diet is not the only source of protein in the intestinal lumen: a significant proportion is reabsorbed from intestinal secretions and desquamated mucosal cells.

Dietary protein is broken down by gastric pepsin, followed by trypsin and other proteolytic enzymes of pancreatic juice; pancreatic trypsinogen is converted to the active trypsin by enterokinase on the brush border. The products of digestion are small peptides and amino acids. Many peptides are further hydrolysed by peptidases on the brush border.

Amino acids are actively absorbed in the small intestine.

Small peptides are actively absorbed into the cell intact, independently of amino acids and, with a few exceptions, are hydrolysed intracellularly.

Protein absorption therefore depends on:

- the presence of *pancreatic proteolytic enzymes* and therefore on normal pancreatic function;
- normal *intestinal mucosa* with normal active transport mechanisms.

Vitamin B_{12} absorption

Vitamin B_{12} can be absorbed only when it has formed a complex with *intrinsic factor*, a glycoprotein secreted by the parietal cells of the stomach: this complex is resistant to proteolytic digestion and binds to specific cell receptors in the distal ileum, from where it is absorbed. Some intestinal bacteria need vitamin B_{12} for growth and may prevent its absorption by competing for it with intestinal cells. Intestinal bacterial overgrowth is associated with intestinal strictures, diverticula and 'blind loops'. Vitamin B_{12} absorption therefore depends on normal:

- *gastric secretion of intrinsic factor;*
- *intestinal mucosa in the distal ileum;*
- *intestinal flora.*

Absorption of other water-soluble vitamins

Most of the other water-soluble vitamins (C and the B group except B_{12}) are absorbed mainly in the upper small intestine, probably by specific transport mechanisms. Clinical deficiencies of all but folate are relatively uncommon in the malabsorption syndromes, probably because absorption of these vitamins does not depend on that of fat. Folate deficiency may be due to intestinal malabsorption, inadequate intake or increased utilization by intestinal bacteria.

Calcium and magnesium absorption

Calcium is actively absorbed in the upper small intestine, particularly in the duodenum and upper jejunum. Normal calcium absorption depends on its presence in the free-ionized form (it is inhibited by the formation of insoluble salts with free fatty acids and with phosphate) and on the presence of the vitamin D metabolite, 1,25-dihydroxycholecalciferol, which influences both active transport and passive diffusion (p. 176). Much of the faecal calcium is endogenous, being derived from intestinal secretions.

Magnesium is also absorbed by an active process which may be shared with calcium.

Normal calcium and magnesium absorption depends on:

- *a low concentration of fatty acids* and of *phosphate* in the intestinal lumen;
- the absorption and metabolism of *vitamin D* and therefore on normal fat absorption;
- normal *intestinal mucosal cells.*

Iron absorption

Iron is absorbed in the duodenum and upper jejunum in the ferrous form (Fe^{2+}). Absorption is increased by anaemia of any kind (p. 390). Some of the iron absorbed into the intestinal cell enters the plasma but some is lost into the lumen with desquamated cells.

Gastric function

The important components of gastric secretion are *hydrochloric acid*, *pepsin* and *intrinsic factor.* All these components are of importance in digestion and absorption. Loss of hydrochloric acid by vomiting in pyloric stenosis is a cause of metabolic alkalosis (p. 97).

Stimulation of gastric secretion occurs by two main pathways:

- through the *vagus nerve* which, in turn, responds to stimuli from the cerebral cortex, normally resulting from the sight, smell and taste of food. Hypoglycaemia can stimulate this pathway and can be used to assess the completeness of vagotomy;
- by *gastrin*, which is carried by the blood stream to the parietal cells of the stomach where it stimulates gastric acid secretion, the action being mediated by histamine. Acid in the pylorus, in turn, inhibits gastrin secretion by negative feedback control. Calcium also stimulates gastrin secretion; this may explain the relatively high incidence of peptic ulceration in patients with chronic hypercalcaemia.

Histamine acts after binding to receptors on the surface of the cells. There are probably two types of receptor:

- those for which antihistamines compete with histamine (H_1 receptors). These are found on smooth muscle cells;

- those on which antihistamines have no effect. These H_2 *receptors* are found on *gastric parietal* cells.

Hypersecretion

Hypersecretion of gastric juice may be associated with *duodenal ulceration*. However, there is overlap between the amount of acid secreted in normal subjects and in those with duodenal ulceration, and gastric acid estimation is of very limited diagnostic value for this condition.

In the rare *Zollinger-Ellison syndrome* (p. 417) acid secretion is very high, due to excessive gastrin production, usually by a pancreatic tumour. The consequent ulceration of the stomach and upper small intestine may cause severe diarrhoea; the low pH inhibits lipase activity and may cause steatorrhoea.

Hyposecretion

Hyposecretion of gastric juice occurs most commonly in *pernicious anaemia*, when it is the result of the formation of antibodies to the parietal cells of the gastric mucosa (p. 283). Extensive *carcinoma of the stomach* and *chronic gastritis* may also cause gastric hyposecretion. However, *estimation of gastric acidity is of no diagnostic value in these conditions*: the diagnosis of pernicious anaemia should be made using haematological investigations, including the Schilling test (p. 261). Results of gastric acid determination in the other two conditions are too variable to be useful.

Tests of gastric function

Tests of gastric function involving measurement of acid secretion have largely been superseded by endoscopy of the stomach and duodenum and by histological examination of the biopsy material obtained.

Stimulation of gastric secretion

Vagal stimulation of gastric secretion and measurement of gastric acidity may be used to test the *completeness of section of the vagus* nerve in patients who remain symptomatic after surgery for duodenal ulceration. The stimulus is stress due to insulin-induced hypoglycaemia (compare the use of insulin to stimulate cortisol secretion). If vagotomy is complete symptomatic hypoglycaemia with a plasma glucose level below 2.5 mmol/litre (45 mg/dl) should produce no increase in acid secretion.

Treatment of hypersecretion

Palliative treatment with antacids and carbenoxolone, or surgery, have largely been replaced by treatment with the drugs *cimetidine* (Tagamet) and *ranitidine*

(Zantac). These drugs compete with histamine for the H_2 receptors, so directly reducing acid secretion. The initial clinical response is usually good, but relapse may occur when treatment is stopped.

The post-gastrectomy syndromes

There may be some degree of malabsorption after gastrectomy (p. 258). Rapid passage of the contents of the small gastric remnant into the duodenum may have two other clinical consequences.

- The '*early dumping syndrome*'. Soon after a meal the patient may experience abdominal discomfort and feel faint and nauseated; these symptoms are thought to be caused by the rapid passage of hypertonic fluid into the duodenum. Before this abnormally large load can be absorbed, water passes along the osmotic gradient from the extracellular space into the intestinal lumen. The reduction in plasma volume causes faintness and the large volume of fluid causes abdominal discomfort.
- *Post-gastrectomy hypoglycaemia (the 'late dumping syndrome').* If a meal containing a high glucose concentration passes rapidly into the duodenum glucose absorption is very rapid. The plasma glucose level rises quickly and causes an outpouring of insulin. The resultant 'overswing' of plasma glucose concentration may cause hypoglycaemic symptoms, typically occurring at about two hours after a meal. This is a form of reactive hypoglycaemia (p. 219).

Both these disabilities can be mitigated if meals low in carbohydrate concentration are taken 'little and often'.

Pancreatic function

Pancreatic secretions can be divided into endocrine and exocrine components. The endocrine secretions control the plasma glucose concentration and are discussed in Chapter 10.

The exocrine secretions are made up of two components, the alkaline pancreatic fluid and the enzymes. The alkaline juice is primarily responsible for neutralizing gastric acid secretions, thus providing an optimum environment for duodenal digestive enzyme activity. The enzymes include proteases (*trypsin and chymotrypsin*), *amylase* and *lipase*. Some of the proteases are secreted as precursors and are converted to the active form within the intestinal lumen.

Pancreatic secretions are controlled by gut peptides, released from the duodenum in response to a fall in pH or the presence of food. *Secretin* stimulates the release of a high volume of alkaline fluid, while *cholecystokinin* stimulates the release of a fluid rich in enzymes.

Tests of exocrine pancreatic function

Plasma enzymes. Plasma *trypsin* can be measured by immunoassay and the finding of *low* levels in adults is occasionally diagnostic of pancreatic insufficiency. Although pancreatic function may initially be normal in the neonate with fibrocystic disease, plasma trypsin levels may be high because blockage of pancreatic ductules by sticky mucus secretion causes trypsin to be regurgitated into the plasma.

In normal subjects plasma *amylase* consists principally of the salivary isoenzyme; the total enzyme activity is therefore not significantly lowered when the secretory cell mass is reduced by chronic pancreatic disease, and measurement of plasma amylase is of limited value in the diagnosis of chronic conditions. However, in acute pancreatitis *total* plasma amylase activity is usually significantly increased by release of this enzyme from damaged cells.

Faecal enzymes. Faecal *trypsin* levels are extremely variable, probably because of bacterial degradation in the intestinal lumen. Their estimation is of no value in the diagnosis of pancreatic hypofunction in adults. In infants with diarrhoea the absence of the enzyme suggests fibrocystic disease of the pancreas, although assay of plasma levels is more reliable.

Duodenal enzymes. Most pancreatic function tests are best performed in special centres. Measurement of pancreatic enzymes, and of the bicarbonate concentration, in duodenal aspirates before and after stimulation of the pancreas with secretin or cholecystokinin is not very suitable for routine use because of the difficulty in positioning the duodenal tube and in quantitative sampling of the secretions.

'Tubeless tests' which avoid the need for intubation, and which overcome the difficulties of sample collection, have been developed. A synthetic *peptide labelled with p-aminobenzoic acid (PABA)* is taken orally and the product of its digestion by chymotrypsin, PABA, is absorbed and excreted in the urine. Urinary excretion of PABA is significantly reduced in chronic pancreatitis. However, abnormal results may also occur if there is renal impairment, liver disease, or malabsorption even with normal pancreatic function. The effect of these conditions is assessed by repeating the test with unconjugated PABA, which eliminates the need for digestion before absorption. A similar, but technically simpler, test uses fluorescein dilaurate (the *pancreolauryl test*).

Disorders of the pancreas rarely associated with malabsorption

Absence of pancreatic secretions may, by impairing digestion, cause malabsorption. However, there are three relatively common pancreatic diseases which are only rarely associated with malabsorption.

- In *acute pancreatitis* necrosis of pancreatic cells results in release of enzymes into the retroperitoneal space and blood stream. The presence of pancreatic juice in the peritoneal cavity causes *severe abdominal pain* and *shock*: this picture is common to many acute abdominal emergencies. A vicious cycle is set up as more pancreatic cells are digested by the released enzymes. Acute pancreatitis is most commonly the result of obstruction of the pancreatic duct, or of regurgitation of bile along this duct. The most important predisposing factors are *alcoholism* and *biliary tract disease. Trauma* to the pancreas, by damaging the cells, may initiate the vicious cycle. *Hypercalcaemia* and *hypertriglyceridaemia* may also precipitate acute pancreatitis.

Typically plasma amylase activity increases five-fold or more in acute pancreatitis. However, it is important to realise that enzyme activities of up to, and even above, this value may be reached in any acute abdominal emergency, but especially after gastric perforation into the lesser sac. Excretion of normal amylase may be impaired in renal glomerular failure: in macroamylasaemia excretion of the high molecular weight enzyme through normal glomeruli is impaired (p. 314); these latter two conditions are usually asymptomatic. Conversely, levels in acute pancreatitis may not be very high and usually fall very fast as the enzyme is lost in the urine. *A high plasma amylase activity is therefore only a rough guide to the presence of acute pancreatitis, and normal or only slightly raised values do not exclude the diagnosis.*

- *Chronic pancreatic failure* is not always associated with steatorrhoea. It may rarely follow a severe attack, but is more likely after repeated acute or subacute attacks.
- *Carcinoma of the pancreas* is very difficult to diagnose by laboratory tests unless the lesion is in the head of the organ and causes obstructive jaundice; extensive gland destruction may cause late-onset diabetes mellitus.

The malabsorption syndromes

Generalized malabsorption associated with steatorrhoea

Generalized malabsorption may result from either intestinal or pancreatic disease.

- In *intestinal* malabsorption fat *digestion is normal*, but *absorption* of the products of digestion is *impaired.*
- *In the rarer pancreatic* steatorrhoea the absorptive capacity is normal but fat *cannot be digested* because there is deficiency of digestive enzymes.

In either case absorption of fat is impaired. However, unequivocal steatorrhoea (a daily fat excretion consistently more than 18 mmol (5 g)) can only be demonstrated when disease is extensive and has usually already been elucidated by radiological or histopathological means.

Reduction of absorptive area or generalized impairment of transport mechanisms

Villous atrophy. Malabsorption may be due to a reduction of the absorptive area because the intestinal villi are flattened. A definitive diagnosis may be made by demonstrating this flattening by microscopic examination of a mucosal biopsy specimen.

- *Gluten-sensitive enteropathy* (coeliac disease) may occur at any age. Sensitivity to gluten and the related compound gliadin, which are found mainly in wheat, causes villous atrophy. Blunting and flattening of the villi is usually reversible following treatment with a gluten-free diet, but the response may be delayed for several months.
- *Tropical sprue* is also associated with villous atrophy, but it does not respond to a gluten-free diet. There may be a bacterial cause for this disorder since it responds to broad-spectrum antibiotics. Folate should also be given.
- The term *idiopathic steatorrhoea* should be reserved for those cases of steatorrhoea with flattened villi, which respond neither to a gluten-free diet nor to broad-spectrum antibiotics.

Extensive surgical resection of the small intestine may so reduce the absorptive area as to cause malabsorption.

Extensive infiltration or inflammation of the small intestinal wall (for example due to Crohn's or Whipple's diseases, or to small intestinal lymphoma) may impair absorption and this may be aggravated by an alteration in the bacterial flora due to reduced intestinal motility.

Increased rate of transit through the small intestine

Food passes through the intestine more rapidly than usual:

- after *gastrectomy*, when normal mixing of food with fluid, acid and pepsin does not occur in the stomach, resulting in impaired enzyme activity within the small intestine. An increased rate of transit through the duodenum may contribute to the malabsorption. However post-gastrectomy malabsorption is rarely severe, although iron deficiency is relatively common;
- *in the carcinoid syndrome* (p. 415), which is associated with excessive production of 5-hydroxytryptamine (5-HT; serotonin) by tumours of argentaffin cells usually arising in the small intestine, or by their metastases, usually in the liver. This is a rare disease which may occasionally present a problem of differential diagnosis of malabsorption. The malabsorption is probably the result of increased intestinal motility induced by 5-HT.

Impaired digestive enzyme activity

Failure of digestive enzyme secretion. *Pancreatic dysfunction* may cause malabsorption by reducing secretion of digestive enzymes.

- *Chronic pancreatitis* may present insidiously. It is most common in alcoholics. Glucose intolerance may develop due to reduced islet cell mass.
- *Fibrocystic disease* (cystic fibrosis) usually presents in early childhood, but may not be diagnosed until the patient is adult. It is transmitted as an autosomal recessive disorder. Thick, viscous pancreatic and bronchial secretions may cause malabsorption and chronic lung disease respectively. Sweat glands are also affected; in children the diagnosis depends on demonstrating a *sweat sodium or chloride concentration of about twice normal.*

If pancreatic disease causes obstructive jaundice (usually carcinoma of the head of the pancreas) it may cause mild steatorrhoea even if enzyme secretion is adequate. This is due to impairment of lipase activity by bile salt deficiency (p. 250).

Inactivation of pancreatic enzymes by acid. The increased intestinal activity in the *Zollinger–Ellison syndrome* (p. 417) inactivates pancreatic lipase and causes precipitation of bile salts within the gut lumen, so resulting in malabsorption.

Differential diagnosis of generalized intestinal and pancreatic malabsorption

The diagnosis of steatorrhoea is usually made by clinical, haematological, histological and radiological means, which may be supplemented by biochemical tests. Malabsorption of fat occurs both in generalized intestinal and in pancreatic disease.

Anaemia is more common in *intestinal* than in pancreatic malabsorption since iron, vitamin B_{12} and folate do not depend on pancreatic enzymes for absorption. In intestinal malabsorption the blood and bone-marrow films typically show a mixed iron deficiency and megaloblastic picture. Malabsorption of iron, vitamin B_{12} and folate may be demonstrable and the absorption of vitamin B_{12} is not increased by the simultaneous administration of intrinsic factor as part of the Schilling test, as it is in pernicious anaemia (p. 261). Anaemia may be aggravated by protein deficiency.

Differences in carbohydrate metabolism. Polysaccharide absorption is impaired in both conditions but hypoglycaemia is rare. In pancreatic malabsorption, in which neither disaccharidase activity nor active monosaccharide absorption is affected, both monosaccharides and disaccharides can be absorbed. Hyperglycaemia suggests that pancreatic disease is the cause of malabsorption, but *it is by no means diagnostic of it.*

Xylose absorption test. Xylose is a pentose which, like glucose, can be absorbed without digestion, but the metabolism of which is not controlled by hormone secretion. For this reason once xylose has been absorbed the plasma levels are, unlike those of glucose, not affected by such endocrine disease as diabetes mellitus.

An oral dose of xylose is absorbed by normally functioning upper small intestinal cells. Metabolism of the absorbed pentose is very slow, and because it is freely filtered by the glomeruli, most of it appears in the urine. In the xylose absorption test either the amount excreted in the urine during a fixed period, or in children the plasma level at a defined time after the dose, is measured. Intestinal disease should impair xylose absorption, and therefore reduce excretion, while pancreatic disease should not affect either. Unfortunately, whether the test is based on plasma or on urinary assays, there are several sources of imprecision (p. 269).

- Xylose is mostly absorbed in the upper small intestine and absorption may therefore be normal despite significant ileal dysfunction (for example in Crohn's disease).

• The urine test depends on accurate collections over two and five hours. Potential errors are even greater than those in tests depending on 24-hour collections, especially in children.

• Even mildly impaired renal function decreases glomerular filtration of xylose and may give falsely low results in the urine test, or a falsely high plasma level. This is a common cause of misleading results in the elderly. The xylose absorption test should not be performed if the plasma urea and/or creatinine levels are even only marginally raised.

• Both plasma levels and urine excretion depend partly on the volume of distribution of the absorbed xylose. In oedematous or very obese patients results may be falsely low.

Failure of absorption of specific substances

An altered bacterial flora causes malabsorption of *vitamin* B_{12} and *fat*.

• *The contaminated bowel ('blind loop') syndrome* is associated with bacterial proliferation due to impaired intestinal motility and stagnation of intestinal contents, and occurs when the 'blind loops' are the result of *surgery* or in the presence of small intestinal *diverticula*. The bacteria may interfere with bile salt metabolism and cause malabsorption of fat and, by utilizing vitamin B_{12} and folate, reduce their absorption. *Megaloblastic anaemia is common.*

• *Treatment with broad-spectrum antibiotics* may, by altering the small intestinal flora, cause a similar syndrome. Abnormal bacterial growth may sometimes be diagnosed by culture of organisms from a specimen collected into a special, anaerobic microbiological medium after intubation of the upper small intestine. *The test is only useful if intubation is performed by an expert.* The *^{14}C-glycocholate breath* test is a relatively simple tubeless investigation for abnormal bacterial growth. ^{14}C-glycocholate is taken orally. The intestinal bacteria split the cholate from the labelled glycine: the latter is absorbed and metabolized, and the $^{14}CO_2$ is measured in the expired air. However, false positive results may be found in patients with disease of the terminal ileum and the test may be normal in some patients in whom the bacteria do not deconjugate bile salts.

Biliary obstruction causes malabsorption of *fat* and of *fat-soluble substances* by preventing the secretion of bile salts into the intestinal lumen; *steatorrhoea* results. The diagnosis is usually clinically obvious due to the presence of jaundice. Retention of bile salts, such as may occur in primary biliary cirrhosis, may cause intense skin irritation.

Local diseases or surgery of the small intestine may cause selective malabsorption of substances absorbed predominantly at those sites. *Diseases involving the terminal ileum*, such as Crohn's disease or tuberculosis, may impair absorption of *bile salts* and *vitamin* B_{12}.

Vitamin B_{12} absorption may also be impaired because of intrinsic factor deficiency due to damage to gastric parietal cells in *pernicious anaemia*, or to *total gastrectomy* or extensive *malignant infiltration of the stomach.*

The Schilling test. Malabsorption of vitamin B_{12} can be demonstrated if a small dose of radiolabelled vitamin is given orally and its excretion in the urine measured. If malabsorption of the vitamin is due to pernicious anaemia administration of the labelled vitamin with intrinsic factor restores normal absorption: if it is due to intestinal disease malabsorption persists. A 'flushing' dose of non-radiolabelled vitamin is given parenterally at the same time as, or just after, the labelled dose to ensure quantitative urinary excretion. Haematological tests such as examination of blood and bone marrow films should have been completed before the vitamin B_{12} is given.

Disaccharidase deficiency may occur in generalized disease of the intestinal wall because disaccharides are localized on the brush border of the enterocyte. This deficiency is relatively unimportant compared with the general malabsorption, and tests for these syndromes are therefore useful only in the absence of steatorrhoea, when selective rather than generalized malabsorption of carbohydrate may be present.

The *symptoms* of disaccharidase deficiency are those of the effects of unabsorbed, osmotically active sugars in the intestinal lumen. They include faintness, abdominal discomfort and severe diarrhoea after ingestion of the offending disaccharide. (Compare the 'dumping syndrome', p. 255). Diarrhoea may be severe enough to cause volume depletion in infants. Stools are typically acid because bacteria metabolize sugars to acids; unabsorbed sugars may be detectable in the faeces, or disaccharides may be absorbed intact and be detectable in the urine.

Lactase deficiency

- *Acquired lactase deficiency* is much commoner than the congenital form and is probably the commonest type of disaccharidase deficiency; it may not present until childhood, or even adult life, possibly in genetically predisposed subjects. It is common in non-Caucasians, especially the Chinese; this may be due to genetic factors, or may be secondary to a low dietary intake of dairy products with consequent failure to induce synthesis of the enzyme after weaning.
- *Lactase deficiency associated with prematurity.* In premature infants lactase may not be present in normal amounts in intestinal cells. Initially these cases resemble the congenital ones. However, the sensitivity to milk usually disappears within a few days of birth.
- *Congenital lactase deficiency* is very rare. Infants present with severe diarrhoea, or colicky abdominal pain, soon after the introduction of milk feeds. The syndrome is treated by removing milk and milk products from the diet.

Sucrase and isomaltase deficiency usually coexist.

- *Congenital sucrase-isomaltase deficiency* is commoner than congenital lactase deficiency.
- *Acquired sucrase-isomaltase deficiency* and *maltase deficiency* of any kind are very rare.

The *diagnosis* of disaccharidase deficiency is made most reliably by estimating the relevant enzymes in intestinal biopsy tissue.

The relevant disaccharide may be given orally and the plasma glucose estimated as in the glucose tolerance test. If the disaccharide cannot be hydrolysed the constituent monosaccharides cannot be absorbed and the curve is flat. The result should usually be compared with that of a glucose tolerance test: if this, too, is flat the abnormal response to disaccharides is not diagnostic. However, if the patient experiences typical symptoms, or if disaccharides are excreted in the urine when the offending sugar is given, the comparison need not be made.

Radiological examination may assist in making the diagnosis of disaccharidase deficiency. Barium is administered first without, and then with, the disaccharide. The osmotic effect of the unabsorbed sugar causes flocculation of the barium.

Protein-losing enteropathy. In many cases of malabsorption protein is lost through the gut due to failure to digest and absorb that entering it in intestinal secretions and by cell desquamation. Isolated protein-losing enteropathy is a very rare syndrome in which absorption is normal, but the intestinal wall is abnormally permeable to large molecules, like the glomerulus in the nephrotic syndrome. It occurs in a variety of conditions in which there is ulceration of the bowel or abnormal lymphatic drainage. The clinical picture, like that of the nephrotic syndrome (p. 342), is due to hypoalbuminaemia.

If the syndrome is suspected, and if there is no evidence of generalized malabsorption, the diagnosis can be made by measuring the loss into the bowel after intravenous injection of a substance of a molecular weight approximating that of albumin; radiolabelled polyvinylpyrrolidone (PVP), dextran or albumin have been used. Typically the serum protein electrophoretic pattern is similar to that found in the nephrotic syndrome (p. 327); the relatively high molecular weight components of the α_2-fraction are retained and all other fractions are reduced.

Metabolic consequences of malabsorption

Impaired fat absorption with steatorrhoea is accompanied by malabsorption of the fat-soluble vitamins (A, D and K). *Vitamin D deficiency* results in impaired *calcium* absorption. The low free-ionized calcium concentration stimulates parathyroid hormone secretion which, by causing phosphaturia, leads to hypophosphataemia (p. 174), and sometimes to osteomalacia and rickets.

Vitamin K is needed for hepatic synthesis of prothrombin and other clotting factors. A *bleeding tendency*, associated with a prolonged prothrombin time, may develop in patients with severe malabsorption. This prothrombin deficiency, unlike that due to liver disease, can be reversed by parenteral administration of vitamin K.

Vitamin A deficiency is rarely clinically evident, although malabsorption of an oral dose of vitamin A can sometimes be demonstrated.

Amino acid and *peptide* malabsorption occur in intestinal disease due to impaired intestinal transport mechanisms: in pancreatic disease the digestion of protein is severely impaired. In either case prolonged disease may cause generalized *muscle and tissue wasting* and *osteoporosis* (p. 186), and a reduced level of all protein fractions in the plasma. The low plasma albumin concentration may cause *oedema* (p. 46) and a reduction of plasma *total calcium* concentration. Reduced *antibody formation* (immunoglobulins) may predispose to infection.

Thus in *severe and long-standing* generalized malabsorption, whether intestinal or pancreatic, the *clinical picture* may include:

- bulky, fatty stools, with or without diarrhoea;
- generalized wasting and malnutrition (protein deficiency);
- osteoporosis and osteomalacia (protein and calcium deficiency);
- oedema (hypoalbuminaemia due to protein deficiency);
- tetany (hypocalcaemia due to vitamin D and calcium deficiency);
- recurrent infections (immunoglobulin deficiency).

The *laboratory findings* may be:

- increased excretion of fat in the stools;
- anaemia (iron and/or folate and/or vitamin B_{12} deficiency);
- hypocalcaemia (protein-bound and free-ionized) with hypophosphataemia;
- raised plasma alkaline phosphatase activity in cases with osteomalacia and rickets;
- low concentrations of all except the α_2 serum protein fractions;
- prolonged prothrombin time (vitamin K deficiency).

The complete picture is only seen in advanced cases of the syndrome. Milder cases present with vague abdominal symptoms, and perhaps with anaemia, but with normal chemical findings.

Laboratory diagnosis of steatorrhoea

Steatorrhoea may be due to generalized or localized intestinal disease, or to pancreatic disease. The important laboratory findings in the differential diagnosis of these conditions are summarized in Table 12.1.

Steatorrhoea is defined as a daily faecal fat excretion consistently more than 18 mmol (5 g) of 'fat', measured as fatty acids. Whatever the aetiology, unequivocal

Table 12.1 Differential diagnosis of steatorrhoea

	Upper small intestinal disease	Pancreatic disease	'Contaminated bowel' syndrome
Anaemia	Mixed megaloblastic and iron deficiency common ('dimorphic anaemia')	Rare	Megaloblastic common
Intestinal biopsy	May show flattened villi or other cause	Normal	Normal
Xylose absorption	Reduced	Normal	Usually normal

steatorrhoea can only be demonstrated when the disease is extensive and has usually already been elucidated by radiological or histopathological means.

Gross steatorrhoea causes foul smelling, bulky, pale, greasy stools which are difficult to flush away in the toilet. There may be diarrhoea.

Faecal fat assay is a relatively insensitive test for small intestinal or pancreatic disease. For example, pancreatic lipase activity can be reduced by over 75 per cent before chemical steatorrhoea occurs. The difficulty of obtaining accurately timed faecal collections often make the assay imprecise (p. 268).

Absorbed fat is ultimately metabolized to carbon dioxide and water. If ^{14}C-labelled triglyceride (triolein) is given by mouth the ratio of $^{14}CO_2$ to unlabelled CO_2 can be measured in expired air. A low ratio usually indicates impaired fat absorption. The breath test eliminates the imprecision of a timed faecal collection. However, various assumptions have to be made when interpreting results, and the assay uses expensive isotopes and needs special expertise. The ^{14}C-triolein breath test does correlate with faecal fat measurements, but the limitations discussed have prevented its adoption as a routine test.

No biochemical test is sensitive enough for the early detection, or precise enough for the differential diagnosis, of generalized malabsorption or of steatorrhoea. Developments such as endoscopy and ultrasound have reduced the need for laboratory tests. Biochemical tests do, however, continue to be important to detect and monitor treatment of those metabolic *effects* of prolonged and severe malabsorption described on p. 263.

A proposed scheme for the investigation of malabsorption is outlined on p. 267.

Summary

Normal intestinal absorption

1. Normal intestinal absorption depends on adequate digestion of food and therefore on normal pancreatic function, and on a normal area of functioning intestinal cells.
2. Normal digestion and absorption of fat depends on the presence of bile salts as well as of lipase.
3. Absorption of cholesterol, phospholipids and fat-soluble vitamins depends on normal triglyceride absorption.

Gastric function

1. Hypersecretion of acid occurs in:

- duodenal ulceration;
- the Zollinger–Ellison syndrome.

2. Hyposecretion of acid occurs in:

- pernicious anaemia;
- extensive gastric infiltration.

3. The vagal stimulation of gastric secretion can be tested by inducing hypoglycaemia with insulin.

Pancreatic function

1. Chemical tests for pancreatic hypofunction are only satisfactory if performed in special centres.

2. Acute pancreatitis is associated with a short-lived rise in plasma amylase activity. This enzyme can also reach a high concentration in many acute abdominal emergencies and in renal glomerular failure. The assay of total amylase activity is useless for detecting chronic pancreatic disease.

The malabsorption syndromes

1. In *intestinal* malabsorption there is malabsorption of small molecules, usually due to a reduced absorptive area.

2. In *pancreatic* malabsorption there is malabsorption of fats, proteins and polysaccharides, but small molecules are usually absorbed normally.

3. An *abnormal bacterial flora* may cause steatorrhoea because of deconjugation of bile salts, and megaloblastic anaemia because of bacterial competition for vitamin B_{12} and folate.

4. There may be steatorrhoea in biliary obstruction, because of lack of bile salts in the intestinal lumen.

5. Selective malabsorption of vitamin B_{12} occurs in pernicious anaemia due to deficiency of intrinsic factor.

6. Selective disaccharidase deficiency causes malabsorption of disaccharides. These deficiencies are more commonly acquired than congenital in origin.

Further reading

Read NW, Corbett CL. The function of the human gastro-intestinal tract and its laboratory assessment. In: Williams DL, Marks V eds. *Biochemistry in Clinical Practice*. London: Heinemann, 1983: 67–95.

Glickman RM. Fat absorption and malabsorption. *Clin Gastroenterol* 1983; **12**: 323–34.

Ravich WJ, Bayless TM. Carbohydrate absorption and malabsorption. In: Sleisenger MH ed. *Clin Gastroenterol* 1983; **12**: 335–56.

Freeman HJ, Sleisenger MH, Kim YS. Human protein digestion and absorption. Normal mechanisms and protein energy malnutrition. In: Sleisenger MH ed. *Clin Gastroenterol* 1983; **12**: 357–78.

Laker MF, Bartlett K. Tubeless tests of small intestinal function. In: Price CP, Alberti KGMM eds. *Recent Advances in Clinical Biochemistry*. London: Churchill Livingstone, 1985; **3**: 195–219.

Scharpe S, Iliano I. Two indirect tests of exocrine pancreatic function evaluated. *Clin Chem* 1987; **33**: 5–12.
Theodossi A, Gazzard BG. Have chemical tests a role in diagnosing malabsorption? *Ann Clin Biochem* 1984; **21**: 153–65.

Investigation of suspected malabsorption

This proposed scheme for the investigation of suspected malabsorption will discuss the use of laboratory tests only. It must be stressed that such tests are relatively insensitive and often unsatisfactory.

Malabsorption may be suspected if the patient presents with one or both of:

- a history of chronic diarrhoea of unknown origin;
- clinical, radiological, haematological and/or biochemical findings suggestive of malnutrition with no obvious cause.

1. Has the patient a *history* which may suggest alteration in the intestinal bacterial flora?

- Has he been on long-term broad-spectrum antibiotics?
- Has he had intestinal surgery which may have resulted in 'blind loops'?
- Has he travelled in the tropics? If so, consider the possibility of tropical sprue.

2. What is the appearance of the *stool*?

- *bulky, pale, greasy stools* suggest steatorrhoea;
- *constipation* with hard dry stools makes the diagnosis of malabsorption highly improbable;
- *watery stools*, especially if bloodstained, suggest colonic disease such as ulcerative colitis. The possibility of purgative abuse must always be considered.

3. *Plasma electrolyte* abnormalities, especially severe hypokalaemia, and signs of extracellular volume depletion without obvious evidence of malnutrition, favour colonic disease as a cause of diarrhoea rather than a small intestinal malabsorption syndrome.

4. *Anaemia* is more likely to be due to intestinal than to pancreatic disease:

- a hypochromic, microcytic picture may be the result of blood loss at any level of the gastrointestinal tract;
- a normochromic, normocytic picture is a non-specific finding in any chronic disease;
- *a dimorphic (mixed iron deficiency and macrocytic) picture is very suggestive of intestinal malabsorption*;
- low red cell folate and/or plasma vitamin B_{12} levels suggest intestinal malabsorption.

5. If any or all of the above investigations suggest steatorrhoea a cause should be sought. The most definitive tests are:

- endoscopy, with histological examination of an intestinal biopsy specimen.
  Flattened villi suggest:
    gluten-sensitive enteropathy;
    idiopathic steatorrhoea;
    tropical sprue.
- radiological examination, which may detect infiltration of the mucosa.

6. If doubt remains faecal fat estimation *may* help. However, because of the difficulty of obtaining accurately timed faecal collection, only very high results are of unequivocal significance; in such cases steatorrhoea will probably be obvious visually.

7. If steatorrhoea is obvious, and if the intestinal histological picture is normal, the most important possibilities are:

- localized intestinal disease, such as Crohn's disease;
- the 'contaminated small bowel' ('blind loop') syndrome;
- pancreatic disease. This is a rare cause of malabsorption unless there is a history of recurrent attacks of acute pancreatitis.

Tests such as abdominal ultrasound or CT scanning may help, especially if pancreatic carcinoma is suspected.

8. The xylose absorption test is rarely helpful. If it is *unequivocally* normal an upper intestinal lesion is unlikely. Such a result does not, however, exclude ileal or pancreatic disease, the 'contaminated bowel' syndrome, or even some cases of gluten-sensitive enteropathy.

9. If malabsorption is proven, and if there is laboratory evidence of malnutrition, blood haemoglobin and plasma calcium, phosphate and albumin concentrations and alkaline phosphatase activity should be monitored to assess the efficacy of treatment.

Faecal fat estimation

Faecal fat excretion should represent the difference between the fat absorbed and that entering the gastrointestinal tract from the diet and from the body (p. 248). Absorption of fat, and the addition of fat to the intestinal contents, occurs throughout the small intestine. A single 24-hour collection of faeces gives very inaccurate results for two reasons.

- the transit time from the duodenum to the rectum is variable;
- rectal emptying is variable and may not be complete.

Consecutive 24-hour collections give faecal fat results which may vary by several hundred per cent. The longer the period of collection the more accurate is the calculated daily mean result; it is desirable to collect at least a five-day specimen of stool. The precision may be increased by collecting between 'markers', usually dyes or radiopaque pellets taken orally: the dye can be detected by visual inspection and the pellets by X-raying the stools.

Procedure
This protocol is for a five-day collection using carmine markers.

Day 0. The first 'marker' (usually two capsules of carmine) is given. As soon as this marker appears in the stool the collection is started. This marker will gradually disappear in subsequent stools.

Day 5 after the first marker. The second marker is given. As soon as this marker appears in the stool the collection is stopped.

Fat is estimated in all the specimens passed between the appearance of the two markers, including only one of the 'marked' stools.

Note 1. It is very important that *all* stools should be collected. To ensure that none is missing it is best to label each specimen with the following information, and a record of this information should be kept on the ward:

Name of patient
Ward
Date of specimen
Time of specimen
Number in the series.

The patient must use only bedpans during the test period, and the importance of obtaining a complete collection should be explained.

2. The time needed for collection of specimens will be about a week (including the time for the appearance of the markers). Before starting the test make sure that during this time the patient is *not to be discharged, is not going to be operated on* (except in an emergency) and *does not receive enemas or aperients*: any patient needing aperients is most unlikely to have steatorrhoea at that time. *Neither a barium enema nor a barium meal* should be performed for a few days before or during this time because barium interferes with the estimation.

The collection and estimation of faecal fat excretion is time-consuming and unpleasant for all concerned, including the patient. It is important that specimen collection is carefully controlled so that the result may be meaningful.

Interpretation

A *mean* daily fat excretion of clearly more than 18 mmol (5 g) indicates steatorrhoea. The result is not affected by diet within very wide limits, because in the normal subject almost all the fat is of endogenous origin. However, if the patient is on a *very* low fat diet mild steatorrhoea may not be detected.

Xylose absorption test

The result of the test is invalidated if there is poor renal function. Oedema also invalidates the results because the volume through which the xylose is distributed may be significantly increased (p. 260).

The test is imprecise because of the problems of urine collection (p. 260).

A dose of 5 g of xylose is preferable to one of 25 g because a high xylose concentration within the intestinal lumen may, like a large dose of glucose (p. 226), have an osmotic effect which causes symptoms and affects the results.

Procedure

The patient fasts overnight.

08.00 h. The bladder is emptied and *the specimen discarded.* 5 g of xylose dissolved in 200 ml of water is given orally.

All specimens passed between 08.00 h and 10.00 h are put into a bottle labelled 'Number 1'.

10.00 h. The bladder is emptied and *the specimen put into Bottle 1,* which is now complete.

All specimens passed between 10.00 h and 13.00 h are put into a bottle labelled 'Number 2'.

13.00 h. The bladder is emptied and *the specimen put into Bottle 2,* which is now complete.

Both bottles are sent to the laboratory for analysis.

Interpretation

In the normal subject more than 23 per cent of the dose (1.15 g) should be excreted during the five hours. Fifty per cent or more of the total excretion should occur during the first two hours. In mild intestinal malabsorption the total five-hour excretion may be normal, but delayed absorption is reflected in a two- to five-hour ratio of less than 40 per cent. In pancreatic malabsorption the result should be normal (p. 259). The limitations of this test are discussed on p. 259.

Principles of intravenous feeding

The principles discussed in Chapters 10, 11 and 12 have an important application in the management of nutrition and of intravenous nutrition in particular.

The metabolism of carbohydrate, fat and, to a lesser extent, the carbon chains of some amino acids, supplies the energy needs of the body by coupling the breakdown of their energy-rich bonds with ATP synthesis. Some of the constituent elements, carbon, hydrogen and oxygen, leave the body as carbon dioxide and water. In a normal adult the daily energy loss as heat is about 120 kJ (30 kcal) per kg body weight. In addition there is a daily protein turnover of about 3 g per kg body weight (about 0.5 g of nitrogen), of which about 0.15 g of nitrogen per kg body weight is excreted (1 g of nitrogen is derived from about 6.25 g of protein). These losses are usually balanced by dietary intake of equivalent amounts of energy as carbohydrate, fat and protein. Excess energy is stored as glycogen and triglyceride. If energy expenditure exceeds intake these energy stores are drawn upon. A well-nourished adult has enough energy stored as hepatic glycogen to last at least a day. Once this store has been depleted energy is derived from triglyceride, and later from structural compounds such as the proteins of cells including those of muscle. This may cause severe ketosis and increase nitrogen turnover and loss.

The daily energy and nitrogen requirements are not constant. They are significantly increased in ill, catabolic patients, in whom stress-induced hormonal responses impair insulin activity, resulting in glucose intolerance and increased protein breakdown. Under such circumstances there is failure efficiently to utilize dietary energy and nitrogen until the factors causing negative balance have been corrected.

If possible all patients should be fed by mouth (enterally); it is simpler than intravenous (parenteral) feeding, and causes fewer complications. However, if enteral feeding is contraindicated for example after major abdominal surgery, or if there is persistent vomiting, parenteral feeding may be essential. It can be given either by peripheral intravenous infusion or through a central venous catheter into a large vessel. The choice depends partly on the length of time for which parenteral feeding is required, but more importantly on the expertise available for the insertion and nursing care of the central catheter. Dietary intake through a peripheral line is restricted because glucose and amino acid solutions are hyperosmolar and cause irritation to small vessel walls, sometimes with thrombophlebitis.

The principles of carbohydrate, lipid and protein metabolism and their interrelationships, and those of fluid and electrolytes, must be fully understood in order to manage patients receiving parenteral nutrition.

Short-term intravenous nutrition (a period of days)

There is rarely any need to provide additional energy and nitrogen to a well-nourished patient during a short-term fast. Energy is derived from hepatic glycogen stores as well as triglyceride in fat; ketosis is usually mild. If, after major trauma such as surgery, the hormonally induced *catabolic phase* is likely to be followed by rapid recovery, amino acid infusion is unnecessary. Moreover, during this phase the body cannot use administered nitrogen efficiently, and much of it is excreted in the urine as urea or free amino acids.

During a short fast, for example in the immediate postoperative period or during an acute gastrointestinal disturbance, maintenance of fluid and electrolyte balance is all that is necessary. When prescribing five per cent glucose (dextrose) as part of a fluid regime it should be considered as a source of water rather than of a significant amount of energy; a litre contains only 850 kJ (205 kcal) and it would require 10 litres to supply the minimum daily energy requirements.

Long-term intravenous nutrition

If prolonged fasting is anticipated, intravenous feeding should be used from the beginning, in order to optimize its benefits. Intravenous feeding should not, if possible, be stopped suddenly; the patient should gradually be weaned on to enteral feeding.

Long-term parenteral nutrition should preferably be given through a central venous catheter inserted into a large vessel. This enables hyperosmolar solutions to be infused with minimal risk of thrombophlebitis. The insertion requires expertise and careful nursing attention to prevent infection, one of the major complications of intravenous nutrition. The regime depends on the clinical condition of the patient and the volume which can safely be infused. The infused energy source can be glucose- or fat-containing fluids ('Intralipid' soybean emulsion, KabiVitrum). The composition of some of these solutions is shown in Table 2.5 (p. 55). At least 50 per cent of the energy requirements should be given as glucose. Severe illness may cause insulin resistance and utilization of large amounts of glucose may be impaired unless exogenous insulin is given. *Hyperglycaemia* may cause a dangerously high plasma osmolality and glycosuria with polyuria. Plasma glucose concentrations must be monitored frequently to ensure that control is adequate. Glucose infusion should not be stopped suddenly; insulin must be stopped first or *hypoglycaemia* may occur. Preferably glucose should be infused throughout the 24 hours.

The fat particles in 'Intralipid' are similar in size to, and are metabolized in the same way as, chylomicrons (p. 235). Provided that some carbohydrate is given at a constant rate, there is no risk of dangerous ketoacidosis. A grossly lipaemic plasma may interfere with some laboratory analyses, particularly that of plasma sodium (p. 446), but this problem is avoided if non-lipid-containing solutions are infused for a few hours before sampling. If lipaemia then persists it suggests that the rate of administration is faster than the rate at which the fat can be metabolized, and the infusion should be slowed.

Giving energy as glucose and fat minimizes the utilization of amino acids for gluconeogenesis and reduces urinary nitrogen loss. However, nitrogen supplementation is needed to replace the obligatory daily loss and to promote tissue healing. Solutions containing the essential amino acids should be infused into patients unable to eat for long periods of time. Estimation of the 24-hour urea nitrogen output can, if glomerular function is normal, be used as a guide to replacement. However, once tissue repair predominates, amino acid utilization increases (the *anabolic phase*) and urinary nitrogen loss may fall suddenly: this indicates an *increased*, not a decreased, *need for nitrogen*.

The assessment of nitrogen requirements is difficult. A 'normal' non-stressed adult patient needs about 9 g of nitrogen a day, but this requirement increases as

Table 12.2 Examples of daily intravenous feeding regimens for adult patients with normal renal function and in fluid and electrolyte balance, requiring either basal or increased nitrogen intake. Additional water, fat soluble vitamins and trace elements should be given

	Volume ml	Non-protein energy kJ	Nitrogen g	Na mmol	K mmol	Ca mmol	P mmol	Mg mmol	Zn μmol
Basal requirements (800 kJ/gN)									
Synthamin 9 (Travenol)	1000		9.1	73	60		30	5.0	
Glucose 20%	1500	5000							
Intralipid 10% (KabiVitrum)	500	2300					7		
Addamel	10					5		1.5	20
	3010	7300	9.1	73	60	5	37	6.5	20
Increased requirements (775 kJ/gN)									
Synthamin 14 (Travenol)	1000		14.0	73	60		30	5.0	
Glucose 20%	1000	3350							
Glucose 50%	500	4200							
Intralipid 10% (KabiVitrum)	500	2300					7		
Addamel	10					5		1.5	20
	3010	9850	14.0	73	60	5	37	6.5	20

the catabolic rate increases in stress and infection to over 20 g of nitrogen a day. About 840 kJ (200 kcal) per g of nitrogen are necessary for a 'normal' adult to meet energy expenditure, and for synthesis of protein from amino acids. This proportion decreases slightly as the catabolic rate and nitrogen requirements increase.

In addition to the nitrogen and energy requirements, vitamin, mineral and trace element supplementation must be given. Tissue destruction leads to loss of intracellular constituents, such as phosphate and some trace elements such as zinc and copper. It may be necessary to add phosphate to the infusion. Urinary trace metal loss is high, with quantitative excretion of supplements during the catabolic phase, but falls as anabolism becomes predominant: if parenteral feeding has been very prolonged deficiencies may become apparent at this time, and weight gain may be impaired if supplements are not given.

Monitoring of long-term intravenous feeding

Initially results of daily plasma sodium, potassium, $T\text{CO}_2$ and urea levels help in the assessment of water and electrolyte needs and renal function. Plasma calcium, phosphate, magnesium and albumin concentrations should be measured weekly to assess the onset of metabolic complications. Blood haemoglobin, MCV and MCHC and the prothrombin time should also be monitored at least weekly. Plasma concentrations of trace elements, in particular zinc and copper, should be measured every two weeks. Daily estimation of the 24-hour output of urinary nitrogen may help to assess nitrogen utilization and the amount which should be replaced: if urea excretion increases quantitatively when nitrogen intake is increased it indicates that the nitrogen cannot be used, and supplements should not be increased; if nitrogen loss falls while supplements are being given it indicates that anabolism is increasing, and supplements should be *increased* until urinary excretion increases. Once the patient has been established on long term feeding the frequency of monitoring can be reduced.

A cholestatic type of liver disorder may develop in some patients receiving intravenous feeding; unless significant symptoms occur, this is not an indication to stop parenteral feeding.

The composition of fluids used as part of an intravenous feeding regime are shown in Tables 2.2, 2.3, and 2.4 (pp. 54 to 55), and examples of feeding regimes for patients *with normal renal function and who are in electrolyte balance* are given in Table 12.2. A more detailed discussion of parenteral feeding, its complications and its monitoring is given in the references below.

Further reading

Blackburn GL, Wolfe RR. Clinical biochemistry and intravenous hyperalimentation, in: Alberti KGMM, Price CP eds. *Recent Advances in Clinical Biochemistry* No 2. Edinburgh: Churchill Livingstone, 1981: 197–228.

Macfie J. Towards cheaper intravenous nutrition. *Br Med J* 1986; **292**: 107–10.

Oxford Parenteral Nutrition Team. Total parenteral nutrition: value of a standard feeding regime. *Br Med J* 1983; **286**: 1323–7.

Marshall WJ, Mitchell, PEG. Total parenteral nutrition and the clinical chemistry laboratory. *Ann Clin Biochem* 1987; **24**: 327–36.

13

Vitamins

Vitamins are organic compounds which are essential dietary constituents (the name 'vitamines' originally meant amines necessary for life); unlike most other such constituents they are required in very small amounts. They must be taken in the diet because the body either cannot synthesize them at all or, under normal circumstances, not in sufficient amounts for its requirements: vitamin D, for example, can be synthesized in the skin from 7-dehydrocholesterol by the action of ultraviolet light, but, for various reasons, the amount of sunlight reaching the skin may be insufficient to provide the required amount.

A normal mixed diet provides adequate amounts of vitamins, and deficiencies are rarely seen in affluent populations except in those people with intestinal malabsorption, on unsupplemented artificial diets or parenteral nutrition, and in food faddists. Vitamin supplementation of a normal diet is unnecessary. Some vitamins (notably A and D) produce toxic effects if taken in excess.

The biochemical functions of many vitamins are known, but it is not always easy to relate this knowledge to the clinical picture seen in deficiency states.

Testing for deficiency should be carried out as soon as the diagnosis is suspected; results of laboratory tests usually revert rapidly to normal once the patient has started eating a normal diet, for example in hospital, and it may then be impossible to confirm the original diagnosis.

Classification of vitamins

Vitamins are classified into two groups on the basis of their solubilities, one group being fat-soluble and the other water-soluble.

The distinction is of clinical importance, because steatorrhoea is associated with deficiency of fat-soluble vitamins, but there is relatively little clinical evidence of lack of most of the water-soluble vitamins except B_{12} and folate.

Fat-soluble vitamins

The fat-soluble vitamins are:

- A (retinol);

- D (calciferol);
- K (2-methyl-1,4-naphthoquinone);
- E (tocopherol).

Each of these has more than one active chemical form, but variations in structure are very slight and in the following discussion each vitamin is considered as a single substance.

Vitamin A (Retinol)

Sources of vitamin A

Precursors of vitamin A (the carotenes) are found in the yellow and green parts of plants and are especially abundant in carrots: for this reason carrots have the reputation of improving night vision, but it is doubtful if they have any effect in someone eating a normal diet. The vitamin is formed by hydrolysis of β-carotene in the intestinal mucosa; each molecule can produce two molecules of vitamin A which are absorbed as retinol esters and stored in the liver. The yield of the vitamin is much less than the theoretical maximum, especially in children. Retinol is transported to tissues bound to retinol-binding protein, an α-globulin.

Vitamin A is stored in animal tissues, particularly in the liver which is an important source of the preformed vitamin; it is also present in milk products and eggs.

Stability of vitamin A

Vitamin A is rapidly destroyed by ultraviolet light and should be kept in dark containers.

Functions of vitamin A

The *retinal pigment*, rhodopsin (visual purple) is necessary for *vision in dim light (scotopic vision)*. Rhodopsin consists of a protein (opsin) combined with vitamin A. In bright light rhodopsin is destroyed. It is partly regenerated in the dark, but, because the regeneration is not quantitatively complete, vitamin A is needed to maintain levels in the retina.

Vitamin A is also essential for normal:

- *mucopolysaccharide synthesis*;
- *mucus secretion*; deficiency causes drying of mucus-secreting epithelia.

Clinical effects of vitamin A deficiency

The clinical effects of vitamin A deficiency are:

- *night blindness*. Vitamin A deficiency is associated with poor vision in dim light, especially if the eyes have recently been exposed to bright light. It is uncommon for the patient to complain of this;
- *drying and squamous metaplasia* of ectodermal tissue;
- *follicular hyperkeratosis*. Skin secretion is diminished and there may be

hyperkeratosis of hair follicles: dry, horny papules, varying in size from a pinhead to a quarter-inch diameter, are found mainly on the extensor surfaces of the thighs and forearms. Squamous metaplasia of the bronchial epithelium has also been reported and may be associated with a tendency to chest infection;

- *xerosis conjunctivae and xerophthalmia*. The conjunctiva and cornea become dry and wrinkled, with squamous metaplasia of the epithelium and keratinization of the tissue, resulting from deficiency of mucus secretion. *Bitot's spots*, seen in more advanced cases, are elevated white patches found in the conjunctivae and composed of keratin debris. Prolonged deficiency leads to *keratomalacia* with ulceration and infection and consequent scarring of the cornea, causing blindness. Keratomalacia is an important cause of blindness in the world as a whole, but is rarely seen in affluent countries;
- *anaemia* which responds to vitamin A, but not to iron therapy.

Causes of vitamin A deficiency

Hepatic stores of vitamin A are so large that clinical signs only develop after many months, or even years, of dietary deficiency. Such prolonged deficiency is very rare in affluent communities. In steatorrhoea clinical evidence of vitamin A is rare although plasma levels may be low. By contrast, deficiency is relatively common in underdeveloped countries, especially in children, and is a common cause of blindness.

Diagnosis of vitamin A deficiency

There is a poor correlation between the eye changes and the biochemical findings. The diagnosis should be made on clinical criteria; very low plasma vitamin A levels usually confirm deficiency. In conditions such as non-cirrhotic liver disease, in which there are low plasma levels of retinol-binding protein, plasma vitamin A levels may be decreased despite normal liver stores. In cirrhosis of the liver the stores may be very low.

Treatment of vitamin A deficiency

Doses of 50 000 to 70 000 international units of vitamin A in fish liver oil should be given daily to treat xerophthalmia. Night blindness and early retinal and corneal changes respond very rapidly to treatment, but corneal scarring is irreversible.

Hypervitaminosis A

Vitamin A in large doses is toxic. Acute intoxication has been reported in Arctic regions as a result of eating polar bear liver, which has a very high vitamin A content, but more commonly overdosage is due to excessive use of vitamin preparations. The symptoms of *acute* poisoning are nausea and vomiting, abdominal pain, drowsiness and headache. In *chronic* hypervitaminosis A there is fatigue, insomnia, bone pain, loss of hair and desquamation and discoloration of the skin.

Vitamin D (Calciferol)

The metabolism and functions of vitamin D, and the effects and treatment of its deficiency, are discussed in Chapter 9.

Overdosage with vitamin D may cause hypercalcaemia and its accompanying dangers (p. 177). In chronic overdosage stores of cholecalciferol are large, and therefore hypercalcaemia may persist, or even progress, for several weeks after stopping ingestion.

Vitamin K

Vitamin K cannot be synthesized by man but, like many of the B vitamins, it can be manufactured by bacteria in the ileum, from which it can be absorbed; dietary deficiency does not occur. In patients with steatorrhoea the vitamin, whether taken in the diet or produced by bacteria, cannot be absorbed normally.

Vitamin K is necessary for the synthesis of prothrombin and coagulation factors VII, IX and X in the liver, and deficiency is accompanied by a bleeding tendency with a prolonged prothrombin time. If these findings are due to deficiency of the vitamin they can be corrected by parenteral administration (p. 292).

Very little vitamin K can be transported across the placenta and the newborn infant may be vitamin K deficient; the neonatal gut is only gradually colonized by bacteria capable of synthesizing vitamin K. Deficiency may be severe enough to cause *haemorrhagic disease of the newborn.*

Vitamin E (Tocopherol)

Vitamin E deficiency may cause haemolytic anaemia, thrombocytosis, oedema and irritability in premature infants. There is some evidence that vitamin E deficiency may cause neurological symptoms in adults.

Water-soluble vitamins

The water-soluble vitamins are:

- the B complex:
  thiamine (aneurin: B_1);
  riboflavine (B_2);
  nicotinamide (pellagra preventive (PP) factor: niacin);
  pyridoxine (B_6);
  folate (pteroylglutamate);
  the vitamin B_{12} complex (cobalamins);
  biotin and pantothenate (probably of no clinical significance in man).
- ascorbate (vitamin C).

The B complex

A group of food factors were originally classified together as the B group, with the exception of vitamin B_{12} and folate, which were discovered later. Most of these act as enzyme cofactors, but it is not easy to relate the clinical findings to the underlying biochemical lesion.

Many are synthesized by colonic bacteria. Opinions vary as to the importance of this source in man, but, because absorption of water-soluble vitamins from the large intestine is poor, probably most of those synthesized within the colon are unavailable to the body.

Clinical deficiency is rare in affluent communities. When deficiency does occur it is usually multiple, involving most of the B group, and is associated with protein malnutrition; for this reason it may be difficult to decide which signs and symptoms are specific for an individual vitamin and which are part of a general malnutrition syndrome.

Thiamine (B_1)

Sources of thiamine and causes of deficiency

Thiamine cannot be synthesized by animals, including man. It is found in most dietary components and wheat germ, oatmeal and yeast are particularly rich in the vitamin. Adequate amounts are present in a normal diet, but the deficiency syndrome is still prevalent in rice-eating areas; polished rice has the husk removed and this is the only source of thiamine in this food. In other areas thiamine deficiency occurs most commonly in alcoholics and in patients with anorexia nervosa.

Functions of thiamine

Thiamine is a component of thiamine pyrophosphate, which is an essential cofactor for *decarboxylation of 2-oxoacids* (cocarboxylase); one such reaction is the conversion of pyruvate to acetyl CoA. In thiamine deficiency pyruvate cannot be metabolized and accumulates in the blood. Thiamine pyrophosphate is also an essential cofactor for *transketolase* in the pentose-phosphate pathway; the keto (oxo) group is transferred from xylulose-5-phosphate to ribose-5-phosphate to produce glyceraldehyde-3-phosphate and sedoheptulose-7-phosphate.

Clinical effects of thiamine deficiency

Deficiency of thiamine causes *beriberi*, in which anorexia and emaciation, neurological lesions (motor and sensory polyneuropathy; Wernicke's encephalopathy) and cardiac arrhythmias may occur; this form is called 'dry' beriberi. In 'wet' beriberi there is nutritional oedema, sometimes with cardiac failure. Some of these findings may be due to protein rather than to thiamine deficiency.

Beriberi may be aggravated by a high carbohydrate diet, possibly because this leads to an increased rate of glycolysis and therefore of pyruvate production.

Laboratory diagnosis of thiamine deficiency

The most reliable test for thiamine deficiency is probably the estimation of *erythrocyte transketolase* activity, with and without added thiamine pyrophosphate. A reduced activity, if due to thiamine deficiency, becomes normal after addition of the cofactor. This test is rarely indicated, and must not be requested after treatment has started, because plasma levels revert rapidly to normal once an adequate diet is started.

Treatment of thiamine deficiency

Beriberi responds to 5 to 10 mg of thiamine daily, although occasionally higher doses may be needed. *In cases in whom multiple deficiency is suspected a mixture of the B complex vitamins should be given.*

Riboflavine (B_2)

Sources of riboflavine

Riboflavine is found in large amounts in yeasts and germinating plants such as peas and beans, and in smaller amounts in fish, poultry and meat, especially offal.

Functions of riboflavine

There are about 15 flavoproteins, mostly enzymes incorporating riboflavine in the form of flavine mononucleotide (FMN) and flavine adenine dinucleotide (FAD). FMN and FAD are reversible *electron carriers* in biological oxidation systems which are, in turn, oxidized by cytochromes (Fig. 13.1, p. 282).

Clinical effects of riboflavine deficiency

Ariboflavinosis causes a rough, scaly skin, especially on the face, cheilosis (red, swollen, cracked lips), angular stomatitis and similar lesions at the mucocutaneous junctions of the anus and vagina, and a swollen, tender, red tongue which is described as magenta coloured. Congestion of conjunctival blood vessels may be visible if the eye is examined with a slit lamp.

Laboratory diagnosis of riboflavine deficiency

Riboflavine acts as a cofactor for *glutathione reductase*. The finding of a low erythrocyte activity of this enzyme, which increases by about 30 per cent after the addition of FAD, suggests riboflavine deficiency.

Nicotinamide (Niacin)

Sources of nicotinamide

Nicotinamide can be formed in the body from nicotinic acid. Both substances are plentiful in animal and plant foods, although much of that in plants is bound in an unabsorbable form. Some nicotinic acid can also be synthesized in man from tryptophan. Probably both dietary and endogenous sources are necessary to provide enough nicotinamide for normal metabolism.

Functions of nicotinamide

Nicotinamide is the active constituent of nicotinamide adenine dinucleotide (NAD), and its phosphate (NADP), which are important cofactors in *oxidation-reduction reactions*. Reduced NAD and NADP are, in turn, reoxidized by flavoproteins, and the functions of riboflavine and nicotinamide are closely linked (Fig. 13.1, p. 282). NAD and NADP and their reduced forms are essential for glycolysis and for oxidative phosphorylation, and for many synthetic processes.

Clinical effects of nicotinamide deficiency

It may be difficult to distinguish between the clinical features due to coexistent deficiencies, especially of pyridoxine, and those specifically due to nicotinamide. However, nicotinamide deficiency is probably the most important factor precipitating the clinical syndrome of *pellagra*, and nicotinic acid has been called 'pellagra-preventive (PP) factor'. The symptoms are often remembered by the mnemonic 'three Ds' – dermatitis, diarrhoea and dementia.

- The *dermatitis* is a sunburn-like erythema, especially severe in areas exposed to the sun, which may progress to pigmentation and to thickening of the skin; pellagra literally means 'rough skin'.
- The *diarrhoea* is due to widespread inflammation of the mucosal membranes of the gastrointestinal tract and may cause anorexia with weight loss.
- *Dementia*, with delusions, may be preceded by irritability and depression.

Other features include achlorhydria, stomatitis and vaginitis.

Causes of nicotinamide deficiency

Dietary deficiency of nicotinamide, like that of the other B vitamins, is rare in affluent communities.

Hartnup disease is due to a rare inborn error of metabolism involving the renal, intestinal and other cellular transport mechanisms for the monoamino monocarboxylic amino acids including tryptophan (p. 373). Subjects with the disease may present with a pellagra-like rash which can be cured by giving between 40 and 200 mg of nicotinamide daily. If the endogenous supply of tryptophan for synthesis is reduced dietary nicotinic acid is probably insufficient to supply the body's needs over long periods of time: under these circumstances only a slight reduction of intake may precipitate pellagra. A similar clinical picture has been reported as a rare complication of the *carcinoid syndrome*, when tryptophan is diverted to the synthesis of large amounts of 5-hydroxytryptamine (p. 416).

Nicotinic acid (but not nicotinamide) may reduce hepatic secretion of VLDL and therefore plasma levels of VLDL and LDL (p. 243).

Laboratory diagnosis of nicotinamide deficiency

Nicotinic acid in body fluids can be measured by biochemical and microbiological assays.

Pyridoxine (B_6)

Sources of pyridoxine and causes of deficiency

Pyridoxine (pyridoxol), its aldehyde (pyridoxal) and its amine (pyridoxamine) are widely distributed in food and dietary deficiency is very rare. The antituberculous drug *isoniazid* (isonicotinic hydrazide) and *L-dopa* have been reported to produce the picture of pyridoxine deficiency, probably by competing with it in metabolic pathways.

Functions of pyridoxine

Pyridoxal phosphate, formed in the liver from pyridoxine, pyridoxal and pyridoxamine, is a cofactor mainly for the *transaminases*, and for *decarboxylation of amino acids*.

Clinical effects of pyridoxine deficiency

Deficiency may cause roughening of the skin, peripheral neuropathy and a sore tongue. A rare hypochromic, microcytic anaemia, with increased iron stores (sideroblastic anaemia) responds to large doses of pyridoxine even when there is no evidence of vitamin deficiency ('pyridoxine-responsive' anaemia).

Laboratory diagnosis of pyridoxine deficiency

Pyridoxal phosphate is needed for conversion of tryptophan to nicotinic acid. In pyridoxine deficiency this pathway is impaired. *Xanthurenic acid* is the excretion product of 3-hydroxykynurenic acid, the metabolite before the 'block'; in pyridoxine deficiency it is found in abnormally high amounts in the urine after an oral *tryptophan load.*

The urinary metabolite of pyridoxal phosphate, 4-pyridoxic acid, may also be measured.

The increase in the activity of *erythrocyte aspartate transaminase* after addition of pyridoxal phosphate may be measured. The more severe the pyridoxine deficiency the greater the increase in enzyme activity after addition of the vitamin.

Biotin and pantothenate

Lack of these two vitamins probably does not produce clinical deficiency syndromes.

Biotin is present in eggs, but large amounts of raw egg white in experimental diets have caused loss of hair and dermatitis thought to be due to biotin deficiency.

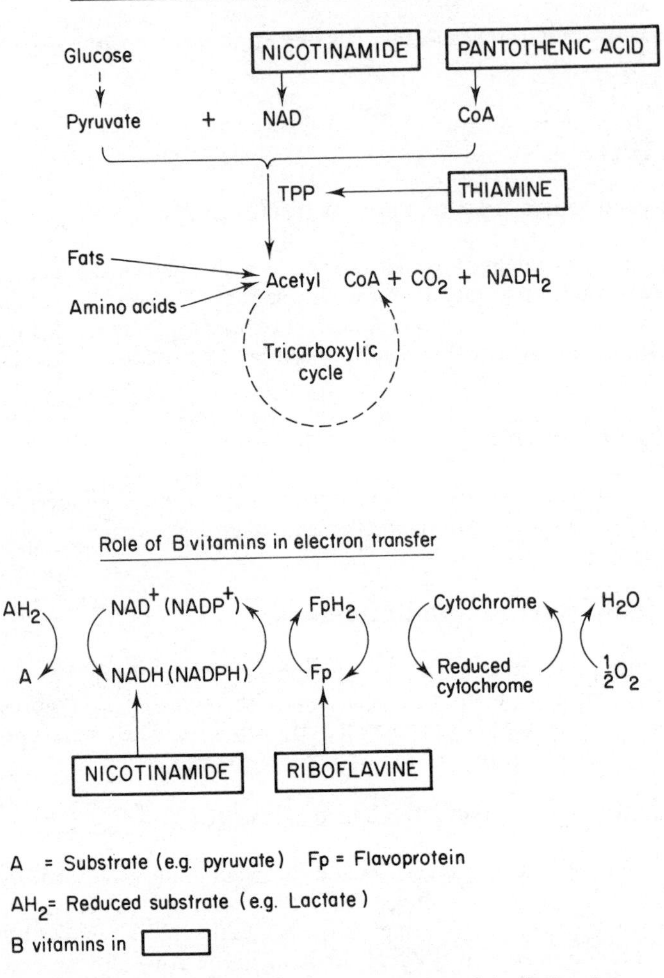

Fig. 13.1 Some biochemical interrelationships of the B vitamins.

Probably the protein *avidin*, present in egg white, combines with biotin and prevents its absorption. Biotin is a cofactor in carboxylation reactions.

Pantothenate is a component of coenzyme A (CoA), which is essential for fat and carbohydrate metabolism. The vitamin is very widely distributed in foodstuffs.

Fig. 13.1 summarizes some of the biochemical interrelationships of the B vitamins discussed so far. Note that, as a general rule, deficiency of this group results in lesions of the skin, mucous membranes and the nervous system.

Folate and B_{12}

These two vitamins are included in the B group and are essential for the normal maturation of erythrocytes; deficiency of either causes *megaloblastic anaemia*. Their effects are so closely interrelated that they are usually discussed together. A fuller discussion of the haematological diagnosis and treatment of such anaemia will be found in haematology textbooks.

Folate is present in green vegetables and some meats. It is easily destroyed during cooking and *dietary deficiency* may rarely occur. It is absorbed throughout the small intestine and, in contrast to most of the other B vitamins (except B_{12}), clinical deficiency is relatively common in intestinal *malabsorption syndromes*, especially in the 'contaminated bowel' syndrome (p. 260). In these conditions, and during *pregnancy* and *lactation*, low red-cell folate levels may be associated with megaloblastic anaemia. The active form of the vitamin is tetrahydrofolate, which is essential for transfer of one-carbon units: it is particularly important in *purine* and *pyrimidine* (and therefore DNA and RNA) synthesis. Methotrexate, a cytotoxic analogue of folate, competes with it for metabolism and therefore inhibits DNA synthesis.

The vitamin B_{12} group includes several cobalamins, found in animal products but not in green vegetables. Dietary deficiency is rare. The cobalamins are transported in plasma by a specific carrier protein, transcobalamin II. Deoxyadenosyl and methylcobalamin have, like folate, coenzyme activity in nucleic acid synthesis. Hydroxycobalamin is the form most commonly used in treatment, and both it and cyanocobalamin are converted to cofactor forms in the body. All forms are absorbed mainly in the terminal ileum, combined with intrinsic factor derived from the gastric parietal cells. In true *pernicious anaemia* antibodies to gastric parietal cells cause malabsorption of vitamin B_{12}. The relation between intestinal dysfunction and vitamin B_{12} deficiency is discussed in Chapter 12.

Deficiency of vitamin B_{12}, like that of folate, causes megaloblastic anaemia; unlike that of folate it can cause *subacute combined degeneration* of the spinal cord. Although the megaloblastic anaemia of vitamin B_{12} deficiency can be reversed by folate, this treatment should never be given in pernicious anaemia because it does not improve, and may even aggravate, the neurological lesions.

Table 13.1 summarizes the synonyms of the B vitamins, their known biological functions and the clinical syndromes associated with deficiencies of each.

Ascorbate (vitamin C)

Sources of ascorbate

Ascorbate is found in fruit and vegetables and is especially plentiful in citrus fruits. A quantitatively significant dietary source is ascorbate added to other foods as a preservative. It cannot be synthesized by man, other primates, or the guinea pig.

Functions of ascorbate

Ascorbate can be reversibly oxidized in biological systems to dehydroascorbate and, although its functions in man are not certain, it probably acts as a hydrogen carrier. It seems to be necessary for normal collagen formation.

Table 13.1 The 'B complex' vitamins

Name	Synonyms	Biochemical function	Clinical deficiency syndrome
Thiamine	Aneurin Vitamin B_1	Cocarboxylase (as thiamine pyrophosphate)	Beriberi (neuropathy) Wernicke's encephalopathy
Riboflavine	Vitamin B_2	In flavoproteins (electron carriers as FAD and FMN)	Ariboflavinosis (affecting skin and eyes)
Nicotinamide	PP factor Niacin	In NAD and NADP (electron carriers)	Pellagra (dermatitis, diarrhoea, dementia)
Pyridoxine	Adermin Vitamin B_6	Cofactor in decarboxylation and deamination (as phosphate)	? Pyridoxine responsive anaemia ? Dermatitis
Biotin		Carboxylation cofactor	Probably unimportant clinically
Pantothenate		In Coenzyme A	
Folate	Pteroyl glutamate	Metabolism of purines and pyrimidines	Megaloblastic anaemia
Vitamin B_{12} group	Cobalamins	Cofactor in synthesis of nucleic acid	Megaloblastic anaemia Subacute combined degeneration of the spinal cord

Causes of ascorbate deficiency

Deficiency of ascorbate causes scurvy and was commonly seen on long sea voyages of exploration in the 16th, 17th and 18th centuries.

Dehydroascorbate is easily and irreversibly oxidized and loses its biological activity in the presence of oxygen; this reaction is catalysed by heat. Scurvy was at one time fairly common in bottle-fed infants, because the ascorbate was often destroyed in the preparation of feeds; it is now rare in this age group. It occurs most commonly in the elderly, especially those living on low incomes who do not eat fresh fruit and vegetables and who tend to cook in frying pans; the combination of heat and the large area of food in contact with air irreversibly oxidizes the vitamin. Ascorbate deficiency can occur in iron overload (p. 398).

Clinical effects of ascorbate deficiency

Many of the signs and symptoms of scurvy can be related to deficient collagen formation. These include:

- fragility of *vascular walls* causing a bleeding tendency, often with a positive Hess test, petechiae and ecchymoses, swollen, tender, spongy, bleeding gums and, occasionally, haematuria, epistaxis and retinal haemorrhages. In infants subperiosteal bleeding and haemarthroses are extremely painful and may lead to permanent joint deformities;

- poor *wound healing*;
- deficiency of *bone matrix* causing osteoporosis and poor healing of fractures. In children bone formation is impaired at the epidiaphyseal junctions, which look 'frayed' radiologically;
- *anaemia*, possibly due to impaired erythropoiesis. This may sometimes be cured by ascorbate alone. Bleeding aggravates the anaemia.

This florid form of scurvy is rarely seen nowadays and the patient most commonly presents complaining that bruising occurs after only minor trauma.

All the signs and symptoms are dramatically cured by the administration of ascorbate.

Diagnosis of ascorbate deficiency

The laboratory confirmation of the clinical diagnosis of ascorbate deficiency can only be made *before* therapy has started. Once ascorbate has been given it is difficult to prove that deficiency was previously present.

Although leucocyte ascorbate assay was said to be more reliable in confirming the diagnosis than that of plasma, levels probably alter in parallel; plasma assay is technically more satisfactory.

Summary

1. Vitamins have important biochemical functions, most of which are now well understood. Unfortunately the relation of these to clinical syndromes is not always obvious.
2. The fat-soluble vitamins, especially vitamin D, may be deficient in steatorrhoea. Both vitamin A and vitamin D are stored in the body and deficiency takes some time to develop.
3. **Vitamin A** is necessary for the formation of visual purple and for normal mucopolysaccharide synthesis. Deficiency is associated with poor vision in dim light and with drying and metaplasia of epithelial surfaces, especially those of the conjunctiva and cornea.
4. **Vitamin D**, as 1,25-dihydroxycholecalciferol, is necessary for normal calcium metabolism and deficiency causes rickets in children and osteomalacia in adults.
5. Both vitamin A and vitamin D are toxic in excess.
6. **Vitamin K** is necessary for prothrombin formation and deficiency is associated with a bleeding tendency.
7. **Thiamine** deficiency causes beriberi.
8. **Riboflavine** deficiency causes ariboflavinosis.
9. **Nicotinamide** can be synthesized from tryptophan in the body, but dietary deficiency causes a pellagra-like syndrome, which may also be seen in Hartnup disease, when tryptophan absorption is deficient, and in the carcinoid syndrome, when tryptophan is used in excess for 5-hydroxytryptamine synthesis.

10. **Pyridoxine-responsive anaemia** may occur.

11. **Folate and vitamin B_{12}** deficiencies produce megaloblastic anaemia and deficiency of vitamin B_{12} can also cause subacute combined degeneration of the cord. Compared with the other B vitamins, deficiencies of these are relatively common in malabsorption syndromes, and vitamin B_{12} and folate deficiency can be features of the 'contaminated bowel' syndrome. Classical pernicious anaemia is due to intrinsic factor deficiency with consequent malabsorption of vitamin B_{12}.

12. **Ascorbate** deficiency causes scurvy.

Further reading

Barker BM, Bender DA eds. *Vitamins in Medicine.* **Vols 1 and II**. London. Heinemann, 1980.

14

Liver disease and gall stones

Liver disease

Outline of the functions of the liver

The liver has important synthetic and metabolic functions. It detoxifies and, like the kidney, excretes end products of metabolism.

The main blood supply to the liver is from the portal vein, which is formed from the superior mesenteric and splenic veins and so drains the intestinal tract. The liver also receives oxygenated blood from the hepatic artery. It is made up of hexagonal lobules of cells. Rows of hepatocytes radiate from the central hepatic vein and are separated by sinusoidal spaces, along the wall of which are interspersed hepatic macrophages, the Kupffer cells, which, as part of the reticuloendothelial system, are phagocytic. At the corners of each lobule are the portal tracts, which contain branches of the hepatic artery, the portal vein and branches of the bile duct. Blood flows from the 'corners' of the lobules towards the central hepatic vein. Hypoxia initially causes damage to the centrilobular area while toxins primarily affect the periphery of the lobule.

All nutrients from the gastrointestinal tract except fat pass through the sinusoidal spaces before entering the systemic circulation. Distortion of this architectural arrangement, by for example cirrhosis of the liver, may allow blood to pass directly from the portal to the hepatic vein, so bypassing the hepatocytes and significantly reducing the detoxifying capacity. This causes many of the features of liver disease described below.

General metabolic functions

Postprandially, when the glucose concentration in the portal vein is high, glycogen is synthesized and stored and glucose is converted into fatty acids which are transported to adipose tissue (Fig. 10.1, p. 203). During *fasting* the systemic plasma glucose concentration is maintained by the breakdown of stored glycogen or by the synthesis of glucose from such substrates as glycerol, lactate and amino acids by gluconeogenesis. Fatty acids reaching the liver from fat stores may be metabolized in the tricarboxylic acid cycle, converted to ketones or incorporated into triglycerides (Fig. 10.2, p. 204).

Synthetic functions

The hepatocytes synthesize:

- *plasma proteins*, except immunoglobulins and complement;
- most *coagulation factors*, including fibrinogen and factors II (prothrombin), V, VII, IX, X, XI, XII and XIII. Of these prothrombin (II) and factors VII, IX and X cannot be synthesized without vitamin K;
- the *lipoproteins*, VLDL and HDL (Chapter 11);
- *primary bile acids*.

The liver has a very large functional reserve. There may be extensive liver disease before deficiencies in synthetic function can be detected.

Extrahepatic causes, such as the loss of protein through the kidney, gut or skin, or the increased capillary permeability which follows inflammation or infection, must be excluded before a fall in plasma albumin concentration is attributed to advanced liver disease.

Prothrombin levels (assessed by measuring the prothrombin time) may be reduced because of impaired hepatic synthesis, whether due to failure to absorb vitamin K or to hepatocellular damage: if hepatocellular function is adequate, parenteral administration of vitamin K will reverse the abnormality (p. 292).

Excretion and detoxification

The excretion of *bilirubin* is considered in more detail below. Other substances which are inactivated and excreted by the liver include:

- *amino acids*, which are deaminated in the liver; the amino groups, and any ammonia produced by intestinal bacterial action which is absorbed into the portal vein, are converted to *urea*;
- *cholesterol*, which is excreted in the bile either unchanged or after conversion to bile acids (p. 300);
- *steroid hormones*, which are metabolized and inactivated by conjugation with glucuronate and sulphate and are excreted in the urine in this water-soluble form;
- many *drugs*, which are metabolized and inactivated by enzymes of the endoplasmic reticulum, including the P_{450} system; some are excreted in the bile.

Efficient excretion of end-products of metabolism and of bilirubin depends on:

- normally functioning liver cells;
- normal blood flow through the liver;
- patent biliary ducts.

Tests for impairment of metabolic, including synthetic and secretory, function are relatively insensitive indicators of liver disease because of the very large hepatic reserve.

Filtering function

The reticuloendothelial Kupffer cells in the hepatic sinusoids are well placed to extract toxic substances which have been absorbed from the gastrointestinal tract. Disruption of the normal architecture, due for example to cirrhosis, enables such toxins to enter the systemic circulation.

Bilirubin metabolism and jaundice

Jaundice becomes clinically apparent when the plasma bilirubin concentration exceeds approximately 35 μmol/litre (2 mg/dl). It occurs when bilirubin production exceeds the hepatic capacity to excrete it. This may be because:

- an increased rate of bilirubin production exceeds normal excretory capacity;
- the normal load of bilirubin cannot be conjugated and/or excreted by damaged liver cells;
- the biliary flow is obstructed, so that conjugated bilirubin cannot flow into the intestine.

Formation of bilirubin

At the end of their lifespan red blood cells are broken down by the reticuloendothelial system, mainly in the spleen. The released haemoglobin is split into globin, which enters the general protein pool and haem, which is converted to bilirubin after removal of iron. The iron is reutilized (p. 390).

About 80 per cent of bilirubin is derived from the breakdown of haem within the reticuloendothelial system. Other sources include breakdown of immature red cells in the bone marrow and of compounds chemically related to haemoglobin, such as myoglobin and the cytochromes.

Unconjugated bilirubin refers to bilirubin which is not water-soluble before conjugation with glucuronate in the liver; it is, however, lipid-soluble and can therefore cross cell membranes and is potentially toxic. It is transported to the liver bound to albumin, and accounts for the bilirubin found in normal plasma. Thus, at physiological concentrations it is all protein-bound and cannot pass through lipid membranes including the 'blood-brain barrier'.

In pathological states the plasma concentration may exceed the protein-binding capacity, and plasma unbound *(free), unconjugated bilirubin* levels increase. The lipid-soluble, unbound, unconjugated bilirubin damages brain cells. This is most likely to occur in the newborn, and in particular in the premature infant, in whom the hepatic conjugating mechanisms are immature (p. 353). Low plasma albumin levels, or displacement of unconjugated bilirubin from binding sites on albumin by high levels of fatty acids, or by drugs such as salicylates or sulphonamides, increase the proportion of the unbound, unconjugated bilirubin and therefore the risk of cerebral damage.

In the liver bilirubin is transferred from plasma albumin, through the permeable vascular sinusoidal membrane, into hepatocytes, where it is bound to *ligandin* (Y protein); from there it is actively transported into the smooth endoplasmic reticulum and is conjugated with glucuronate by a process catalysed by uridyl diphosphate (UDP) glucuronyl transferase. This sequence of events can only continue if there is a free flow of bile. Other anions, including drugs, may compete for binding to ligandin and so impair bilirubin conjugation and therefore excretion.

In the adult about 300 μmol (18 mg) of bilirubin reaches the liver and is conjugated daily: the hepatocyte can handle a much greater load than this. Jaundice due to unconjugated hyperbilirubinaemia only occurs if there is:

- a marked increase in the bilirubin load as a result of haemolysis, or of the breakdown of large amounts of blood after haemorrhage into the gastrointestinal tract or, for example, under the skin with extensive bruising;
- impaired binding of bilirubin to ligandin or impaired conjugation with glucuronate.

Unconjugated bilirubin is normally all protein-bound and is not water-soluble and therefore cannot be excreted in the urine. Patients with unconjugated hyperbilirubinaemia do not have bilirubinuria ('acholuric jaundice').

Biliary excretion of bilirubin

Bilirubin monoglucuronide passes to the canalicular surface of the hepatocyte where, after addition of a second glucuronate molecule, it is secreted by active processes into the bile canaliculi. Further transport depends on normal bile flow, which, in turn, is largely dependent on active secretion of bile acids from the hepatocyte. These energy-dependent steps are the ones *most likely to be impaired by liver damage (and by hypoxia and septicaemia) and by increased pressure in the biliary tract.* An increase in plasma conjugated bilirubin concentration is the earliest manifestation of impaired excretion. *In most cases of jaundice in adults both conjugated and unconjugated fractions of plasma bilirubin are elevated,* but conjugated bilirubin predominates. *Conjugated bilirubin* is water-soluble and is less strongly protein-bound, and can be *excreted in the urine. Bilirubinuria is always pathological* because it reflects the presence of *conjugated* bilirubin in the plasma. Dark urine may be an early sign of some forms of hepatobiliary disease.

Urobilin

The conjugated bilirubin enter the gut in bile. It is broken down by bacteria in the distal ileum and in the colon into a group of products, known collectively as *stercobilinogen (or faecal urobilinogen).* Some is absorbed into the portal circulation and most of it is reexcreted in bile (*enterohepatic circulation*). A small fraction enters the systemic circulation and is excreted in the urine as *urobilinogen,* which can be oxidized to a coloured pigment, *urobilin.*

Urobilinogen, in contrast to bilirubin, is often detectable in the urine of normal people by testing with commercial strips (such as Urobilistix, Ames), particularly if the urine, and therefore the urobilinogen, is concentrated. Urinary urobilinogen is increased if:

- *haemolysis* is severe. Large amounts of bilirubin enter the bowel lumen and are converted to stercobilinogen. An increased amount of urobilinogen is formed and absorbed and, if the hepatic capacity to resecrete it is exceeded, it is passed in the urine;
- *liver damage* impairs reexcretion of normal amounts of urobilinogen into the bile.

Urinary urobilinogen excretion is so variable in the normal subject that only the very high urinary levels found during acute haemolytic episodes are sometimes of diagnostic significance.

Unabsorbed stercobilinogen is oxidized to stercobilin, a pigment which contributes to the brown colour of faeces. Pale stools may, therefore, suggest biliary obstruction.

Urobilinogen and stercobilinogen are *colourless.*

Bilirubin, urobilin and *stercobilin are coloured* ('bile pigments').

The metabolism and excretion of bilirubin are summarized in Fig. 14.1.

Biochemical tests for liver disease

The mechanisms underlying hepatic diseases can be divided into three main groups; these often coexist, but one usually predominates in any condition.

- *Liver-cell damage is typified by release of enzymes* from damaged hepatocytes, which causes a rise in their plasma activities.
- *Cholestasis* is typified by retention of conjugated bilirubin and of alkaline phosphatase (ALP), and by increased ALP synthesis at the sinusoidal surface, which cause *conjugated hyperbilirubinaemia* and *increased plasma alkaline phosphatase activity.*
- *A considerably reduced mass of functioning cells* is typified by a reduction in prothrombin and albumin synthesis, which cause a *prolonged prothrombin time and hypoalbuminaemia.*

The plasma enzymes most commonly measured as indicators of liver-cell

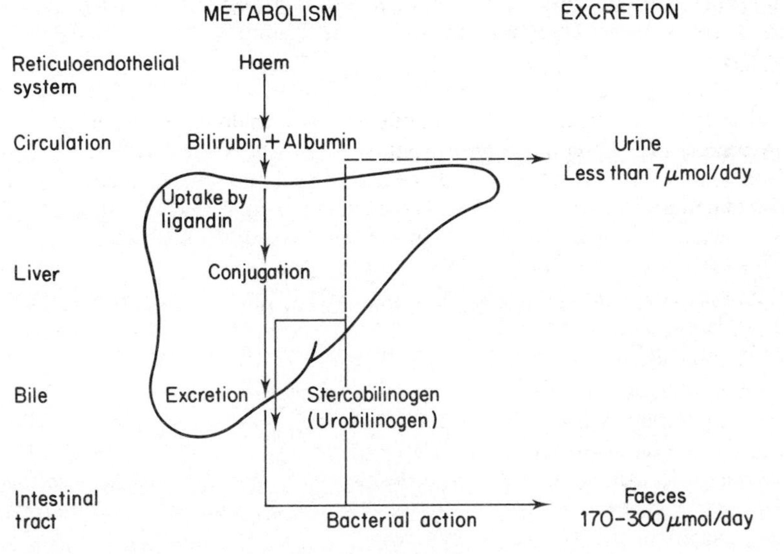

Fig. 14.1 Metabolism and excretion of bilirubin.

damage are *aspartate (AST; GOT) and alanine (ALT; GPT) transaminases;* these are often included in a group of tests misleadingly called 'liver function tests'. A rise in plasma transaminase activities is a sensitive indicator of *damage to cytoplasmic and/or mitochondrial membranes* even if there is no detectable impairment of function; the hepatic synthetic and secretory capacities are large; only severe and usually prolonged liver disease demonstrably impairs prothrombin and albumin synthesis. Hypoalbuminaemia is a common finding in many severe illnesses and is a less specific indicator of impaired synthetic capacity than a prolonged prothrombin time. The reserve secretory capacity is even larger, and hyperbilirubinaemia is rare unless there is cholestasis. By contrast, plasma enzyme activities rise when the membranes of very few cells are damaged.

Liver cells contain more AST than ALT and in most conditions damage to *both mitochondrial and cytoplasmic membranes leads to a relatively greater increase in plasma AST activity than in that of ALT.* However, ALT is confined to the cytoplasm in which its concentration is higher than that of AST. In conditions such as viral hepatitis the *cytoplasmic membrane* sustains the main damage and mitochondria are much less affected: leakage of cytoplasmic contents leads to a relatively *greater increase in plasma ALT than AST.* Therefore, relative plasma activities of AST and ALT may help to indicate the type of cell damage.

A raised plasma activity of *γ-glutamyltransferase (GGT),* derived from the endoplasmic reticulum of the cells of the hepatobiliary tract, does not necessarily indicate hepatocellular damage. The endoplasmic reticulum proliferates, and synthesis of the enzyme is *induced, in response to prolonged intake of alcohol, and of drugs such as phenobarbitone and phenytoin.*

A prolonged prothrombin time may be due to *cholestasis*; fat-soluble vitamin K cannot be absorbed normally if fat absorption is impaired due to intestinal bile salt deficiency (p. 262). The abnormality is then corrected by parenteral administration of the vitamin. A prolonged prothrombin time may also result from severe impairment of synthetic ability if the *liver cell mass is greatly reduced*; in such cases it is *not* corrected by parenteral administration of vitamin K.

Diseases of the liver

Cholestasis

Cholestasis may be:

- *intrahepatic,* in which bile secretion from the hepatocytes into the canaliculi is impaired due to:
  - viral hepatitis;
  - drugs such as chlorpromazine or toxins such as alcohol;
  - inflammation of the biliary tract (cholangitis);
  - autoimmune disease (primary biliary cirrhosis).
- *extrahepatic,* due to obstruction to the flow of preformed bile through the biliary tract by stones, inflammation, or pressure from outside by malignant tissue, usually of the head of the pancreas.

It is essential to distinguish between extra- and intrahepatic causes of cholestasis.

Surgery is often indicated for the former, but is usually contraindicated for intrahepatic lesions. The biochemical findings are similar: unless the cause is clinically obvious, evidence for the dilated ducts of extrahepatic obstruction should be sought using ultrasound, CT scanning or cholangiography.

A recent history of acute hepatitis suggests intrahepatic cholestasis.

There may be no hyperbilirubinaemia if only part of the biliary system is involved by intrahepatic lesions such as *cholangitis, early primary biliary cirrhosis* or primary or secondary *tumours* (p. 298). Bilirubin can be secreted by the unaffected areas. By contrast the increased synthesis of alkaline phosphatase in the affected ducts increases the activity of this plasma enzyme: therefore *measurement of plasma alkaline phosphatase activity is the most sensitive test for cholestasis*. If this is the only abnormal finding, and if the cause is not obvious on other grounds, it must be shown to be of hepatic origin before it is assumed to indicate liver disease (p. 316).

Cases with prolonged and more extensive cholestasis may present with severe jaundice (the plasma bilirubin concentration may be as high as 850 μmol/litre; 50 mg/dl), and sometimes with itching due to deposition of retained bile salts in the skin; more rarely there is bleeding due to malabsorption of vitamin K, with consequent prothrombin deficiency. Cholesterol retention may cause hypercholesterolaemia. Dark urine and pale stools suggest biliary retention of conjugated bilirubin, although steatorrhoea may contribute to the latter finding.

The jaundice of extrahepatic obstruction by malignant tissue is typically painless and progressive, but there may be a history of vague persistent back pain and weight loss. By contrast, intraluminal obstruction by a gall stone may cause severe pain which, like the jaundice, is intermittent. Gall stones may not always cause such symptoms, and if a large stone lodges in the lower end of the common bile duct the picture may be indistinguishable from that of malignant obstruction.

Although most of the findings are directly attributable to cholestasis, back pressure may damage hepatocytes and plasma transaminase activities are usually slightly increased.

Primary biliary cirrhosis is a rare disease which occurs most commonly in middle-aged women. Destruction and proliferation of the bile ducts produces a predominantly *cholestatic* picture, with itching and a very high plasma ALP activity. However, there is a variable degree of hepatocellular destruction, and in the early stages the picture may resemble that of chronic active hepatitis (p. 296). Jaundice develops late in most patients. *Mitochondrial antibodies* are detectable in the serum of over 96 per cent of cases, and the *IgM concentration is usually raised.*

Acute hepatitis

The biochemical findings in acute hepatitis are usually predominantly those of cell membrane damage (an increase in plasma ALT activity greater than that of AST), although there may be cholestasis and, in very severe cases, impaired prothrombin synthesis.

Viral hepatitis may be associated with many infections, such as infectious mononucleosis, rubella and cytomegalovirus. The term is most commonly used to describe three types of infection, all transmissible by the faecal-oral route.

- *Hepatitis A ('infectious hepatitis')* is a food-borne infection relatively common in schools.
- *Hepatitis B ('serum hepatitis')* is transmitted by blood products or other body fluids, and occurs more sporadically than hepatitis A.
- *Non-A, non-B hepatitis* is usually the result of transfusion of blood or blood products.

There may be a three or four day history of anorexia, nausea and tenderness or discomfort over the liver before the onset of jaundice. Some patients remain anicteric. *Plasma transaminase activities are very high* from the onset of symptoms and peak about four days later, when jaundice becomes detectable.

In the early stages there is often a cholestatic element, with *pale stools,* due to reduced intestinal bilirubin, and *dark urine* due to a *rise in plasma conjugated bilirubin* concentration: *unconjugated bilirubin levels also increase* due to impaired hepatocellular conjugation. Unless cholestasis is severe, in which case the biochemical picture may resemble that described on p. 292, *plasma bilirubin levels rarely exceed 350 μmol/litre (about 20 mg/dl) and the plasma ALP activity is only moderately raised or even normal.* If hepatocellular damage is severe and extensive the *prothrombin time may be increased,* and, in cholestatic cases, this finding may be contributed to by malabsorption of vitamin K.

Plasma transaminase activities (especially that of ALT) may remain elevated for several months.

Serological findings. Testing for viral antigens, or for antibodies synthesized in response to the virus, can be used to test for hepatitis A or B. Hepatitis A viral (HAV) antibodies of the IgG class are detectable in the serum of many normal subjects, the prevalence increasing with age, and testing for them is therefore less useful as a diagnostic tool than that for hepatitis B. The finding in plasma of IgM specific hepatitis A antibodies, or of a rising titre, strongly suggests that infection is present.

In the early stage of *hepatitis B* a viral surface antigen (HB_sAg) is detectable in serum. It is short-lived, but during the next few weeks an antibody to the viral core, anti-HB_c, can be detected. At the end of this period the surface antibody, anti-HB_s, develops (Fig. 14.2). HB_sAg may persist, especially in patients with an impaired immune response, in whom the early manifestations are usually mild: this finding may indicate that the patient is a carrier of the disease, and that he should not therefore be allowed to donate blood for transfusion. Serological tests for post-transfusion hepatitis (non-A, non-B) are usually negative and cannot be used to test for the carrier state.

Most cases of hepatitis A recover completely. However, a patient with hepatitis B may not only become a carrier of the disease, but may develop chronic active hepatitis (p. 296), and later cirrhosis: chronicity is indicated by persistently raised plasma transaminase activities, an increase in γ-globulin levels and a positive test for HB_s antigen.

Acute alcoholic hepatitis occurs in heavy drinkers, often after a period of increased alcohol intake. Although the clinical features may mimic acute viral hepatitis, the plasma transaminase activities and the plasma bilirubin concentration are not usually much elevated. The finding of a raised plasma γ-glutamyltransferase (GGT) activity, especially with macrocytosis, hyperuricaemia and

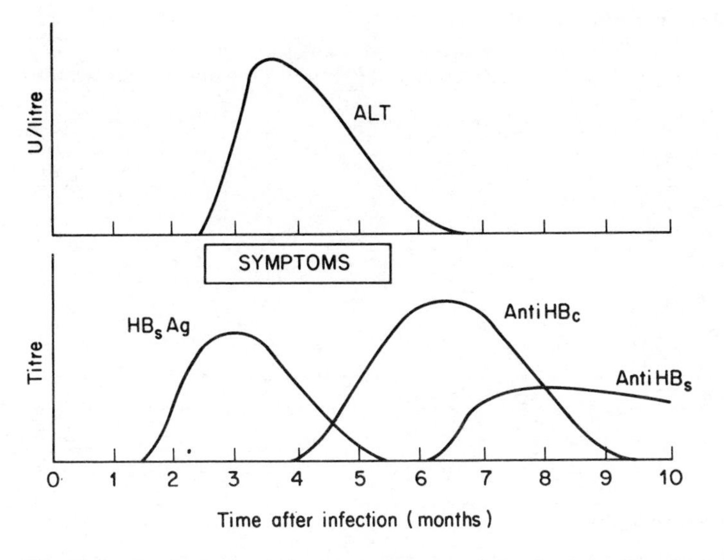

Fig. 14.2 Serological and biochemical changes following infection with hepatitis B virus.

hypertriglyceridaemia, may suggest that there has been chronic alcohol ingestion and that recently increased intake has caused the hepatitis. *None of these findings is diagnostic of alcoholism*; for example, GGT synthesis is also induced by several drugs. Moreover, an alcoholic has an equal chance of developing viral hepatitis as the rest of the population.

Alcoholic hepatitis may progress to cirrhosis.

Other drugs and toxins are hepatotoxic, sometimes directly and sometimes due to a hypersensitivity reaction; in the latter case the damage is not dose-related. The clinical picture may resemble that of viral hepatitis. Table 14.1 lists some of these drugs. *A drug history is an essential part of the assessment of a patient presenting with liver disease.*

Chronic hepatitis

The finding of persistent, usually only slightly, raised plasma transaminase activities, sometimes with chronic or recurrent symptoms suggesting liver disease, may be caused by several disorders. The plasma *ALT activity is usually relatively higher than that of AST*, and may be the only abnormal biochemical finding.

Chronic persistent hepatitis is the term used to describe the finding of slightly raised plasma transaminase activities without clinical signs or symptoms, and without a significant change in these activities over many years. They rarely exceed three times the upper reference limits and jaundice is unusual. The biochemical findings may be discovered by chance or may persist after acute non-A, non-B hepatitis. The condition is probably benign.

Table 14.1 Some drug effects on the liver

	Hepatic necrosis	Hepatitis-like reaction	Chronic hepatitis	Cholestasis
α-Methyldopa	+	+	+	+
Carbamazepine				+
Chlorambucil		+		+
Chlordiazepoxide		+		
Chlorpromazine				+
Chlorpropamide				+
Chlortetracycline	+			
Cytotoxic drugs	+ *			
Erythromycin				+
Ferrous sulphate	+ *			
Halothane	+	+		+
Indomethacin				+
Isoniazid		+	+	
Monoamine oxidase inhibitors (MAO)		+		
Methotrexate			+	
Nitrofurantoin		+	+	+
Oxyphenisatin		+	+	
Para-amino salicylic acid		+		+
Paracetamol (acetaminophen)	+ *		+	
Phenothiazines				+
Phenylbutazone		+		+
Phenytoin		+		
Salicylates (aspirin)	+ *			
17 α-alkylated steroids (in oral contraceptives)				+
Tolbutamide				+
Valproate	+			+

An asterisk indicates that damage is dose-dependent and predictable

Chronic active hepatitis is due to active and continuous hepatocellular destruction, and may progress to cirrhosis. It occurs at any age, but is most common in young women. It may:

- be associated with, or perhaps be caused by, HB_sAg;
- be part of an autoimmune process which sometimes involves more than one organ;
- have no obvious cause.

The earliest findings which differentiate it from chronic persistent hepatitis are an *increasing serum IgG level*, perhaps detected by a rising γ-globulin, and the presence of *smooth muscle and antinuclear antibodies*. As the disease progresses more cells are destroyed, and the *AST may rise to or above that of the ALT*; slight jaundice may develop. If there is much cell destruction plasma albumin levels fall.

Cirrhosis

Cirrhosis is the end result of many inflammatory and metabolic diseases involving the liver, including prolonged toxic damage most usually due to alcohol. In 'cryptogenic cirrhosis' the cause is unknown.

The fibrous scar tissue distorts the hepatic architecture and regenerating nodules of hepatocytes disrupt the blood supply, sometimes increasing the pressure in the portal vein (portal hypertension); blood may be shunted from the portal into the hepatic vein, bypassing the liver.

In the early stages there may be no abnormal biochemical findings. During phases of active cellular destruction the plasma *AST, and sometimes ALT* activities rise slightly. The biochemical findings in advanced cases are mostly associated with a reduced functioning cell mass (p. 291). The vascular shunting may allow antigenic substances which have been absorbed from the intestine to bypass the normal hepatic sinusoidal filtering process, and to stimulate increased synthesis of IgG and IgA, producing the typical serum protein electrophoretic pattern of *β-γ fusion* (p. 327).

Portal hypertension and impaired lymphatic drainage lead to accumulation of fluid in the peritoneal cavity (*ascites*); this may be aggravated by hypoalbuminaemia, which may also cause peripheral *oedema*. In advanced cirrhosis the findings of hepatocellular failure develop.

Primary hepatocellular carcinoma may develop in a cirrhotic liver.

Hepatocellular failure

Liver damage severe enough to cause obvious signs of impaired hepatocellular function may be due to severe hepatitis, advanced cirrhosis, or follow an overdose of a liver toxin such as paracetamol (acetaminophen). The biochemical picture may include any or all of the findings of acute hepatitis. Jaundice is progressive. Plasma enzyme activities may fall terminally, when very few cells remain. Other features may include:

- *hypovolaemia and hypotension* which are due to loss of circulating fluid in ascites and in the oedema fluid formed because of hypoalbuminaemia (p. 34), and which are aggravated by vomiting. The resultant low renal blood flow may have two consequences:
    - secondary hyperaldosteronism, causing *electrolyte disturbances, especially hypokalaemia*, and sometimes *dilutional hyponatraemia* (p. 46);
    - renal circulatory insufficiency (p. 13), *causing oliguria, a high plasma creatinine level, and usually uraemia* despite reduced urea synthesis;
- *impaired hepatic deamination of amino acids*, causing accumulation of amino acids in the plasma and overflow *aminoaciduria*. If the formation of urea from amino acids is not balanced by renal retention due to a reduced GFR, the *plasma urea concentration may be low*;
- *impairment of hepatic gluconeogenesis* may occasionally cause *hypoglycaemia*.

Metabolic liver disease

A group of metabolic disorders, most of which are inherited, is associated with liver disease, especially cirrhosis.

α_1-antitrypsin deficiency (p. 336) is associated with neonatal hepatitis which progresses to cirrhosis in childhood. Often there is basal emphysema.

Galactosaemia (p. 221), if untreated, may cause cirrhosis.

Haemochromatosis (p. 397). Iron deposition in the liver may cause cirrhosis, and may be due either to a primary defect in iron metabolism, with increased intestinal absorption, or be secondary to an excessive intake of iron, usually parenterally. As in any form of cirrhosis a primary liver carcinoma may develop. Other findings may be those of pancreatic insufficiency with diabetes mellitus, cardiac failure, arthritis and hypogonadism. Reducing iron stores by venesection may be effective treatment.

Wilson's disease (p. 374) may cause hepatitis and cirrhosis in young adults and is due to excessive accumulation of copper.

Reye's syndrome. This rare cause of acute fatty infiltration of the liver in children is associated with marked encephalopathy and severe metabolic acidosis with hypoglycaemia. The plasma transaminase activities are high, but plasma bilirubin levels are only slightly raised. The aetiology is uncertain, but there seems to be an association with aspirin ingestion; in some countries it has been recommended that children are not given this drug.

Hepatic invasion or infiltration

Invasion of the liver by secondary carcinoma, or infiltration by lymphoma or granulomas such as sarcoidosis may be associated with abnormal biochemical tests; *sometimes the only abnormal finding is a raised plasma AST activity*; the ALT may also be raised to a lesser extent. The picture may reflect cholestasis, with or without jaundice. Metabolic function is rarely demonstrably impaired. If a primary hepatocellular carcinoma develops, either in a cirrhotic liver or *de novo*, the plasma transaminase and ALP activities usually rise rapidly, and plasma α-fetoprotein levels are often very high; this latter finding is *not diagnostic* of primary hepatic malignancy (p. 423).

Haemolytic jaundice

Haemolytic jaundice is not due to liver disease, and has been discussed more fully on p. 290. Unconjugated hyperbilirubinaemia is usually mild in adults (plasma bilirubin concentration less than 70 μmol/litre; 4 mg/dl) because of the large reserve hepatic secretory capacity. Erythrocytes contain a high concentration of AST and lactate dehydrogenase (HBD isoenzyme) (p. 312). After or during severe haemolysis a rise in plasma activities of these enzymes should not be misinterpreted as evidence of myocardial damage or of liver disease.

Jaundice in the newborn

Red-cell destruction, together with immature hepatic handling of bilirubin, may cause a high plasma level of unconjugated bilirubin in the newborn infant; so-called 'physiological jaundice' is common (p. 353). Usually as the result of haemolytic disease, the level of unconjugated bilirubin may be as high as 400 to 500 μmol/litre (25 to 30 mg/dl) or more and may exceed the plasma protein-binding capacity; free, unconjugated bilirubin may be deposited in the brain and cause *kernicterus*. Neonatal jaundice and its treatment are discussed more fully on p. 352.

The course and severity of neonatal hyperbilirubinaemia may be influenced by drugs.

- Several drugs, including sulphonamides and salicylates, *displace bilirubin from plasma albumin* and increase the risk of deposition in the brain, with resultant cerebral damage. Salicylates should not, in any case, be prescribed for children because of the association with Reye's syndrome.
- Novobiocin *inhibits glucuronyl transferase* and complicates unconjugated hyperbilirubinaemia.
- Any drug causing *haemolysis* aggravates the condition.

The inherited hyperbilirubinaemias

There is a group of inherited disorders in which hyperbilirubinaemia (unconjugated or conjugated) is the only detectable abnormality.

Unconjugated hyperbilirubinaemia

Gilbert's disease is a relatively common condition characterized by plasma unconjugated bilirubin levels of between 20 and 40 μmol/litre (1.2 and 2.5 mg/dl) and rarely exceeding 80 μmol/litre (5 mg/dl). Values fluctuate, and may rise during intercurrent illness or during fasting. This familial condition may be noticed at any age, but usually presents after the second decade. It is often discovered when plasma bilirubin levels fail to return to normal after an attack of hepatitis, or during any mild illness which, because of the jaundice, may be misdiagnosed as hepatitis. The condition is harmless but must be differentiated from haemolysis and from hepatitis. Although some patients do have shortened red-cell survival, the reason for the hyperbilirubinaemia is not clear and may be due to several factors involved in the hepatic uptake and conjugation of bilirubin.

The Crigler-Najjar syndrome, due to deficiency of hepatic glucuronyl transferase, is more serious. It usually presents at birth. The plasma unconjugated bilirubin may increase to levels that exceed the binding capacity of plasma albumin and so cause *kernicterus*. The defect may be complete (Type I) and inherited as an autosomal recessive condition, or partial (Type II) and inherited as an autosomal dominant condition. In the latter the plasma bilirubin concentration may be reduced by drugs that induce enzyme synthesis, such as phenobarbitone.

Conjugated hyperbilirubinaemia

The Dubin-Johnson syndrome is due to defective excretion of conjugated bilirubin, but not of bile acids, and is characterized by slightly raised plasma conjugated bilirubin levels that tend to fluctuate. Because the bilirubin is conjugated it may be detectable in the urine. Plasma alkaline phosphatase activities are normal. There may be hepatomegaly and the liver is dark brown due to the presence in the cells of a pigment with the staining properties of lipofuscin. The condition is harmless and the diagnosis may be confirmed by the characteristic staining in a specimen obtained by liver biopsy.

Rotor syndrome is similar in most respects to the Dubin-Johnson syndrome, except that there is no pigmentation of the liver cells.

Bile and gall stones

Bile acids and bile salts

Four bile acids are produced in man. Two of these, *cholic acid* and *chenodeoxycholic acid* are synthesized in the liver from cholesterol and are called *primary bile acids*. They are secreted in the bile as sodium salts, conjugated with the amino acids glycine or taurine (*primary bile salts*). These are converted by bacterial action within the intestinal lumen to the *secondary* bile salts, *deoxycholate* and *lithocholate* respectively.

Secondary bile salts are partly absorbed from the terminal ileum and are reexcreted by the liver (enterohepatic circulation of bile salts). Bile therefore contains a mixture of primary and secondary bile salts.

Deficiency of bile salts in the intestinal lumen leads to impaired micelle formation and malabsorption of fat (p. 250). Such deficiency may be caused by cholestatic liver disease (failure of bile salts to reach the intestinal lumen) or by ileal resection or disease (failure of reabsorption causing a reduced bile-salt pool).

Formation of bile

About one to two litres of bile are produced by the liver daily. This *hepatic bile* contains bilirubin, bile salts, phospholipids and cholesterol, as well as electrolytes in concentrations similar to those in plasma. Small amounts of protein are also present.

In the gall bladder there is active reabsorption of sodium, chloride and bicarbonate, together with an isosmotic amount of water. The end result is *gall-bladder bile* which is ten times more concentrated than hepatic bile and in which sodium is the major cation and bile salts the major anions. The concentrations of other non-absorbable molecules, conjugated bilirubin, cholesterol and phospholipids, also increase.

Gall stones

Most gall stones contain all biliary constituents, but consist predominantly of one. Only about 10 per cent contain enough calcium to be radiopaque and in this way they differ from renal calculi.

Pigment stones

Pigment stones are found in such *chronic haemolytic states* as hereditary spherocytosis. Increased breakdown of haemoglobin increases bilirubin formation and therefore biliary secretion. The stones consist mostly of bile pigments, with variable amounts of calcium. They are small, hard and dark green or black, and are usually multiple. Rarely they contain enough calcium to be radiopaque.

Cholesterol-containing stones

Cholesterol is most likely to precipitate in bile already supersaturated with the steroid, and further precipitation on a nucleus of crystals causes progressive enlargement. Not all patients with a high biliary cholesterol concentration form bile stones. Changes in the relative concentrations of different bile salts may favour precipitation. The stones may be single or multiple. They may be mulberry-shaped, and are white or yellowish; the cut surface appears crystalline.

There is *no* association between hypercholesterolaemia and the formation of cholesterol gall stones.

Mixed stones

Most gall stones contain a mixture of bile constituents, usually with a cholesterol nucleus as a starting point. They are multiple, faceted, dark brown stones, with a hard shell and a softer centre. They may contain enough calcium to be radioopaque.

Cholesterol and mixed stones are said to be commonest in multiparous women and the incidence may be increased by oral contraceptives. Although this may suggest a hormonal factor, the disease is not uncommon in men and in young women not taking oral contraceptives.

Consequences of gall stones

Gall stones may remain silent for an indefinite length of time and be discovered only at laparotomy for an unrelated condition. They may, however, lead to several clinical consequences.

• *Acute cholecystitis.* Obstruction of the cystic duct by a gall stone causes chemical irritation of the gall bladder mucosa by trapped bile and secondary bacterial infection.
• *Chronic cholecystitis* may also be associated with gall stones.
• *Obstruction of the common bile* duct occurs if a stone lodges in the bile duct. The patient may present with biliary colic, obstructive jaundice (usually intermittent) or acute pancreatitis if the pancreatic duct is also occluded.
• Extremely rarely gall stones may be associated with *carcinoma of the gall bladder*.

Summary

Liver disease

1. The liver has a central role in many metabolic processes.
2. Bilirubin derived from haemoglobin is conjugated in the liver and excreted in the bile. Conversion to stercobilinogen (faecal urobilinogen) takes place in the intestinal lumen. Some reabsorbed urobilinogen is excreted in the urine.
3. Bilirubin metabolism may be assessed by measuring plasma levels of bilirubin and by visual inspection of the stool and urine.
4. Jaundice is due to a raised plasma bilirubin level, either of the unconjugated fraction only, or, in adults, most commonly of both fractions.
5. The results of initial biochemical tests may be characteristic of one or more of three underlying pathological processes:

• liver cell damage (high transaminases);
• cholestasis (high alkaline phosphatase);
• reduced functioning tissue mass (low albumin or prolonged prothromin time).

Jaundice may or may not be present with any of these processes.

6. Unconjugated without conjugated hyperbilirubinaemia is usually due to haemolysis. In the newborn it may, if severe, exceed the plasma protein-binding capacity, and free unconjugated bilirubin may enter brain cells and cause kernicterus.
7. A group of inherited conditions is characterized by hyperbilirubinaemia. Most are relatively harmless, but the Crigler-Najjar syndrome may cause kernicterus.

Bile and gall stones

1. Bile secreted by the liver is concentrated in the gall bladder before passing into the intestinal lumen.
2. Pigment stones may occur in chronic haemolytic states.
3. Cholesterol stones may occur when cholesterol crystals precipitate from supersaturated bile. The initiating factors are unknown.

Further reading

Sherlock S. *Diseases of the liver and biliary system*. Oxford, Blackwell Scientific, **7th ed.** 1985.

Sherlock S, Virus hepatitis. *Clin Gastroenterol* London: WB Saunders, 1980; **9**.

Czaja AJ. Serologic markers of hepatitis A and B in acute and chronic liver disease. *Mayo Clin Proc* 1979; **54**: 721–32.

Czaja AJ, Davis GL. Hepatitis non-A, non-B. Manifestations and implications of acute and chronic disease. *Mayo Clin Proc* 1982; **57**: 639–52.

Advisory Committee on Dangerous Pathogens. *LAV/HTLV III - the causative agents of AIDS and related conditions - Revised guidelines*. London: DHSS Health Publications, 1986.

Warnes TW. Investigation of the jaundiced patient. *Br J Hosp Med* 1982; **28**: 385–91.

Marks V. Clinical pathology of alcohol. *J Clin Pathol* 1983; **36**: 365–78.

Isherwood DM, Fletcher KA. Neonatal jaundice: investigation and monitoring. *Ann Clin Biochem* 1985; **22**: 109–28.

Ludwig J. Drug effects on the liver. A tabular compilation of drugs and drug-related hepatic diseases. *Dig Dis Sci* 1979; **24**: 785–96.

Handling of samples from patients with possible hepatitis or acquired immune deficiency syndrome

Samples from all patients with viral hepatitis, undiagnosed jaundice, a positive test for HB_sAg, known or suspected HIV (AIDS), or at risk because they are in a dialysis unit, should be considered to be infectious. Anyone handling such a specimen, whether medical, nursing, portering or laboratory staff, is at risk. It is *the responsibility of the clinician sending the blood to identify it clearly as potentially dangerous*. The sample must be sent to the laboratory in leak-proof tubes in a sealed plastic bag and be clearly labelled as a biohazard. Strict adherence to the local safety policy is mandatory. In the UK national guidelines have been produced for the handling of infectious specimens: it is strongly recommended that investigations should not be carried out if these guidelines are contravened.

Investigation of suspected liver disease

The most commonly available laboratory tests for the diagnosis of liver disease include measurement of plasma levels of:

- bilirubin — excretory function
- transaminases (AST and/or ALT) — hepatocellular damage
- alkaline phosphatase (ALP) — cholestasis
- albumin and/or prothrombin time — synthetic function

The initial selection of investigations may depend on the clinical features. The earlier part of this chapter considered the biochemical changes and clinical course of specific disorders. Different diseases may, however, present in a similar way and a single disorder may present in more than one way. In this section we will consider the differential diagnosis of clinical problems.

Jaundice as a presenting feature

While a patient with chronic liver disease may present with jaundice, the differential diagnosis of jaundice with bilirubinuria (due to conjugated bilirubin) in a previously well patient is usually between acute hepatocellular damage and cholestasis. A marked increase in urinary urobilinogen *without bilirubin* in the presence of jaundice suggests haemolysis as a cause.

1. Pay special attention to a history of:

- recent exposure to hepatitis or infectious mononucleosis;
- recent receipt of blood or blood-products;
- drug and alcohol intake;
- associated symptoms such as abdominal pain, pruritus, weight loss or anorexia and nausea;
- changes noted by the patient in the colour of the urine or stools;
- the presence of hepatomegaly.

2. Inspect the colour of the urine and stools. Dark yellow or brown urine and pale stools suggest biliary obstruction. Fresh urine containing only urobilinogen is initially of normal colour.

If considered necessary, test the urine; this is most useful to confirm the presence of bilirubin, and so of conjugated hyperbilirubinaemia. Reagent strips (for example, those manufactured by Ames) are available for testing for bilirubin and urobilinogen. It is essential to test *fresh* urine.

Ictostix (incorporated in Multistix and Bili-Labstix) includes stabilized, diazotized 2,4-dichloraniline, which reacts with *bilirubin* to form azobilirubin. The test will detect about 3 μmol/litre (0.2 mg/dl) of bilirubin. Drugs, such as large doses of chlorpromazine, may give *false positive* results.

Urobilistix (also incorporated in Multistix) includes paradimethylaminobenzaldehyde, which reacts with *urobilinogen*. Urobilistix does *not* react with porphobilinogen. This test will detect urobilinogen in urine from some normal subjects. *False positive* results may occur after taking drugs such as *p*-aminosalicylic acid and some sulphonamides.

All reagent strips must be stored, and the tests performed, strictly according to the manufacturer's instructions.

3. Whatever the results of the urine and stool inspection and testing:

   (a) request plasma transaminase and alkaline phosphatase assays;
   (b) if either hepatitis or infectious mononucleosis is suggested by the history, or if there is a predominant increase in transaminases, and especially if the ALT is higher than the AST, request serological tests (p. 294).

4. If there is predominant elevation of the plasma ALP activity determine if there is bile duct dilatation using ultrasound or radiological tests.

- If the bile ducts are dilated there is obstruction that may require surgery.
- If dilated ducts are not demonstrated there is probably intrahepatic cholestasis. Request serological tests for hepatitis, for smooth muscle, mitochondrial and nuclear antibodies and immunoglobulin levels.

If the diagnosis is still in doubt, and if the prothrombin time is normal (a prolonged prothrombin time increases the chance of bleeding), a liver biopsy may be indicated.

5. If acute alcoholic hepatitis is suspected contributory evidence may be the finding of a γ-glutamyltransferase activity which is disproportionately high compared with that of the transaminases, and of macrocytosis, hypertriglyceridaemia and hyperuricaemia. *None of these findings is diagnostic.*

Suspected chronic liver disease (with or without jaundice)

1. Relevant points in the clinical evaluation are:

- a previous history of hepatitis;
- alcohol intake;
- the presence, or history of, other autoimmune disorders;
- pruritus or features of malabsorption.

2. Request plasma transaminases and alkaline phosphatase assays.

- A plasma ALT activity higher than that of the AST may be due to reversible alcoholic hepatitis, to chronic persistent hepatitis, or to early chronic active hepatitis;
- A plasma AST activity higher than that of ALT may be due to cirrhosis or severe chronic active hepatitis;
- A high ALP activity suggests cholestasis.

3. A high blood alcohol level despite denial of drinking by the patient suggests, *but does not prove*, an alcoholic aetiology.

4. Detectable plasma mitochondrial or smooth-muscle antibodies suggest a non-alcoholic cause.

5. Serum protein electrophoresis and immunoglobulin assay may help. High serum IgG and IgA levels, causing β-γ fusion on the electrophoretic strip, suggest cirrhosis. A high serum IgG and normal IgA suggest chronic active hepatitis. A high serum IgM is often found in primary biliary cirrhosis.

Hepatic invasion and infiltration

1. Significant infiltration of the liver by tumour cells, or by granulomas such as sarcoidosis, may occur without detectable biochemical abnormality, or there may be any of the changes described on p. 298. In this situation measurement of the plasma AST activity is usually the most sensitive test: it may be high despite a normal ALT.

2. If primary hepatocellular carcinoma is suspected, the plasma ALP activity and α-feto-protein level may also be high.

3. Radionuclide scans or other imaging procedures, or a liver biopsy, may be indicated.

Unconjugated hyperbilirubinaemia

Jaundice in which plasma conjugated bilirubin levels are normal, or are less than about 10 per cent of the total, and in which there is therefore little or no bilirubinuria, may be due to:

- an increased bilirubin load. If there is no obvious cause, such as extensive bruising, haematological tests for haemolysis are indicated;
- inherited defects of bilirubin uptake or conjugation (p. 299). Other tests for liver disease are usually normal, and the diagnosis is made by the exclusion of evidence for haemolysis.

A greatly increased bilirubin load may result in an increased urinary urobilinogen excretion (p. 290).

15

Plasma enzymes in diagnosis

An enzyme is a protein which catalyses one or more specific biochemical reactions; its presence at very low concentrations results in a large increase in the rate of substrate utilization and product formation *in vitro*. Therefore the measurement of enzyme *activity* in body fluids by monitoring changing substrate or product concentrations is usually relatively easy, while the concentration of enzyme *protein* is so low that it is often difficult to measure by simple methods.

Most enzymes are present in cells at much higher concentrations than in plasma. Some enzymes occur predominantly in cells of certain tissues, where they may be located in different compartments, such as the cytoplasm or the mitochondria. 'Normal' plasma enzyme levels reflect the balance between the *rate of synthesis* and *release into plasma* during cell turnover, and the *rate of clearance* from the circulation. *Proliferation of cells, an increase in the rate of cell turnover, cell damage* or an *increase in the rate of enzyme synthesis (induction),* or *reduced clearance* result in increased plasma enzyme concentrations. Changes in plasma enzyme activities may sometimes help to detect and localize tissue cell damage or proliferation, or to monitor treatment and progress of disease.

Very occasionally plasma enzyme activities may be *lower than normal,* either due to *reduced synthesis,* or to *congenital deficiency* or the presence of *inherited variants of relatively low biological activity:* an example of the latter are the cholinesterase variants (p. 321).

Assessment of cell damage and proliferation

Plasma enzyme levels depend on:

- the rate of release from damaged cells which, in turn, depends on the rate at which damage is occurring;
- the extent of cell damage.

In the absence of cell damage the rate of release depends on:

- the rate of cell proliferation;
- the degree of induction of enzyme synthesis.

These factors are balanced by:

- the rate of enzyme clearance from the circulation.

Acute cell damage, such as occurs, for example, in viral hepatitis, may cause very high plasma enzyme activities, which fall as the condition resolves. By contrast, the liver may be much more extensively involved in advanced cirrhosis, but the *rate* of cell damage is often low, and plasma enzyme activities may be only slightly raised or be within the reference range. In very severe liver disease plasma enzyme activities may even fall terminally, when the number of hepatocytes is grossly reduced.

It is not known how most enzymes are removed from, or their action inhibited in, the circulation. Relatively small peptides, such as α-amylase, can be cleared by the kidney, but most enzymes are large proteins and are probably catabolized by plasma proteases before being taken up by the reticuloendothelial system. In health each enzyme has a fairly constant biological half-life which is characteristic of that enzyme; a knowledge of this half-life may be of help in assessing the time since the onset of an acute illness. After a myocardial infarction, for example, plasma levels of aspartate transaminase and creatine kinase (with short half-lives) fall to normal before those of lactate dehydrogenase (with a longer half-life). The half-life may be lengthened if there is circulatory failure.

Renal glomerular impairment may delay the rate of fall of those plasma enzymes cleared through the kidney. Plasma amylase activity may be high due to renal glomerular impairment alone.

Localization of the damage

Most of the enzymes commonly measured to assess tissue damage are present in nearly all cells, although their relative concentrations in certain tissues may differ. Measurement of the plasma activity of an enzyme known to be in high concentration within cells of a particular tissue may indicate an abnormality of those cells, but the results will rarely enable a specific diagnosis to be made. For example, if there is circulatory failure after a cardiac arrest very high plasma levels of enzymes from many tissues may occur because of hypoxic damage to cells and reduced rates of clearance; the raised plasma levels of 'cardiac' enzymes do not necessarily mean that a myocardial infarction caused the arrest. The possibility of *malignancy* should be considered if the cause of persistently high plasma enzyme activities cannot be explained: some malignant cells contain very high concentrations of enzymes such as lactate dehydrogenase. *In vivo* or *in vitro* haemolysis is often associated with increased plasma activities of lactate dehydrogenase and other enzymes released from damaged erythrocytes.

Diagnostic precision may be improved by:

- *isoenzyme determination.* Some enzymes may exist in more than one form, and these isoenzymes may be separated by their different physical or chemical properties. If they originate in different tissues such identification will give more information than the measurement of plasma total enzyme activity: for example, creatine kinase may be derived from skeletal or from cardiac muscle,

but one of its isoenzymes is found predominantly in the myocardium;
- *estimation of more than one enzyme.* Many enzymes are widely distributed, but their *relative concentrations may vary in different tissues.* For instance, although both alanine and aspartate transaminases are abundant in the liver, the concentration of aspartate transaminase is much greater than that of alanine transaminase in heart muscle.

The distribution of enzymes within cells may differ. Alanine transaminase and lactate dehydrogenase are present only in cytoplasm; glutamate dehydrogenase is found only in mitochondria, while aspartate transaminase occurs in both of these cellular compartments. Different disease processes may affect the cell in different ways, causing an alteration in the relative plasma enzyme activities (p. 292).

Non-specific causes of raised plasma enzyme activities

Before attributing a change in plasma enzyme activity to a specific disease process it is important to exclude the presence of factitious or non-specific causes. Some, such as the effect of circulatory failure, have already been mentioned.

Slight rises in plasma aspartate transaminase activities are common, non-specific, findings in many illnesses. Moderate exercise, or a large intramuscular injection, may lead to a rise in plasma creatine kinase activity, but isoenzyme determination may identify skeletal muscle as the tissue of origin (p. 313).

Some drugs, such as the anticonvulsants phenytoin and phenobarbitone, may induce the synthesis of the microsomal enzyme, γ-glutamyltransferase, and so increase its plasma activity even without disease.

Plasma enzyme activities may be raised if the rate of clearance from the circulation is reduced. In the absence of liver or renal disease this may occur if the enzyme forms macromolecules in plasma as, for example, in macroamylasaemia (p. 314), or if it forms complexes with immunoglobulins; for example, lactate dehydrogenase, alkaline phosphatase or creatine kinase may be bound to IgG.

Factors affecting results of plasma enzyme assays

Analytical factors affecting results. The total concentration of all plasma enzyme protein is less than 1 g/litre. Results of enzyme assays are not usually expressed as concentrations, but as activities. Changes in concentration may give rise to proportional changes in catalytic activity, but the results of such measurements depend on many analytical factors. These include the concentrations of the substrate and product, the pH and temperature at which the reaction is carried out, the type of buffer, and the presence of activators or inhibitors. *Because the definition of 'international units' does not take these factors into account, results from different laboratories, apparently expressed in the same units, may not be directly* *comparable.* Because of these variations, reference ranges are not quoted in this chapter.

Physiological factors affecting results. Physiological factors affect plasma enzyme activities. For example:

- plasma aspartate transaminase activity is moderately higher during the *neonatal period than in adults;*
- *plasma alkaline phosphatase activity is higher in children* than in adults, and peaks during the *pubertal growth spurt, before falling to adult levels. It is also raised during the last trimester of pregnancy* because of the presence of the placental isoenzyme;
- *during* and *immediately after labour* there is a moderate rise in several enzymes, such as the transaminases and creatine kinase;
- plasma γ-glutamyltransferase activity is higher in men than in women.

Plasma enzyme activities must be interpreted in relation to the sex- and age-matched reference range of the issuing laboratory.

The actual concentrations of some plasma enzymes can now be measured by immunoassay and results are being compared with activity measurements.

Abnormal plasma enzyme activities

In the following section individual enzymes of clinical importance will be considered. Applications of their assays in defined clinical situations will be discussed later in the chapter.

Transaminases

The transaminases are enzymes which need the cofactor, pyridoxal phosphate, and which are involved in the transfer of an amino group from a 2-amino- to a 2-oxoacid. They are widely distributed in the body.

Aspartate transaminase (AST)

AST (glutamate oxaloacetate transaminase, GOT) is present in high concentrations in the heart, liver, skeletal muscle, kidney and erythrocytes. Damage to any of these tissues may increase plasma AST levels.

Causes of raised plasma AST activities

- *Artefactual:*
  due to *in vitro* haemolysis.

- *Physiological:*
  during the neonatal period (about 1.5 times the upper adult reference limit).
- *Markedly raised levels* (10 to 100 times the upper adult reference limit):
  circulatory failure with 'shock' and hypoxia (p. 308);
  myocardial infarction (p. 318);
  acute viral or toxic hepatitis (p. 293).
- *Moderately raised levels:*
  cirrhosis (may be normal, but may rise to twice the upper adult reference limit);
  infectious mononucleosis (due to liver involvement);
  cholestatic jaundice (up to 10 times the upper adult reference limit);
  malignant infiltration (may be normal, but may rise to twice the upper reference limit);
  skeletal muscle disease (p. 319);
  after trauma or surgery (especially after cardiac surgery);
  severe haemolytic anaemia.

Alanine transaminase (ALT)

ALT (glutamate pyruvate transaminase, GPT) is present in high concentrations in liver and, to a lesser extent, in skeletal muscle, kidney and heart.

Causes of raised plasma ALT activities

- *Markedly raised levels* (10 to 100 times the upper adult reference range):
  circulatory failure with 'shock' and hypoxia;
  acute viral or toxic hepatitis.
- *Moderately raised levels:*
  cirrhosis (may be normal, but may rise to twice the upper adult reference limit);
  infectious mononucleosis (due to liver involvement);
  liver congestion secondary to congestive cardiac failure;
  cholestatic jaundice (up to 10 times the upper adult reference limit);
  extensive trauma and muscle disease (much less affected than AST).

Lactate dehydrogenase (LD)

LD catalyses the reversible interconversion of lactate and pyruvate. The enzyme is widely distributed in the body, with high concentrations in heart and skeletal muscle, liver, kidney, brain and erythrocytes, and measurement of plasma total LD is therefore a non-specific marker of cell damage.

Causes of raised plasma LD activities

- *Artefactual:*
  due to *in vitro* haemolysis or delayed separation of plasma from whole blood.

- *Marked increase* (more than 5 times the upper adult reference limit)
  circulatory failure with 'shock' and hypoxia;
  myocardial infarction (p. 318);
  some haematological disorders. In blood diseases such as megaloblastic anaemias, acute leukaemias and lymphomas, very high levels (up to 20 times the upper adult reference limit) may be found. Smaller increases occur in other disorders of erythropoiesis such as thalassaemia, myelofibrosis and haemolytic anaemias;
  renal infarction, or, sometimes, during rejection of a renal transplant.
- *Moderate increase:*
  viral hepatitis;
  malignancy of any tissue;
  skeletal muscle disease;
  pulmonary embolism;
  infectious mononucleosis.

Isoenzymes of LD

Five isoenzymes can be detected by eletrophoresis and are referred to as LD_1 to LD_5. LD_1, the fraction which migrates fastest towards the anode, predominates in heart muscle and the kidney. The slowest moving isoenzyme, LD_5, is the most abundant form in the liver and in skeletal muscle. While in many conditions there is an increase in all fractions, the finding of certain patterns is of diagnostic value.

- Predominant elevation of LD_1 and LD_2 (LD_1 greater than LD_2) occurs after myocardial infarction, in megaloblastic anaemia, and after renal infarction.
- Predominant elevation of LD_2 and LD_3 occurs in acute leukaemia; LD_3 is the main isoenzyme elevated due to malignancy of many tissues.
- Elevation of LD_5 occurs after damage to the liver or skeletal muscle.

The finding of a rise in LD_1 is most significant in the diagnosis of myocardial infarction. LD_1 and, to a lesser extent, LD_2 and LD_3 can use 2-hydroxybutyrate, as well as lactate, as substrate, while LD_4 and LD_5 cannot. Some laboratories make use of this fact and assay hydroxybutyrate dehydrogenase (HBD) as an index of LD_1 activity. Immunological methods for the specific measurement of LD_1 are now available, but are expensive.

Creatine kinase (CK)

CK is most abundant in heart and skeletal muscle and in brain, but also occurs in other tissues such as smooth muscle.

Causes of raised plasma CK activities

- *Artefactual:*
  due to *in vitro* haemolysis, using most methods.
- *Physiological:*
  neonatal period (slightly raised);
  during and for a few days after parturition.

- *Marked increase:*
  shock and circulatory failure;
  myocardial infarction (p. 318);
  muscular dystrophies (p. 319) and rhabdomyolysis (breakdown of skeletal muscle).
- *Moderate increase:*
  muscle injury;
  after surgery (for about a week);
  physical exertion. There may be a significant rise in plasma activity after only moderate exercise, muscle cramp or following an epileptic fit;
  after an intramuscular injection;
  hypothyroidism (thyroxine may influence the catabolism of the enzyme);
  alcoholism (possibly partly due to alcoholic myositis);
  some cases of cerebrovascular accident and head injury;
  some patients predisposed to malignant hyperpyrexia (p. 376).

Plasma CK activity is raised in all types of muscular dystrophy, but not usually in neurogenic muscle diseases such as poliomyelitis, myasthenia gravis, multiple sclerosis or Parkinson's disease.

Isoenzymes of CK

CK consists of two protein subunits, M and B, which combine to form three isoenzymes, BB (CK-1), MB (CK-2) and MM (CK-3). CK-MM is the predominant isoenzyme in skeletal and cardiac muscle and is detectable in the plasma of normal subjects. CK-MB accounts for about 35 per cent of the total CK activity in cardiac muscle and less than five per cent in skeletal muscle, and its plasma activity is always high after myocardial infarction. The use and limitations of CK-MB estimation are considered on p. 319. CK-MB may be detectable in the plasma of patients with a variety of other disorders in whom the total CK activity is raised, but the CK-MB activity then accounts for less than six per cent of the total.

CK-BB is present in high concentrations in the brain and in the smooth muscle of the gastrointestinal and genital tracts. Raised plasma activities may occur during parturition. Although they have also been reported after brain damage and in association with malignant tumours of the bronchus, prostate and breast, measurement is not of proven value for diagnosing these conditions. In malignant disease total plasma CK activity is usually normal.

α-Amylase

Amylase breaks down starch and glycogen to maltose. It is present at a high concentration in pancreatic juice and in saliva and may be extracted from such other tissues as the gonads, Fallopian tubes, skeletal muscle and adipose tissue. Most plasma amylase in normal subjects is derived from the pancreas and salivary glands and, being of relatively low molecular weight, is excreted in the urine.

Estimation of plasma amylase activity is mainly requested to help in the diagnosis of acute pancreatitis, in which the plasma activity may be very high. However, it may also be raised in association with other intra- and extraabdominal conditions which cause similar acute abdominal pain, and a high result is not a specific diagnostic marker for acute pancreatitis.

Causes of raised plasma amylase activity

- *Marked increase* (5 to 10 times the upper reference limit):
  - acute pancreatitis;
  - severe glomerular impairment;
  - severe diabetic ketoacidosis;
  - perforated peptic ulcer.
- *Moderate increase* (up to 5 times the upper reference limit):
  - other acute abdominal disorders:
    - perforated peptic ulcer (may be markedly raised);
    - acute cholecystitis;
    - intestinal obstruction;
    - abdominal trauma;
    - ruptured ectopic pregnancy.
  - salivary gland disorders:
    - mumps;
    - salivary calculi;
    - after sialography.
  - morphine administration (spasm of the sphincter of Oddi);
  - severe glomerular dysfunction (may be markedly raised);
  - myocardial infarction (occasionally);
  - acute alcoholic intoxication;
  - diabetic ketoacidosis (may be markedly raised);
  - macroamylasaemia.

Macroamylasaemia. In some patients a high plasma amylase activity is accompanied by low renal excretion of the enzyme despite normal glomerular function. The condition is symptomless; it is thought that either the enzyme is bound to a high molecular weight plasma component such as protein, or that the amylase molecules form high molecular weight polymers which cannot pass through the glomerular membrane. This harmless condition may be confused with other causes of hyperamylasaemia.

Pancreatic pseudocyst. If the plasma amylase activity fails to fall after an attack of acute pancreatitis there may be leakage of pancreatic fluid into the lesser sac (a pancreatic pseudocyst). Urinary amylase levels are high, differentiating it from macroamylasaemia. This is one of the few indications for estimating urinary amylase excretion, which is only inappropriately low for the plasma activity if there is glomerular impairment or macroamylasaemia.

Low plasma amylase activity may be found in infants up to one year of age.

Isoenzymes of amylase

Plasma amylase is derived from the pancreas and from the salivary glands. It is rarely necessary to identify the isoenzyme components in plasma, but they can be distinguished by electrophoresis, or by using an inhibitor derived from wheat germ. Possible indications for isoenzyme determination include:

- the possibility of chronic pancreatic disease, in which *low* levels may be found;
- the coexistence of mumps or renal failure, which complicate the interpretation of *high* levels due to acute pancreatitis.

Alkaline phosphatase (ALP)

The alkaline phosphatases are a group of enzymes which hydrolyse phosphates at high pH. They are present in most tissues but are in particularly high concentration in the osteoblasts of bone, the hepatobiliary tract, the intestinal wall, the renal tubules and the placenta. The exact metabolic function of ALP is unknown, but it is probably important for calcification of bone.

Plasma ALP in adults is mainly derived from bone and liver; the proportion due to the bone fraction is increased when there is increased osteoblastic activity. In the preterm infant plasma total ALP activity is up to five times the upper adult reference limit, and consists predominantly of the bone isoenzyme. In children the total activity is about 2.5 times, and increases to up to five times, this upper limit during the pubertal growth spurt. There is a gradual increase in the proportion of liver ALP to that of bone, and in the adult the liver isoenzyme contributes to just over half the plasma total activity. In the elderly the plasma bone isoenzyme activity increases slightly. During the last trimester of pregnancy the plasma total ALP activity rises due to the contribution of the heat-stable placental isoenzyme.

Causes of raised ALP activity

- *Physiological:*
  - preterm infants (up to 5 times upper adult reference limit);
  - children, until puberty (up to 2 to 2.5 times adult reference limit);
  - puberty. During the pubertal growth spurt levels may be 5 or 6 times the upper adult reference limit;
  - pregnancy, during the last trimester.
- *Bone disease:*
  - osteomalacia and rickets (p. 183);
  - Paget's disease of bone (may be very high):
  - secondary deposits of carcinoma in bone;
  - extensive osteogenic sarcoma;
  - primary hyperparathyroidism *with bone disease* (usually normal but may be slightly elevated) (p. 179).
- *Liver disease:*
  - intra- or extrahepatic cholestasis (p. 293);
  - space-occupying lesions, tumours, granulomas, and other causes of hepatic infiltration;
- *Malignancy:*
  - bone or liver involvement or direct tumour production.

A placental-like, so-called 'Regan', isoenzyme may be identified in plasma in patients with malignant disease, especially carcinoma of the bronchus.

Very high levels of ALP have been recorded transiently in children under three years of age, but the clinical significance of this finding is unknown.

Plasma total alkaline phosphatase activity is not usually increased in myelomatosis despite the X-ray appearance of multiple 'punched-out' osteolytic lesions. However, it may be raised if there is liver involvement, or, more rarely, if there is healing of very extensive pathological fractures.

Low levels of plasma ALP activity may be associated with:

- *arrested bone growth:*
    - achondroplasia;
    - cretinism;
    - ascorbate deficiency.
- *hypophosphatasia,* an autosomal recessive disorder in which there is a low plasma alkaline phosphatase activity and rickets or osteomalacia.

Isoenzymes of alkaline phosphatase

Bone disease with increased osteoblastic activity, or liver disease with involvement of the biliary tract, are the commonest causes of an increased alkaline phosphatase activity.

Rarely, the cause of a raised plasma total alkaline phosphatase activity is not apparent, and further tests may be helpful. The isoenzymes originating from bone, liver, intestine and the placenta may be separated by electrophoresis, but interpretation may be difficult if the total activity is only marginally raised. The placental and 'Regan' isoenzymes are more stable at 65 °C than the bone, liver and intestinal isoenzymes, and heat inactivation may help to differentiate the heat-stable from the heat-labile fraction.

The placental isoenzyme does not cross the placenta and is therefore not detectable in the plasma of the newborn.

Acid phosphatase

Acid phosphatase is found in the prostate, liver, erythrocytes, platelets and bone. The main indications for estimation are to help in the diagnosis of prostatic carcinoma and to monitor its treatment.

Normally acid phosphatase drains from the prostate, through the prostatic ducts, into the urethra, and very little can be detected in plasma. In extensive prostatic carcinoma, particularly if it has metastasized, plasma acid phosphatase activity rises, probably because of the increased number of prostatic acid phosphatase-containing cells. If the tumour is small, or is too undifferentiated to synthesize the enzyme, plasma activities may be normal. For this reason the assay is more useful for monitoring treatment of a known case of disseminated prostatic carcinoma than for making the diagnosis.

Sampling for acid phosphatase assay

Opinions differ about whether rectal examination increases the serum activity by pressure on prostatic cells. The effect is certainly relatively rare if the examination is performed by an experienced clinician, especially if, because he knows that the estimation is to be performed, he exerts minimum pressure. However, we have found that the serum ACP activity can rise to two or three times the upper reference limit in some cases, and that it only falls to its basal level after several days. An example of this effect is given, and a sampling procedure suggested, on p. 447. It is obvious that prostatectomy will release large amounts of the enzyme into the plasma and the assay should not be requested for at least a week after the operation.

Heparin inhibits the activity of the enzyme, and clotted rather than heparinized blood must be used.

Acid phosphatase is unstable and specimens for assay should be sent to the laboratory without delay.

Haemolysed specimens must not be assayed, partly because the enzyme is released from blood cells *in vitro.*

Acid phosphatase isoenzymes

Release of acid phosphatase from blood cells *in vitro* may occur even in unhaemolysed samples, and many methods have been devised in an attempt to measure only the prostatic fraction, without complete success. One method makes use of the fact that L-tartrate inhibits prostatic acid phosphatase; the assay is performed with and without the addition of L-tartrate, and the difference in activity between the results, the *tartrate-labile* fraction, is mainly *prostatic* acid phosphatase. It is now possible to measure the prostatic enzyme protein concentration by immunoassay, but the diagnostic value of the assay remains to be proved.

Causes of raised serum acid phosphatase activity

- *Tartrate-labile:*
  - *artefactually* following rectal examination, acute retention of urine or passage of a catheter, due to pressure on prostatic cells;
  - disseminated carcinoma of the prostate.
- *Total:*
  - *artefactually* in a haemolysed specimen, or following rectal examination, acute retention of urine or passage of a catheter;
  - disseminated carcinoma of the prostate;
  - Paget's disease of bone;
  - some cases of metastatic bone disease, especially with osteosclerotic lesions;
  - Gaucher's disease (probably from Gaucher cells);
  - occasionally in thrombocythaemias.

The assay is not of diagnostic value in the last four conditions.

γ-Glutamyltransferase (GGT)

γ-Glutamyltransferase occurs mainly in the liver, kidney, pancreas and prostate. Plasma GGT activity is higher in men than in women.

Causes of raised plasma GGT activity

- Induction of enzyme synthesis, without cell damage, by drugs or alcohol. Many drugs, most commonly the anticonvulsants phenobarbitone and phenytoin, and alcohol induce proliferation of the endoplasmic reticulum.
- Cholestatic liver disease, when changes in GGT activity usually parallel those of alkaline phosphatase. In the cholestatic jaundice of pregnancy plasma GGT activities do not increase.
- Hepatocellular damage, such as that due to infectious hepatitis; measurement of plasma transaminase activities is a more sensitive indicator for such conditions.

Slightly or moderately raised activities (up to about three times the upper reference limit) are particularly difficult to interpret. Very high plasma GGT activities, out of proportion to those of the transaminases, may be due to:

- alcoholic hepatitis or gross alcohol abuse;
- induction by anticonvulsant drugs;
- cholestatic liver disease.

A patient should never be labelled alcoholic because of a high plasma GGT activity alone.

Plasma enzyme patterns in disease

Myocardial infarction

The plasma enzyme estimations of greatest value in the diagnosis of myocardial infarction are AST, LD (or HBD) and CK. The choice of estimations depends on the time which has elapsed since the suspected infarction. An approximate guide to the sequence of changes is given in Table 15.1.

All plasma activities (including that of CK-MB) may be normal until at least four hours after infarction: *blood should not be taken for enzyme assay until this time has elapsed after the onset of chest pain.*

Table 15.1 The time sequence of changes in plasma enzymes after myocardial infarction

Enzyme	Starts to rise (hours)	Time after infarction of peak elevation (hours)	Duration of rise (days)
CK (total)	4–8	24–48	3–5
AST	6–8	24–48	4–6
LD (HBD)	12–24	48–72	7–12

Plasma enzyme levels are raised in about 95 per cent of cases of myocardial infarction, and are sometimes very high. The degree of rise is a very rough indicator of the size of the infarct, but is of limited prognostic value. The prognosis often depends more on the site than on the size of the infarct. A second rise of plasma enzyme activities after their return to normal indicates extension of the infarct or the development of congestive cardiac failure; in the latter case plasma HBD and CK activities do not rise.

Plasma enzyme activities do not rise significantly after an episode of angina pectoris.

Even a small myocardial infarct causes some hepatic congestion due to right-sided heart dysfunction, and there is therefore a slight rise of ALT activity. This is rarely a diagnostic problem, because the increase in activity of AST is much greater than that of ALT, and that of HBD (LD_1) is unequivocally raised. If there is primary hepatic dysfunction, congestive cardiac failure without infarction or pulmonary embolism (which, by impairing pulmonary blood flow, usually causes some hepatic congestion), the rises in AST and ALT are more comparable, and HBD (but not total LD) activity is normal.

The sequence of changes in plasma CK activity after myocardial infarction are similar to those of AST (Table 15.1), although they rise more above the upper reference limit. If the plasma total CK activity is high when the AST and HBD (LD_1) activities are normal, the elevation is more likely to be due entirely to the MM isoenzyme derived from skeletal muscle; this may follow recent intramuscular injection, exercise or surgery. Detection of an elevated plasma CK-MB fraction by isoenzyme electrophoresis may occasionally help to detect a myocardial infarction which has occurred between 4 and 24 hours earlier, but in most cases nothing is lost by taking a sample later for AST and LD_1 (HBD) measurement. Any patient suspected of having a myocardial infarction should be kept under observation and the delay of a few hours will not be detrimental to his management.

Most of the CK released after myocardial infarction is the MM isoenzyme, which is found in both skeletal and myocardial muscle; this has a longer half-life than the MB fraction and *after about 24 hours* the finding of a high MM and undetectable MB does *not exclude myocardial damage* as a cause of high total CK activities: by this time the plasma transaminase and HBD patterns are usually characteristic. *In most cases of suspected myocardial infarction measurement of plasma AST and HBD (LD_1) activities, together with the clinical and electrocardiographic findings, are adequate to make a diagnosis, and plasma total CK activity may be misleading.*

Liver disease

Plasma enzyme changes in liver disease are discussed in Chapter 14.

Muscle disease

In the muscular dystrophies plasma levels of the muscle enzymes, CK and the transaminases, are increased, probably because of leakage from the diseased cells. Results of CK estimation are the more specific. Points to consider in interpretation are:

- activities are *highest* (up to 10 times the upper reference limit or more) in the *early stages* of the disease. *Later*, when much of the muscle has wasted, they are lower and *may even be normal*;
- activities are *higher following muscular activity which has been taken immediately after rest* (because of release of CK built up in muscle during rest) than after prolonged activity;
- activities are *higher in the newborn than in adults.*

Similar, but less marked, changes are found in many patients with myositis.

Carriers of the Duchenne type of muscular dystrophy can often be detected by finding raised plasma CK activities. The rise is only moderate, and non-specific causes of raised enzyme activities (p. 309) must be excluded.

The specimen should be collected:

- at a time when normal plasma activities would be expected to be highest, that is *late in the day after ordinary physical activity* (levels may be normal in known carriers in the morning);
- *not during pregnancy* (levels may be low in early pregnancy);
- at a time when release from skeletal muscle is not abnormally high, that is:
  - *not for 48 hours after severe or prolonged exercise;*
  - *not for 48 hours after an intramuscular injection.*

The assay must be performed on a fresh or appropriately stored specimen, preferably on three separate occasions, to minimize errors of interpretation due to random variations in plasma levels. The laboratory should be informed *before* the blood is taken.

Although plasma enzyme activities are usually *normal* in *neurogenic muscular atrophy*, the number of false positives makes such tests unreliable in differentiating these conditions from primary muscle disease.

Enzymes in malignancy

Plasma total enzyme activities may be raised, or an abnormal isoenzyme detected, in several neoplastic disorders.

- Serum prostatic (tartrate-labile) acid phosphatase activity rises in some cases of malignancy of the prostate gland.
- Any malignancy may be associated with a non-specific increase in plasma LD, HBD and, occasionally, transaminase activity.
- Plasma transaminases and alkaline phosphatase may be of value to monitor treatment of malignant disease. Raised levels may indicate secondary deposits in liver or, of alkaline phosphatase, in bone. Liver deposits may also cause an increase in plasma LD or GGT.
- Tumours occasionally produce a number of enzymes, such as the 'Regan' ALP isoenzyme, LD (HBD) or CK-BB, assays of which may be used as an aid to diagnosis or for monitoring treatment.

Haematological disorders

The very high activities of LD (HBD) found in megaloblastic anaemias and leukaemias have been mentioned (p. 312). Similar rises may be found in other conditions in which marrow activity is abnormal. Typically there is much less change in the plasma AST than in the LD (HBD) activities.

Severe *in vivo* haemolysis produces changes in both AST and LD (HBD) activities which mimic those of myocardial infarction.

Plasma cholinesterase and suxamethonium sensitivity

There are normally two cholinesterases, one found predominantly in erythrocytes and nervous tissue (acetylcholinesterase). The other, found in plasma (*cholinesterase* or 'pseudocholinesterase'), is synthesized mainly in the liver.

Causes of decreased plasma cholinesterase activity

- hepatic parenchymal disease (reduced synthesis);
- ingestion, or absorption through the skin, of such anticholinesterases as organophosphates;
- inherited abnormal cholinesterase variants, with a low biological activity.

Causes of increased plasma cholinesterase activity

- recovery from liver damage;
- nephrotic syndrome.

Suxamethonium sensitivity

The muscle relaxant suxamethonium (succinyl choline; 'scoline') is usually broken down by plasma cholinesterase, and this limits the duration of its action. Giving suxamethonium to patients with a low cholinesterase activity, usually due to an enzyme variant, is often followed by a prolonged period of apnoea ('scoline apnoea'); such patients may need ventilatory support after operation. The abnormal cholinesterase variants may be classified by measuring the percentage inhibition of the enzyme activity by dibucaine (dibucaine number) or by fluoride (fluoride number). A normal dibucaine number despite a low plasma activity suggests that synthesis has been impaired, but does not completely exclude a genetic abnormality. A low dibucaine number with a very low activity is usually due to the presence of an abnormal gene. Intermediate dibucaine numbers result from other genetic abnormalities, whether homozygous or heterozygous, and the variants may have normal or low activities.

Identification of patients susceptible to suxamethonium, and of their affected relatives, is important. All blood relatives should be traced and investigated to identify their genotype, and so to predict the chance of later anaesthetic risk. All affected individuals should carry a warning card, or should wear some other form of warning (for example, a 'Medic Alert' bracelet).

Summary

1. Enzyme concentrations are high in cells. Natural decay of these cells releases enzymes into the plasma. Plasma activities are usually measurable, but low.

2. Plasma enzyme assays are most useful in the detection of raised levels due to cell damage.

3. Assays of a few enzymes may help to identify the damaged tissue, and isoenzyme studies may increase the specificity. In general, a knowledge of the patterns of enzyme changes, together with the clinical and other findings, are needed if a useful interpretation is to be made.

4. Non-specific causes of raised enzyme activities include peripheral circulatory insufficiency, trauma, malignancy and surgery.

5. Artefactual increases may occur in haemolysed samples.

6. Enzyme estimations may be of value in the diagnosis and monitoring of:

- myocardial infarction (AST, LD and its isoenzymes such as HBD, and sometimes CK);
- liver disease (transaminases, ALP and sometimes GGT);
- bone disease (ALP);
- prostatic carcinoma (tartrate-labile ACP);
- acute pancreatitis (α-amylase);
- muscle disorders (CK).

Further reading

Wilkinson JH ed. *The Principles and Practice of Diagnostic Enzymology*. London: Edward Arnold, 1976.

(The chapters dealing with myocardial infarction (8), enzymes in disease of skeletal muscle (9) and enzyme tests in diseases of the liver and hepatobiliary tract (10) are particularly useful)

Lee TH, Goldman L. Serum enzyme assays in the diagnosis of acute myocardial infarction. *Ann Int Med* 1986; **105**: 221–33.

Posen S, Doherty E. The measurement of serum alkaline phosphatase in clinical medicine. *Adv Clin Chem* 1981; **22**: 163–245.

Yam LT. Clinical significance of the human acid phosphatases: a review. *Am J Med* 1974; **56**: 604–16.

Penn R, Worthington DJ. Is serum γ-glutamyltransferase a misleading test? *Br Med J* 1983; **286**: 531–35.

Brown SS, Kalow W, Pilz W, Whittaker M, Woronick CL. The plasma cholinesterases: a new perspective. *Adv Clin Chem* 1981; **22**: 2–123.

16

Proteins in plasma and urine

Plasma proteins

Plasma contains a mixture of proteins which differ in origin and function.

Metabolism of plasma proteins

The amount of protein in the vascular compartment depends on the balance between the rate of synthesis and that of catabolism or loss; the *concentration* depends on the relative amount of protein and water in the compartment, and abnormal levels do not necessarily reflect abnormalities in protein metabolism.

- *Synthesis.* Hepatocytes synthesize most of the plasma proteins; those of the complement system are synthesized both in the liver and by macrophages. Immunoglobulins are the exception, and are derived only from the B cells of the immune system.
- *Catabolism and loss.* Most plasma proteins are taken up by pinocytosis into *capillary endothelial cells or mononuclear phagocytes*, where they are *catabolized.* Small proteins are *lost passively* through the *renal glomerulus* and *intestinal wall.* Some are reabsorbed, either directly by renal tubular cells or after digestion in the intestinal tract; some are catabolized by renal tubular cells.

Functions of plasma proteins

The following is an outline of the main functions of the plasma proteins.

Inflammatory response and control of infection. The immunoglobulins and the complement proteins form part of the immune system and the latter, together with a group of proteins known as 'acute-phase reactants', are involved in the inflammatory response.

Transport. Albumin and specific binding proteins transport many hormones, vitamins, lipids, bilirubin, calcium, trace metals and some drugs. Combination with protein allows poorly water-soluble substances to be transported in plasma. The protein-bound fraction of many of these is less physiologically active than the unbound fraction.

Control of extracellular fluid distribution. Distribution of water between the intra- and extravascular compartments is influenced by the colloid osmotic effect of plasma proteins, especially albumin (p. 34).

Peptide hormones and *blood clotting factors* contribute quantitatively relatively small, but physiologically important, amounts of plasma protein. Only a few circulating enzymes are functional; most originate from cell breakdown.

This list is by no means complete. The function of many plasma proteins which have been identified is unknown.

Methods of assessing plasma proteins

Proteins may be quantitated, either in groups (for example, total protein) or, more often, as individual proteins (most commonly albumin). Changes in the relative proportions of groups of proteins may be visually assessed after electrophoresis of serum.

The levels of plasma proteins may be expressed as concentrations (for example, g/litre) or as activities of those proteins with defined functions (for example, clotting times for prothrombin). The distinction is important when abnormal forms of protein are present at normal concentration, but with impaired function (for example, C_1 inhibitor).

Total protein

Total protein estimation is of limited clinical value. Acute changes in concentration, like those of all proteins, reflect both the protein and the fluid content of the vascular compartment. Acute changes are more likely to be due to loss from, or gain by, the vascular compartment of protein-free fluid than of protein (compare sodium, p. 51). Only marked changes of the major constituents, albumin and immunoglobulins, are likely to alter total protein levels significantly.
Low total protein levels may be due to:

- dilution, for example if blood is taken near the site of an intravenous infusion;
- hypoalbuminaemia (p. 329);
- profound immunoglobulin deficiency.

Raised total protein levels may be due to:

- loss of protein-free fluid, or excessive stasis during venepuncture (p. 447);
- a major increase in one or more of the immunoglobulins.

Total protein levels may be misleading and may be normal in the face of quite marked changes in the constituent proteins. For example:

- a fall in plasma albumin concentration may roughly be balanced by a rise in immunoglobulin levels. This is quite a common combination;
- most individual proteins, other than albumin, make a relatively small contribution to total protein: quite a large *percentage* change in the concentration of one of them may not cause a detectable change in total protein concentration.

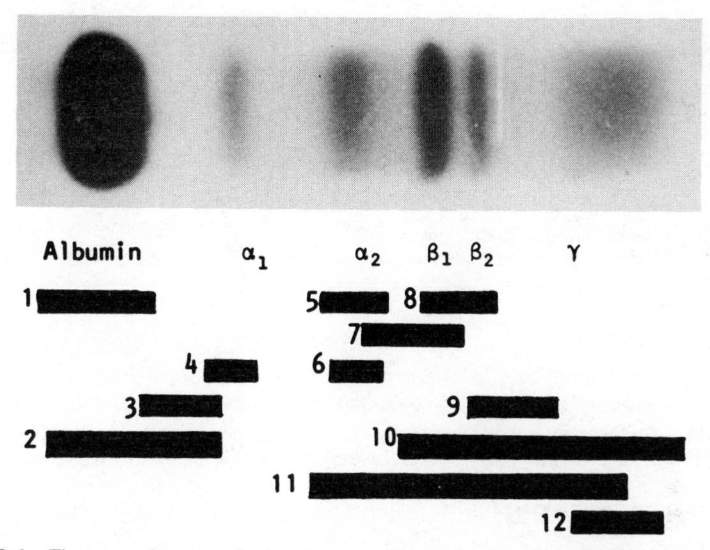

Fig. 16.1 The normal serum electrophoretic pattern. In this example the β-globulin has separated into β_1 and β_2 fractions. This finding is not invariable, especially in stored specimens.

1. Albumin
2. α-lipoprotein (HDL)
3. α_1-acid glycoprotein
4. α_1-antitrypsin
5. α_2-macroglobulin
6. haptoglobin
7. β-lipoprotein (LDL)
8. transferrin
9. C_3 fraction of complement
10. IgG
11. IgA
12. IgM

Electrophoresis

Electrophoresis, which separates proteins according to their different electrical charges, is usually performed by applying a small amount of serum to a strip of cellulose acetate or agarose and passing a current across it for a standard time. In this way five main groups of proteins, albumin and the α_1-, α_2-, β-, and γ-globulins, may be distinguished after staining, and may be compared with those in a normal control serum. Each of the globulin fractions contains several proteins (Fig. 16.1). Changes in electrophoretic patterns are most obvious when the level of a protein normally present in high concentration (for example, albumin) is abnormal, or when there are parallel changes in several proteins in the same fraction.

The following description applies to the normal appearance, in adults, of cellulose acetate strips: **Albumin**, usually a single protein, makes up the most obvious band. The **α_1-globulin** band consists almost entirely of *α_1-antitrypsin.* The **α_2-globulin** band consists mainly of *α_2-macroglobulin* and *haptoglobin. The* **β-globulin** *often separates into two bands: β_1* consists mainly of *transferrin*, with a contribution from *LDL*, and β_2 consists of the *C_3 fraction of complement. The* **γ-globulins** *are immunoglobulins.* Some immunoglobulins are also found in the α_2 and β regions.

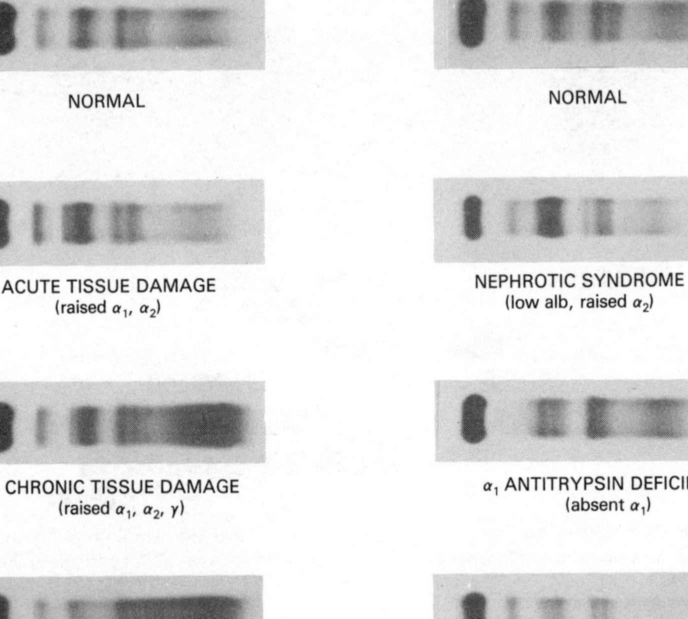

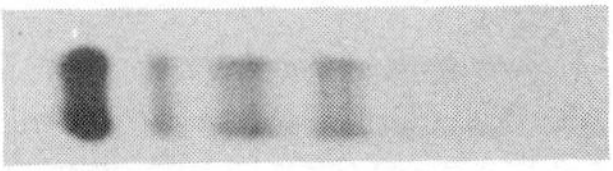

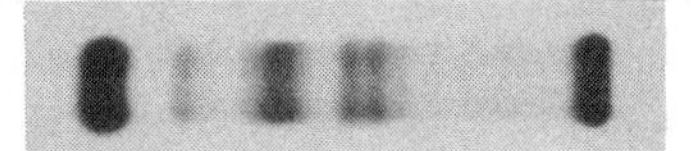

Fig. 16.2 Electrophoretic patterns in disease.
Note:

1. *Nephrotic syndrome*. In this typical example the level of all fractions except that of α_2-globulin is low. Occasionally the γ fraction may be normal or increased.
2. *Paraproteinaemias*. Immune paresis is not invariable (p. 338).

If *plasma* rather than serum is used a sixth band, *fibrinogen*, appears in the β-γ region. This may make interpretation difficult and blood should be allowed to clot if electrophoresis is requested.

Electrophoretic patterns in disease (Fig. 16.2)

Some abnormal electrophoretic patterns are characteristic of a particular disorder or group of related disorders, while others indicate non-specific pathological processes.

Parallel changes in all fractions (not shown in Fig. 16.2). This is a *normal pattern with an abnormal total protein concentration.* An *increase* in all fractions (including immunoglobulins) may be found if there is *volume depletion due to loss of protein-free fluid*, or if there has been *stasis* during venepuncture (p. 447), and a reduction in *overhydration* or in specimens taken from a 'drip arm'. A *reduction* also occurs in *severe malnutrition*, sometimes due to *malabsorption*, unless accompanied by infection.

The acute-phase pattern. Tissue damage of any kind triggers the sequence of biochemical and cellular events associated with inflammation (p. 329). The biochemical changes include stimulation of synthesis of the so-called acute-phase proteins, although secondary utilization of synthesized complement causes a fall in its concentration: the plasma levels of these proteins reflect the activity of the inflammatory response, and their presence is responsible for the rise in the erythrocyte sedimentation rate (ESR) and the increased plasma viscosity characteristic of such a response.

Chronic inflammatory states. In chronic inflammation immunoglobulin synthesis is often increased, and may be evident as a diffuse rise in γ-globulin. If there is still an active inflammatory reaction this increased density of γ-globulin will be associated with the increase in the α_1- and α_2-fractions of the acute phase response.

Cirrhosis of the liver. The changes in the plasma proteins in liver disease are considered more fully on p. 306. They are usually 'non-specific', but in cirrhosis a characteristic pattern is sometimes seen. Albumin and often α_1-globulin levels are reduced and the γ-globulin concentration is markedly raised, with apparent fusion of the β and γ bands because of an increase in plasma IgA levels.

Nephrotic syndrome. Plasma protein changes depend on the severity of the renal lesion (p. 342). In early cases a low albumin level may be the only abnormality, but the typical pattern in established cases is a reduced albumin, α_1- and γ-globulin, with an increase in α_2-globulin because of an increase in the high molecular-weight α_2-macroglobulin; β-globulin levels are usually normal. If the syndrome is due to systemic lupus erythematosus the γ-globulin may be normal or raised.

α_1-antitrypsin deficiency. Because the α_1-band seen after cellulose acetate electrophoresis consists almost entirely of α_1-antitrypsin, absence or an obvious reduction in density of this band suggests α_1-antitrypsin deficiency (p. 336).

Paraproteinaemia and hypogammaglobulinaemia are discussed in the sections on immunoglobulins (pp. 337 and 335).

Although the changes in the electrophoretic patterns usually indicate disease, they are rarely pathognomonic.

Albumin

Albumin, with a molecular weight of 65 000, is synthesized by the liver. It has a normal plasma biological half-life of about 20 days. About 60 per cent of albumin in the extracellular fluid is in the large interstitial compartment. However, the *concentration* of albumin in the smaller plasma compartment is much higher because of the relative impermeability of the blood-vessel wall. This concentration gradient is important in maintaining plasma volume (p. 34).

There are several inherited variants of albumin synthesis including the *bisalbuminaemias*, in which two chemical types of albumin are present; these are curiosities only, because there are usually no clinical consequences. In another, *analbuminaemia*, there is deficient synthesis of the protein. Clinical consequences are slight, and oedema, although present, is surprisingly mild.

An abnormally high plasma albumin level is found only after loss of protein-free fluid or, artefactually, in a sample taken with prolonged venous stasis (p. 447).

Causes of hypoalbuminaemia. A low plasma albumin level despite a normal total body albumin may be due to dilution by an excess of protein-free fluid, or to redistribution into the interstitial fluid due to increased capillary permeability. There may be true albumin deficiency due to a decreased rate of synthesis, or to an increased rate of catabolism or loss from the body.

Dilutional hypoalbuminaemia may be the result of:

- taking blood from the arm into which an infusion is flowing (artefactual, p. 448);
- administration of an excess of protein-free fluid;
- fluid retention, usually in oedematous states or during late pregnancy.

Redistribution of albumin from plasma to interstitial fluid may be the result of:

- recumbency: plasma albumin levels may be 5 to 10 g/litre lower in the recumbent than in the upright posture, perhaps because of redistribution of fluid;
- increased capillary permeability: this is probably the cause of the rapid fall in plasma levels found in many conditions, such as postoperatively and in severe illnesses such as septicaemia.

The slight fall in albumin level found in even mild illness may be due to a combination of the above two factors.

Decreased synthesis of albumin. Normally about 4 per cent of the body albumin is replaced each day by hepatic synthesis. Hepatic impairment causes hypoalbuminaemia if the rate of synthesis of new amino acids is inadequate to replace those deaminated during metabolism; most of the amino nitrogen is then lost as urinary urea. Hypoalbuminaemia may therefore be due to:

- impairment of synthesis due to chronic liver dysfunction;
- malnutrition, which may be due to malabsorption, resulting in an inadequate supply of dietary amino acids.

Increased catabolism of albumin. Catabolism, and therefore nitrogen loss, is increased in many illnesses. This may aggravate the hypoalbuminaemia due to other causes.

Increased loss of albumin from the body. Because of its relatively low molecular weight, significant amounts of albumin are lost in conditions in which permeability of membranes separating plasma from the outside of the body is increased. The plasma concentration of albumin is higher than that of other low-molecular-weight plasma proteins, and its loss is therefore more obvious. Such protein-losing states include loss through:

- the glomerulus in the nephrotic syndrome (p. 342);
- the skin because of extensive burns or skin diseases such as psoriasis. A large part of the interstitial fluid is subcutaneous;
- the intestinal wall in protein-losing enteropathy (p. 262).

Consequences of hypoalbuminaemia

Fluid distribution. Albumin is quantitatively the most important protein contributing to the plasma colloid osmotic pressure (p. 34). *Oedema* can occur in severe hypoalbuminaemia.

Binding functions. About half the plasma calcium is bound to albumin (p. 173) and hypoalbuminaemia is accompanied by *hypocalcaemia.* Because this involves only the protein-bound (physiologically inactive) fraction tetany does not develop and calcium and vitamin D supplementation is contraindicated.

Albumin also binds *bilirubin, free fatty acids*, and a number of *drugs* such as salicylates, penicillin and sulphonamides. The albumin-bound fractions are physiologically and pharmacologically inactive. A marked reduction in plasma albumin, by reducing the binding capacity, may increase free levels of these substances and may cause toxic effects if drugs are given in their normal dosage (p. 435). Drugs which are albumin-bound may, if administered together, compete for binding sites, also increasing free concentrations (p. 436): an example of this is simultaneous administration of salicylates and the anticoagulant warfarin, with potentiation of the effect of the latter.

Proteins of inflammation and the immune system

The body responds to tissue damage, and to the presence of infecting organisms or other foreign substances, by a complex, interrelated series of cellular and chemical responses. The cells and humoral factors act together to initiate and control the inflammatory reaction and so to remove damaged tissues and foreign substances.

The inflammatory response and the ability to kill organisms may be impaired if there is deficiency of either cellular or humoral components. In autoimmune disease an inappropriate response damages host tissue. This chapter describes briefly some of the chemical responses to tissue damage, including the response of the complement system, and some of the changes in immunoglobulins (synthesized by B lymphocytes) which may be associated with antigenic challenge. Further details, and an account of the cellular response (polymorphonuclear leucocytes, monocytes and T lymphocytes), will be found in textbooks of immunology.

Acute-phase proteins

The non-specific changes in the electrophoretic pattern which occur in response to acute or chronic tissue damage have already been described. The rise in the α_1- and α_2-globulin fractions is due to an increase in several proteins which are synthesized in the liver in response to peptide mediators, such as interleukin 1, released from inflammatory cells. These acute-phase reactants include:

- **activators** of other inflammatory pathways such as *C-reactive protein* (so-called because it reacts with the C-polysaccharide of pneumococci). This protein combines with bacterial polysaccharides or phospholipids released from damaged tissue to become an activator of the complement pathway; plasma levels of C-reactive protein rise rapidly in response to acute inflammation and its assay is particularly useful in the early detection of acute infection;

- **inhibitors** such as *α_1-antitrypsin*, which control the inflammatory response and so minimize damage to host tissue;

- **haptoglobin**, which binds haemoglobin released by local haemolysis during the inflammatory response.

The concentrations of plasma *fibrinogen* and of several *complement* fractions also *increase*. The plasma albumin concentration falls (p. 328) and secondary decreases of complement and haptoglobin occur if there is excessive utilization. Haptoglobin may be undetectable if much haemoglobin is released from red cells due to intravascular haemolysis or haemorrhage into tissues.

The non-specific nature of the response means that measurement of individual acute-phase proteins is rarely helpful as an aid to diagnosis. However, their rates of synthesis and half-lives differ, and measurements of the relative changes in the levels of more than one may be useful to monitor the progress of inflammatory disease such as rheumatoid arthritis.

Complement

The complement proteins are synthesized by macrophages or hepatocytes and, because of the presence of inhibitors in the plasma, usually circulate in an inactive form. Sequential activation with utilization of complement proteins during the

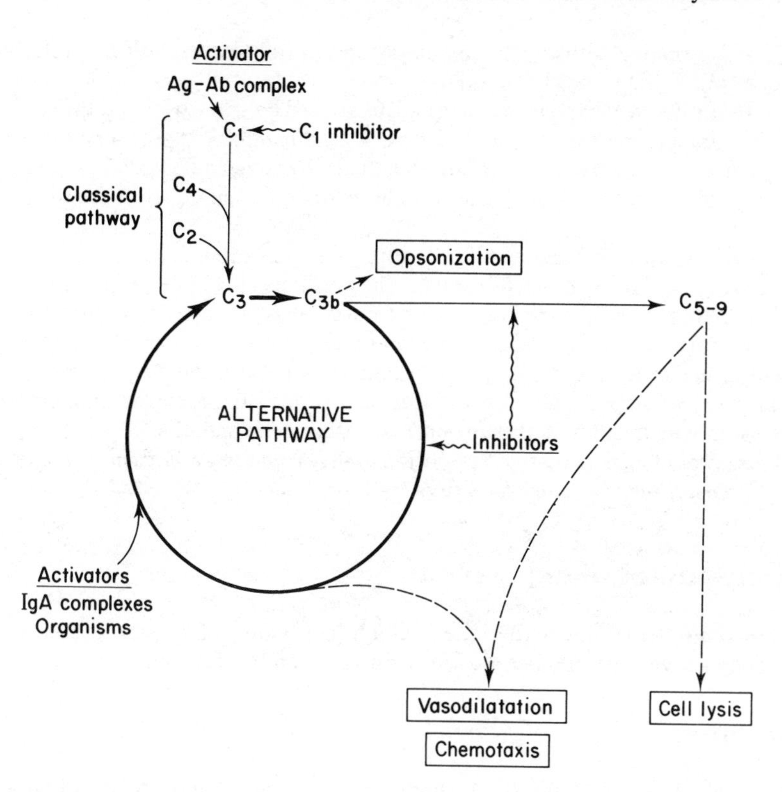

Fig. 16.3 The complement pathway (much simplified).

inflammatory process reduces their plasma levels. The products of activation attract phagocytes to the area of inflammation (*chemotaxis*) and, by increasing the permeability of the capillary wall to both cellular and chemical components, allow them to reach affected cells; acting together with immunoglobulins they opsonize and lyse foreign cells. Other acute-phase proteins, synthesized in response to mediators released from the inflammatory cells, help to control and limit the response (see earlier).

Clinically the most important of the groups of complement proteins is C_3, activation of which results in chemotaxis. Two main pathways are concerned with activation of C_3 (Fig. 16.3). The names of these pathways reflect the order in which they were discovered, rather than their importance. Activation of either results in low C_3 levels.

- In the *alternative pathway*, which is the more important of the two, C_3 is activated by IgA immune complexes, lipopolysaccharides on the surfaces of invading organisms, or by the products of activation of the classical pathway. C_{3b} is formed, which in turn activates more C_3. Other products of C_{3b} cause vasodilatation and cell lysis. The cyclical process is self-perpetuating and *C_3 levels fall*. If the initial stimulus is removed, inhibitors, such as those formed in the acute-phase reaction, control the reaction and C_3 levels become normal.

The alternative pathway is most important in individuals with no preformed antibodies and in newborn infants.

- In the *classical pathway* C_1 is usually activated by antigen-bound IgG or IgM (immune complexes). C_4 (and C_2) are used during the resultant sequence of events, which activates the alternative pathway. *C_3 and C_4 levels therefore fall.* Once the formation of immune complexes stops C_3 and C_4 levels are restored.

In immune complex disease (for example, systemic lupus erythematosus, SLE) circulating immune complexes persist. The products of C_3, released by continued activation of the classical pathway, may damage blood vessels, joints and kidneys. Both plasma C_3 and C_4 levels are low, and persistently low C_3 levels may help to distinguish chronic mesangiocapillary glomerulonephritis, with a poor prognosis, from the less serious and self-limiting acute post-streptococcal glomerulonephritis, in which C_3 levels become normal within a few months.

This account is, of necessity, simplified. *Before requesting complement studies you should contact the laboratory for advice.*

Immunoglobulins

It has long been known that the γ-globulin fraction of plasma proteins has antibody activity. Antibodies also occur in the β- and α_2-fractions.

Structure

The basic immunoglobulin is a Y-shaped molecule depicted schematically in Fig. 16.4.

- Usually four polypeptide chains are linked by disulphide bonds. There are two heavy (H) and two light (L) chains in each unit. The *H chains* in a single unit are similar and determine the immunoglobulin *class* of the protein. H chains γ, α, μ, δ and ϵ occur in IgG, IgA, IgM, IgD and IgE respectively. *L chains* are of two *types*, κ or λ. In a single molecule the L chains are of the same type, although the Ig class as a whole contains both types.

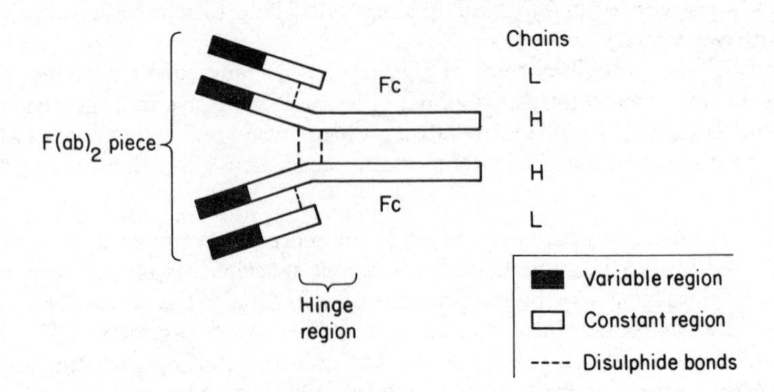

Fig. 16.4 Schematic representation of Ig unit.

- There are two antigen-combining sites per unit, together known as the $F(ab)_2$ piece. These lie at the ends of the arms of the Y: both H and L chains are necessary for full antibody activity. The amino acid composition of this part of the chain varies in different units (*variable region*). When both of these sites combine with antigen, conformational changes are transmitted through the hinge region (Fig. 16.4) and the Fc portion of the molecule becomes activated.
- The rest of the H and L chains is less variable (*constant or Fc region*). The constant region of the H chains is responsible for such properties of the Ig unit as the ability to bind complement or actively to cross the placental barrier. The H chains are associated with a variable amount of carbohydrate; IgM has the highest content.

Some immunoglobulin molecules may contain more than one basic unit bound together by 'J' chains: for example, the IgM molecule consists of 5 units. Because of this variation in size and therefore in density, the classes can be separated by ultracentrifugation. They are classified by their Svedberg coefficient (S), the S value of a protein increasing with increasing size.

The S values, most important functions and other properties of immunoglobulins are given in Table 16.1. As we have seen, they act synergistically with the acute-phase proteins, including complement.

Table 16.1 Properties and functions of plasma immunoglobulins

	IgG	IgA	IgM	IgE	IgD
Molecular weight	160 000	160 000 (polymers occur)	1 000 000	200 000	190 000
Sedimentation coefficient	7S	7S	19S	8S	7S
% total plasma Ig	73	19	7	0.001	1
Complement activation	Yes	Yes (as complexes)	Yes	No	No
Placental transfer	Yes	No	No	No	No
Approx. mean normal adult concentration (g/litre)	10	2.0	1.0	0.0003	0.03
Adult levels reached by:	3 to 5 years	15 years	9 months	? 15 years	? 15 years
Major function	Protects extravascular tissue spaces Secondary response to antigen. Neutralizes toxins	Protects body surfaces as secretory IgA (11S)	Protects blood stream Primary response to antigen Lyses bacteria	Mast-cell bound antibodies of immediate hyper-sensitivity reactions	Not known

Normal immunoglobulin response to infection

Immunoglobulins are synthesized by B lymphocytes.

IgM is synthesized first in response to particulate antigens such as blood-borne organisms. Because of their large size they are almost confined to the intravascular compartment, and this fact, together with the speed of synthetic response, makes them the first line of defence amongst immunoglobulins against invading organisms. The fetus can synthesize IgM and high levels at birth usually indicate that there has been intrauterine infection (p. 357).
IgG levels rise slightly later in response to soluble antigens such as bacterial toxins. Because of their relatively low molecular weight they can diffuse fairly freely into the interstitial fluid and act in tissue against infection. Within a few weeks of the initial infection raised levels of all immunoglobulins may be demonstrated and may be evident as a diffusely increased γ band on the electrophoretic strip.
Much **IgA** is synthesized submucosally and is present in intestinal and respiratory secretions, sweat, tears and colostrum. It is affected more than other immunoglobulins in diseases of the gastrointestinal and respiratory tracts. Secretory IgA is a dimer in which two subunits are joined by a peptide J chain and, in addition, have a 'secretory piece' synthesized by epithelial cells.

Most infections produce a general *polyclonal* immunoglobulin response. In such circumstances immunoglobulin estimation adds little to the observation of an increase in γ-globulin on routine electrophoresis. In certain conditions, some of which are listed in Table 16.2, one or more immunoglobulin classes predominate. Although there is considerable overlap, individual Ig estimation may help in the diagnosis of such cases.

The immunoglobulin response to allergy

IgE is synthesized by plasma cells beneath the mucosae of the gastrointestinal and respiratory tracts and by those in the lymphoid tissue of the nasopharynx. It is present in nasal and bronchial secretions. Circulating IgE is rapidly bound to cell surfaces, particularly those of mast cells and circulating basophils and plasma levels are therefore very low. Combination of antigen with this cell-bound antibody results in the cells releasing mediators and accounts for immediate hypersensitivity reactions such as occur in hay fever. Desensitization therapy of allergic

Table 16.2 Some abnormalities of plasma immunoglobulins in disease

Predominant Ig	Examples of clinical conditions
IgG	Autoimmune diseases, such as SLE or chronic active hepatitis
IgA	Diseases of the intestinal tract, for example Crohn's disease
	Diseases of the respiratory tract, for example tuberculosis, bronchiectasis
IgM	Primary biliary cirrhosis
	Viral hepatitis
	Parasitic infestations, especially when there is parasitaemia
	At birth, indicating intrauterine infection

disorders aims at stimulating production of circulating IgG against the offending antigen, to prevent it reaching cell-bound IgE and/or at suppressing IgE synthesis. Raised *concentrations* are found in several diseases with an *allergic* (atopic) component such as in some cases of eczema, asthma and parasitic infestations.

Deficiencies of the proteins of the inflammatory response

Deficiencies of any of the proteins described above may lead to an increased susceptibility to infection or to a modification of the inflammatory response.

Immunoglobulin deficiency

It is important to remember that the plasma immunoglobulins reflect only the humoral phase of the immune system. Cellular immunity is not assessed and *normal immunoglobulin levels do not exclude an immune deficiency state.* For example, plasma immunoglobulin concentrations are usually raised in the acquired immune deficiency syndrome (AIDS). Severe reduction in immunoglobulin levels may produce obvious hypogammaglobulinaemia in the electrophoretic pattern, but usually measurement of individual proteins is needed to make a diagnosis.

The effects of deficiency of individual immunoglobulins are related to their functions and distribution. In *IgM deficiency* septicaemia is common. *IgG deficiency* may result in recurrent pyogenic infections of tissue spaces, especially in the lungs and skin, by toxin-producing organisms such as staphylococci and streptococci. *IgA deficiency* may be symptomless, or may be associated with recurrent, mild respiratory tract infections or intestinal disease.

Primary immunoglobulin deficiency is less common than deficiency secondary to other disease.

Several classifications have been proposed for these deficiences and three categories presenting with Ig deficiency will be discussed.

Transient immunoglobulin deficiency. In the newborn infant circulating IgG is derived from the mother by placental transfer. Levels decrease over the first 3 to 6 months of life and then gradually rise as endogenous IgG synthesis increases. In some subjects onset of synthesis is delayed and 'physiological hypogammaglobulinaemia' may persist for several months.

Most of the IgG transfer across the placenta takes place in the last three months of pregnancy. Severe deficiency may therefore develop in very premature babies, because the level of IgG derived from the mother falls before the endogenous level rises.

Primary immunoglobulin deficiency. Primary *IgA deficiency* in plasma, saliva and other secretions is relatively common, with an incidence of about 1 in 500 of the population. Most are symptom-free.

Several rare syndromes, usually familial, have been described. In one, *infantile sex-linked agammaglobulinaemia* (Bruton's disease), which occurs only in males, there is almost complete absence of B cells and circulating immunoglobulins, while cellular (T cell) immunity seems normal. Other syndromes have varying degrees of immunoglobulin deficiency and impaired cellular immunity, and can occur in either sex.

Secondary immunoglobulin deficiency. Low serum immunoglobulin levels are most common in patients with malignant disease, particularly of the haemopoietic and immune systems, and are often precipitated by chemo- or radiotherapy; they are an almost invariable finding in patients with myelomatosis. In severe protein-losing states, such as the nephrotic syndrome, low immunoglobulin levels (especially of IgG) are partly due to the loss of relatively low-molecular-weight Ig and partly due to increased catabolism.

Complement deficiency

Inherited deficiences of most of the proteins of the complement system have been described. These may be associated with repeated infection or immune-complex disease or may have no clinical consequences. Most are extremely rare.

In **hereditary angioneurotic oedema**, C_1 inhibitor (Fig. 16.3, p. 331) is deficient or non-functioning. The condition is characterized by episodically increased capillary permeability and consequent oedema of the subcutaneous tissues and mucous membranes of the upper respiratory or gastrointestinal tract. Laryngeal oedema may be fatal. The level, or sometimes the activity, of the inhibitor may be measured.

Alpha$_1$ antitrypsin deficiency

The α_1-antitrypsins normally control the proteolytic action of enzymes from phagocytes (protease inhibitors; Pi).

There are about 30 genetic variants of α_1-antitrypsin, which are inherited as autosomal codominant alleles. The normal allele is PiM and the normal genotype MM.

The most important abnormal alleles which cause α_1-antitrypsin deficiency are called *null*, in which none of the protein is synthesized, and *Z*, in which protein accumulates in the liver because it cannot be secreted after synthesis. The condition may be suspected after serum protein electrophoresis if the α_1 band is much reduced or absent (Fig. 16.2, p. 326). The unopposed action of proteases from phagocytes in the lung may, by destroying elastic tissue, cause basal emphysema in young adults who are homozygous for either of these abnormal alleles; the condition may be exacerbated by cigarette smoking or infection. Hepatic damage occurs in 10 to 20 per cent of subjects, such as those with PiZZ, in whom the protein cannot be secreted by hepatocytes; the condition may present as hepatitis in the neonatal period or as cirrhosis in children or young adults.

The diagnosis can usually be confirmed by demonstrating that the

α_1-antitrypsin level is low, but sometimes the phenotype should also be identified. Blood relations should be investigated and all those with abnormal levels advised not to smoke.

B cell disorders

The normal B cell response to antigenic stimulation is *polyclonal*. Many different groups (clones) of B cells synthesize a range of different immunoglobulins which cause the appearance of diffuse hypergammaglobulinaemia in the electrophoretic pattern.

Paraproteinaemia

Each B cell is highly specialized and synthesizes a single class and type of immunoglobulin. If, by contrast to the usual polyclonal response to an antigenic challenge such as infection, one of these cells proliferates to form a clone of like cells, a single protein will be produced in excess. This *monoclonal* proliferation of B cells is often, but not always, malignant.

The term 'paraprotein' refers to the appearance of an abnormal narrow, dense band on the electrophoretic strip, most commonly in the γ-region, but which may be anywhere from the α_2 to the γ regions inclusive (Fig. 16.2, p. 326). A paraprotein can often be shown to be monoclonal.

Causes of paraproteinaemia. Although the presence of a paraprotein is strongly suggestive of a malignant process, this is not always so.

Paraproteins may be found in the following malignant conditions:

- myelomatosis, which accounts for most of the cases of malignant paraproteinaemia;
- macroglobulinaemia;
- B cell lymphomas, including chronic lymphatic leukaemia.

Immune paresis

The synthesis of immunoglobulins from other clones may be suppressed by monoclonal proliferation. In such cases the γ band, other than the narrow paraprotein, is reduced or absent and levels of the other immunoglobulins can be shown to be reduced (Fig. 16.2, p. 326).

Bence Jones protein

Bence Jones protein (BJP) is usually, but not invariably or exclusively, found in the urine of patients with malignancy of B cells, but can rarely be found without malignant disease. It consists of free monoclonal light chains, or fragments of them, which have been synthesized much in excess of H chains, implying a degree of dedifferentiation. Because of its relatively low molecular weight (20 000 to 40 000) the protein is filtered at the glomerulus and only accumulates in the plasma if there is glomerular impairment or if it polymerizes. BJP may damage renal tubular cells and may form large casts, producing the 'myeloma kidney'. It may also form amyloid deposits in tissues.

The coexistence of paraproteinaemia, immune paresis and Bence Jones protein is very suggestive of malignancy of B cells.

Results of malignant B cell proliferation

Some of the clinical and laboratory findings are similar in all malignant B cell tumours. Whether they occur depends on the level of paraprotein, the presence or absence of immune paresis and the presence of BJP. It should be remembered that malignant tumours of B cells can exist without all, or rarely any, of these findings.

Consequences of the presence of paraprotein. Unless the level of paraprotein is very high these findings are not present. Very high levels (which are suggestive of malignancy) may be associated with a very high ESR and may cause:

- *in vivo* effects of increased plasma viscosity with sluggish blood flow in small vessels; these effects include *retinal-vein thrombosis* with impairment of vision, *cerebral thrombosis*, or even *peripheral gangrene* (hyperviscosity syndrome);
- an obviously increased blood viscosity during venepuncture: the blood may clot in the syringe, and it may be difficult to prepare blood films;
(Hyperviscosity is most common in macroglobulinaemia, but may sometimes occur in myelomatosis).
- a high plasma total protein despite a normal or low albumin level;
- spurious hyponatraemia due to the space-occupying effect of the protein (p. 446).

Consequences of immune paresis. Because of the abnormal spectrum of immunoglobulins there may be increased susceptibility to infection.

Effects of Bence Jones protein. Cases with large amounts of Bence Jones protein are especially likely to be associated with:

- *renal failure*, due to deposition of BJP in tubular cells;
- *amyloidosis*.

Other common findings in cases with malignant B cell tumours are:

- *normochromic, normocytic anaemia*, a common presenting feature of any malignant disease;
- small *haemorrhages*, perhaps due to complexing of coagulation factors by the paraprotein;
- *Raynaud's phenomenon* if the paraprotein is a cryoglobulin (p. 340).

β_2-microglobulin is a low molecular weight (11 800) protein that forms part of the HLA antigen on the surface of all nucleated cells. The protein is readily filtered at the glomerulus and plasma levels are normally low. Raised plasma levels occur in a number of haematological malignancies. In myelomatosis, for example, the plasma β_2-microglobulin level is an index of the extent of the disease.

Myelomatosis (multiple myeloma; plasma-cell myeloma)

Myelomatosis is a condition which becomes increasingly frequent after the age of 50; it is rare before the age of 30. It occurs equally in both sexes. In the commonest form there is malignant proliferation of plasma cells throughout the bone marrow. In such cases the clinical features are due to:

- malignant proliferation of plasma cells;
- disordered immunoglobulin synthesis and/or secretion from the cell.

Malignant proliferation of plasma cells. *Bone pain* may be severe and is due to pressure from the proliferating cells. *X-rays* may show discrete *punched-out areas* of radiotranslucency, most frequently in the skull, vertebrae, ribs and pelvis. There may be generalised osteoporosis. Histologically there is little osteoblastic activity around the lesion and plasma alkaline phosphatase activity is therefore normal.

Pathological fractures may occur.

The effects of paraproteinaemia, immune paresis, BJP (which occurs in about 70 per cent of cases) and the other possible findings have been discussed above.

Other laboratory findings. The immunoglobulin usually increased is IgG, less commonly IgA (about 2.5:1) and rarely Bence Jones protein (if renal failure is present). Occasionally IgD, IgM or IgE are found, the last two being very rare.

In about 20 per cent of cases, usually those with BJP production, no paraprotein can be detected in the plasma. Rarely, neither a paraprotein nor BJP can be found. In either case there is usually immune paresis.

In IgD myelomatosis an increase in γ-globulin may not be detectable by routine electrophoresis.

Hypercalcaemia may be present (p. 181). High levels are usually suppressed by hydrocortisone (see steroid suppression test, p. 187) and steroids can be used to treat such hypercalcaemia.

As there is little osteoblastic activity, the *alkaline phosphatase activity is normal* unless there is liver involvement, when the raised level can be shown to be of hepatic, not bony, origin. A normal alkaline phosphatase activity in cases with bone lesions suggest myelomatosis rather than bony metastases.

Bone marrow appearance. The proportion of plasma cells in the bone marrow is increased, and many of these cells are atypical ('myeloma cells'). *Examination of a bone marrow film must be carried out before myelomatosis is diagnosed or excluded.*

Soft tissue plasmacytoma. Rarely myeloma involves soft tissues, without marrow changes (extramedullary plasmacytoma). Although the protein abnormalities of myelomatosis are often found in these cases, the behaviour and prognosis of the two conditions are different. Spread of soft-tissue plasmacytoma is slow and tends to be local. Local excision of the solitary tumour is often effective. The paraprotein should be typed, and the level monitored to follow progress.

Waldenström's macroglobulinaemia

Like myelomatosis, macroglobulinaemia usually occurs in the older age group (between 60 and 80 years), but unlike myelomatosis is commoner in men than in women. It is due to malignancy of B cells, but the malignant cells resemble lymphocytes rather than plasma cells. Symptoms of the 'hyperviscosity syndrome' are commoner than in myelomatosis, probably because of the large size of the IgM molecule, but skeletal manifestations are rare. There is lymphadenopathy.

Laboratory findings and diagnosis. *Serum protein changes*. The paraprotein in the γ region can be identified as monoclonal IgM. The serum IgA concentration is usually reduced, but that of IgG *may* be raised.

The *bone marrow* aspirate or lymph node biopsy contains atypical lymphocytoid cells.

Heavy-chain diseases

The heavy-chain diseases are a rare group of disorders characterized by the presence of an abnormal protein in plasma or urine identifiable as part of the H chain (α, γ or μ). The clinical picture is that of generalized lymphoma (γ-chain disease), intestinal lymphomatous lesions, with malabsorption (α-chain disease) or chronic lymphatic leukaemia (μ-chain disease). In some cases a paraprotein is detectable in the serum.

Cryoglobulinaemia

Proteins which precipitate when cooled below body temperature are called cryoglobulins. They may be associated with diseases known to produce paraproteins. About half of them can be shown to consist of a monoclonal immunoglobulin (usually IgM or IgG). The patient usually presents with other symptoms of the underlying disease and the cryoglobulin is found during investigation. Occasionally, especially if the concentration of protein is high and if it precipitates at temperatures above 22°C, intravascular precipitation may cause such skin lesions as purpura and Raynaud's phenomenon.

In some cases the protein can be shown to be polyclonal and sometimes to include complement; these cases may be associated with immune complex disease (p. 332), but occasionally no underlying abnormality can be found (*essential cryoglobulinaemia*).

Paraproteinaemia without obvious cause

If a paraprotein is found, investigation for one of the diseases discussed above should be initiated. In between 10 and 30 per cent of cases in hospital (and probably more in the 'well population') no cause can be found. The condition, which may be transient, has been called '*benign*' or '*essential*' *paraproteinaemia* or *monoclonal gammopathy of undetermined significance.* The diagnosis should be made provisionally and the patients followed up; they may later develop obvious myelomatosis or macroglobulinaemia.

Proteinuria

The loss of most plasma proteins through the glomerulus is restricted by the size of the pores in, and by a negative charge on the basement membrane that repels negatively-charged protein molecules. Alteration of either of these factors by glomerular disease may allow albumin and larger proteins to enter the filtrate. Low-molecular-weight proteins are filtered even under normal conditions. Most are absorbed and metabolized by tubular cells, but normal subjects excrete up to 0.08 g of protein a day in the urine, amounts undetectable by usual screening tests. Proteinuria of more than 0.15 g a day almost always indicates disease.

Significant proteinuria may be due to renal disease or, more rarely, may occur because large amounts of low-molecular-weight proteins are circulating and therefore being filtered. *Blood and pus in the urine give positive tests for protein.*

Renal proteinuria

Glomerular proteinuria is due to increased glomerular permeability *(nephrotic syndrome)*: this is discussed more fully below. *Albumin* is usually the predominant protein in the urine.

'*Orthostatic (postural) proteinuria.*' Most proteinuria is more severe in the upright than in the prone position. The term 'orthostatic' or 'postural' has been applied to proteinuria, often severe, which disappears at night. It appears to be glomerular in origin and is commonest in adolescents and young adults. Although it is often harmless, evidence of renal disease may occur after some years.

Microalbuminuria. Sensitive immunological assays have shown the normal daily excretion of albumin to be less than 50 mg. Patients with diabetes mellitus who excrete more than this, but whose total urinary protein excretion is 'normal', are said to have microalbuminuria. Such patients have a greater risk of developing progressive renal disease than those whose albumin excretion is normal.

Tubular proteinuria may be due to renal tubular damage from any cause, especially pyelonephritis. If glomerular permeability is normal, proteinuria is usually less than 1 g a day, and consists mainly of low molecular weight globulins. One of these, β_2-microglobulin (p. 338) has been measured as a sensitive marker of renal tubular damage.

Proteinuria with normal renal function

Proteinuria can be due to production of *Bence Jones protein,* to severe haemolysis with *haemoglobinuria,* or to severe muscle damage with *myoglobinuria.* In the latter two cases the urine may be red or brown in colour.

Apparent proteinuria has occasionally been found because the patient has added egg white or animal blood to the urine.

Bence Jones proteinuria can be inferred by comparison of urinary and serum electrophoretic patterns. BJP is the only protein of a molecular weight lower than albumin likely to be found in significant amounts in unconcentrated urine (in the absence of haemoglobinuria or myoglobinuria). The presence in the urine, especially if it is not present in the serum, of a band which is denser than that of albumin suggests BJP.

Nephrotic syndrome

In the nephrotic syndrome *increased glomerular permeability* causes a protein loss, by definition, of *more than 6 g a day,* with consequent hypoalbuminaemia and oedema, and with hyperlipidaemia. The renal disease may be primary, or secondary to other pathology. It has been reported:

- *in most types of glomerulonephritis,* usually due to deposition of circulating immune complexes in the glomerulus (about 80 per cent of nephrotics). In children 'minimal change' glomerulonephritis is the commonest cause;
- *secondary to:*
  - diabetes mellitus;
  - systemic lupus erythematosus (SLE) (due to immune complexes);
  - inferior vena-caval or renal-vein thrombosis;
  - amyloidosis;
  - malaria due to P. *malariae* (due to immune complexes).

Laboratory findings

Protein abnormalities. *Proteinuria* in the nephrotic syndrome ranges from 6 to 50 g a day. The proportion of different proteins lost is an index of the severity of the glomerular lesion. In mild cases albumin (MW 65 000) and transferrin (MW about 80 000) are the predominant urinary proteins, and α_1-antitrypsin (MW 50 000) is also present. With increasing glomerular permeability IgG (MW 160 000) and larger proteins appear.

The *serum protein electrophoretic picture* reflects the pattern of loss and has been described on p. 327.

Electrophoresis of urine on cellulose acetate gives some idea of the severity of the glomerular lesion. The *differential protein clearance* is a more precise measure of the selectivity of the lesion. The clearance of a low MW protein, such as transferrin or albumin, is compared with that of a larger one, such as IgG. The result is usually expressed as a ratio, obviating the need for timed collections. A ratio of IgG to transferrin or albumin clearance of less than 0.2 indicates high selectivity (predominant loss of small molecules), with a more favourable prognosis than those cases in whom the ratio is higher: such cases usually respond well to steroid or cyclophosphamide therapy.

The consequences of the protein abnormalities are:

- *oedema* due to hypoalbuminaemia (p. 329);

- *reduction in the concentration of protein-bound substances* due to loss of carrier proteins. It is important not to misinterpret low total levels of calcium, thyroxine, cortisol and iron.

Lipoprotein abnormalities. In mild cases LDL increases, with consequent *hypercholesterolaemia.* In more severe cases a rise in VLDL *(triglycerides)* may cause plasma turbidity. The reason for these abnormalities is not understood.
Fatty casts may occur in the urine.

Renal function tests. In the early stages glomerular permeability is high and the plasma urea concentration normal. Later glomerular failure may develop, with uraemia. At this stage, protein loss is reduced and plasma levels of protein and lipid may revert to normal. In the presence of uraemia this does *not* indicate recovery.

The student should read the indications for protein estimation on p. 345.

Summary

1. Albumin is the plasma protein most often measured specifically. Its functions include control of fluid distribution between plasma and the extracellular compartment and binding and consequent inactivation of many endogenous and exogenous substances.
2. Groups of proteins may be separated by electrophoresis. In normal serum five bands can be seen – albumin and α_1-, α_2-, β-and γ-globulins. In plasma a sixth band, fibrinogen, is present.
3. Clinically helpful electrophoretic patterns may be found in the nephrotic syndrome, the paraproteinaemias, hypogammaglobulinaemia, α_1-antitrypsin deficiency, and some cases of cirrhosis of the liver.
4. The acute-phase reaction occurs in many inflammatory states and in tissue damage. It is characterized by increased density of the α_1 and α_2 bands on electrophoresis.
5. Measurement of specific acute-phase proteins, such as α_1-antitrypsin, C-reactive protein and haptoglobin, is occasionally helpful in monitoring inflammatory disease. The level of the C_3 fraction of complement may fall in such disease.
6. The immunoglobulins, complement and other acute-phase proteins act synergistically in the presence of invading organisms.
7. The immunoglobulins (Ig) are a group of proteins that are structurally related. Five classes are described. The main ones are IgG, IgA and IgM. Paraproteins are immunoglobulins and Bence Jones protein is related to them.
8. The functions and properties of the Ig classes and their response to antigenic stimuli differ. Estimation of IgG, IgA and IgM may be helpful in a few conditions.
9. Immunoglobulin deficiency is only one aspect of immunological deficiency. *Normal Ig levels do not exclude immunological deficiency.* Ig deficiencies may be primary or secondary and may involve one or all Ig classes.
10. A paraprotein is a narrow band most commonly found in the γ-globulin

region of the electrophoretic strip. It usually consists of monoclonal immunoglobulins, and almost always, but not invariably, indicates malignant proliferation of B cells.

11. Paraproteins are most commonly associated with myelomatosis. Myelomatosis presents clinically in a variety of ways, reflecting bone marrow replacement and abnormal plasma protein levels. The laboratory diagnosis is made by bone marrow examination, and by finding protein abnormalities in serum and/or urine.

12. Bence Jones protein (usually found only in urine) consists of monoclonal free light chains. Its presence usually indicates malignancy of B cells.

13. Cryoglobulins are abnormal proteins which precipitate when cooled below body temperature. They may cause symptoms on exposure to cold. They may occur in any of the diseases associated with paraproteinaemia.

Proteinuria

1. Proteinuria may be due to glomerular or tubular disease. Glomerular proteinuria is the commoner form: massive proteinuria is always of glomerular origin.

2. Proteinuria may occur with normal renal function if abnormally large amounts of low-molecular-weight proteins are being produced.

3. The nephrotic syndrome is characterized by proteinuria of at least 6 g a day, with a low serum albumin, oedema and hyperlipidaemia. The proteinuria is glomerular in type. The severity of the lesion may be assessed by differential protein clearance.

Further reading

Whicher JT. The interpretation of electrophoresis. *Br J Hosp Med.* 1980; **24:** 348–60.

Whicher JT. Serum protein zone electrophoresis – An outmoded test? *Ann Clin Biochem* 1987; **24:** 133–9.

Powell RJ. Serum complement levels. *Br J Hosp Med* 1984; **32:** 104–10.

Whicher JT. The role of immunoglobulin assays in clinical medicine. *Ann Clin Biochem* 1984; **21:** 461–66.

Rosen FS, Cooper MD, Wedgwood RJP. The primary immunodeficiencies. *New Engl J Med* 1984; **311:** 235–42 and 300–8.

Whicher JT. The laboratory investigation of paraproteinaemia. *Ann Clin Biochem* 1987; **24:** 119–32.

Kyle RA. Multiple myeloma. Review of 869 cases. *Mayo Clin Proc* 1975; **50:** 29–40.

Parfrey PS. The nephrotic syndrome. *Br J Hosp Med.* 1982; **27:** 155–62.

Roitt IM, Brostoff J, Male DK. *Immunology.* Edinburgh: Churchill Livingstone, 1985.

Blood sampling for protein estimations

Blood for protein estimations, including those for immunoglobulins, should be taken with a *minimum of stasis,* or falsely high results may be obtained.

Clotted blood is essential for routine electrophoresis, because the presence of fibrinogen may mask, or be interpreted as, an abnormal protein.

Samples for *complement* estimations must be taken and processed so as to minimize *in vitro* activation. *It is essential to contact your laboratory for details before taking the blood.*

Blood for *cryoglobulin* measurement should be collected in a syringe *warmed to 37°C, and should be maintained at this temperature until it has been tested.* Failure to observe this precaution may result in false negative findings, because the cryoprecipitate is incorporated into the blood clot on cooling.

Indications for protein estimations

We have discussed the changes that may occur in plasma protein levels in disease. Demonstrating these changes does not always aid diagnosis. Before making a request, some assessment should be made as to whether the result of the estimation will aid diagnosis or treatment. The following are the commonest indications for protein estimation.

To assess changes in hydration

Changes in plasma *total protein* or *albumin* levels over short periods of time are almost certainly due to changes in hydration (p. 324), or to changes in capillary permeability (p. 328).

To evaluate apparently abnormal levels, or changes in level, of a protein-bound substance

Plasma *albumin* estimation must always accompany that of calcium (pp. 194 and 196). If changes in other protein-bound substances parallel those of albumin they are probably due to changes in protein levels.

To investigate oedema (see p. 46)

Very low plasma *albumin* levels suggest that hypoalbuminaemia is the cause of the oedema.

In the presence of hypoalbuminaemia a *serum electrophoretic pattern* typical of nephrotic syndrome or hepatic cirrhosis (Fig. 16.2, p. 326) may help to confirm the cause: the diagnosis of nephrotic syndrome depends on finding gross proteinuria. If oedema is due to causes other than hypoalbuminaemia the dilution leads to a fall in the levels of albumin and all other protein fractions.

To investigate suspected immune complex disease

Plasma C_3 determination is indicated if immune complex disease is suspected. If such disease is in an active phase, the *level will be low:* if high it suggests another cause for tissue damage.

The *electrophoretic pattern* may show a diffuse rise in γ-globulin in immune complex disease. It should be remembered that this finding may be due to a number of other types of disease.

To investigate the cause of recurrent infection

Serum immunoglobulin concentrations should be measured as part of the assessment of the adequacy of the immune system. Note that these tests assess only *humoral* immunity.

Serum electrophoresis may indicate the cause of low immunoglobulins (for example, there may be a protein-loss pattern (p. 327) or a paraprotein). Estimation of the 24-hour *urinary protein loss* may be indicated if this cause is suspected.

To investigate suspected myelomatosis or macroglobulinaemia

Electrophoresis should be carried out on *serum,* to detect a paraprotein, *and on urine,* to detect Bence Jones protein. The diagnosis of myelomatosis must be confirmed by inspection of a bone marrow aspirate or biopsy.

Plasma immunoglobulin levels should be measured to assess the degree of immune paresis.

To investigate liver disease

In chronic liver disease *serum albumin* levels may be low.

Serum electrophoresis may show β-γ fusion in cirrhosis or a diffuse increase in γ-globulin in chronic active hepatitis.

Serum immunoglobulins may help in the differential diagnosis. In cirrhosis there is a predominant increase of IgA, in biliary cirrhosis of IgM, and in chronic active hepatitis of IgG.

Protein estimations are usually unhelpful in acute liver disease.

Testing for urinary protein

Several rapid screening tests are in routine use. There are two *limitations* of screening tests.

1. The tests were developed to detect albumin and may be negative in the presence of other proteins, such as BJP.
2. Because the test depends on protein *concentration*, very dilute urine may give negative results despite significant proteinuria.

It is *essential* that the sample should be *fresh*.

Albustix (Ames). The test area of the reagent strip is impregnated with an indicator, tetrabromphenol blue, buffered to pH 3. At this pH it is yellow in the absence of protein; because protein forms a complex with the dye, stabilizing it in the blue form, it is green or bluish-green if protein is present. The colour after testing is compared with the colour chart provided, which indicates the approximate protein concentration. The strips should be kept in the screwtop bottles, in a cool place. The instructions on the container should be carefully followed.

False positive results occur:

- if the specimen is contaminated with vaginal or urethral secretions;
- in strongly alkaline (infected or stale) urine, when buffering capacity is exceeded. A green colour in this case is a reflection of the alkaline pH;
- if the urine container is contaminated with disinfectants such as chlorhexidine.

False negative results occur if acid has been added to the urine as a preservative (for example, for the estimation of urinary calcium).

17

The clinical chemistry of the newborn

The survival rate of very small premature infants has increased because of the improved specialized medical and nursing techniques for treating the newborn, but often at the expense of serious short- and longterm morbidity. This improved survival rate has increased the need for laboratory methods adapted for assaying very small specimens.

Diseases occurring during the neonatal period can be divided into two main groups:

- those of the ill, full-term infant;
- those of the infant born before term, in whom immaturity contributes to the severity of the disease.

The most common disorders in both groups are perinatal asphyxia, the respiratory distress syndrome and infection.

With the exception of some tests for inborn errors of metabolism, the same biochemical investigations are used to diagnose and monitor disease during the neonatal period as in adults. However, analysis is more labour intensive than it is in adult chemical pathology. Requesting is inevitably less selective because of the non-specificity of the presenting clinical signs and the inability of the infant to give a history; by contrast, the number of investigations that can be performed may be limited by the small volume of the samples. The blood volume of a premature infant weighing 1 kg is about 90 ml, compared with about 5 litres in a 70 kg adult: only a small amount of blood can be taken without causing anaemia. Venous (or arterial) samples, although sometimes difficult to obtain, are preferable to capillary ones, since the latter are more prone to contamination from the interstitial and cellular fluids. *Before requesting tests on small samples the clinician should contact the laboratory to discuss the tests needed in order of priority.*

Interpretation of results is influenced by the different reference ranges at different ages. Some of these, such as the relatively high plasma alkaline phosphatase activity found during childhood, are discussed in other chapters: some will be mentioned in this chapter.

Many disorders of infancy and childhood are discussed in the relevant chapters. Some inborn errors of metabolism are described in Chapter 18. In this chapter we will discuss those diseases particularly affecting the newborn infant and their pathophysiology. A list of references giving a more detailed account of paediatric chemical pathology is given at the end of this chapter.

Renal function

By the 36th week of gestation the kidney has a full complement of nephrons, but renal function is not fully mature until the age of about two years. Glomerular function develops more rapidly than that of the tubules.

The glomerular filtration rate (GFR) doubles during the first two weeks of life because of an increase in, and redistribution of, renal blood flow. Consequently the plasma *creatinine* concentration, which is higher at birth than that of the adult, decreases rapidly at first: the fall is slower in the preterm infant. The GFR *related to the child's surface area* and the plasma creatinine level reach adult values by about six months of age.

Plasma *urea* levels are low in newborn infants, despite the relatively low GFR; the high anabolic rate results in more nitrogen being incorporated into protein, rather than into urea, when compared with the adult. Plasma urea levels fluctuate markedly with varying nutritional state, metabolic rate and state of hydration. Neither plasma urea nor creatinine level is a very sensitive indicator of renal function in this age group. Such function is difficult to assess in the neonatal period.

The control of secretion of the hormones acting on the kidney, ADH and aldosterone, is fully developed even in premature neonates. However, the kidney cannot respond fully to these hormones: this is similar to the case of renal dysfunction in adults, but in the newborn the impaired response is due to renal immaturity, not disease.

The renal concentrating capacity is lower in the newborn than in the adult; the maximum urinary osmolality which can be achieved in response to water deprivation, even with stimulation by exogenous ADH, is between 500 and 700 mmol/kg. This is partly because the loops of Henle do not penetrate as deeply into the renal medulla as in the adult, and partly due to the relatively low rate of urea production and therefore excretion. Reabsorption of urea by the distal nephron contributes to the interstitial medullary osmolality, so facilitating water reabsorption in response to ADH (p. 6). Both the short loops and the low urea excretion render the countercurrent mechanism relatively inefficient. Renal concentrating ability may rise if urea production is increased by giving the infant a high protein diet.

Plasma renin activity and aldosterone concentration are relatively high in the newborn infant. Nevertheless, the premature infant cannot conserve sodium efficiently because renal tubular function is immature. Excretion of a sodium load is impaired, perhaps because of the low glomerular filtration rate. The total body sodium content of a 1 kg premature neonate is only about 45 mmol, compared with about 3000 mmol in the adult, and sodium balance must be carefully controlled.

Hydrogen ion secretion is also inefficient in neonates. The urinary phosphate concentration, important for renal buffering in older children and in adults, may be low, especially if dietary phosphate intake is inadequate; formation of ammonia by tubular cells (p. 85) is also inefficient. The resultant inability fully to 'reabsorb' all the filtered bicarbonate and to regenerate enough to replace that used in buffering may contribute to the relatively low neonatal plasma $T\text{CO}_2$ concentration.

Neonatal renal function is adequate to maintain homeostasis in the normal newborn infant, but is unable to respond adequately to illness or other stress. It is difficult to determine if there is renal impairment because of the unsatisfactory nature of renal function tests at this age; both plasma urea and creatinine concentrations may fluctuate considerably. A plasma urea level above 8 mmol/litre suggests retention due to glomerular impairment, whether parenchymal or prerenal in origin, especially if the urinary output can be shown to be low. Management depends to a large extent on clinical criteria.

Water and electrolytes

Water

Water probably constitutes about 80 per cent of the weight of a premature infant weighing 1 kg, compared with about 60 per cent of that of an adult (the relative amount of water is related more to the proportion of fat to lean tissue than to total body weight). Probably relatively more of the water is in the extracellular than in the intracellular compartment at birth. The extracellular fluid compartment contracts during the first week of life: the loss of this fluid in the urine may account for the relatively high total amount of sodium excreted by preterm infants in the absence of pathology.

'Insensible' water loss is proportionately much higher in infants than in adults because:

- the ratio of surface area to body volume is high;
- there is a high metabolic and respiratory rate.

Daily fluid requirements are therefore up to five times higher per kg body weight than in adults.

Sodium

A normal neonate requires about 2 to 4 mmol/kg of sodium daily; the premature infant may need up to 6 mmol/kg a day, because of the high losses. The plasma sodium concentration should be monitored to ensure that the proportion of sodium to water is correct.

The total body sodium and the plasma sodium concentration (and therefore osmolality) may fluctuate because renal function is immature. The neonate, unlike children and adults, is unable to make its need for water clear, and it is difficult for the doctor accurately to assess requirements. Rapid changes in extracellular osmolality due to inappropriate intake of water cause significant shifts of water between the intra- and extracellular compartments, and signs varying from listlessness to convulsions, or even coma, may be caused by the changes in cerebral hydration.

Hypernatraemia develops if water loss exceeds that of sodium, or if excess of sodium relative to water is infused or fed.

If replacement is inadequate the high 'insensible' water loss, aggravated by urinary loss due to inefficient renal concentrating ability, may cause rapid extracellular fluid depletion with hypotension and, if water depletion is predominant, hypernatraemia: this is particularly likely if insensible loss is increased by pneumonia, diarrhoea or phototherapy for jaundice. Infants given sodium bicarbonate are even more likely than adults to develop hypernatraemia because of the low body sodium content, and because excretion of a sodium load is impaired.

Hyponatraemia is a common finding, as it is in adults. It may either be dilutional or due to high renal loss with replacement with fluid of low sodium concentration. Causes include:

- prolonged maternal infusion of oxytocin ('Syntocinon') in 5 per cent glucose, or hypoosmolal fluid, to induce labour. Oxytocin has an antidiuretic action similar to that of ADH and may cause dilutional hyponatraemia in both mother and infant;
- inappropriate ADH secretion:
  following post-partum intracranial haemorrhage;
  in severe pulmonary disease, such as the respiratory distress syndrome (hyaline membrane disease) or pneumonia.
- infusion of water in excess of sodium, either as 5 per cent dextrose or as hypoosmolal sodium solutions;
- acute renal disease, especially if hypoosmolal fluid is given to a child with glomerular dysfunction;
- diuretic treatment, particularly in the premature infant and if hypoosmolal fluid has been given.

Potassium

The normal neonate needs about 2 to 4 mmol/kg of potassium a day.

As in the adult, artefactual causes of abnormal potassium levels must be excluded. Pseudohyperkalaemia is especially likely if capillary samples are used, because cells may be damaged if the skin is squeezed. It may also be due to *in vitro* haemolysis, or to withdrawal of blood from a cannula through which a potassium solution is being infused: artefactual hypokalaemia occurs if the infusion contains no potassium.

Hyperkalaemia. The total body potassium is only about 75 mmol, compared with about 3000 mmol in an adult. Since only about 2 per cent (4 mmol) is in the extravascular compartment overzealous potassium treatment can easily cause hyperkalaemia. Other common causes of hyperkalaemia include glomerular dysfunction and tissue damage due to hypoxia.

Hyperkalaemia is better tolerated than in adults, particularly in premature infants. Plasma levels of up to 8 mmol/litre only need urgent treatment if there are significant changes in the electrocardiogram.

Hypokalaemia may be caused by diarrhoea or diuretic treatment.

Perinatal asphyxia

Renal complications and disturbances of electrolyte balance are especially likely to develop in infants with perinatal asphyxia.

Cerebral oedema or haemorrhage may stimulate ADH secretion from the posterior pituitary gland, causing *oliguria and a dilutional hyponatraemia* with hypoosmolality, accompanied by a *high urinary sodium concentration due to plasma volume expansion* (p. 49).

The hypotension occurring during asphyxia may reduce renal blood flow enough to cause acute oliguric renal failure (acute tubular necrosis). In addition to the oliguria and hyponatraemia, there will be *uraemia* and *hyperkalaemia*, with proteinuria and often haematuria.

Hydrogen ion homeostasis

Plasma $T\text{CO}_2$ ('bicarbonate') levels should be interpreted with caution in the preterm infant: they are particularly likely to be artefactually low in very small samples (p. 104).

Normal plasma $T\text{CO}_2$ levels are about 3 mmol/litre lower in infants than in adults. This is partly due to renal immaturity and partly to the low concentration of urinary buffers (p. 348); both these factors impair bicarbonate 'reabsorption' and regeneration.

Disturbances of hydrogen ion homeostasis are common in the newborn infant.

Metabolic acidosis

Causes of metabolic acidosis include:

- renal dysfunction;
- lactic acidosis due to:
  - tissue hypoxia resulting from hypotension and the low $P\text{O}_2$ accompanying asphyxia or sepsis;
  - some inborn errors of metabolism, such as glucose-6-phosphatase deficiency (p. 220).
- inborn errors of amino acid metabolism with aminoacidaemia, such as maple syrup urine disease (p. 371).

Unless the cause is obvious inborn errors should be considered. Simple screening tests (Table 17.1) may give the clue, but, whether they are positive or negative, further tests must be carried out, if necessary in a specialist centre.

Mixed respiratory and metabolic acidosis

The respiratory distress syndrome (hyaline membrane disease) is common in premature neonates and is due to immaturity of the enzymes responsible for intra-

Table 17.1 Screening tests for inborn errors of metabolism in the neonatal period

Possible disorder or group of disorders	Tests	Remarks
May suggest inborn error	Blood pH, P_{CO_2}, P_{O_2}	Detects hydrogen ion disturbances, and type
Congenital adrenal hyperplasia (p. 129)	Plasma Na^+↓, K^+↑	Non-specific finding Cause must be confirmed
Renal tubular acidosis (p. 93)	Plasma T_{CO_2}↓, Cl^-↑	More commonly acquired, or due to other inborn error, than congenital
Urea cycle disorders	Plasma urea↓, ammonia↑	Unable to form urea from ammonia
	Urinary amino acids	Specific amino acids high in some forms
Disorders of carbohydrate metabolism (p. 220)	Plasma glucose↓ Urinary reducing substances, ketones and amino acid chromatography	Non-specific finding Suspected disorder must be confirmed and typed
Disorders of organic acid metabolism	Urinary amino acids and ketones	

uterine synthesis of pulmonary surfactant: the likelihood of its occurrence may be monitored prenatally by performing the relevant assays on amniotic fluid (p. 145).

The condition presents with pulmonary collapse (atelectasis) and secondary lung infection is common. The blood $P\text{CO}_2$ is high, causing respiratory acidosis. The low blood $P\text{O}_2$, but mainly the reduced blood flow due to hypotension, cause tissue hypoxia with lactic acidosis. Renal failure may aggravate this metabolic acidosis. The combination of respiratory and metabolic components may cause very severe acidosis, with a blood pH well below 7.0.

The infant usually needs positive pressure ventilation to expand the lungs and so correct blood gas abnormalities. If this is successful, the improved general condition increases tissue perfusion, correcting lactic acidosis and improving renal function.

The $P\text{O}_2$ and $P\text{CO}_2$ of cutaneous capillary blood can be monitored continuously using electrodes placed on the skin. These bring the capillary $P\text{O}_2$ and $P\text{CO}_2$ to near arterial levels by heating the skin to about 44°C and so increasing cutaneous blood flow. The electrodes must be repositioned every four hours to prevent local burns and they need to be recalibrated frequently. In case of litigation results of transcutaneous monitoring may not be legally acceptable as evidence. *Transcutaneous blood gas monitoring supplements, but does not replace, arterial blood gas analysis.*

Bilirubin metabolism

Bilirubin metabolism is discussed in Chapter 14.

More unconjugated bilirubin reaches the liver in the newborn infant than in the adult because:

- the red cell half-life is shorter in infants, and the blood haemoglobin level falls rapidly during the first week of life, even in normal infants;
- delayed clamping of the umbilical cord may significantly increase red cell mass;
- bruises may occur during birth, and resorption of haemoglobin breakdown products from these increases the plasma bilirubin concentration.

The mature liver can conjugate very large amounts of bilirubin and jaundice due to increased bilirubin production is rarely severe in adults. In the newborn, even at term, the conjugating enzymes are not fully developed and even a marginally increased load causes jaundice.

Mild jaundice *not present at birth*, but developing during the first few days and continuing during the *first week* of life, and for which there is no obvious pathological reason, is known as *'physiological jaundice'*. Such jaundice is very common in normal newborn infants.

Unconjugated hyperbilirubinaemia

Plasma unconjugated bilirubin levels may be very high in premature infants because of *hepatic immaturity*. If the bilirubin concentration exceeds the albumin-binding capacity, the unbound, fat-soluble unconjugated bilirubin may cross cell membranes and be deposited in the brain, causing *kernicterus* (p. 299). This is a serious complication and may cause permanent brain damage or death.

The risk of kernicterus is increased:

- the more premature the infant;
- if the plasma unconjugated bilirubin concentration is rising rapidly, perhaps due to haemolysis;
- if the bilirubin-binding capacity is low, due to:
  - hypoalbuminaemia;
  - displacement of bilirubin from albumin by some drugs;
  - displacement of bilirubin from albumin by hydrogen ions in acidosis due to hypoxia or other serious illness.

Jaundice *during the first 24 hours of life is more likely to be pathological than physiological.* Causes include:

- *maternofetal rhesus or ABO blood group incompatibility;* this is particularly likely in infants born to multiparous mothers, who may have developed antibodies during previous pregnancies;
- *inherited erythrocyte abnormalities associated with haemolysis,* such as glucose-6-phosphate dehydrogenase deficiency or hereditary spherocytosis.

Bilirubin should be measured in blood taken from the umbilical cord of all infants known to be at risk for one of the above reasons: blood group typing and Coombs' testing for red cell antibodies can also be performed on cord blood.

Management. If the plasma bilirubin level is rising rapidly or exceeds about 340 μmol/litre (20 mg/dl) exchange transfusion may be needed. Biochemical complications of exchange transfusions are usually due to the anticoagulant used in the infused blood, and are transitory. They include:

- hyperkalaemia;
- hypocalcaemia;
- metabolic acidosis;
- hypoglycaemia.

Bilirubin is destroyed by ultraviolet light, and lesser degrees of unconjugated hyperbilirubinaemia may be treated by phototherapy. Water loss from the skin may be high and fluid balance must be carefully monitored.

Prolonged unconjugated hyperbilirubinaemia, not in itself requiring treatment, may be associated with *chronic infections* such as that of cytomegalovirus, or with *hypothyroidism*. It is more common in breast-fed infants than in those given formula feeds: the reason for this is unknown.

Conjugated hyperbilirubinaemia

Impaired excretion of bilirubin which has been conjugated may be due to:

- intrahepatic cholestasis in cystic fibrosis;
- congenital biliary atresia; it is important to make this diagnosis early because it is usually amenable to surgical treatment;
- biliary obstruction by pressure on the bile ducts by, for example, extrabiliary tumours.

Other causes of neonatal jaundice

Neonatal hepatitis rarely presents clinically until after the first week of life. Causes include intrauterine infection with organisms such as cytomegalovirus and rubella.

Metabolic causes of jaundice include galactosaemia (p. 221) and α_1-antitrypsin (p. 336) deficiency.

Other *inherited abnormalities* associated with jaundice include the Dubin-Johnson syndrome (p. 300).

Investigation of neonatal jaundice

The same investigations are used as in adults, but interpretation must allow for the different reference ranges. Plasma transaminase activities are up to twice the upper limit of the adult reference range during the first three months of life and reach adult levels by the age of about one year. The plasma total alkaline phosphatase activity is derived mainly from growing bone: it is significantly higher in infancy and childhood and falls to adult levels after the pubertal growth spurt (p. 315).

Glucose metabolism

Hypoglycaemia in the neonatal period is discussed on p. 220. Hepatic glycogen stores increase about threefold, and adipose tissue (another source of energy) is laid down, during the last 10 weeks of pregnancy. Very premature infants therefore have little liver glycogen, and are very prone to hypoglycaemia. Full-term infants may become hypoglycaemic if initially adequate stores are drawn on more rapidly than normal, for example during perinatal asphyxia.

Plasma glucose levels as low as 1.7 mmol/litre (about 30 mg/dl) during the first 72 hours in the premature infant, or 2.2 mmol/litre (about 40 mg/dl) in the later neonatal period, may not be associated with any clinical signs. When such signs do occur they include tremors and apnoeic attacks with convulsions.

Calcium, phosphate and magnesium metabolism

Calcium metabolism in adults is discussed in Chapter 9. Calcium and phosphate are actively transported across the placenta and total and free-ionized calcium concentrations are higher in fetal than in maternal plasma. 25-hydroxycholecalciferol, but not its active metabolite, 1,25-dihydroxycholecalciferol, nor parathyroid hormone (PTH), can pass the placental membrane. Calcium and phosphate accumulate rapidly between the 30th and 36th week of fetal life: a premature infant may therefore be calcium and phosphate deficient and may need to be given supplements.

During intrauterine life the high fetal plasma free-ionized calcium concentration suppresses PTH release from its own parathyroid glands. The glands may not recover immediately after birth and there may be transient hypoparathyroidism. Plasma total and free-ionized calcium levels fall by up to 30 per cent after birth, but, in the normal infant, return to normal spontaneously within about three days: this fall rarely needs to be treated.

The neonatal reference range for plasma total calcium concentration is wider than that in adults. Plasma phosphate values are higher in actively growing infants and in children than in adults.

Neonatal hypocalcaemia

Hypocalcaemia during the first two weeks of life can be divided into two groups.

- *Hypocalcaemia of early onset* occurs during the first three days of life and is commonest:
  - in low birth-weight, preterm infants for the reasons given above;
  - following perinatal asphyxia;
  - in infants of diabetic mothers.

  It may also occur if the mother was hypercalcaemic during pregnancy, for example due to primary hyperparathyroidism: the maternal hypercalcaemia is reflected in the fetus, and the suppressed fetal parathyroid glands may take some time to recover. This condition is a more severe form of the 'physiological' condition described above.

- *Hypocalcaemia of late onset* is less common than it used to be. It usually occurred following the introduction of high phosphate-containing feeds.

Rickets of prematurity

Bone demineralization (osteopenia) is common in small preterm infants and often resolves spontaneously before obvious rickets develop. The plasma alkaline phosphatase activity may rise to more than six times the upper adult reference limit, the plasma calcium levels are usually low normal, and those of phosphate low.

Rickets of prematurity usually becomes clinically evident between the fourth and 12th weeks of life, usually in very premature, low birth-weight infants. Longitudinal growth slows, and the decalcification of the bones predisposes to pathological fractures: respiration is impaired by the soft ribs.

X-ray changes may be minimal or absent in early cases. In more advanced disease the classical radiological changes appear at the ends of long bones.

Rickets of prematurity may be due to:

- calcium and phosphate depletion. In premature infants unsupplemented breast milk may not contain enough of these minerals to replace that which should have accumulated *in utero* during the last 10 weeks of gestation. Phosphate depletion is probably the commonest cause and is indicated by very *low plasma phosphate concentration and urinary phosphate excretion.* Glomerular dysfunction reduces phosphate excretion even if there is no deficiency (p. 174), and the diagnosis must not be made on the basis of this finding alone;
- maternal vitamin D deficiency during pregnancy, or in the infant after birth. In very premature infants such deficiency may be due to low activity of the renal 1-α-hydroxylase needed to convert 25-hydroxycholecalciferol to the active 1,25-dihydroxycholecalciferol (1,25-DHCC) (p. 176). Whatever the cause of 1,25-DHCC deficiency, calcium absorption may be impaired. In such cases the *urinary phosphate excretion will be inappropriately high* due to secondary hyperparathyroidism, provided that renal function is normal;
- drugs such as frusemide increase urinary calcium loss;
- renal tubular disorders of phosphate reabsorption may cause phosphate depletion (p. 186). In such cases the plasma calcium is usually normal, but may even be high because it cannot be deposited in bone without phosphate.

Nutritional rickets is now uncommon during later childhood in affluent societies, but remains a problem in some countries. In the UK it is commonest in Asian children (p. 184) and may develop in those receiving anticonvulsants (p. 184).

Treatment. Vitamin D and calcium and phosphate supplementation must be monitored by measuring serial plasma alkaline phosphatase activities. If treatment is successful these may continue to rise for several weeks while bone is being actively laid down, before falling once the bone is adequately calcified.

Many premature infants are given prophylactic vitamin D supplementation. This is unlikely to prevent rickets unless accompanied by adequate calcium and phosphate intake.

Magnesium

Hypomagnesaemia often accompanies hypocalcaemia and may be caused by dietary deficiency, or by increased intestinal or urinary loss. Low plasma magnesium levels may impair the release and action of PTH, and so delay correction of plasma calcium levels.

Hypomagnesaemia should be considered if the infant has convulsions despite normocalcaemia.

Plasma proteins

At term the mean plasma total protein concentration is about 12 g/litre lower than in adults. By contrast, that of albumin is about the same, although, as in adults, it may fall during illness. The acute phase proteins, reflected in the α- and β-globulins on the electrophoretic strip, reach adult levels by about the age of 6 months, but are affected by illness, again as they are in the adult.

The immunoglobulin pattern differs significantly from that of adults. During normal pregnancy placental transfer of maternal IgG leads to a gradual increase in fetal plasma immunoglobulin concentrations. At birth these maternally derived IgG levels fall and endogenous immunoglobulin synthesis starts. Adult levels of plasma IgM are reached by the age of about 9 months, IgG by about three years and IgA only by the age of 15 (Table 16.1, p. 333).

Plasma immunoglobulin levels may need to be measured to diagnose immune deficiency states, whether primary, secondary or transient. They may also be used to detect infection, whether it occurred during the intrauterine period or has developed after birth. Results must be compared with the age-matched reference range, allowance being made for both the time since conception (gestational age) and age since birth (postnatal age).

A high plasma IgM concentration in blood obtained from the umbilical cord, or within the first four weeks of life, may indicate intrauterine or neonatal infections, such as syphilis, rubella, toxoplasmosis or cytomegalovirus. Allowance must be made for the normal rise in plasma IgM which starts after about 6 weeks of age.

Specific inborn errors, such as α_1-antitrypsin deficiency, may be diagnosed by measuring the appropriate plasma protein.

Neonatal thyroid function

The fetal hypothalamic-pituitary-thyroid axis develops independently of maternal hormones.

Immediately after birth plasma TSH levels rise rapidly, probably in response to

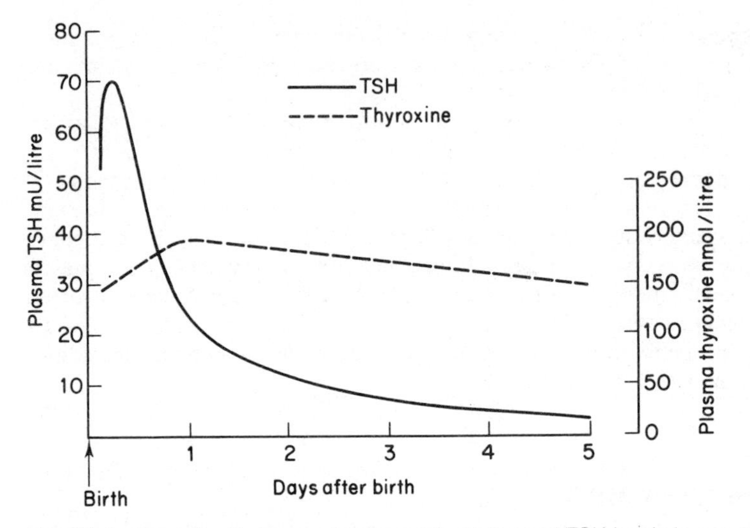

Fig. 17.1 Example of changes in plasma thyroxine and TSH levels in a normal full-term infant during the first five days after birth.

the stress of birth, to about 15 times the upper adult reference limit. They reach a peak within the first hour before falling, rapidly at first, during the next week. Plasma total thyroxine levels peak within the first 24 hours and then fall gradually (Fig. 17.1). Screening tests for neonatal hypothyroidism should be delayed for a week after birth to allow the levels to stabilize.

There is a similar pattern of secretion in the healthy preterm infant, but the peak levels of both hormones are lower. This reduced response is more marked in ill preterm infants, in whom it may be very difficult to interpret results of thyroid function tests: plasma total thyroxine levels are low, those of TSH 'normal' and of thyroxine-binding globulin normal or low. This pattern resembles that in ill adults. Treatment with thyroxine is usually not indicated.

The incidence and diagnosis of neonatal hypothyroidism is discussed on p. 167. Thyroid function tests should be repeated in infants found to have a positive screening test at a week after birth; if the diagnosis is confirmed thyroid replacement should be started immediately. Thyroid function should be reassessed at the age of one year because neonatal hypothyroidism is sometimes transient. Results of screening tests may be misleading in ill, premature infants and may need to be repeated before discharge from hospital.

Hypothyroidism may be suspected clinically, for example because of failure to thrive or persistent jaundice. In such cases thyroid function should be fully investigated.

Summary

1. In preterm infants immaturity contributes to the pathogenesis of disease.
2. Biochemical results in infants and children must be interpreted using matched sex and age reference ranges.
3. 'Insensible' water loss is much higher in newborn infants than in adults.
4. A mixed respiratory and metabolic acidosis, usually due to the respiratory distress syndrome, is relatively common in preterm infants.
5. Hypoglycaemia is common in preterm infants because of inadequate glycogen stores.
6. 'Physiological' jaundice is very common in newborn infants. Severe jaundice, needing treatment, occurs in premature neonates and those with maternofetal blood group incompatibility.
7. Metabolic bone disease, including rickets, may occur in very small preterm infants.
8. Plasma IgM levels measured in blood taken from the umbilical cord may help to diagnose infections which have developed *in utero*.
9. Thyroid function tests may be very difficult to interpret in very premature and ill infants.

Further reading

Clayton BE, Round JM, eds. *Chemical pathology and the sick child.* Oxford: Blackwell Scientific, 1984.

Brook CGD, ed. *Clinical paediatric endocrinology.* Oxford: Blackwell Scientific, 1981.

Roberton NRC, ed. *Textbook of neonatology.* Edinburgh: Churchill Livingstone, 1986.

Holton JB. Diagnosis of inherited metabolic diseases in severely ill children. *Ann Clin Biochem* 1982; **19**: 389–95.

Isherwood DM, Fletcher KA. Neonatal jaundice: investigation and monitoring. *Ann Clin Biochem* 1985; **22**: 109–28.

18

Inborn errors of metabolism

The inherited characteristics of an individual are determined by about 50 000 gene pairs, arranged on 23 pairs of chromosomes, one of each pair coming from the father and one from the mother. Genotype diversity is introduced by random selection and recombination during meiosis, as well as by occasional mutation. These genetic variants may, at one extreme, be incompatible with life, or at the other, produce biochemical differences detectable only by special techniques, if at all. Between the two extremes there are many variations which produce functional abnormalities.

Genetic disorders fall into three main categories.

- *Chromosomal disorders* due to the absence, or abnormal arrangement, of chromosomes affecting many genes and therefore many gene products. Examples include Down's syndrome (trisomy for chromosome 21) and Klinefelter's syndrome (XXY).
- *Multifactorial disorders* due to the interaction of multiple genes with environmental or other exogenous factors, such as diabetes mellitus.
- *Monogenic disorders* due to an abnormality of a single gene which is the primary determinant of the disorder and which is inherited in a predictable pattern.

General principles

The sequence of bases in the genes codes for protein structure through the medium of messenger RNA.

The abnormalities in monogenic disorders may be due to an abnormal *structural* gene, with production of one abnormal protein: in this case all the biochemical abnormalities can be explained by defective synthesis of a single peptide. In other cases an abnormal *control* gene alters the rate at which one or more structural genes function and consequently alters the *amounts* of one or more proteins which are, nevertheless, structurally normal.

In most of the examples discussed in this chapter the affected protein is an enzyme, but in other conditions it may, for instance, be a receptor, transport or structural protein, a peptide hormone, an immunoglobulin or a coagulation factor. In many inherited disorders the abnormal protein is not yet known, and the defect is only recognized by its characteristic clinical presentation, as in, for example, cystic fibrosis.

Patterns of inheritance

Every inherited characteristic is governed by a pair of genes on homologous chromosomes, one gene being received from each parent. Different genes governing the same characteristic are called alleles. If an individual has two identical alleles he is *homozygous* for that gene or inherited characteristic; if he has two different alleles he is *heterozygous*. Genes may be carried on the sex chromosomes (X and Y) or on the autosomes (similar in both sexes): the patterns of inheritance differ.

Autosomal inheritance

If one parent (Parent 1 in the example below) is heterozygous for an abnormal gene (*A*) and the other parent is homozygous for the normal gene (N), the possible gene combinations in the offspring are shown in the square.

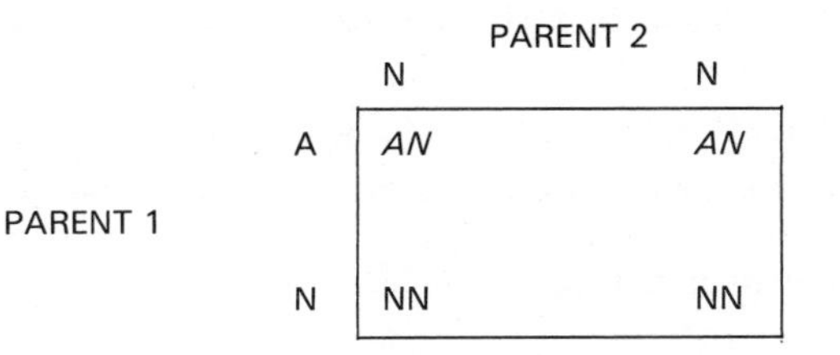

On a *statistical basis,* half the offspring will be *heterozygous (AN)* for gene *A*, like Parent 1. None will be homozygous for the abnormal gene (*AA*).

If both parents are heterozygous (*AN*), in a large series a quarter of the offspring will be homozygous (*AA*) and half heterozygous (*AN*).

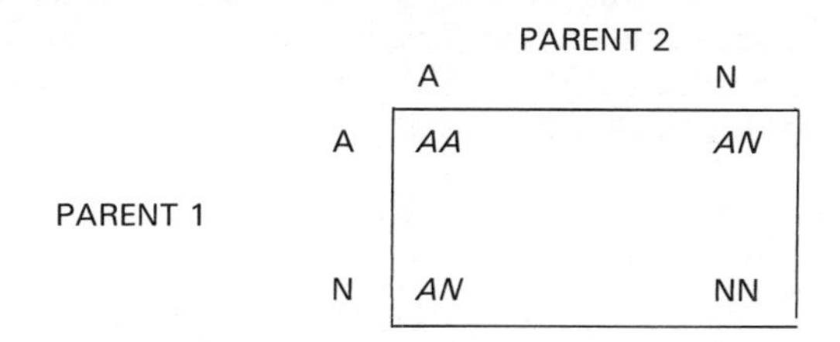

If one parent is homozygous (*AA*) and the other normal (NN) all the offspring will be heterozygous.

The metabolic consequences of an abnormal gene depend on the effectiveness of that gene compared with the normal one.

Dominant abnormal genes affect both heterozygotes (*AN*) and homozygotes (*AA*), although homozygotes may be more severely affected. In the first example,

Parent 1 (*AN*) and half the offspring would be affected, and in the second example both parents and three out of four of the offspring (*AA* and *AN*) would be affected. Characteristically, if there is autosomal dominant inheritance:

- every affected individual has at least one affected parent;
- offspring in successive generations are affected;
- clinically normal offspring are not carriers of the abnormal gene;
- statistically, three in four children are affected if both parents are heterozygous.

Recessive abnormal genes only affect homozygous offspring (*AA*). In the first example neither parent, nor the offspring, would be affected, and in the second example the parents would both appear normal, but statistically one in four of the offspring would be affected.

Therefore, in autosomal recessive inheritance:

- heterozygous parents are not clinically affected;
- clinical consequences may miss offspring in succeeding generations;
- clinically normal children may be heterozygous and therefore be carriers of the abnormal gene;
- statistically, one in four children of heterozygous parents are clinically affected.

Disorders inherited as a recessive trait have a lower expression frequency in affected families than dominant disorders, but usually tend to be more severe.

Sex-linked inheritance

Some abnormal genes are carried only on the sex chromosomes, almost always on the X chromosome.

X-linked recessive inheritance. Women have two X chromosomes and men one X and one Y. In X-linked recessive inheritance an abnormal X chromosome (*Xa*) is latent when combined with a normal X, but active when combined with a Y, chromosome. If the mother carries *Xa* she will appear to be normal but, statistically, half her sons will be affected (*YXa*). Half her daughters will be carriers (*XXa*), but all her daughters will be clinically unaffected.

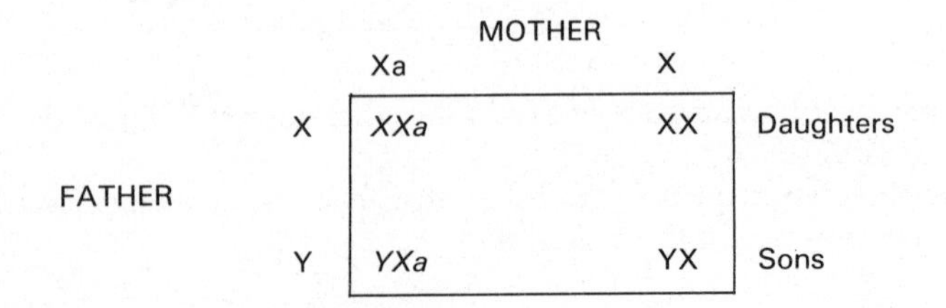

If the father is affected and the mother carries two normal genes none of the sons will be affected, but all the daughters will be carriers.

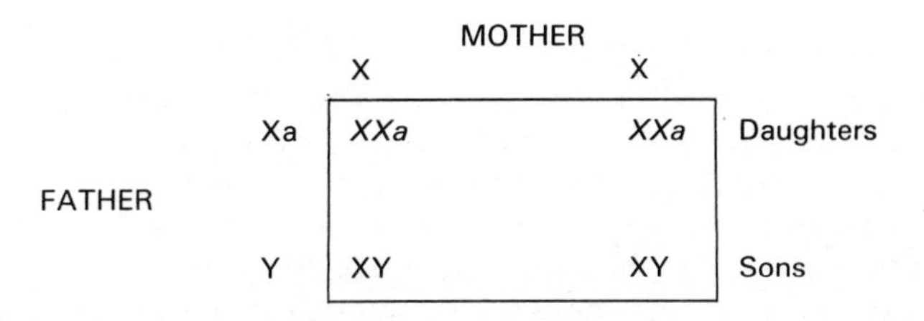

Inherited disease manifesting in male offspring and carried by females is typical of X-linked inheritance. The female is only clinically affected in the extremely rare circumstance when she is homozygous for the abnormal gene. This would only occur if she inherited the abnormal genes from her affected father and from her carrier mother.

Haemophilia is the classical example of an X-linked recessive disorder.

X-linked dominant inheritance. In this type of inheritance both *XXa* women and *YXa* men are affected. An example of this type of disorder is familial hypophosphataemia (p. 186).

Multiple alleles

Occasionally there may be several alleles governing the same characteristic. Different pair combinations may then produce different disease patterns such as, for example, some of the haemoglobinopathies, or the variant may only be detectable by biochemical testing, such as, for example, some of the plasma protein variants.

The rules outlined above may not apply in all cases, because new mutations can occur at any time and may produce dominant disorders in unaffected families.

The terms 'dominant' and 'recessive' are relative. A dominant gene may fail to manifest itself (*incomplete penetrance*) and may therefore appear to skip a generation. A gene may vary in its *degree of expression,* and therefore in the degree of abnormality that it produces. Finally, a recessive gene, which produces disease only in homozygotes (*AA*) may, nevertheless, be detectable by laboratory tests in heterozygotes.

Possible metabolic consequences

As stated above, although monogenic disorders may involve any peptide, they are usually due to an enzyme abnormality.

Deficiency of a single enzyme in a metabolic pathway may produce its effects in several ways. Suppose that substance A is acted on by enzyme X to produce substance B, and that substance C is on an alternative pathway.

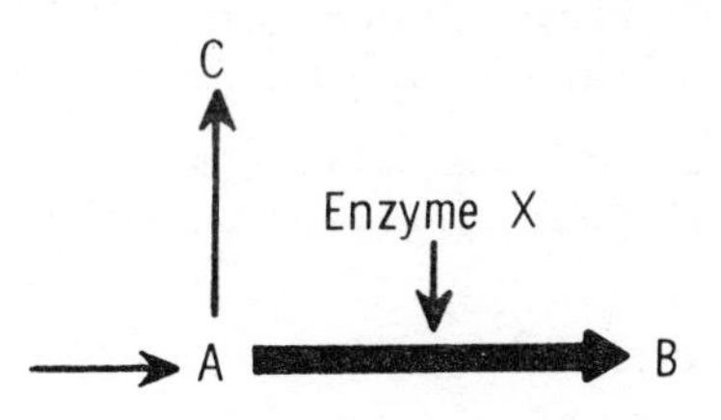

The consequences of a deficiency of X may be due to:

- *deficiency of the products of the enzyme reaction* (B). Examples include:
  cortisol deficiency in congenital adrenal hyperplasia (p. 129);
  the hypoglycaemia of some forms of glycogenosis.
- *accumulation of the substance acted on by the enzyme* (A). Examples include:
  phenylalanine in phenylketonuria (p. 369);
  complex structural molecules in lysosomal storage diseases, such as Gaucher's disease, in which substances that are usually hydrolysed accumulate in lysosomes because one of the metabolizing enzymes is deficient.
- *diversion through an alternative pathway.* Some product(s) of the latter, (C), may accumulate and produce effects, as, for example, in congenital adrenal hyperplasia when accumulation of androgens causes virilization (p. 129).

The effects of the last two types of abnormality will be aggravated if the whole metabolic pathway is controlled by negative feedback from the final product. For example, in congenital adrenal hyperplasia cortisol deficiency stimulates steroid synthesis, and therefore the accumulation of androgens, and consequent virilization is accentuated.

The clinical effects of some inborn errors may be modified by, or depend entirely on, environmental factors. For example, iron loss occurs during menstruation and pregnancy: women with idiopathic haemochromatosis accumulate iron less rapidly than men with the same condition, and rarely present with clinical features before the menopause. Patients with cholinesterase variants only develop symptoms if the muscle relaxant, suxamethonium, is given (p. 321).

Clinical importance of inborn errors of metabolism

The recognition of many inborn errors of metabolism is only of academic interest, because the abnormality produces no clinical effects. In others, even though no effective treatment is available, it may be important to make a diagnosis so that genetic counselling can be undertaken and, if termination is acceptable, prenatal diagnosis offered during subsequent pregnancies.

There is a group of diseases in which recognition in early infancy is of great importance, because *treatment may prevent irreversible clinical consequences or death.* Some of the more important of these are:

- phenylketonuria (p. 369);
- galactosaemia (p. 221);
- congenital hypothyroidism (p. 166);
- maple-syrup urine disease (p. 371).

Examples of conditions which should be sought in *relatives of affected patients*, either because *further ill effects may be prevented*, or because a *precipitating factor should be avoided* are:

- cholinesterase abnormalities (p. 321);

- glucose-6-phosphate dehydrogenase deficiency (p. 375);
- acute porphyrias (p. 405);
- haemochromatosis (p. 397);
- cystinuria (p. 373);
- Wilson's disease (p. 374).

Many other conditions can be *treated symptomatically.* Examples are:

- hereditary nephrogenic diabetes insipidus (p. 44);
- congenital disaccharidase deficiency (p. 261);
- Hartnup disease (p. 373).

Some inborn errors are completely, or almost completely, harmless. They are important because they produce *effects which may lead to misdiagnosis* or which may alarm the patient. Examples are:

- renal glycosuria (p. 210);
- alkaptonuria (p. 371);
- Gilbert's disease (p. 299).

Finally, clinical effects of some inborn errors of metabolism may not appear until child-bearing age is reached: in these cases *genetic counselling* of blood relatives is desirable. Examples of this type of disease are:

- Wilson's disease (p. 374);
- haemochromatosis (p. 397).

Screening for inborn errors of metabolism

Screening newborn infants

Many countries have instituted programmes for screening all newborn infants for certain metabolic disorders. The criteria for instituting such a programme should depend on the following characteristics of the disorder or of the test.

- The disease should be not be clinically obvious and should have a relatively high incidence in the population screened.
- The disease should be treatable, and it must be possible to obtain the result of the screening test before irreversible damage is likely to have occurred.
- The screening test should be simple and reliable, and the cost of the programme should ideally be, at least partly, offset by the cost-savings resulting from early treatment; for example, such treatment may sometimes eliminate the need for prolonged institutional care.

Not all these criteria are necessarily fulfilled in all available screening programmes, which include those for:

- phenylketonuria;
- galactosaemia;
- hypothyroidism;
- cystic fibrosis.

Such screening tests may be performed on blood or urine during the neonatal period. Blood obtained from a heel prick is often collected on filter paper, which is easy to transport to the laboratory.

The timing of sample collection is important if false results are to be avoided. Substances on the metabolic pathway before the enzyme block, such as galactose in galactosaemia, only accumulate once the infant starts ingesting the precursor, in this case milk or milk-products. For this reason, blood samples for screening apparently well infants are usually collected at between the sixth and ninth day of life, although abnormal metabolites may not be detectable in the urine until four to six weeks after birth if the 'renal threshold' is relatively high.

A positive result of screening should be confirmed by quantitative analysis or by repeating the testing. Many abnormalities are transient, and do not necessarily indicate the presence of an inborn error.

Prenatal screening

Prenatal screening of high-risk groups only may be performed for some disorders. This is most commonly done by culturing fibroblasts obtained by amniocentesis early in the second, or by chorionic villus sampling during the first, trimester and by demonstrating the metabolic defect in the cultured cells. Examples of those in whom such screening may be indicated are:

- women with a previously affected infant;
- ethnic groups in whom the carrier state is more common than in the general population. For example, there is a relatively high incidence of Tay-Sach's disease in Ashkenazi Jews.

The many pitfalls of interpreting results of pre- or postnatal screening for inborn errors, and the responsibility of those who undertake to do so, are reviewed in the references quoted at the end of this chapter.

When to suspect an inborn error of metabolism

The possibility of an inherited metabolic defect should be considered if there are bizarre and inexplicable clinical or laboratory findings in infancy or early childhood, especially if more than one infant in the family has been affected. The following features are particularly suggestive:

- vomiting;
- failure to thrive;
- hepatosplenomegaly;
- prolonged jaundice;
- retarded mental development, fits or spasticity;
- a peculiar smell, or staining, of the napkins;
- hypoglycaemia;
- metabolic acidosis;

- refractory rickets;
- renal calculi.

In older infants the symptoms may be more specific.

Laboratory diagnosis of inborn errors of metabolism

The diagnosis of an inborn error should always be confirmed in a centre specializing in such disorders.

If the symptoms are strongly suggestive of a particular disorder, specific tests, such as measurement of plasma caeruloplasmin concentration and urinary copper excretion in Wilson's disease (p. 375), may be performed.

In the neonatal period symptoms are often non-specific. Disorders presenting at this age usually progress rapidly and a screening policy must be agreed. Inborn errors presenting acutely during the neonatal period are usually due to an enzyme abnormality. This abnormality may be demonstrated:

- *indirectly* by detecting a high concentration of the substance normally metabolized by the enzyme, or a low concentration of the product;
- *directly* by demonstrating a low enzyme activity in the appropriate tissue or in blood cells. These assays may only be available at special centres. If possible all cases should be confirmed in this way.

Examples of indirect screening methods include:

- estimation of plasma ammonia for disorders of the urea cycle, in which ammonia cannot be converted to urea as efficiently as normal, and accumulates in the blood stream;
- chromatography of plasma and urine for amino acids for detection of many disorders of amino acid metabolism;
- detection of organic acids in the urine in disorders of branched-chain amino acid metabolism.

Screening tests should be interpreted with caution and a suspected diagnosis confirmed by more specific techniques. For example, aminoaciduria, although a common finding in inborn errors of metabolism, is not specific for such a disorder.

Aminoaciduria. Amino acids are usually filtered by the glomerulus and reach the proximal tubule at concentrations equal to those in plasma: they are almost completely reabsorbed by tubular cells. Aminoaciduria may therefore be of two types:

- *overflow aminoaciduria* in which, because of raised plasma levels, amino acids reach the proximal tubules at concentrations higher than the reabsorptive capacity of the cells;
- *renal aminoaciduria* in which plasma levels are low because of urinary loss due to defective tubular reabsorption.

Aminoaciduria may also be subdivided according to the pattern of excreted amino acids.

- *Specific aminoaciduria* is due to increased excretion of either a single amino acid, or a group of chemically related amino acids. It may be overflow or renal in type.
- *Non-specific aminoaciduria,* in which there is increased excretion of a number of unrelated amino acids, is almost always due to an acquired lesion. It may be overflow in type, as in severe hepatic disease, when impaired deamination of amino acids causes raised plasma levels: more commonly renal aminoaciduria results from non-specific proximal tubular damage (p. 16), and other substances usually almost completely reabsorbed by the proximal tubule are also lost (phosphoglucoaminoaciduria; the Fanconi syndrome). If it occurs in inborn errors of metabolism it is usually secondary to tubular damage caused by deposition of the substance not metabolized normally, such as copper in Wilson's disease, than directly to a genetic effect.

Principles of treatment of inborn errors of metabolism

Some inborn errors can be treated by limiting the dietary intake of precursors in the affected metabolic pathway, such as phenylalanine in phenylketonuria or lactose in galactosaemia, or by supplying the missing metabolic product, such as cortisol in congenital adrenal hyperplasia. Accumulated products, such as iron in haemochromatosis or copper in Wilson's disease, may be removed or their accumulation reduced. In disorders such as cystic fibrosis only the complications can be treated, and there is no treatment yet available for many inborn errors. Enzyme replacement by bone marrow transplantation has been tried for some disorders with a grave prognosis, but the results are still being assessed.

Diseases due to inborn errors of metabolism

Only a very few of the known inborn errors of metabolism will be discussed. The choice must be biased by what the authors feel to be important and others might disagree. Some of the more clinically important abnormalities are listed on p. 366, and on p. 377 a fuller, but by no means complete, list, with the mode of inheritance when known, is included. Many of these conditions are mentioned briefly in the relevant chapters. For more detailed and comprehensive reviews the student should consult the references listed at the end of this chapter.

Disorders of amino acid metabolism

Most disorders of amino acid metabolism are characterized by raised plasma levels of an amino acid or amino acids, with overflow aminoaciduria.

Disorders of aromatic amino acid metabolism

The main metabolic pathway for aromatic amino acids is outlined in Fig. 18.1, and indicates the known enzyme defects. *Tyrosine*, normally produced from phenylalanine, is the precursor of several important substances. The inherited defects of thyroid hormone synthesis are considered briefly on p. 166.

Phenylketonuria

Phenylketonuria is an autosomal recessive disorder caused by an abnormality of the *phenylalanine hydroxylase* system. The enzyme itself is most usually affected, but in about three per cent of cases the enzymes responsible for the synthesis of the cofactor, tetrahydrobiopterin, are abnormal. Therefore several different inherited deficiencies may have very similar biochemical and clinical consequences. Because phenylalanine cannot be converted to tyrosine, it accumulates in the plasma and is excreted in the urine with its metabolites such as phenylpyruvic acid: the disease acquired its name from the detection of the latter 'phenylketone' in the urine.

The clinical features are:

- irritability, feeding problems, vomiting and fits during the first few weeks of life;
- mental retardation developing at between four and six months, with psychomotor irritability;
- often generalized eczema;
- a tendency to reduced melanin formation because of reduced production of tyrosine. Many patients are pale-skinned, fair-haired and blue-eyed.

Diagnosis. The *phenylalanine* concentration may be measured in *blood* taken from a heel prick. The microbiological Guthrie test is suitable for mass screening, and should be performed at about 6 days after birth. Blood phenylalanine levels may be raised in conditions other than phenylketonuria; in the newborn and especially in premature infants the enzyme system may not be fully developed and false positive results are likely if the test is performed too early. If a positive result is found the test should be repeated later, to allow time for development of the enzyme. Measurement of blood tyrosine levels may identify other causes of false positive results; tyrosine levels do not rise in phenylketonuria, but are high in many of the other conditions.

Heterozygotes may be clinically normal, but can be detected by biochemical tests.

Urinary phenylpyruvic acid reacts with Phenistix (Ames), but this should not be used as a screening test because it may not detect the amino acid until about six weeks after birth, by which time there may be permanent cerebral damage.

A variant, persistent hyperphenylalaninaemia, without mental retardation has been described.

Babies exposed *in utero* to the high phenylalanine levels of unrecognized phenylketonuric mothers may be mentally retarded, although they themselves do not have detectable phenylketonuria (*maternal phenylketonuric syndrome*).

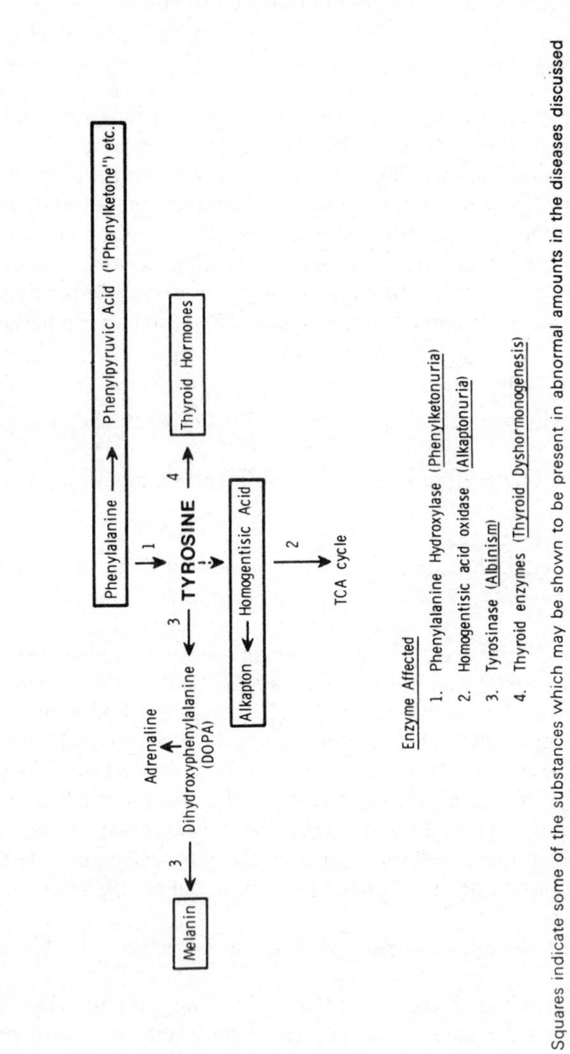

Squares indicate some of the substances which may be shown to be present in abnormal amounts in the diseases discussed

Fig. 18.1 Some inborn errors of the aromatic amino acid pathway.

Management. The aim of management is to lower plasma phenylalanine levels by giving a low phenylalanine diet. Such treatment is difficult, expensive and tedious for both the patient and the parents, and should be monitored carefully, especially if the patient becomes pregnant. It is still unknown whether strict dietary restriction can ever safely be discontinued.

Deficiency of phenylalanine, an essential amino acid, has deleterious effects which include an impaired growth rate, eczema and mental retardation. Some of these problems can be overcome by supplementing the diet with the immediate metabolic product of phenylalanine metabolism, tyrosine; tyrosine is not usually an essential amino acid, but becomes so if it cannot be produced from phenylalanine.

The form of hyperphenylalaninaemia caused by cofactor deficiency cannot adequately be treated by diet.

Alkaptonuria

Alkaptonuria is an autosomal recessive disorder due to a deficiency of *homogentisic acid oxidase*. Homogentisic acid accumulates in the blood and tissues, and is passed in the urine. Oxidation and polymerization of homogentisic acid produces the pigment alkapton in much the same way as polymerization of DOPA (Fig. 18.1) produces melanin. Deposition of alkapton in cartilages, with consequent darkening, is called *ochronosis*, which may cause *arthritis* in later life and result in visible darkening of the cartilages of the ears. Conversion of homogentisic acid to alkapton is accelerated in alkaline conditions, and the most obvious abnormality in alkaptonuria is *darkening of the urine as it becomes more alkaline on standing*: this finding may, however, not always be present. If it is, the condition is often first noticed by the mother, who is worried by the black nappies which become even blacker when washed in alkaline detergents.

The condition is compatible with a normal life span and, despite the tendency for patients to develop arthritis in later life, treatment is considered unnecessary.

Homogentisic acid is a reducing substance which reacts with Clinitest tablets (p. 230).

Albinism

A deficiency of *tyrosinase* in melanocytes causes one form of albinism, and it is inherited as a recessive character. Pigmentation of the skin, hair and iris is reduced, and the eyes may be pink. Reduced pigmentation of the iris causes photosensitivity, and decreased skin pigmentation is associated with an increased incidence of skin cancer. The tyrosinase involved in catecholamine synthesis is a different isoenzyme, controlled by a different gene, and adrenaline metabolism is normal in albinos.

Disorders of other amino acids

Maple syrup urine disease

In maple syrup urine disease, which is inherited as an autosomal recessive condition, there is deficient decarboxylation of the oxoacids resulting from deamination of the three *branched-chain amino acids*, leucine, isoleucine and valine. These amino acids accumulate in the plasma and are excreted in the urine with their corresponding oxoacids. The smell of the urine is like that of maple syrup, and gives the condition its name.

The disease presents during the first week of life, and if not treated severe neurological lesions develop which cause death within a few weeks or months. If a diet low in branched-chain amino acids is given, normal development seems possible.

The *diagnosis* is made by demonstrating raised levels of branched-chain amino acids in plasma and urine. It may be confirmed by demonstrating the enzyme defect in leucocytes.

Histidinaemia

Histidinaemia is associated with deficiency of *histidinase*, an enzyme needed for normal histidine metabolism, and is probably inherited as an autosomal recessive trait. About half the described cases have mental retardation and speech defects, but the other half seem to be normal. The results of dietary treatment are inconclusive.

The *diagnosis* is made by demonstrating raised plasma levels of histidine, and by finding histidine and a metabolite, *imidazole pyruvic acid*, in the urine.

Like phenylpyruvic acid, imidazole pyruvic acid reacts with Phenistix (Ames).

Inherited disorders of transport mechanisms

Groups of chemically similar substances are often transported by shared or interrelated pathways. Such group-specific mechanisms usually affect transport across all cell membranes, and defects often involve both the renal tubule and intestinal mucosa. Inborn errors of the following amino and imino acid group pathways have been identified:

- the *dibasic* amino acids (with two amino groups), cystine, ornithine, arginine and lysine (*cystinuria*) (COAL is a useful mnemonic);
- many *neutral* amino acids (with one amino and one carboxyl group) (*Hartnup disease*);
- the *imino acids*, proline and hydroxyproline, which probably share a pathway with glycine (*familial iminoglycinuria*).

Cystinuria

Cystinuria is due to an autosomal recessive inherited abnormality of tubular reabsorption, with excessive urinary excretion, of the dibasic amino acids, cystine, ornithine, arginine and lysine. A similar transport defect has been demonstrated in the intestinal mucosa, but although dibasic amino acid absorption is reduced, they can be synthesized in the body, and deficiency does not occur. Cystine is relatively insoluble and, because of the high urinary concentrations in homozygotes, may precipitate and form calculi in the renal tract. Although increased excretion can be demonstrated in heterozygotes, concentrations are rarely high enough to cause precipitation in these cases.

The *diagnosis* of cystinuria is made by demonstrating excessive urinary excretion of the characteristic amino acids. All the amino acids must be identified to distinguish the stone-forming homozygotes from heterozygous cystine-lysinuria and from cystinuria occurring as part of generalized aminoaciduria.

The *management* of cystinuria aims to prevent calculus formation by reducing urinary concentration. The patient should drink plenty of fluid *day and night.* Alkalinizing the urine increases the solubility of cystine. If these measures prove inadequate D-penicillamine may be given: it forms a chelate which is more soluble than the cystine alone.

Cystinosis. This very rare but serious disorder of cystine metabolism is characterized by intracellular accumulation and storage of cystine in many tissues. It must be distinguished from cystinuria, which is a relatively harmless condition. Renal tubular damage causes the Fanconi syndrome. Aminoaciduria is non-specific and of renal origin. Death occurs at a young age.

Hartnup disease

Hartnup disease, named after the first patient described, is a rare autosomal recessive disorder in which there are renal and intestinal transport defects involving neutral amino acids.

Most, if not all, the clinical manifestations can be ascribed to the reduced intestinal absorption and increased urinary loss of *tryptophan.* This amino acid is normally partly converted to nicotinamide, the conversion being especially important if dietary intake of nicotinamide is marginal (p. 280). The clinical features of Hartnup disease are intermittent and resemble those of pellagra, namely:

- a red scaly rash on exposed areas of skin;
- reversible cerebellar ataxia;
- mental confusion of variable degree.

In spite of the generalized defect of amino acid absorption there is no evidence of protein malnutrition; this may be because intact peptides can be absorbed by a different pathway.

Excessive amounts of *indole* compounds, originating from bacterial action on unabsorbed tryptophan, are absorbed from the gut and excreted in the urine.

The *diagnosis* is made by demonstrating the characteristic amino acid pattern in the urine. Heterozygotes are difficult to detect by available techniques.

Familial iminoglycinuria

Increased urinary excretion of the imino acids, proline and hydroxyproline, and of glycine, despite normal plasma levels, is due to a transport defect for these three compounds. It is inherited as an autosomal recessive trait. The condition is apparently harmless, but must be differentiated from other more serious causes of iminoglycinuria, such as the defect of proline metabolism, hyperprolinaemia.

Storage defects

A variety of disorders affect storage in the body. Some, such as the glycogen storage disorders (p. 220) and the mucopolysaccharidoses, present in infancy and childhood: in others, such as haemochromatosis (p. 397) and Wilson's disease, the abnormal accumulation only becomes severe enough to cause clinical features by adult life.

The mucopolysaccharidoses (MPS)

These rare conditions are caused by defects of any of the several enzymes that hydrolyse mucopolysaccharides (glycosaminoglycans), which therefore accumulate in such tissues as the liver, spleen, eyes, central nervous system, cartilage and bone. *Hurler's syndrome (MPS I H)* is the least rare, and is inherited as an autosomal recessive disorder. Patients present in infancy or early childhood with the characteristic coarse features of gargoylism, short stature, mental retardation and clouding of the cornea. They usually die young of cardiorespiratory disease. *Scheie's syndrome (MPS I S)* is difficult to distinguish clinically from Hurler's syndrome at the time of diagnosis, but has a much better prognosis, and there is little mental retardation. *Hunter's syndrome (MPS II)*, in contrast to all the other mucopolysaccharidoses, is inherited as a sex-linked recessive trait.

The mucopolysaccharidoses can initially be diagnosed biochemically by demonstrating increased urinary excretion of sulphated glycosaminoglycans, such as dermatan, heparan and keratan sulphates, the excretion pattern being characteristic for each syndrome. The diagnosis should be confirmed by direct enzyme assay.

There is no effective treatment at present, although enzyme replacement by bone marrow transplantation is being evaluated.

Wilson's disease

Some plasma copper is loosely bound to albumin, but most is usually incorporated in the protein *caeruloplasmin.* Copper is mostly excreted in the bile.

There are two defects of copper metabolism in Wilson's disease:

- *impaired biliary excretion* leads to *copper deposition in the liver;*
- *deficiency of caeruloplasmin* results in *low plasma copper* levels; most of this copper is in a loosely bound form and is *deposited in tissues*; more than normal is filtered at the glomerulus and *urinary copper excretion is increased.*

Excessive deposition of copper in the basal ganglia of the brain, and in the liver, renal tubules and eyes, produces:

- *neurological symptoms* due to degeneration of the basal ganglia;
- *liver damage* leading to cirrhosis;
- *renal tubular damage* with any or all of the associated biochemical features, including aminoaciduria (Fanconi syndrome);
- *Kayser-Fleischer rings* at the edges of the cornea due to deposition of copper in Desçemet's membranes.

Diagnosis. Most patients have low plasma caeruloplasmin and copper concentrations and a high urinary copper excretion.

Low plasma caeruloplasmin levels may also occur during the *first few months of life*, due to *malnutrition*, and in the *nephrotic syndrome* due to urinary loss.

Raised plasma levels are found in *active liver disease*, in women taking oral contraceptives, during the *last trimester of pregnancy* and non-specifically when there is *tissue damage* due for example to inflammation or neoplasia. These may account for the rare finding of 'normal' plasma caeruloplasmin levels in patients with Wilson's disease.

The clinical condition has a recessive mode of inheritance, but heterozygotes may have reduced caeruloplasmin levels. The distinction between presymptomatic homozygotes and heterozygotes is important because the former can be treated.

Treatment with copper-chelating agents such as D-penicillamine may reduce tissue copper concentrations.

Drugs and inherited metabolic disorders

The variation in individual response to drugs may partly be due to genetic variation. There are a number of well-defined inherited disorders that are aggravated by, or which only become apparent after, administration of certain drugs. These disorders may be classified into two groups.

Disorders resulting from deficient metabolism of a drug

The muscle relaxant, suxamethonium (succinyl choline; 'scoline') normally has a very brief action because it is rapidly broken down by plasma cholinesterase. In *suxamethonium sensitivity* (p. 321) a cholinesterase variant of low biological activity impairs breakdown of the drug, and prolonged postoperative respiratory paralysis may result ('scoline apnoea').

Two other inherited disorders are characterized by defective metabolism of the drugs *isoniazid* and *phenytoin*. In both toxic effects occur more frequently, and at lower dosages, than in normal individuals.

Disorders resulting from an abnormal response to a drug

Deficiency of *glucose-6-phosphate dehydrogenase* (G-6-PD) may cause haemolytic anaemia, and is relatively common in some ethnic groups, such as those of Mediterranean origin. This enzyme catalyses the first step in the hexose monophosphate pathway, and is needed for the formation of NADP, which is probably essential for the maintenance of an intact red cell membrane. Numerous variants of G-6-PD deficiency have been described. Haemolysis may be precipitated by certain antimalarial drugs such as primaquine, and by sulphonamides and vitamin K analogues.

In the inherited *hepatic porphyrias* (p. 405) acute attacks may be precipitated by several drugs, particularly barbiturates.

Some people react to general anaesthetics (most commonly halothane with suxamethonium) with a rapidly rising temperature, muscular rigidity and acidosis (*malignant hyperpyrexia*): most of them die as a result. Many, but not all, susceptible subjects in affected families have a high plasma creatine kinase activity.

This short section should remind readers to remember the possibility of an inborn error when an abnormal reaction to a drug is encountered. A reference to the field of *pharmacogenetics* is given at the end of the chapter.

Summary

1. Inborn errors of metabolism may cause diseases due to inherited defects of protein synthesis. Most cases presenting with early clinical symptoms are due to abnormalities of enzyme synthesis.
2. Inborn errors of metabolism may produce no clinical effects, may only produce them under certain circumstances or, at the other extreme, may produce severe diseases incompatible with life.
3. Recognition of some inherited abnormalities is only of academic interest. Diagnosis is important if the condition is serious but treatable, if precipitating factors can be avoided, if confusion with other diseases is possible or if genetic counselling is considered.
4. Inheritance may be autosomal or sex-linked, dominant or recessive. In diseases producing severe clinical effects inheritance is most often autosomal recessive and they are most common in the offspring of consanguineous marriages.
5. In many cases in which the clinical disease is inherited as a recessive trait lesser degrees of abnormality can be detected by laboratory testing.
6. Some inborn errors of metabolism not mentioned elsewhere in this book are discussed in this chapter.

Further reading

Holton JB. Diagnosis of inherited metabolic diseases in severely ill children. *Ann Clin Biochem* 1982; **19**: 389–95.

Holtzman NA. Newborn screening for inborn errors of metabolism. *Paediatr Clin N Am* 1980; **25**: 411–21.

Scriver CR, Clow CL. Phenylketonuria: epitome of human biochemical genetics. *New Engl J Med* 1980; **303**: 1336–42, 1394–1400.

Vesell ES. Pharmacogenetics: multiple interactions between genes and environment as determinants of drug response. *Am J Med* 1979; **66**: 183–7.

For reference

Stanbury JB, Wyngaarden JB, Fredrickson DS, Goldstein JL, Brown MS eds. *The Metabolic Basis of Inherited Disease*. **5th ed**. New York: McGraw-Hill, 1983.

Sinclair L. *Metabolic Disease in Childhood.* Oxford: Blackwell Scientific, 1979.

Table 18.1 Some inborn errors and their inheritance

The following list of inborn errors of metabolism is far from complete. It is meant for reference only and the student should not attempt to learn it. Most of the abnormalities have been discussed in this book, and a page reference is given. Where it is known, the mode of inheritance is stated, unless the heading applies to a group of diseases of different modes of inheritance

D = Autosomal Dominant
R = Autosomal Recessive
X-linked D = X-linked Dominant
X-linked R = X-linked Recessive

	Inheritance	Page
1. **Disorders of cellular transport**		
Most of these are recognized as renal tubular transport defects, and in some, defective intestinal transport can also be demonstrated		
Amino and imino acids		
Dibasic amino acids. Cystinuria	R	372
Neutral amino acids. Hartnup disease	R	373
Familial iminoglycinuria	R	373
Glucose		
Renal glycosuria	D	210
Water (failure to respond to ADH)		
Hereditary nephrogenic diabetes insipidus	X-linked R	44
Potassium (all cells)		
Familial periodic paralysis	D	60
Calcium (failure to respond to PTH)		
Pseudohypoparathyroidism	X-linked D	185
Phosphate		
Familial hypophosphataemia	X-linked D	186
Hydrogen ion		
Renal tubular acidoses	D	93
Bilirubin (liver cells)		
Congenital hyperbilirubinaemias		
Crigler-Najjar syndrome Type I	R	299
Crigler-Najjar syndrome Type II	D	299
Dubin-Johnson syndrome	?R	300
Gilbert's disease	?	299
Rotor syndrome	?R	300
2. **Disorders of amino and imino acid metabolism**		
Aromatic amino acids		
Phenylketonuria	R	369
Alkaptonuria	R	371
Albinism	R	371
Thyroid dyshormonogenesis	All R	166
Sulphur amino acids		
Cystinosis	R	373
Homocystinuria	R	
Branched-chain amino acids		
Maple syrup urine disease	R	371
Histidinaemia	R	372
3. **Disorders of carbohydrate metabolism**		
Glycogenoses	R	220
Galactosaemia	R	221
Hereditary fructose intolerance	R	221
Essential fructosuria	R	230
Diabetes mellitus	?	211

Table 18.1 (continued)

	Inheritance	Page
4. **Disorders of lipid metabolism**		
Hyperlipoproteinaemias	—	240
Hypolipoproteinaemias	R	242
Plasma LCAT deficiency	R	242
5. **Disorders of lysosomal metabolism**		
Mucopolysaccharidoses		
Hurler's syndrome (MPS I H)	R	374
Hunter's syndrome (MPS II)	X-linked R	374
Gaucher's disease	R	364
6. **Abnormalities of plasma proteins**		
Immunoglobulin deficiencies	—	335
Carrier protein abnormalities		
Transferrin deficiency	?R	396
Thyroxine-binding globulin deficiency	X-linked	162
Cholinesterase variants	R	321
α_1-Antitrypsin deficiency	—	336
Bisalbuminaemia	D	328
Analbuminaemia	R	328
7. **Disorders of metal metabolism**		
Haemochromatosis	?R	397
Wilson's disease	R	374
8. **Erythrocyte abnormalities** (see haematology textbooks)		
Haemoglobinopathies	—	—
Glucose-6-phosphate dehydrogenase deficiency	X-linked	375
NADP methaemoglobin reductase deficiency	R	—
9. **Disorders of porphyrin metabolism**		
Porphyrias	—	403
10. **Disorders of steroid metabolism**		
Congenital adrenal hyperplasia	All R	129
11. **Disorders of purine metabolism**		
Primary gout	?	384
Xanthinuria	R	386
Lesch-Nyhan syndrome	X-linked R	385
12. **Disorders of digestion**		
Disaccharidase deficiencies	R	261
Cystic fibrosis	R	258
13. **Disorders of oxalate metabolism**		
Primary hyperoxaluria	R	22
14. **Disorders precipitated by drugs**		
Suxamethonium sensitivity	R	321
Slow inactivation of isoniazid	R	375
Phenytoin toxicity	D	375
Glucose-6-phosphate dehydrogenase deficiency	X-linked	375
Malignant hyperpyrexia	D	376

19

Purine and urate metabolism

Hyperuricaemia and gout

Hyperuricaemia may be due to a familial primary abnormality of purine metabolism or be secondary to a variety of other conditions. It may be asymptomatic, or may be associated with the clinical syndrome of gout.

At plasma pH most urate is ionized at position 8 of the purine ring (Fig. 19.1). This anionic group is associated with the predominant extracellular cation, sodium. Unequivocally high urate concentrations, even if without symptoms, should be treated because the relatively insoluble monosodium urate may, like calcium, precipitate in tissues. If this takes place in the kidney it may cause renal damage. Ionization of uric acid decreases as the pH falls, and it therefore becomes less soluble: at a urinary pH below about 6 uric acid may form renal calculi.

Normal urate metabolism

Urate is the end product of purine metabolism in primates, including man. In most other mammals it is further metabolized to the more soluble allantoin; it is because of the poor solubility of urates that man is prone to clinical gout and renal damage by urate. The purines adenine and guanine are constituents of both types of *nucleic acid* (DNA and RNA). The purines used by the body for nucleic acid synthesis may be derived from the breakdown of ingested nucleic acid, mostly taken in cell-rich meat, or they may be synthesized *de novo* from small molecules.

Synthesis of purines

The purine synthetic pathway involves the incorporation of many small molecules into the relatively complex purine ring. The upper part of Fig. 19.1 summarizes some of the more important synthetic steps. Some cytotoxic drugs inhibit various stages of the pathway, so preventing DNA formation and cell growth.

The following stages shown in Fig. 19.1 should especially be noted.

- The first step in purine synthesis is condensation of pyrophosphate with phosphoribose to form phosphoribose diphosphate (phosphoribosyl pyrophosphate; PRPP).
- The amino group of glutamine is incorporated into the ribose phosphate

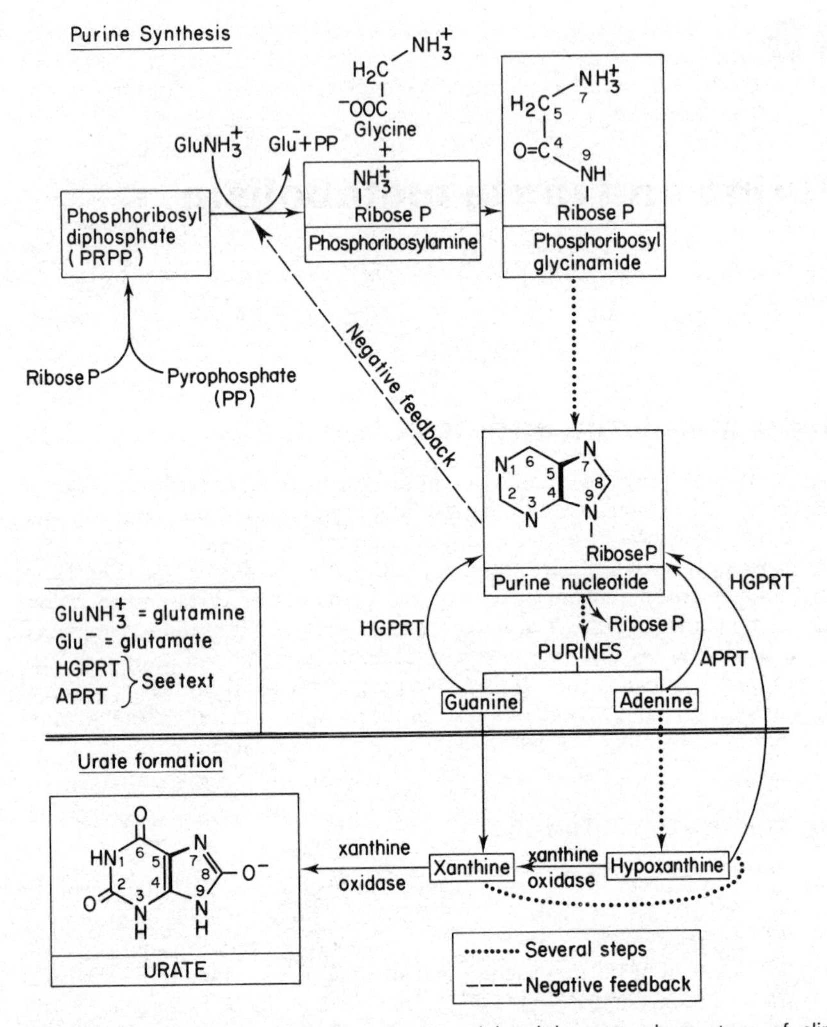

Fig. 19.1 Summary of purine synthesis and breakdown to show steps of clinical importance.

molecule and pyrophosphate (PP) is released. Amidophosphoribosyl transferase catalyses this *rate-limiting* or controlling step. The enzyme is subject to feedback inhibition by increasing concentrations of purine nucleotides: thus the rate of synthesis is slowed when its products increase. The control of this rate-limiting step may be impaired in primary gout.

• Glycine is added to phosphoribosylamine. By using labelled glycine it has been shown that the rate of purine synthesis is increased in primary gout. In Fig. 19.1 the atoms in the glycine molecule have been numbered to correspond with those in the purine and urate molecules, and the heavy lines further indicate the final position of the amino acid in these molecules.

After many further steps purine ribonucleotides (purine ribose phosphates) are formed and, as previously mentioned, control the second step in the synthetic pathway – the formation of phosphoribosylamine. Ribose phosphate is split off, thereby releasing the purines.

Fate of purines

Purines synthesized in the body, those derived from the diet, and those liberated by endogenous catabolism of nucleic acids may follow one of two pathways. They may be:

- reused for nucleic acid synthesis;
- oxidized to urate.

Formation of urate from purines. As shown in the lower part of Fig. 19.1, some adenine is oxidized to hypoxanthine, which is further oxidized to xanthine. Guanine can also form xanthine. Xanthine, in turn, is oxidized to form urate. The oxidation of both hypoxanthine and xanthine is catalysed by *xanthine oxidase* in the liver. Thus the formation of urate from purines depends on xanthine oxidase activity: gout may be treated using an inhibitor of this enzyme (allopurinol).

Reutilization of purines. Some xanthine, hypoxanthine and guanine can be resynthesized to purine nucleotides by pathways involving, among other enzymes, hypoxanthine-guanine phosphoribosyl transferase (HGPRT) and adenine phosphoribosyl transferase (APRT).

Excretion of urate. Urate is filtered through the glomerulus and most is reabsorbed from the tubular lumen. Over 80 per cent of the urinary urate is derived from subsequent active tubular secretion.

Urinary excretion is slightly lower in males than in females, and this may contribute to the higher incidence of hyperuricaemia in men. Renal secretion may be enhanced by uricosuric drugs used to treat hyperuricaemia, which may block tubular reabsorption. Tubular secretion of urate is inhibited by such organic acids as lactic and oxoacids, and by thiazide diuretics.

75 per cent of excreted urate enters the urine. The remaining 25 per cent passes into the *intestinal lumen*, where it is broken down by intestinal bacteria, the process being known as *uricolysis*.

Causes of hyperuricaemia

Fig. 19.2 summarizes the factors which may contribute to hyperuricaemia. These are:

- increased rate of urate formation:
  - increased synthesis of purines (a);
  - increased intake of purines (b);
  - increased turnover of nucleic acids (c).
- reduced rate of excretion (e).

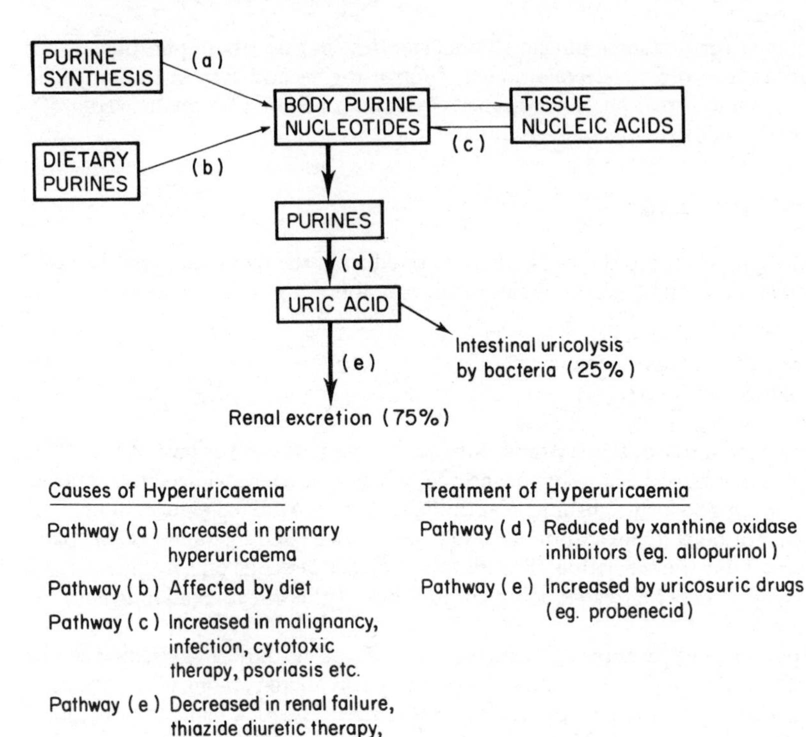

Fig. 19.2 Origin and fate of urate in normal subjects.

Abnormalities of steps (b), (c) and (e) are causes of secondary hyperuricaemia. Increased synthesis (Step a), due to impaired feedback control, is probably the most important mechanism causing primary hyperuricaemia.

Dangers of hyperuricaemia

Urate is poorly soluble in plasma. Precipitation in tissues may be favoured by a variety of local factors of which the most important are probably pH and trauma.

Crystallization in *joints*, especially those in the feet, produces the classical picture of gout first recorded by Hippocrates in 469 BC. Urate precipitation at these sites causes an inflammatory response with leucocytic infiltration and it is thought that lactic acid production by these white cells causes a local fall in pH: this converts urate to uric acid, which is less soluble than urate, and a vicious circle is set up in which further precipitation, and therefore further inflammation, occur. In *acute attacks* of gouty arthritis local factors are more important than the *plasma urate* concentration, which is usually *normal* during the attack.

Precipitation may occur in subcutaneous tissues, especially on the ears, and in the olecranon and patellar bursae and tendons. Such deposits are called *gouty tophi.*

A potentially serious effect of hyperuricaemia is precipitation of urate in the kidneys, causing progressive renal disease. For this reason it has been recommended that even asymptomatic cases should be treated if the plasma urate concentration is consistently higher than 0.6 mmol/litre (10 mg/dl).

Secondary hyperuricaemia

High plasma urate concentrations may be secondary to:

increased turnover of nucleic acids ((c) in Fig. 19.2):

- in rapidly growing malignant tissue, especially in leukaemias, lymphomas and polycythaemia rubra vera;
- in psoriasis, when turnover of skin cells is increased;
- increased tissue breakdown:
  - after treatment of large malignant tumours;
  - after tissue damage due to trauma;
  - during acute starvation.

reduced excretion of urate ((e) in Fig. 19.2) due to:

- renal glomerular dysfunction;
- treatment with thiazide diuretics;
- acidosis;
- low doses of salicylates.

Breakdown of tumour tissue during *treatment of large masses by radiotherapy or cytotoxic drugs* can cause massive, sudden release of urate which, by crystallizing in and blocking renal tubules, may cause acute oliguric renal failure. During such treatment allopurinol should be given and, if glomerular function is not impaired, fluid intake should be kept high.

Tissue damage and starvation. Tissue breakdown is increased after trauma, including that of surgery; during starvation or stringent dieting the patient's own tissues may be used as an energy source. In both cases endogenous urate release is increased. Starvation may be associated with mild ketoacidosis, and protein catabolism releases acidic amino acid residues (p. 76): these acids, by inhibiting secretion of urate, may aggravate the hyperuricaemia due to overproduction. During complete fasting plasma concentrations may be 0.9 mmol/litre (15 mg/dl) or more.

Glomerular dysfunction causes urate retention. Before the diagnosis of primary hyperuricaemia is made plasma urea should be assayed on the same specimen as urate, to exclude renal disease as a cause. It may be difficult to decide which abnormality is the primary one because hyperuricaemia can *cause* renal dysfunction. As a rough guide the plasma urate concentration would be expected to be about 0.6 to 0.7 mmol/litre (10 to 12 mg/dl) due to glomerular retention alone at a plasma urea concentration of about 50 mmol/litre (300 mg/dl); if it is much higher than this hyperuricaemia should be suspected to be the primary cause of the renal dysfunction. Clinical gout is rare in the secondary hyperuricaemia of renal disease.

Thiazide diuretics inhibit renal urate excretion. Although hyperuricaemia is relatively common during diuretic treatment, clinical gout is a rare complication. It may, however, be precipitated in those patients with a gouty tendency.

Association with hypercalcaemia. For unknown reasons hyperuricaemia, and even clinical gout, are relatively common in patients with hypercalcaemia from any cause, and in patients with recurrent renal calculi, even when not associated with hypercalcaemia.

Glucose-6-phosphatase deficiency (p. 220). The tendency to hyperuricaemia in patients with glucose-6-phosphatase deficiency may be related directly to the inability to convert glucose-6-phosphate to glucose. More G-6-P is available for metabolism through intracellular pathways, including:

- the pentose-phosphate pathway, thus increasing ribose phosphate (phosphoribose) synthesis. This may accelerate the first step in purine synthesis, with consequent *urate overproduction;*
- glycolysis, thus increasing lactic acid production (p. 208). Lactic acid may *reduce renal urate excretion.*

Primary hyperuricaemia and gout

Familial incidence

In AD 150 Galen said that gout was due to 'debauchery, intemperance and an hereditary trait'. As we shall see, 'intemperance' may aggravate the condition. The striking familial incidence of hyperuricaemia confirms that there is probably an 'hereditary trait', but in this respect we know little more than Galen did; the mode of inheritance, like that of diabetes mellitus, is still not fully established.

Sex and age incidence

Primary hyperuricaemia and gout are very rare in children, and rare in women of child-bearing age. The higher incidence in males cannot be due to sex-linked inheritance because the syndrome can be transmitted by males. Plasma urate concentrations are low in children and rise in both sexes at puberty, more so in males than females. Women become more prone to hyperuricaemia and gout in the post-menopausal period (compare plasma cholesterol and iron levels). It may be that renal excretion of urate is affected by sex-hormone levels.

Precipitating factors

The classical image of the gouty subject is the red-faced, good-living, hard drinking squire depicted in novels and paintings of the 18th century. Galen mentioned 'debauchery and intemperance' as causes of gout. Two factors probably account for the high incidence of clinical gout in this type of subject.

- *Alcohol* has been shown to decrease renal excretion of urate. This may be because it increases lactic acid production, which inhibits urate secretion.
- *A high meat diet* contains a relatively high proportion of *purines.*

Neither of these factors is likely to precipitate gout in a normal person, but, like thiazide diuretics, may do so in a subject with a hyperuricaemic tendency.

Biochemical lesion of primary hyperuricaemia

Purine synthesis is increased in about 25 per cent of cases of primary hyperuricaemia due to overactivity of amidophosphoribosyl transferase which controls the formation of phosphoribosylamine (p.380).

Reduced renal tubular secretion of urate has also been demonstrated in other cases of primary hyperuricaemia. Many subjects may have both increased synthesis and decreased excretion.

Principles of treatment of primary hyperuricaemia

Treatment may be based on:

- **reducing dietary purine intake** ((b) in Fig. 19.2). This treatment is not very effective by itself.

- **increasing renal excretion of urate** with *uricosuric drugs,* such as probenecid and salicylates in large doses ((e) in Fig. 19.2). These drugs are very effective if renal function is normal, but are useless if there is renal failure. Fluid intake must be kept high. *Low doses* of most uricosuric drugs *reduce* urate secretion;

- **reducing urate production** by using drugs which inhibit xanthine oxidase activity, such as *allopurinol* (hydroxypyrazolopyrimidine) ((d) in Fig. 19.2), which is structurally similar to hypoxanthine and acts as a competitive inhibitor of the enzyme. *De novo* synthesis may also be decreased by this drug;

- **colchicine,** which has an anti-inflammatory effect in acute gouty arthritis, does not affect urate metabolism.

Juvenile hyperuricaemia (Lesch-Nyhan syndrome)

This is an *exceedingly rare* inborn error of urate metabolism, probably carried on an X-linked recessive gene. Severe hyperuricaemia occurs in young male children. A deficiency of the enzyme *hypoxanthine-guanine phosphoribosyl transferase (HGPRT)* has been demonstrated in affected subjects. Hypoxanthine and other purines cannot be recycled to form purine nucleotides, and probably produce more urate. The syndrome is associated with mental deficiency, a tendency to self-mutilation, aggressive behaviour, athetosis and spastic paraplegia.

Pseudogout

Pseudogout is not a disorder of purine metabolism. The clinical picture is, however, similar to that of gout. Calcium pyrophosphate precipitates in joint cavities, and calcification of cartilages is demonstrable radiologically. The plasma urate concentration is normal. The crystals of calcium pyrophosphate may be identified in joint fluid using a polarizing microscope.

Hypouricaemia

Hypouricaemia, unless the result of treatment of hyperuricaemia, is rare. It is an unimportant finding associated with proximal renal tubular damage, in which reabsorption of urate is reduced.

Xanthinuria is a very rare inborn error of purine metabolism in which there is a deficiency of xanthine oxidase in the liver. The mode of inheritance is probably autosomal recessive. Purine breakdown stops at the xanthine-hypoxanthine stage, and plasma and urinary urate levels are very low. Increased xanthine excretion may lead to formation of xanthine stones: this does not occur during treatment of gout with xanthine oxidase inhibitors, perhaps because the drugs also inhibit purine synthesis.

Summary

1. Urate is the end product of purine metabolism.
2. Hyperuricaemia may result from:

- an increased rate of nucleic acid turnover (malignancy, tissue damage, starvation);
- an increased rate of synthesis of purines (primary gout);
- a reduced rate of renal excretion of urate (glomerular dysfunction, thiazide diuretics, acidosis).

3. Hyperuricaemia may be aggravated by:

- high purine diets;
- acidosis or a high alcohol intake.

4. Primary hyperuricaemia and gout have a familial incidence. Both are rare in women of child-bearing age.

5. Severe hyperuricaemia may cause renal damage and should be treated even if asymptomatic.

6. Hypouricaemia is rare and usually unimportant. It occurs in the very rare inborn error, xanthinuria.

Further reading

Cameron JS, Simmonds HA. Uric acid, gout and the kidney. *J Clin Path* 1981; **34**: 1245–54.

Emmerson BT. *Hyperuricaemia and gout in clinical practice.* Bristol: John Wright and Sons. 1983.

20

Iron metabolism

The iron-containing pigment, haemoglobin, carries oxygen from the lungs to metabolizing tissues, and myoglobin in muscle increases the local supply of oxygen. Oxygen-carrying ability depends mainly on the presence of ferrous iron in the haem molecule; iron deficiency is associated with deficient haem synthesis, and the symptoms of anaemia are due to tissue hypoxia. Cytochromes and some enzymes needed for electron-transfer reactions also contain iron, but probably only very severe iron deficiency affects these.

Normal iron metabolism

Distribution of iron in the body

Fig. 20.1 represents diagrammatically the distribution of the 50 to 70 mmol (3 to 4 g) of total iron in the body.

- About 70 per cent of the total iron is circulating in *erythrocyte haemoglobin*.
- Up to 25 per cent of the body iron is stored in the reticuloendothelial system, in the liver, spleen and bone marrow: bone marrow iron is drawn on for haemoglobin synthesis. Iron is stored as the protein complexes, *ferritin* and *haemosiderin*. Ferritin iron is more easily released from protein than that in haemosiderin. Haemosiderin, which may be an aggregate of ferritin, can be seen by light microscopy in unstained tissue preparations. Ferritin and haemosiderin, but not haem iron, stain with potassium ferrocyanide (Prussian blue reaction), and this staining characteristic may be used to assess the size of iron stores.

  Iron deficiency only becomes haematologically evident when *no stainable iron* is detectable in the *reticuloendothelial cells* in *bone marrow* films.
  Iron overload is likely when, because reticuloendothelial storage capacity is exceeded, *stainable iron* is demonstrable in *parenchymal cells* in *liver* biopsy specimens.

Histological assessment is more reliable for detecting iron deficiency than overload (p. 397).

- *Only about 50 to 70 μmol (3 to 4 mg), or about 0.1 per cent of the total body iron is*

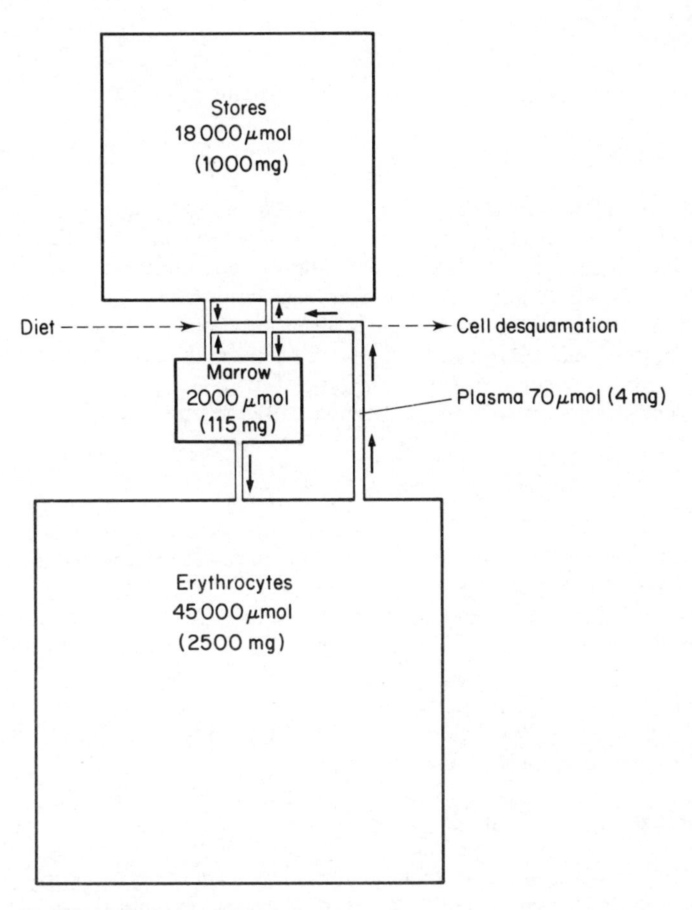

Fig. 20.1 Body iron compartments.

circulating in the plasma: this fraction, which is bound to the protein transferrin, is measured in *plasma iron* assays.

• The rest of the body iron is incorporated in myoglobin, cytochromes and iron-containing enzymes.

Iron can only cross cell membranes by active transport in the ferrous (Fe^{2+}) form: it is in this reduced state in both oxyhaemoglobin and 'reduced' haemoglobin. It is in the ferric (Fe^{3+}) form in ferritin and haemosiderin, and when bound to transferrin.

The control of iron distribution in the body is poorly understood. However, even in normal subjects, there is considerable interchange between stores and plasma; *plasma iron concentrations vary by 100 per cent or more for purely physiological reasons, and are also affected by many pathological factors other than the amount of iron in the body.*

Iron balance

Fig. 20.1 shows that iron, once in the body, is in a virtually closed system.

Iron absorption

The control of body iron content depends upon control of absorption.

Iron is absorbed by an active process in the upper small intestine. Within the intestinal cell some of the iron combines with the protein apoferritin to form ferritin, which, as elsewhere in the body, is a storage compound.

Normally about 18 μmol (1 mg) of iron is absorbed each day and this just replaces loss. This amounts to about ten per cent of that taken in the diet, although the proportion depends to some extent on the type of food.

Iron absorption seems to be influenced by any or all of the following factors:

- oxygen tension in the intestinal cells;
- marrow erythropoietic activity;
- the size of the body iron stores.

It is important to remember that *iron absorption is increased in many non-iron deficiency anaemias.*

Most normal women taking an adequate diet probably absorb slightly more iron than men and so replace their higher losses in menstrual blood and during pregnancy.

Iron requirements for growth during childhood and during adolescence are similar to, or slightly higher than, those of menstruating women. This need, too, can be met by increased absorption from a normal diet.

Iron excretion

There is probably no control of iron excretion, and loss from the body probably depends on the ferritin iron content of cells lost by desquamation, mostly into the intestinal tract and from the skin. The total daily loss by these routes is about 18 μmol (1 mg). Urinary loss is negligible, reflecting the fact that all circulating iron is protein-bound.

Women lose a mean of about 290 μmol (16 mg) a month in menstrual fluid; this is equivalent to a mean of 10 to 18 μmol (0.5 to 1 mg) a day above the basal 18 μmol. Losses can be much higher if there is menorrhagia, and may then cause iron deficiency. During pregnancy the mean extra daily loss to the fetus and placenta is about 27 μmol (1.5 mg). As a comparison, a male blood donor losing a unit (pint) of blood every 4 months averages an extra loss of 36 μmol (about 2 mg) daily above the basal 18 μmol (Table 20.1).

Normal iron loss is so small, and normal iron stores are so large, that it would take about 3 years to become iron-deficient on a completely iron-free diet. Of course, this period is much shorter if there is any abnormal blood loss.

Iron transport in plasma

Iron is transported in the plasma in the ferric form, attached to the specific binding

Table 20.1 Comparison of iron losses

	Source of loss	Extra loss	Daily extra loss	Daily total loss
Men and non-menstruating women	Desquamation	—	—	18 μmol (1 mg)
Menstruating women (mean value)	Desquamation + menstruation	290 μmol (16 mg)/month	9 μmol (0.5 mg)	27 μmol (1.5 mg)
Pregnancy	Desquamation + loss to fetus and in placenta	7000 μmol (380 mg)/ 9 months	27 μmol (1.5 mg)	45 μmol (2.5 mg)
Male blood donors	Desquamation + 1 unit of blood	4500 μmol (250 mg)/ 4 months	36 μmol (2.0 mg)	54 μmol (3.0 mg)

protein, *transferrin*, at a concentration of about 18 μmol/litre (100 μg/dl). Transferrin is normally capable of binding about 54 μmol/litre (300 mg/dl) of iron and is therefore about a third saturated. Transferrin-bound iron is carried to stores and to bone marrow cells: in stores it is laid down as ferritin and haemosiderin, and in the marrow some may pass directly from transferrin into the developing erythrocyte to form haemoglobin.

Free iron is toxic. In normal subjects iron is all protein-bound; in the plasma it is bound to transferrin, in the stores to protein in ferritin and haemosiderin, and in the erythrocytes it is incorporated in haemoglobin.

Factors affecting the plasma iron concentration

Plasma iron estimation is often requested, but is rarely of clinical value, and results are often misinterpreted. The plasma iron concentration is likely to be a poor index of the total body content because only a very small proportion is in this compartment; it has no function, except as a protein-bound transport fraction. The concentration is not tightly controlled, and varies greatly even under physiological conditions.

Physiological factors affecting the plasma iron concentration

The causes of physiological changes in plasma iron concentrations are not well understood, but alterations can be very rapid, and almost certainly represent shifts between plasma and stores, not changes in total body iron. The following factors are known to affect levels and, in an individual subject, the last three can cause changes of 100 per cent or more.

Sex and age differences

Plasma iron levels, like those of haemoglobin and the erythrocyte count, are higher in men than in women, probably for hormonal reasons. The difference is first evident at puberty, before significant menstrual loss has occurred, and disappears at the menopause. Androgens tend to increase the plasma iron concentration and oestrogens to lower it.

Cyclical variations

Circadian (diurnal) rhythm. The plasma iron concentration is higher in the morning than in the evening. If subjects are kept awake at night this difference is less marked; it is reversed in night workers.

Monthly variations in women. The plasma iron may reach very low concentrations just before or during the menstrual period. The reduction is probably due to hormonal factors rather than blood loss.

Random variations

Day-to-day variations, which may be as much as three-fold, occur in plasma iron concentrations and these usually overshadow cyclical changes. They may be associated with physical or mental stress, but more usually a cause cannot be found.

Effects of pregnancy and oral contraceptives

In the first few weeks of pregnancy the plasma iron may rise to concentrations similar to those found in men. A similar rise occurs in women taking some oral contraceptives.

Pathological factors affecting the plasma iron concentration

- Iron deficiency and iron overload usually cause low and high plasma iron concentrations respectively.

    Iron deficiency is associated with a hypochromic, microcytic anaemia, and with reduced amounts of stainable marrow iron. Plasma ferritin concentrations are usually, but not always, low.
    Iron overload is associated with increased amounts of stainable iron in liver biopsy specimens, and plasma ferritin concentrations are high.

- *Any illness*, whether acute or chronic, causes hypoferraemia: even a bad cold can lead to a fall in plasma iron concentration. Chronic conditions such as malignancy, renal disease, rheumatoid arthritis and chronic infecions are often associated with normochromic, normocytic anaemia. Iron stores and plasma ferritin levels are normal or even increased, and the anaemia does not respond to iron therapy.

Iron deficiency may be superimposed on the anaemia of chronic illness: this is especially likely in such conditions as rheumatoid arthritis in which salicylates, or other anti-inflammatory drugs used to treat the primary condition, may cause low-grade bleeding from small gastric erosions. Ferritin levels are variable, and the finding of hypochromic erythrocytes is the most sensitive index of this complication. Low plasma iron concentrations occur *whether or not* there is any iron deficiency.

- In conditions in which the *marrow cannot use iron*, either because it is hypoplastic, or because some other essential erythropoietic factor such as vitamin B_{12} or folate is deficient, plasma iron levels are often high. Blood and marrow films may show a typical picture, but, for example, in pyridoxine-responsive anaemia (p. 281) and in thalassaemia, the findings in the blood film may resemble those of iron deficiency; in the last two conditions the presence of stainable marrow iron stores excludes the diagnosis of iron deficiency.
- In *haemolytic anaemia* the iron liberated from the haemoglobin released from destroyed erythrocytes enters the plasma and reticuloendothelial system. The plasma iron concentration may be high during a haemolytic episode and is usually normal during quiescent periods. Marrow iron stores and plasma ferritin levels are usually increased in chronic haemolytic conditions.
- In *acute liver disease* disruption of hepatocytes may release ferritin iron into the blood stream and cause a transient rise in plasma iron concentration. *Cirrhosis* may be associated with a similar finding, perhaps due to increased iron absorption and intake.

Transferrin and total iron-binding capacity (TIBC)

Plasma iron concentrations alone give no information about the state of iron stores. In rare situations in which doubt remains after haematological investigation, diagnostic precision may sometimes be improved by measuring the plasma transferrin concentration at the same time as the plasma iron. It is completely uninformative to estimate only plasma iron.

If direct plasma transferrin assay is not available its concentration can be assessed by measuring the *iron-binding capacity* of the plasma. An excess of inorganic iron is mixed with the plasma and any not bound to protein is removed, usually with an exchange resin. The iron remaining in the plasma is assayed and the result expressed as the total iron-binding capacity (TIBC). This is usually a valid measure of the transferrin concentration. In rare circumstances, of which the most common is severe liver disease, plasma ferritin levels are high enough to bind significant amounts of iron, and results of iron-binding capacity measurement are then misleading as an assessment of the transferrin concentration.

Physiological changes in the transferrin concentration

The plasma transferrin concentration is less labile than that of iron. However, it rises:

- after about the 28th week of *pregnancy even if iron stores are normal*;
- in women on some *oral contraceptive preparations*;
- in any patient treated with *oestrogens.*

Pathological changes in transferrin concentration

- The transferrin concentration and TIBC *rise in iron deficiency* and *fall in iron overload.*
- The transferrin concentration and TIBC fall in those chronic illnesses associated with low plasma iron concentrations.

In the nephrotic syndrome both plasma iron and transferrin concentrations may be very low because the relatively low molecular weight transferrin is lost in the urine together with the iron.

Transferrin concentration and TIBC are unchanged in acute illness.

Thus the low plasma iron concentration of uncomplicated iron deficiency is associated with a high transferrin level and TIBC. That of anaemia not due to iron deficiency is associated with a low TIBC.

If iron deficiency coexists with the anaemia of chronic illness the opposing effects of the two conditions on the transferrin concentration make it difficult to interpret transferrin, as well as plasma iron, concentrations.

Plasma ferritin concentrations

Circulating plasma ferritin is usually in equilibrium with that in stores. However, it is an 'acute phase' protein (p. 330) and its synthesis is increased in many inflammatory conditions.

The normal plasma ferritin concentration is about 100 μg/litre. *A level below about 10μg/litre almost certainly indicates iron deficiency*, although the assay is rarely necessary to make the diagnosis. Results can be misleading if there is coexistent inflammatory disease, since accelerated synthesis may lead to normal or even high levels despite very low iron stores: in this situation results of plasma iron and transferrin assays are also difficult to interpret, and haematological parameters remain the most reliable diagnostic indicators of iron deficiency.

High levels of plasma ferritin always occur in significant iron overload, but may also be due to:

- inflammatory conditions;
- malignant disease;
- liver disease.

Thus the finding of a *normal or low plasma ferritin level almost certainly excludes the diagnosis of iron overload, but a high one does not necessarily confirm it.*

The laboratory findings in conditions which may affect plasma iron concentrations are summarized in Table 20.2.

The student should read the section on 'Investigation of anaemia' on p. 400.

Table 20.2 Laboratory changes associated with plasma iron abnormalities

	Plasma concentrations			Marrow stores
	Fe	Transferrin	Ferritin	
Low iron concentration				
Before menstruation	↓	Normal	Normal	Normal
Iron deficiency	↓	↑	Usually ↓	Absent or ↓↓
Acute illness	↓	Normal	Normal or ↑	Normal
Chronic illness	↓	↓	Normal or ↑	Usually ↑
High iron concentration				
Early pregnancy	↑*	Normal	Normal	Normal
Late pregnancy or oral contraceptives	variable	↑	Normal	Normal
Iron overload	↑	↓	↑	Usually ↑
Liver disease	↑	↓	↑	May be ↑
Impaired marrow utilization	↑	Normal or ↓	↑	↑
Haemolysis	↑	Normal or ↓ (acute)	↑ (chronic)	↑ (chronic)

* = *to male levels*

Iron therapy

Because body iron content is determined by control of absorption, not excretion, *parenteral iron therapy*, which by-passes absorption, may cause iron overload. A unit of blood contains 4.5 mmol (250 mg) of iron, and repeated blood transfusions carry the same danger. In anaemias other than those of iron deficiency stores are normal, or even increased: parenteral iron should only be given if the diagnosis of iron deficiency is beyond doubt, and even then the oral route is preferable. Repeated blood transfusion may be needed to correct severe non-iron deficiency anaemia, for example if the marrow is hypoplastic, but the danger of overload should be remembered.

Iron absorption is stimulated by anaemia even if iron stores are increased. Treatment of non-iron deficient anaemia with oral iron supplements is not only ineffective, but can lead to iron overload: this is especially likely in haemolytic anaemia, in which the iron released from destroyed erythrocytes remains in the body.

The control of absorption is relatively inefficient if large amounts of oral iron are taken. Overload has been reported in a non-anaemic woman who continued to take oral iron for many years against medical advice.

Iron therapy is potentially dangerous, and should be prescribed only when iron deficiency is proven.

Iron overload

As emphasized on p. 390, the only route of iron loss is cell desquamation. Iron absorbed from the gastrointestinal tract or administered parenterally in excess of daily loss accumulates in body stores. If such 'positive balance' is maintained over long periods, iron stores may exceed 350 mmol (20 g) (about five times the normal amount).

Causes of iron overload

Increased intestinal absorption:

- idiopathic haemochromatosis;
- anaemia with increased, but ineffective, erythropoiesis;
- liver disease (rare cause);
- dietary excess;
- inappropriate oral therapy.

Parenteral administration:

- multiple blood transfusions;
- inappropriate parenteral iron therapy.

A very rare cause of iron overload is an inherited deficiency of transferrin.

Consequences of iron overload

The effect of the accumulated iron depends on its distribution in the body. This, in turn, is influenced partly by the route of entry. Two main patterns are seen at postmortem or in biopsy specimens.

Parenchymal iron overload occurs in idiopathic haemochromatosis and in patients with ineffective erythropoiesis. Iron accumulates in the parenchymal cells of the liver, pancreas, heart and other organs. There is usually associated functional disturbance or tissue damage.

Reticuloendothelial iron overload is seen after excessive *parenteral administration of iron* or *multiple blood transfusions.* The iron accumulates initially in the reticuloendothelial cells of the liver, spleen and bone marrow. There are few harmful effects but under certain circumstances (p. 398) the distribution may change so that parenchymal damage occurs.

In dietary iron overload both hepatic reticuloendothelial and parenchymal overload may occur, associated with scurvy and osteoporosis (p. 398). Whatever the cause of *massive* iron overload, there may be parenchymal accumulation and tissue damage.

- *Haemosiderosis* is defined as an increase in iron stores as haemosiderin and is a histological definition. It does not necessarily mean that there is an increase in total body iron; for example, in many types of anaemia there is reduced haemoglobin iron (less haemoglobin) but increased storage iron.
- *Haemochromatosis* describes the clinical disorder due to parenchymal iron-induced damage.

Syndromes of iron overload

Idiopathic haemochromatosis

Idiopathic haemochromatosis is a genetically determined disease in which increased intestinal absorption of iron over many years produces large iron stores of parenchymal distribution. It presents, usually in middle age, as cirrhosis with diabetes mellitus, hypogonadism and increased skin pigmentation. Because of the darkening of the skin, due to an increased in melanin rather than to iron deposition, the condition has been referred to as 'bronzed diabetes', although the colour is grey rather than bronze. Cardiac manifestations may be prominent, particularly in younger patients, many of whom die of cardiac failure. In about 10 to 20 per cent of cases hepatocellular carcinoma develops.

Idiopathic haemochromatosis has an autosomal recessive mode of inheritance. The gene for this disorder is closely associated with the HLA gene locus, a fact of importance in family studies. Those members of a family with an HLA haplotype identical with that of the patient are likely to develop the disease.

Factors such as alcohol abuse may hasten the accumulation of iron and the development of liver damage.

It may be difficult to distinguish between idiopathic haemochromatosis and alcoholic cirrhosis. In both conditions diabetes mellitus and hypogonadism may occur and although the incidence of these complications is higher in idiopathic haemochromatosis this does not help in the diagnosis of the individual case. Examination of liver biopsy specimens may further confuse the issue. The liver in cases of alcoholic cirrhosis not infrequently contains increased stainable iron. Not only do some alcoholic drinks, notably wines, contain significant amounts of iron, but there is evidence that in cirrhosis there may be increased iron absorption due possibly to the effect of alcohol. However, the majority of patients with cirrhosis do not have increased iron stores as shown by chemical testing, and the liver biopsy specimen shows that the iron is mainly present in the portal tracts. The clinical, histological and biochemical changes of cirrhosis are usually more obvious than those of iron accumulation: this contrasts with the picture in haemochromatosis with an apparently equivalent load.

Rarely, patients with cirrhosis or a portocaval shunt may have true iron overload and the distinction from idiopathic haemochromatosis may be extremely difficult. A family history, or investigation of near relatives, may help in diagnosis. The treatment of iron overload is the same in either case.

The diagnosis of idiopathic haemochromatosis *must* be followed by investigation of other members of the family, and treatment of those in whom increased iron stores are found.

Treatment of iron overload. The excess iron is usually removed from patients with idiopathic haemochromatosis by weekly venesection until the haemoglobin level falls. Each unit of blood removes about 4.5 mmol (250 mg) of iron.

Anaemia and iron overload

Several types of anaemia may be associated with iron overload. In some, such as aplastic anaemia and the anaemia of chronic renal failure, the cause is multiple blood transfusions and the iron initially accumulates in the reticuloendothelial system. With massive overload (over 100 units of blood), parenchymal overload and haemochromatosis may develop.

In anaemias characterized by erythroid marrow hyperplasia, but with ineffective erythropoiesis, such as thalassaemia major and sideroblastic anaemia, there is, in addition, increased absorption of iron. Haemochromatosis develops at a lower transfusion load than in aplastic anaemia.

Iron overload associated with anaemia can obviously not be treated by venesection, since this would aggravate the anaemia by removing more haemoglobin. The tendency for transfusion further to aggravate overload can be minimized by giving the iron-chelating agent, desferrioxamine, each time: this can be excreted in the urine with any non-haemoglobin iron.

Dietary iron overload

Increased iron absorption due to excessive intake is rare. One well-described form is, however, relatively common in the rural black population of southern Africa. The source is beer brewed in iron containers. Usually the excess is confined to the reticuloendothelial system and the liver (both portal tracts and parenchymal cells), and there is no tissue damage. In a small number of cases deposition in the parenchymal cells of other organs occurs and the clinical picture may resemble that of idiopathic haemochromatosis: it may be distinguished by the high concentration of iron in the reticuloendothelial system seen in the bone marrow and spleen (at autopsy).

Scurvy and osteoporosis may occur in this form of iron overload. The ascorbate deficiency may be due to its irreversible oxidation in the presence of excessive amounts of iron, and osteoporosis sometimes accompanies scurvy. Ascorbate deficiency also interferes with normal mobilization of iron from the reticuloendothelial cells; plasma iron levels may be low, and the response to chelating agents poor, despite iron overload.

The student should read the section on 'Investigation of suspected iron overload' on p. 400.

Summary

1. There is no significant iron excretion. Body stores are determined by control of absorption. Parenteral iron therapy should be given with care.

2. Anaemia, even if not due to iron deficiency, increases absorption. Prolonged oral iron therapy should only be given if iron deficiency is proven.

3. Plasma iron concentrations vary considerably for physiological reasons, and fall in many cases of non-iron deficiency anaemia. They give no information about the state of body iron stores.

4. Plasma iron is transported bound to the protein transferrin. Transferrin concentrations may be measured directly, or indirectly as the total iron-binding capacity (TIBC) of the plasma.

5. The transferrin level rises in iron deficiency and falls in iron overload.

6. The transferrin level falls in many cases of non-iron deficiency anaemia associated with a low plasma iron concentration. A low plasma iron with a high transferrin concentration is more suggestive of iron deficiency than a low plasma iron level alone.

7. Plasma ferritin levels are affected by the state of iron stores, and by an increased rate of synthesis in inflammatory states. A very low plasma ferritin concentration almost certainly confirms the diagnosis of iron deficiency, but a normal or high level does not exclude it. A normal or low plasma ferritin concentration almost certainly excludes a diagnosis of iron overload, but a high one does not necessarily confirm it.

8. The factors governing the distribution of excessive iron are not fully understood. A feature common to all forms of parenchymal overload is a high percentage saturation of transferrin.

9. Iron overload may be the result of excessive intestinal absorption or of parenteral iron administration, usually as blood. The distribution of iron in the body differs in the various forms of iron overload.

10. Iron overload can be demonstrated by the response to repeated venesection or to chelating agents. The diagnosis of idiopathic haemochromatosis should only be made if massive iron overload is present.

Further reading

Cavill I, Jacobs A, Worwood M. Diagnostic methods for iron status. *Ann Clin Biochem* 1986; **23**: 168–71.

Tavill AS, Bacon BR. Hemochromatosis: how much iron is too much? *Hepatology* 1986; **6**: 142–5.

Gordeuk VR, Devee Boyd R, Brittenham GM. Dietary iron overload persists in rural sub-Saharan Africa. *Lancet* 1986; **1**: 1310–3.

Investigation of disorders of iron metabolism

Investigation of anaemia

Anaemia may be due either to iron deficiency or, more commonly, to a variety of other conditions, sometimes associated with high iron stores. The subject of the diagnosis of anaemia is covered more fully in textbooks of haematology. However, so that we may see the value of chemical estimations in perspective, it is worth considering the order in which anaemia may usefully be investigated.

1. The clinical impression of anaemia should be confirmed by *haemoglobin* estimation. Iron deficiency can, however, exist with haemoglobin concentrations within the reference range.

2. The *mean corpuscular haemoglobin (MCH)* and *mean corpuscular volume (MCV)* should be noted and, if necessary, a *blood film* should be examined. Iron deficiency anaemia is hypochromic and microcytic in type, and these findings may be evident before the haemoglobin concentration has fallen below the reference range. Normochromic, normocytic anaemia is a non-specific finding, usually associated with other chronic disease; it is only due to iron deficiency if there has been very recent blood loss. Typical appearances of other types of anaemia may be seen on the blood film.

In most cases of anaemia consideration of these findings, together with the clinical picture, will give the cause. Anaemias such as those due to thalassaemia or of the pyridoxine-responsive type, although rare, are most likely to confuse the picture, since they too are hypochromic, but are not due to iron deficiency.

3. A *marrow film* may be needed to confirm a diagnosis of, for example, megaloblastic anaemia. If such a film is available, staining with potassium ferrocyanide may indicate the state of the iron stores.

4. In the *rare cases* in which the diagnosis is not yet clear, and if a bone marrow aspiration is felt to be unjustified, biochemical investigations may occasionally help. Plasma iron estimation without an assessment of transferrin levels is uninformative. An unequivocally low plasma ferritin concentration confirms iron deficiency, but a normal or high one should not be assumed to exclude it.

If a patient with *proven iron deficiency* shows no response to oral iron treatment within a few weeks he is probably not taking the tablets; normal coloured, rather than black, faeces, detectable if necessary by rectal examination, confirms this suspicion. If iron has been taken malabsorption, usually part of a general absorption defect, is a possible explanation for a poor response.

Investigation of suspected iron overload

Initial tests. Measure plasma iron, percentage saturation of transferrin and plasma ferritin. The *plasma iron* concentration is almost invariably high in idiopathic haemochromatosis, often above 36 μmol/litre (200 mg/dl). This is associated with a reduced transferrin level (as shown by a lowered TIBC) and the *percentage saturation* is usually over 80 per cent, and is often 100 per cent: in the presence of infection or malignancy, however, the plasma iron level and percentage saturation may be lower than expected; the TIBC remains low. The lowering effect of ascorbate deficiency on plasma iron levels has already been mentioned.

Plasma ferritin levels are high in most patients with iron overload (whether reticuloendothelial or parenchymal). Only a few families with idiopathic haemochromatosis despite normal plasma ferritin levels have been described.

If all these values are normal it is unlikely that the patient has iron overload.

Demonstration of increased iron stores. The diagnosis of iron overload can only be made after proof has been obtained of increased iron stores.

Response to venesection. The lack of response of the patient to a therapeutic course of venesection offers the most convincing proof of increased iron stores, albeit retrospectively. Removal of a unit of blood (4.5 mmol, or 250 mg, of iron), repeated weekly, produces a rapid fall in plasma iron, soon followed by iron-deficiency anaemia, in a subject with normal iron stores. In patients with idiopathic haemochromatosis however, 350 mmol (20 g) or more of iron may be removed in this way before evidence of iron deficiency develops.

Use of chelating agents. An alternative method uses the chelating action of substances, such as desferrioxamine, which bind iron and which are subsequently excreted in the urine. Following administration of desferrioxamine, subjects with increased iron stores excrete more iron in the urine than do normal people.

Plasma ferritin levels are high in most cases of reticuloendothelial iron overload, but *normal concentrations have been found in cases of idiopathic haemochromatosis.* Raised levels may also occur in many hepatic diseases, including cirrhosis.

Liver biopsy specimens contain large amounts of stainable iron, which may be mainly in parenchymal or mainly in reticuloendothelial cells. Chemical estimation of iron is more reliable than histochemical evaluation.

Marrow iron content is *usually normal* in haemochromatosis, in which overload is predominantly parenchymal, but may be greatly increased in reticuloendothelial overload. A similar loading of the reticuloendothelial cells is found when there is deficient utilization of marrow iron for haemoglobin synthesis, as in many haematological, neoplastic and chronic inflammatory diseases.

Family studies. Relatives of a patient found to have haemochromatosis should be investigated. Plasma iron and percentage saturation of transferrin are the most sensitive tests. Plasma ferritin may be normal if iron stores are not significantly increased. If available, HLA typing will identify those most likely to have inherited the disorder (p. 397).

21

The porphyrias

The porphyrias are a group of disorders, usually inherited, of haem synthesis. Although most of them are uncommon and some are very rare, it is important to recognize them and to investigate relatives of known cases. For example, drugs which may precipitate acute attacks, sometimes with fatal consequences, must be avoided in the inherited hepatic porphyrias. It is equally important not to confuse the commoner, secondary, causes of abnormal porphyrin excretion with true porphyrias.

Physiology

Haem is synthesized in most tissue of the body. In the bone marrow it is incorporated into *haemoglobin.* In other cells it is used for the synthesis of *cytochromes* and related compounds. The cytochromes are constituents of the electron transport chain, which harnesses the energy of metabolic processes. Impaired cytochrome synthesis can therefore have serious consequences. The liver is quantitatively the largest non-erythropoietic haem-producing organ.

Biosynthesis of haem

The main steps are outlined below and in Fig. 21.1.

- *5-aminolaevulinate (ALA)* is formed by condensation of glycine and succinate. The reaction requires pyridoxal phosphate and is catalysed by *ALA synthase.*
- Two molecules of ALA condense to form a monopyrrole, *porphobilinogen (PBG).*
- Four molecules of PBG combine to form a tetrapyrrole, *uroporphyrinogen* (Fig. 21.2).
- Two isomers are formed, I and III. The major pathway involves the III isomer and leads to the formation of haem by the successive production of *coproporphyrinogen and protoporphyrin,* followed by the incorporation of iron.

Each step is controlled by a specific enzyme. Normally the *rate-limiting step* is that catalysed by *ALA synthase and this step is regulated by feedback inhibition by the final product, haem.*

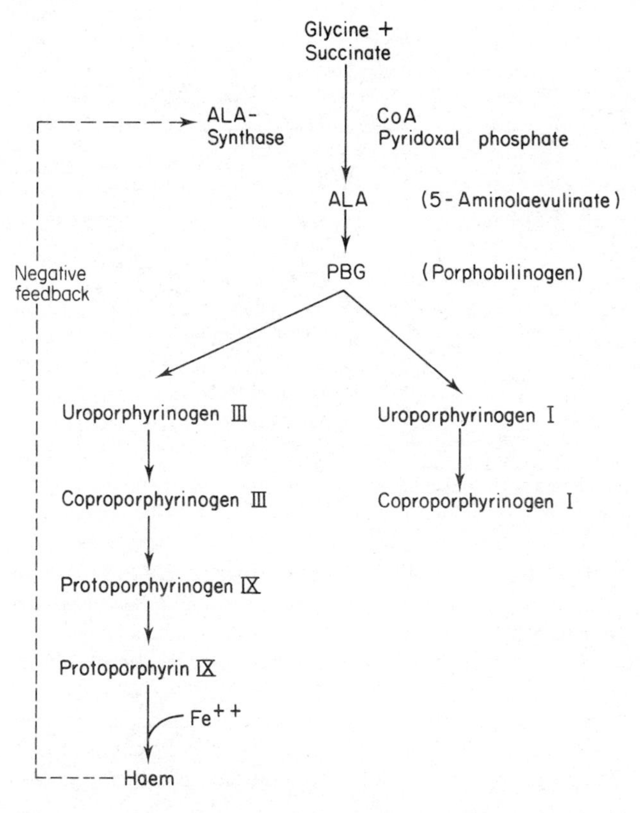

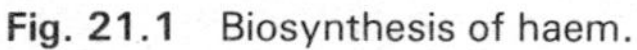
Fig. 21.1 Biosynthesis of haem.

The *porphyrinogens* and their precursors, *ALA* and *PBG*, are *colourless* compounds. Porphyrinogens, however, oxidase spontaneously to the corresponding *porphyrins* which are *dark red* and *which fluoresce* in ultraviolet light. PBG, too, may spontaneously form uroporphyrin when exposed to air and light. A urine specimen containing large amounts of porphyrinogens or their precursors will gradually darken if left standing.

Excretion

Any excess of the intermediates on the haem pathway is excreted. ALA, PBG and uroporphyrin(ogen) are water-soluble and appear in the *urine*. Protoporphyrin is excreted in bile and appears in the *faeces*. Coproporphyrin(ogen) may be excreted by either route. Renal excretion of coproporphyrinogen, like that of urobilinogen, rises with increasing alkalinity.

The porphyrias

The porphyrias result from a deficiency of one of the enzymes on the haem

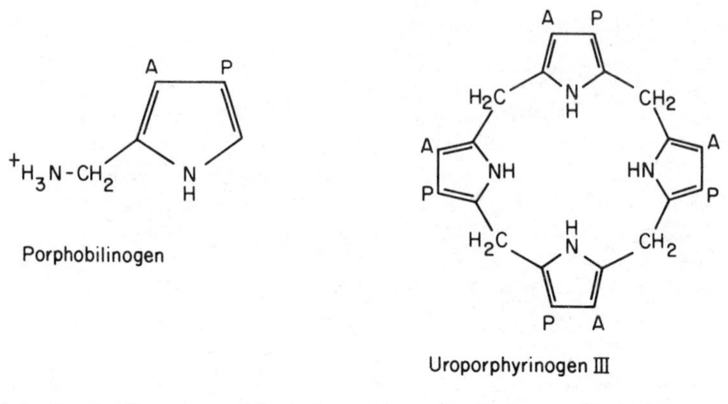

Fig. 21.2 Porphobilinogen, and the tetrapyrrole uroporphyrinogen III which incorporates four porphobilinogen units. Uroporphyrinogen I differs only in the order of the side-chains on one of the rings. (Side-chains: A = acetate; P = propionate)

synthetic pathway: haem production is therefore impaired. *Reduced feedback inhibition of ALA synthase may maintain adequate haem levels but at the expense of overproduction of porphyrins or their precursors.*

The *symptoms* of porphyria correlate well with the biochemical abnormalities.

Skin lesions, varying from mild photosensitivity to severe blistering, occur when *porphyrins* are produced in excess. The lesions typically occur in the exposed areas where sunlight activates porphyrin in the skin to release energy which damages tissue.

Neurological disturbances, such as *peripheral neuritis*, *abdominal pain*, or both in the serious *acute attack*, occur only in those porphyrias in which the *precursors*, ALA and PBG, are produced in excess. It is not known whether the neurological damage is due to haem deficiency in the nervous system or to a direct toxic effect of ALA or PBG.

Many different forms of porphyria have been described. They are usually classified according to whether the main site of porphyrin accumulation is in the liver or the erythropoietic system. This does not necessarily mean that the enzyme defect is confined to that system.

The main porphyrias are:

- **hepatic porphyrias**

acute intermittent porphyria;
hereditary coproporphyria;
porphyria variegata; } acute porphyrias

porphyria cutanea tarda:
- genetic predisposition;
- acquired.

- **erythropoietic porphyrias**

  congenital erythropoietic porphyria;
  protoporphyria.

The features are outlined in Table 21.1 and discussed briefly below.

The acute hepatic porphyrias

- Acute intermittent porphyria.
- Hereditary coproporphyria.
- Porphyria variegata.

These three *dominantly* inherited disorders have latent and acute phases. The symptoms and biochemical abnormalities of the *latent phases* differ and reflect the nature of the enzyme defect. In the *acute phases*, however, the biochemical and clinical picture characteristic of excessive ALA and PBG production, associated with neurological and abdominal symptoms, develops. The similarities and differences are best explained by referring to a simplified scheme of the haem synthetic pathway (Fig. 21.3).

Latent phase

The enzyme defect tends to reduce haem levels which in turn increase ALA synthase activity by the negative feedback loop. Increased ALA synthase activity has been demonstrated in all the porphyrias. This maintains haem levels at the expense of accumulation and excretion of the substance immediately before the block. The main biochemical abnormalities are predictable (Fig. 21.3).

Acute intermittent porphyria, in which there is increased *urinary ALA and PBG* (not detectable in all patients).

Hereditary coproporphyria, in which there is increased *faecal coproporphyrin.*

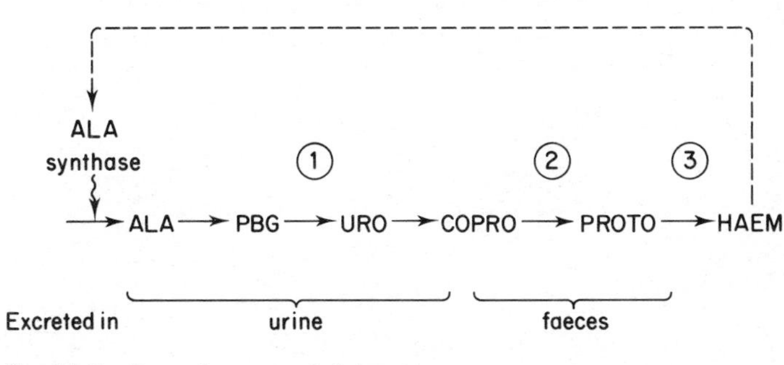

Fig. 21.3 Sites of enzyme deficiencies in:
1. acute intermittent porphyria
2. hereditary coproporphyria
3. porphyria variegata.

Table 21.1 The major clinical and biochemical features of the porphyrias

	Hepatic porphyrias							Porphyrias involving the erythropoietic system	
	Acute intermittent porphyria		Porphyria variegata		Hereditary coproporphyria		Porphyria cutanea tarda	Congenital erythropoietic porphyria	Protoporphyria
	acute	latent	acute	latent	acute	latent			
Clinical features									
Abdominal and neurological symptoms	+	−	+	−	+	−	−	−	−
Skin lesions	−	−	+	+	rarely		+	+	+
Chemical Abnormalities									
Urine PBG and ALA	+	+	+	−	+	−	−	−	−
Urine porphyrins	+	−	+	−	+	−	+	+	−
Faecal porphyrins	−	−	+	+	+	+	−	+	+
	Acute attacks precipitated by many drugs (for instance barbiturates, oestrogens, sulphonamides)						Symptoms may be relieved by venesection	Erythrocyte porphyrins increased	

Porphyria variegata, in which there is increased *faecal protoporphyrin.*

In acute intermittent porphyria the latent phase is usually asymptomatic. In porphyria variegata, and less commonly in hereditary coproporphyria, the increase in porphyrins may produce skin lesions.

Acute phase

ALA synthase activity may be further increased by a number of drugs, particularly barbiturates, oestrogens, sulphonamides and griseofulvin, and by acute illness: this may be a direct effect or may be due to an increased demand for haem. It results in a marked increase in ALA and PBG production. In acute intermittent porphyria this increase is due to the block imposed by an inherited deficiency of porphobilinogen deaminase; in hepatic coproporphyria and porphyria variegata, this enzyme becomes rate-limiting and is unable to respond normally to the increased demand. *The increase in urinary ALA and PBG is the hallmark of the acute porphyric attack* and, in hereditary coproporphyria and porphyria variegata, is superimposed on all the other biochemical abnormalities. The accelerated activity of the pathway and the spontaneous conversion of the precursors to porphyrin lead to increased urinary porphyrin excretion.

Acute attacks occur only in a small proportion of patients exposed to the provoking agents. They occur usually only after puberty and are commoner in women than men. Colicky abdominal pain (due to involvement of the autonomic nervous system) and neurological symptoms ranging from peripheral neuritis to quadriplegia are usually the presenting features. Death may result from respiratory paralysis. Hyponatraemia is a common finding.

The acute attack closely resembles serious acute intraabdominal conditions, and if the diagnosis is not made the patient may be subjected to surgery, with the use of a barbiturate anaesthetic; barbiturates and the stress of operation may aggravate the condition.

Diagnosis of latent porphyria

It is essential to investigate the blood relatives of any patient with porphyria. Screening tests for excess urinary PBG and ALA are inadequate to diagnose latent acute intermittent porphyria, and even quantitative estimation may fail to detect all carriers. The activity of the enzyme, porphobilinogen deaminase, should be measured in erythrocytes (the most easily available cells).

Porphyria cutanea tarda

In patients with porphyria cutanea tarda the skin is unduly sensitive to minor trauma, particularly in sun-exposed areas: the commonest presenting feature is blistering on the backs of the hands. Less commonly the lesions appear on the face. Increased facial hair and hyperpigmentation occur in chronic cases. Acute attacks do not occur.

The basic defect is an inability to convert uroporphyrinogen to coproporphyrinogen due to a deficiency of uroporphyrinogen decarboxylase. This may be a dominantly inherited disorder, but most cases are sporadic. Factors which produce clinical disease, possibly by aggravating an underlying genetic deficiency, include alcohol abuse, iron overload or high-dose oestrogen therapy. Symptoms improve when the offending substance is withdrawn. Some liver toxins such as hexachlorbenzene directly inhibit the activity of the enzyme.

The impaired conversion leads to an accumulation of uroporphyrinogen and porphyrins intermediate between it and coproporphyrinogen. These deposit in the skin and are excreted in the urine in increased amounts. Faecal porphyrins are not increased but the abnormal pattern of intermediate porphyrins may be detected by chromatography; this finding is of diagnostic value.

It is important not to confuse this disorder with the coproporphyrinuria of liver disease (see below).

Erythropoietic porphyrias

Two rare inherited disorders are associated with accumulation of porphyrins in erythrocytes. Acute porphyric attacks do not occur and ALA and PBG excretion are normal.

Congenital erythropoietic porphyria, unlike all the other porphyrias discussed, is inherited as a *recessive* characteristic. Usually from infancy onwards *erythrocyte* and *plasma uroporphyrin I* levels are very high and these is severe *photosensitivity*. Porphyrin is also deposited in bones and teeth, which fluoresce in ultraviolet light, and the teeth may be brownish-pink. Hirsutism, especially of the face, also occurs and there is a haemolytic anaemia.

Urinary porphyrin levels are grossly increased, and faecal levels less so.

Protoporphyria. In this *dominantly inherited* disorder *protoporphyrin* levels are increased in *erythrocytes* and *faeces*. There is mild photosensitivity, and hepatocellular damage may lead to liver failure.

Other causes of excessive porphyrin excretion

Porphyria is not the only cause of disordered porphyrin metabolism and positive screening tests *must be confirmed* by quantitative analysis, with *identification* of the porphyrin. Three causes must be considered.

Lead poisoning inhibits several of the enzymes involved in haem synthesis and eventually causes anaemia. The *urine* contains increased amounts of *ALA* (an early and sensitive test), and *coproporphyrin*. Some of the symptoms of lead poisoning, such as abdominal pain, are similar to those of the acute porphyric attack, and may cause difficulty in diagnosis. PBG excretion, however, is not usually increased.

Liver disease may increase *urinary coproporphyrin*, possibly due to decreased biliary excretion. This is probably the commonest cause of porphyrinuria. Occasionally there is mild photosensitivity (in porphyria cutanea tarda the more severe skin lesions are due to uroporphyrin excess).

Ulcerative lesions of the upper gastrointestinal tract may produce raised levels of *faecal porphyrin* by degradation of haemoglobin. If there is bleeding from the lower part of the tract the blood reaches the rectum before there is time for conversion: this may help roughly to localize the site of bleeding.

Summary

1. Porphyrins are by-products of haem synthesis. 5-aminolaevulinate (ALA) and porphobilinogen (PBG) are precursors.
2. The porphyrias are diseases associated with disturbed porphyrin metabolism. Most are inherited. The main clinical and biochemical features are outlined in Table 21.1 (p. 406).
3. Acute attacks with abdominal or neurological symptoms are a feature of the inherited hepatic porphyrias. Such attacks are potentially fatal and may be provoked by a number of drugs. The diagnosis of porphyria in the acute phase depends on the demonstration of ALA and PBG in the urine.
4. The diagnosis of inherited porphyria must be followed by investigation of all blood relatives to detect asymptomatic cases. Screening tests may be negative in some types and quantitative estimations are necessary. Both urine and faeces should be examined.
5. Other causes of abnormalities in porphyrin excretion are lead poisoning, liver disease and upper gastrointestinal bleeding.
6. The very rare erythropoietic porphyrias cause excessive accumulation of porphyrin in erythrocytes.

Further reading

Hindmarsh JT. The porphyrias: recent advances. *Clin Chem* 1986; **32**: 1255–63.

Sweeney GD. Porphyria cutanea tarda, or the uroporphyrinogen decarboxylase deficiency diseases. *Clin Biochem* 1986; **19**: 3–15.

Moore MR. International review of drugs in acute porphyria. *Int J Biochem* 1980; **12**: 1089–97.

Investigation of suspected porphyria

Check with the laboratory for the type of samples required. Samples should not be sent simply requesting a 'porphyrin screen'. Indicate which type of porphyria is suspected, so that the laboratory can select the appropriate tests.

Suspected acute attack

Fresh urine should be tested immediately for porphobilinogen. If PBG is not present it is highly unlikely that the patient is suffering from an acute porphyric attack.

Note however that:

1. patients with acute intermittent porphyria who have once had an acute attack may continue to excrete increased amounts of PBG for many years.
2. a negative test does not exclude the diagnosis of *latent* porphyria.

Suspected latent porphyria

A patient with a history of repeated attacks of abdominal pain or neurological symptoms may have acute intermittent porphyria, porphyria variegata or hereditary coproporphyria.

1. Measure PBG and ALA in a 24 hour urine collection. Increased excretion suggests acute intermittent porphyria.
2. Measure porphyrins in a random sample of faeces. Raised values suggest porphyria variegata (proto- and coproporphyrin) or hereditary coproporphyria (coproporphyrin).
3. If the assay is available, measure the activity of porphobilinogen deaminase in red blood cells. Decreased values are found in patients with acute intermittent porphyria. There is often overlap in levels between affected and normal people but, within a family, carriers can be shown to have levels about half those of unaffected members. This test will also detect affected children.

Suspected porphyria with skin lesions

Skin lesions may occur in any of the porphyrias other than acute intermittent porphyria. Blood, urine and faeces should be sent for testing.

1. Increased erythrocyte porphyrins suggest protoporphyria or congenital erythropoietic porphyria (very rare). High values may also occur in iron deficiency anaemia and lead poisoning.
2. Increased urinary porphyrins suggest porphyria cutanea tarda or congenital erythropoietic porphyria. The increased uroporphyrin excretion in these conditions must be distinguished from the coproporphyrinuria of liver disease.
3. Increased faecal porphyrins occur in protoporphyria, porphyria variegata and hereditary coproporphyria. These may be distinguished by chromatographic separation of porphyrins; this will also demonstrate the abnormal pattern of porphyria cutanea tarda.

Investigation of family members

If one of the acute porphyrias is diagnosed it is essential to investigate blood relatives. Those found to have the condition must be counselled regarding drug usage.

Carriers of porphyria variegata and hereditary coproporphyria can be identified, after puberty, by demonstration of clearly increased faecal porphyrin excretion. Normal excretion before puberty does not exclude the diagnosis.

Acute intermittent porphyria is detected by measuring red cell porphobilinogen deaminase activity.

Test for porphobilinogen

1. Equal parts of *fresh* urine and Ehrlich's reagent (2 per cent *p*-dimethylaminobenzaldehyde in 5M HCl) are mixed. If PBG is present a red colour develops.
2. If a red colour develops, 3 ml of *n*-butanol is added, the tube shaken and the phases allowed to separate. If the red colour does *not enter* the butanol (upper) layer it is PBG.

It is important to use fresh urine because PBG disappears on standing. A number of other substances, especially urobilinogen, may also give a red colour with Ehrlich's reagent. All such known substances are, however, extracted into the butanol layer. Chloroform has been used instead of butanol, but may not extract all non-PBG reactants and so give a false positive result.

Tests for porphyrins

The tests described below are only suitable for *screening* suspected cases of porphyria. *Positive results must be confirmed by quantitative analysis* and by *typing* the porphyrins, *preferably in a centre specializing in testing for porphyrias.* Increased porphyrin excretion may occur in conditions other than porphyria.

If screening tests are performed the following points must be remembered.

1. The *correct type of sample for the type of porphyria suspected must be tested* (urine or faeces).
2. During the *latent phase of acute intermittent porphyria screening tests are usually negative.* If the tests are correctly performed, *negative results for faecal porphyrins* almost certainly *exclude the diagnosis of porphyria variegata.*
3. *Before puberty tests are negative,* even in those children who will suffer from porphyria.

Urine. About 10 ml of fresh urine is mixed with 2 ml of a mixture of equal parts of ether, glacial acetic acid and amyl alcohol: the aqueous and solvent phases are allowed to separate. A red fluorescence under ultraviolet light in the *upper* layer indicates the presence of porphyrin.

Faeces. A pea-sized piece of faeces is mixed with the solvent described for urine. A red fluorescence under ultraviolet light indicates either porphyrin *or* chlorophyll (which has a similar chemical structure, and which is derived from the diet). *1.5M HCl extracts porphyrin,* but *not* chlorophyll.

22

Biochemical effects of tumours

Some rare syndromes are associated with neoplasia of cells which normally produce hormones, but, because of their scattered nature, are not usually thought of as endocrine organs. In addition tissues which usually do not seem to secrete hormones may do so if they become malignant. Both these syndromes have been explained on the basis of the so-called APUD system. Although this is by no means the only possible explanation, we will start with a brief outline of the APUD concept.

The diffuse endocrine (APUD) system

APUD cells are widely dispersed, have common cytological characteristics, and may originate in embryonic ectoblast. Most have endocrine or neurotransmitter properties. The acronym APUD is derived from their ability for *A*mine *P*recursor *U*ptake and *D*ecarboxylation to produce amines. Tumours of APUD cells have been called APUDomas. Some of these cells secrete physiologically active *amines*. In many the amine production appears to be associated with the synthesis and secretion of *peptide hormones*. Others do not normally seem to secrete hormones.

Many of the peptide-hormone-producing cells form such recognizable *endocrine organs* as the pituitary and parathyroid glands, and the calcitonin-producing cells of the thyroid. Other APUD cells, both peptide and amine secreting, are found in *specialized neurological tissues*; examples include the hypothalamic cells which synthesize the trophic hormones and ADH, and the adrenaline- (epinephrine-) and noradrenaline- (norepinephrine-) secreting cells of the sympathetic nervous system, including those of the adrenal medulla. Abnormalities of most of these tissues are discussed elsewhere in this book. In this chapter we shall describe the secreting tumours of the sympathetic nervous system, phaeochromocytoma and neuroblastoma.

Hormone-secreting tumours of APUD cells are scattered through tissues of non-ectodermal origin. Many occur in the *gastrointestinal tract* and *pancreas*. Insulinomas are discussed in Chapter 10, but there are many other types of peptide-hormone secreting APUD cells in the pancreas; many of these also occur in the gastrointestinal tract, and the amine-secreting, or amine-precursor-secreting, carcinoid tumours occur mainly in the intestine.

Probably all APUD cells have the potential to secrete any of the APUD hormones. A common site for apparently non-secretory cells is the bronchial tree. Bronchial carcinomas are the commonest of the tumours, apparently originating from non-endocrine tissue, which sometimes secrete hormones and other peptides normally foreign to that tissue. Some of these syndromes may be due to malignancy of APUD cells in the region, rather than of the host tissue, but this is not the only possible explanation. The second part of this chapter is a brief account of these not uncommon syndromes.

Catecholamine-secreting tumours

The sympathetic nervous tissue, comprising the adrenal medulla and sympathetic ganglia, is derived from the embryonic neural crest and is composed of two types of cell, the *chromaffin cells* and the *nerve cells*, both of which can synthesize active catecholamines (dihydroxylated phenolic amines). Adrenaline (epinephrine) is almost exclusively a product of the adrenal medulla, while noradrenaline (norepinephrine) is predominantly formed at sympathetic nerve endings.

Metabolism of the catecholamines

Adrenaline and noradrenaline are formed from the amine precursor, tyrosine, via dihydroxyphenylalanine (DOPA) and dihydroxyphenylethylamine (DOPamine). DOPA, DOPamine, adrenaline and noradrenaline are all catecholamines. Adrenaline and noradrenaline are both metabolized to the inactive 4-hydroxy-3-methoxymandelate (HMMA), often called vanillyl mandelate (VMA), each by similar pathways on which metadrenaline and normetadrenaline respectively are intermediates (Fig. 22.1). Adrenaline, noradrenaline and the metadrenalines, their conjugates, and HMMA, can be measured in the urine.

Action of catecholamines

Both adrenaline and noradrenaline act on the cardiovascular system. Noradrenaline causes generalized vasoconstriction, with hypertension and pallor, while adrenaline may dilate blood vessels in muscles, with variable effects on blood pressure.

Adrenaline may cause hyperglycaemia due to stimulation of glycogenolysis and other anti-insulin effects.

A more detailed discussion of the action of these two hormones will be found in textbooks of pharmacology.

Catecholamine-secreting tumours

Although tumours of the sympathetic nervous system, whether adrenal or extra-adrenal, can produce an excess of catecholamines, the two main types have

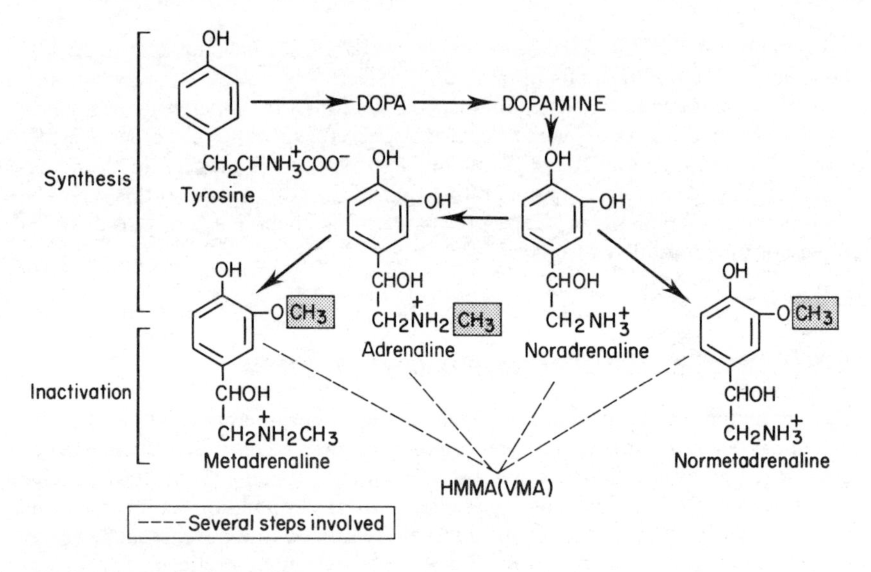

Fig. 22.1 Synthesis and metabolism of the catecholamines.

completely different clinical manifestations.

- *Phaeochromocytomas* occur in *chromaffin tissue:*
  about 90 per cent are in the adrenal medulla;
  about 10 per cent are extraadrenal.
- *Neuroblastomas* are tumours of *nerve cells*:
  about 40 per cent are in the adrenal medulla;
  about 60 per cent are extraadrenal.

Phaeochromocytoma occurs mainly in *adults* and is usually a *non-malignant tumour*. The symptoms and signs can be related to very high levels of catecholamines and include *paroxysmal hypertension* accompanied by *anxiety*, *sweating*, a throbbing *headache* and either *facial pallor* or *flushing*. There may also be hyperglycaemia, and even glycosuria, during the attack. Occasionally there is persistent, rather than paroxysmal, hypertension. Although this tumour accounts for only a very small proportion of cases of hypertension, it is potentially curable by surgery: for this reason its presence should be sought if a young adult presents with hypertension for no apparent reason. Table 22.1 summarizes some relatively rare causes of hypertension which should be excluded before 'essential hypertension' is diagnosed.

Neuroblastoma is a very *malignant* tumour of sympathetic nervous tissue occurring in *children*. The plasma levels of catecholamines are often as high as, or higher than, those in patients with phaeochromocytoma, but the clinical syndrome described above is rare. It is not known why this should be so, but the hormones may be released into the circulation in an inactive form. Some neuroblastomas secrete DOPamine.

Diagnosis of catecholamine-secreting tumours

Chemical diagnosis of the tumours can usually be made by measuring the daily

Table 22.1 Some metabolic causes of hypertension

	First line investigations
Renal disease	Plasma urea or creatinine assay Examination of the urine for protein, blood and casts
Primary hyperparathyroidism	Plasma calcium, phosphate etc assays (p. 194)
Cushing's syndrome	Cortisol (p. 123)
Primary hyperaldosteronism	Plasma potassium and T_{CO_2} (p. 74)
Phaeochromocytoma	Urinary HMMA (VMA) assays

urinary excretion of HMMA (VMA), the major catabolic product of catecholamines. An excretion of more than twice the upper limit of the reference range is diagnostic of such a tumour, but slightly increased excretion may be found in cases of essential hypertension. Many dietary components and drugs affect the analysis and it is important to consult the laboratory before starting a urine collection.

In rare cases the HMMA excretion is normal, but plasma levels of metadrenaline or of total catecholamines are increased. These estimations usually need only be performed if the patient has signs and symptoms highly suggestive of a phaeochromocytoma despite a normal excretion of HMMA. However, estimation of catecholamines in samples obtained by selective venous catheterization may help to localize a tumour before surgery. *It is very important to contact the laboratory before taking blood specimens for such estimations.*

If results of 24 hour urinary assays are equivocal, and if the hypertension is paroxysmal, urine should be collected during and immediately after the attack: this may be the only time at which increased excretion can be demonstrated.

The carcinoid syndrome

Normal metabolism of 5-hydroxytryptamine

Some APUD cells are called *argentaffin* because they reduce, and therefore stain with, silver salts. They are normally found in tissues derived from embryonic gut. They are most abundant in the ileum and appendix, but some are found in the pancreas, stomach and rectum. They synthesize the biologically active amine, *5-hydroxytryptamine* (5-HT; serotonin) from the amine precursor tryptophan, the intermediate product being *5-hydroxytryptophan* (5-HTP). 5-HTP is inactivated by deamination, and oxidation by monoamine oxidases, to *5-hydroxyindole acetic acid* (5-HIAA) (Fig. 22.2); the oxidases, and aromatic amino acid decarboxylase, are present in other tissues, as well as argentaffin cells. 5-HIAA is usually the main urinary excretion product of argentaffin cells.

Argentaffin cells may also excrete a peptide, *substance P*, an excess of which causes flushing, tachycardia, increased bowel motility and hypotension.

Causes of the carcinoid syndrome

Tumours of the argentaffin system, which may be benign or malignant, are most

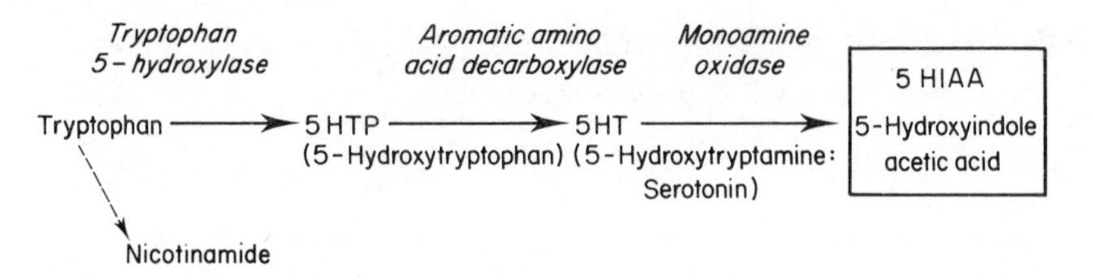

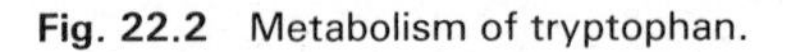

Fig. 22.2 Metabolism of tryptophan.

common in the ileum and appendix; those in the appendix rarely metastasize and may differ histochemically from the other small intestinal argentaffin tumours. More rarely the neoplasm is bronchial, pancreatic or gastric in origin: argentaffin tumours are very rarely found at any other site.

The carcinoid syndrome is usually associated with abnormally high concentrations of plasma 5-HT. Ileal and appendiceal tumours only produce the typical clinical syndrome when they have metastasized, usually to the liver, but primary tumours at other sites do cause symptoms: perhaps at least some of the products of intestinal tumours are inactivated in the liver, but those from other sites may be released directly into the systemic circulation in an active form.

Clinical picture of the carcinoid syndrome

Symptoms and signs of the carcinoid syndrome include *flushing*, *diarrhoea*, *bronchospasm* and *right-sided fibrotic lesions of the heart*: the heart lesions do not occur if the primary tumour is in the bronchus. Diarrhoea may be severe enough to cause the malabsorption syndrome. The signs and symptoms are more likely to be due to the presence of substance P than to 5-HT. Histamine has also been suggested as a cause of the flushing. A *pellagra-type syndrome* may occasionally develop, because tryptophan is diverted from nicotinamide to 5-HT synthesis (Fig. 22.2).

Diagnosis of the carcinoid syndrome

Urinary 5-HIAA secretion is usually very high in the carcinoid syndrome. A daily excretion of more than 130 μmol (25 mg) is a diagnostic finding if walnuts and bananas, which are high in hydroxyindoles, have been excluded from the diet for 24 hours before the urinary collection is started. Less elevated levels have been reported in association with small bowel disease and intestinal obstruction.

In very rare cases, usually of bronchial or gastric tumours, the argentaffin cells lack aromatic amino acid decarboxylase. In such cases, despite increased secretion of 5-HTP, urinary 5-HIAA excretion may not be increased to diagnostic levels. If there is a strong clinical suspicion of the carcinoid syndrome despite the finding of a normal 5-HIAA excretion, estimation of *total* 5-hydroxyindole (5-HTP, 5-HT and 5-HIAA) excretion may rarely be indicated.

Peptide-secreting tumours of the enteropancreatic system

All tumours in this group are rare and no attempt has been made to describe them all. They are usually found in the pancreatic islets.

Gastrinomas may cause the *Zollinger–Ellison syndrome.* G cell tumours, usually in the pancreatic islets, secrete large quantities of gastrin despite high gastric hydrogen ion levels, which should suppress secretion by feedback (p. 253). About 60 per cent of gastrinomas are malignant. Of the remaining 40 per cent only about a third (13 per cent of the total) are single, resectable adenomas, the rest being multiple. More rarely the syndrome is due to hyperplasia of G cells in the gastric antrum. The very high rate of gastric acid secretion causes ulceration in the stomach and proximal small intestine with severe diarrhoea: inhibition of pancreatic lipase activity by the low pH may cause steatorrhoea. The syndrome may be associated with benign, and usually non-functioning, adenomas, for example, in the anterior pituitary, parathyroid, and thyroid glands and in the adrenal cortex.

In the Zollinger–Ellison syndrome the *fasting plasma gastrin* concentration is 5 to 30 times the upper reference limit. Hypochlorhydria is associated with high gastrin levels, but the simultaneous finding of a very low gastric pH indicates autonomous hormone secretion, and therefore confirms the diagnosis.

Glucagonomas usually arise from the α-cells of the pancreatic islets. The presenting clinical feature is a bullous rash, known as *necrolytic migratory erythema*, which is often accompanied by psychiatric disturbances, thromboembolism, glossitis, weight loss, impaired glucose tolerance, anaemia and a raised erythrocyte sedimentation rate. High glucagon levels could explain the impaired glucose tolerance, but the cause of the other features, all of which are cured by resection of the tumour, is not clear. Diagnosis can be made by finding a very high plasma glucagon concentration.

VIPomas are *extremely rare* tumours, usually of the pancreatic islet cells, secreting large amounts of vasoactive intestinal peptide (VIP), a hormone which increases intestinal motility. They cause a syndrome in which there is very profuse, watery diarrhoea: earlier names were the *Verner-Morrison* or *WDHA* (*W*atery *D*iarrhoea, *H*ypokalaemia and *A*chlorhydria) syndrome. It is not certain if all cases of the Verner-Morrison syndrome are due to VIPomas.

Multiple endocrine adenopathy (MEA)

In the rare syndromes of MEA (pluriglandular syndromes), two or more endocrine glands secrete inappropriately high amounts of hormones, usually from adenomas. There are two main groups of syndromes.

MEA I may involve two or more of the following endocrine tissues; the glands involved are listed in order of decreasing incidence of involvement:

- parathyroid gland (hyperplasia or adenoma);
- pancreatic islet cells:
  gastrinomas;
  insulinomas;
- anterior pituitary gland;
- adrenal cortex;
- thyroid.

MEA II includes:

- medullary carcinoma of the thyroid gland;
- phaeochromocytoma;
- adenoma or carcinoma of the parathyroid gland.

Other combinations are very rare. The reason for the grouping is not clear, but may be associated with lines of development of the APUD system.

Hormonal effects of tumours in non-endocrine tissue

This section will be best understood after reading the chapters concerned with the hormones discussed.

'Ectopic' hormone secretion occurs at sites other than the normal tissue of origin; *inappropriate* secretion may or may not be ectopic (p. 48), but symptomatic ectopic secretion, because it is not under normal feedback control, is always inappropriate.

Many tumours of non-endocrine tissues may secrete hormonal substances very similar to, and often identical with, the natural hormone: some such tumours have been shown to secrete two or more hormones.

Mechanism of ectopic hormone production

It remains uncertain why ectopic hormone secretion should occur. Many theories have been propounded, but we, like the authors of the second reference given at the end of the chapter, believe the one discussed below to be the most likely. The interested student should consult this and other references.

DNA, through RNA production, codes for peptide and protein synthesis, the sequence of bases in the DNA molecule determining the structure of the peptide synthesized. The DNA complement of the fertilized ovum is determined by that of the unfertilized ovum and that of the sperm fertilizing it; all other cells are derived from the fertilized ovum. After fertilization the DNA in the ovum is replicated during repeated cell division and every daughter cell has an identical genetic complement. There is no evidence for mutation during differentiation of tissues, and so every cell must have the potential to produce any peptide coded for in the fertilized ovum.

It is thought that, during functional and histological cell differentiation, various parts of the DNA molecule are consecutively 'repressed' (stopped from functioning) and 'derepressed': for example, during the early stages of blastula formation DNA replication is dominant, while at other stages of development different proteins are manufactured. Much of the genetic information is probably repressed in the fully differentiated cell, and only those peptides essential for its metabolism, and those concerned with the specialized function of the organ, are synthesized. The histological changes associated with neoplasia reflect altered chemical function, probably due to a changed pattern of repression of the DNA molecule. It is possible to see how a cell could revert to synthesizing significant quantities of a peptide normally foreign to the fully differentiated tissue.

If this explanation were correct, the ectopic hormone should be identical with that secreted by normal endocrine tissues. In several, but not all, cases evidence based on biological activity and chemical, physical and immunological properties suggests, or even confirms, that this is true. As might also be expected if the above theory were correct, tumours have been described which secrete two or more peptide hormones.

Some of these syndromes may be due to malignancy of the APUD cells normally present in the tissue. In normal numbers these cells might secrete undetectable amounts of hormone, which, if the cells multiply, reach concentrations high enough to produce clinical effects: alternatively, derepression of normally non-secretory APUD cells might lead to peptide hormone production.

It seems to us that derepression of tissue, rather than of APUD, cells is the likely explanation for most of these syndromes. Derepression cannot be non-specific, because some hormones are secreted more often by one type of tumour than others. This may reflect a differing chemical background in different parts of the body.

No theory is yet proven. We hope that we have stimulated the student to read more about the subject.

Hormonal syndromes

Many of these syndromes have been discussed in the relevant chapters. The following may be due to hormone secretion by the tumour:

- hypercalcaemia (PTH or a PTH-like substance);
- hyponatraemia (ADH);
- hypokalaemia (ACTH);
- polycythaemia (erythropoietin);
- hypoglycaemia (insulin or an insulin-like substance);
- gynaecomastia (gonadotrophin);
- hyperthyroidism (TSH);
- carcinoid syndrome (5-HT and 5-HTP).

All the hormones listed are peptides except 5-HT and 5-HTP, excess secretion of which could be associated with overproduction of a single enzyme (see later).

Hormones produced at ectopic sites, like those secreted by overactivity of endocrine glands, are not under normal feedback control, and secretion continues under conditions in which it should be suppressed: such secretion is therefore inappropriate.

It may be difficult to prove conclusively that tumour cells are secreting hormones, but in many cases strong presumptive evidence enables the clinician to diagnose the syndrome with a high degree of confidence.

Hypercalcaemia due to secretion of PTH or PTH-like substances

Tumour types and incidence. In our experience hypercalcaemia with hypophosphataemia is common in most types of malignant disease. Parathyroid hormone, or a peptide very similar to it, has been extracted from a wide variety of tumours, and circulating levels have occasionally been shown to be inappropriately high.

Biochemical syndrome. The syndrome has been described on p. 180. The possibility of other causes of hypercalcaemia in malignancy has been discussed on p. 181. The student should keep an open mind about the cause, and note the level of plasma phosphate *in relation to that of urea.*

With improved treatment of malignant disease it is important to control potentially lethal hypercalcaemia in these patients. The onset of hypercalcaemia can be very rapid and plasma calcium concentrations should be estimated frequently, and especially in any patient with malignant disease who complains of nausea or other symptoms which might be associated with high plasma calcium levels (p. 177). Treatment is discussed on p. 187.

Hypokalaemia is commonly associated with hypercalcaemia (p. 177). Its presence does not usually indicate simultaneous ACTH production.

Hyponatraemia due to ADH secretion

Tumour types and incidence. Although ectopic ADH production is most commonly associated with the relatively rare oat-cell carcinoma of the bronchus, the syndrome of *inappropriate* ADH secretion has been reported in a wide variety of tumours; in our experience, if evidence is sought, it can be shown to be a common cause of mild hyponatraemia in patients with malignant disease or other illness: in most cases other than those of oat-cell carcinoma of the bronchus the hormone is thought to originate in the hypothalamic-pituitary region and therefore not to be ectopic in origin. The biochemical and clinical syndrome has been described on p. 48.

Diagnosis. Severe volume depletion can stimulate ADH secretion which is inappropriate to the plasma osmolality. The criteria for diagnosing ADH secretion inappropriate both to the extracellular osmolality and to the circulating volume, whether ectopic or not, are that:

- the patient is well-hydrated and normotensive;
- the plasma osmolality is low;
- the urinary osmolality is higher than that of plasma, suggesting that ADH secretion is not subject to normal osmotic feedback control.
- the urinary sodium concentration is relatively high, due to plasma volume expansion

The diagnosis should only be made if all the last three findings are present. The plasma osmolality is low, and the urinary sodium concentration may be high when, for example, plasma volume has been expanded by infusing fluid of low sodium concentration, but the urine is then, appropriately, of lower osmolality than the plasma. The urinary osmolality is appropriately high in water depletion, but the plasma osmolality is then high normal or high, and the urinary sodium concentration low. If there is inappropriate ADH secretion the plasma urea concentration is usually normal or low, due to the high GFR associated with volume expansion.

The usually gradual fall in plasma osmolality allows time for osmotic equilibrium to be established across cell membranes, and the patient may not experience symptoms of hypoosmolality.

Treatment. In asymptomatic cases no treatment other than fluid restriction is needed. If the plasma osmolality has fallen rapidly enough to cause symptoms of cerebral oedema more active treatment may be indicated.

- Infusion of a small amount of hyperosmolar fluid with a diuretic will increase plasma osmolality without further volume expansion. The following two methods depend on this principle:
  - *hyperosmolar saline* may be infused with a diuretic such as frusemide;
  - *mannitol infusion* directly increases plasma osmolality, and causes an osmotic diuresis (p. 8): more water than sodium is lost in the urine.
- *Demeclocycline*, a tetracycline, directly inhibits the action of ADH on the renal tubules. This drug may be given to patients with severe symptoms when simpler methods have failed, or are contraindicated.

Hypokalaemia due to ACTH secretion

Tumour types and incidence. *Symptomatic* inappropriate ACTH secretion is rarer than the two syndromes already discussed, although immunoreactive hormone may often be detected in the plasma of patients without signs or symptoms. It is most commonly associated with the relatively uncommon oat-cell carcinoma of the bronchus, but has been described less frequently in association with a variety of other malignant lesions, especially pulmonary carcinoid tumours, or those of thymic or pancreatic origin.

Clinical and biochemical syndrome. ACTH stimulates the secretion of all adrenocortical hormones except aldosterone (p. 122). In spite of this the picture of ectopic secretion often resembles that of primary hyperaldosteronism, and the patient presents with severe hypokalaemic alkalosis. However, ectopic overproduction has been claimed to account for 20 per cent of *ACTH-dependent* Cushing's syndrome.

Treatment. In cases with hypokalaemic alkalosis, potassium replacement is only effective if administered with spironolactone, amiloride or triamterene (p. 66), or an inhibitor of steroid synthesis such as metyrapone.

Polycythaemia due to erythropoietin secretion

The well-recognized association between polycythaemia and renal carcinoma is thought to be due to excessive secretion of an erythropoietin-like molecule by the tumour; erythropoietin stimulates bone marrow erythropoiesis. Because this hormone is a normal product of the kidney this is not an example of ectopic hormone secretion: however, the syndrome has been reported in association with other tumours, especially hepatocellular carcinoma (primary hepatoma).

Hypoglycaemia due to insulin or an insulin-like hormone

Severe hypoglycaemia, although a rare complication, has been reported in association with many tumours. Only a few cases, usually with carcinoid tumours, have been shown to have inappropriately high plasma immunoreactive insulin concentrations during the attack, and have therefore been proved to be due to ectopic hormone secretion. Severe hypoglycaemia with appropriately *low* immunoreactive insulin levels, usually in cases with very large retroperitoneal or thoracic mesenchymal tumours resembling fibrosarcomas, or with hepatocellular carcinoma, is a more common association. The syndrome has not been satisfactorily explained.

Gynaecomastia due to gonadotrophin secretion

Various types of tumour, including carcinoma of the bronchus, breast and liver, may cause gynaecomastia associated with high circulating concentrations of chorionic gonadotrophin or luteinizing-hormone: increased follicle-stimulating hormone activity is rare. In children with hepatoblastoma this may be a cause of precocious puberty. The syndrome is rarely reported, but a mild degree of gynaecomastia may be overlooked.

Hyperthyroidism due to TSH secretion

Some tumours of trophoblastic cells (choriocarcinoma, hydatidiform mole and testicular teratoma) have been shown to secrete a TSH-like substance. Clinical hyperthyroidism is extremely rare even when plasma hormone concentrations are very high.

Carcinoid syndrome due to 5-HT and 5-HTP secretion

All the syndromes so far discussed have been due to peptide hormones, and could be caused by derepression of DNA. A possible explanation for the overproduction of non-peptide hormones in the carcinoid syndrome, sometimes associated with oat-cell carcinoma of the bronchus, is that the tumour is one of bronchial APUD cells.

A different theory is more speculative and assumes derepression of bronchial cells. In this form of carcinoid syndrome the urinary excretion of 5-HTP and 5-HT, compared with that of true carcinoid tumours, is proportionally higher than that of 5-HIAA. Only *one enzyme*, tryptophan-5-hydroxylase, is needed to convert tryptophan to 5-HTP (Fig. 22.2, p. 416), while the decarboxylase and monoamine oxidase are found in normal non-argentaffin tissues. Derepression of the production of this one enzyme would lead to excessive synthesis of 5-HTP. 5-HT and 5-HIAA could be produced in other tissues, but in relatively smaller amounts than in the usual carcinoid syndrome.

Non-hormonal peptides as indicators of malignancy

The production of non-hormonal peptides by derepression of DNA would not necessarily be clinically obvious. Such circulating peptides have been demonstrated by immunological techniques: the hopes that these assays could be useful as early tests for malignancy have not fully been realized. Three such proteins are discussed briefly here: the first two are normally present in the fetus, but production seems to be largely suppressed in the normal adult.

α-Fetoprotein levels may be very high in the plasma of patients with *hepatocellular carcinoma* (primary hepatoma) and *teratoma.* Moderately raised levels may be due to non-malignant liver disease.

Carcinoembryonic antigen (CEA) may be produced by many *malignant tumours, especially of the gastrointestinal tract.* Plasma levels may also rise due to non-malignant disease of the gastrointestinal tract. Results of this assay may therefore be misleading if used as an aid to diagnosis of malignancy. If the initial concentration is raised in a case of known gastrointestinal malignancy, serial CEA estimations may sometimes help to monitor the effectiveness of, or recurrence after, treatment.

Human chorionic gonadotrophin (HCG) can be used as a marker for malignancy of the gonads (p. 146).

Summary

Catecholamine-secreting tumours

1. Tumours of sympathetic nervous tissue are associated with increased urinary excretion of the catecholamines, adrenaline and noradrenaline, and their metabolic products, the metadrenalines and hydroxymethoxymandelate (HMMA; VMA).
2. Phaeochromocytoma is a rare tumour, usually occurring in adults. It originates most commonly in the adrenal medulla, and much more rarely in extra-adrenal sympathetic nervous tissue. It is associated with hypertension and other symptoms of increased catecholamine secretion.
3. Neuroblastoma is a tumour of childhood, occurring either in extraadrenal

sympathetic nervous tissue, or in the adrenal medulla. Catecholamine secretion is increased, but symptoms are rarely referable to this chemical abnormality.

The carcinoid syndrome

1. Argentaffin cells synthesize 5-hydroxytryptamine (5-HT) which, after conversion to 5-hydroxyindoleacetate (5-HIAA), is excreted in the urine.
2. Tumours of argentaffin tissue are usually found in the intestine, and at this site do not cause typical symptoms of the carcinoid syndrome until they have metastasized to the liver.
3. The carcinoid syndrome is usually associated with increased urinary 5-HIAA excretion.

Peptide-secreting tumours of the enteropancreatic system

1. Gastrinomas cause the Zollinger–Ellison syndrome. Gastric hyperacidity causes peptic ulceration, diarrhoea and sometimes steatorrhoea.
2. Glucagonomas are associated with necrolytic migratory erythema and non-specific symptoms and signs.
3. VIPomas (tumours secreting vasoactive intestinal peptide) may cause the Verner-Morrison syndrome – very severe watery diarrhoea which often causes hypokalaemia (WDHA).

All these tumours are very rare.

Multiple endocrine adenopathy

In these syndromes two or more endocrine glands secrete excessive amounts of hormones.

Hormonal effects of tumours of non-endocrine tissue

1. Many tumours secrete hormonal substances normally foreign to them, which are sometimes identical with the hormones produced by normal endocrine glands.
2. Symptomatic hypercalcaemia, probably often due to secretion of PTH or a PTH-like substance, should be sought and treated in all cases of malignancy.

Non-hormonal peptides as indicators of malignancy

Some non-hormonal peptides may be detected by immunological techniques and have been used in an attempt to diagnose and monitor malignancy of specific tissues.

Further reading

The APUD system and ectopic hormone production

Pearse AGE. The diffuse endocrine system and the implications of the APUD concept. *Int Surg* 1979; **64** No 2: 5–7.

Stevens RE, Moore GE. Inadequacy of APUD concept in explaining production of peptide hormones by tumours. *Lancet* 1983; **1**: 118–9.

Phaeochromocytoma

Bravo EL, Gifford RW. Pheochromocytoma: diagnosis, localization and management. *N Engl J Med* 1984; **311**: 1298–1303.

The carcinoid syndrome

Wareing TH, Sawyers JL. Carcinoids and the carcinoid syndrome. *Am J Surg* 1983; **145**: 769–72.

Oates, JA. The carcinoid syndrome. *N Engl J Med* 1986; **315**: 702–4.

Ectopic ACTH secretion

Howlett TA, Rees LH. Is it possible to diagnose pituitary-dependent Cushing's disease? *Ann Clin Biochem* 1985; **22**: 550–8.

Tumour markers

Begent RHJ. The value of carcinoembryonic antigen measurement in clinical practice. *Ann Clin Biochem* 1984; **21**: 231–8.

23

The cerebrospinal fluid

Cerebrospinal fluid (CSF) is formed from plasma by the filtering and secretory activities of the choroid plexus in the lateral ventricles. It passes through the third and fourth ventricles into the subarachnoid space between the pia mater and subarachnoid mater, and completely surrounds the brain and spinal cord. CSF is reabsorbed into the blood stream by the arachnoid villi which project into the subarachnoid space. In the adult the total volume of the CSF is about 140 ml.

The CSF supports the brain and protects it against injury. It also has a similar function to lymph and removes waste products of metabolism. It is essentially a plasma ultrafiltrate and has a very low protein and lipid content. Some active secretion of, for example, chloride may occur. The circulation of CSF is very slow, allowing long contact with cells of the central nervous system (CNS); the uptake of glucose by these cells may account for its low concentration in the CSF relative to plasma. The composition of CSF obtained by lumbar puncture differs from that of CSF obtained by cisternal or ventricular puncture.

Concentrations of analytes in the CSF should always be compared with those in plasma, because alterations in the latter are reflected in the CSF even when CNS metabolism is normal.

Examination of the cerebrospinal fluid

Biochemical investigation of the CSF is usually less important diagnostically than simple inspection and microbiological and cytological examination of the fluid. Textbooks of microbiology should be consulted for further details.

Taking the sample

CSF is usually collected by lumbar puncture. This procedure is usually contra-indicated if the intracranial pressure is raised and the clinician must confirm the absence of papilloedema before proceeding. If possible a total of about 6 ml of CSF should be collected as 2 ml aliquots into sterile containers and sent first for micro-biological examination: any remaining specimen can then be used, if necessary, for chemical investigations, but if the tests are performed in the reverse order bacterial contamination may occur. If indicated, a further 0.5 ml should be collected into a tube containing fluoride for glucose estimation and should be sent to the laboratory with a blood sample taken at the same time.

CSF is potentially highly infectious and, like all specimens, must be handled and transported with care.

Appearance

Normal CSF is completely clear and colourless and should be compared visually with water; slight turbidity is most easily detected by this method.

Colour

Bright red blood may be due to:

- a recent haemorrhage involving the subarachnoid space;
- damage to a blood vessel during lumbar puncture.

If CSF is collected as three separate aliquots, all three will be equally blood-stained in the first case, but progressively less so in the second.

Xanthochromia is defined as yellow coloration of the CSF, and may be due to:

- *altered haemoglobin*, the colour appearing several days after a subarachnoid haemorrhage, and, depending on the extent of the bleeding, lasting for up to a week or more;
- *large amounts of pus*. The cause will be obvious from the gross turbidity of the sample and from the presence of pus cells detectable by microscopy;
- a very high protein content, as found, for example, if there is a *cerebral or spinal tumour* near the surface which impairs the circulation of CSF. Specimens from these cases tend to clot spontaneously due to the presence of fibrinogen;
- *jaundice*, which will be clinically obvious, may impart a yellow colour to the CSF.

Turbidity

Turbidity is usually due to an *excess of white blood cells (pus)*. Slight turbidity will, of course, occur after haemorrhage, but the cause can be differentiated by microscopical examination of the fluid.

Spontaneous clotting

Clotting occurs when there is an excess of fibrinogen in the specimen, usually associated with a very high protein concentration. This finding occurs classically in association with tuberculous meningitis or with tumours of the CNS.

The following are the most frequently requested biochemical estimations.

Glucose

The CSF glucose concentration is slightly lower than that in plasma but, under normal circumstances, is rarely less than 50 per cent of the plasma level. Provided that CSF for glucose assay has been preserved with fluoride, an abnormally low glucose concentration occurs in:

- *infection*. If there is an increased polymorphonuclear leucocyte count or a bacterial infection, CSF glucose levels may be very low because of increased utilization of glucose. If obvious pus is present estimation of CSF glucose will yield little additional information. The assay is most useful when the CSF is clear and *tuberculous meningitis* is suspected: CSF glucose may then be low, but not as low as in pyogenic meningitis. In viral meningitis the CSF glucose concentration is often normal. *However, CSF glucose estimation does not reliably distinguish between different forms of infective meningitis, because the result may be normal in any form.*
- *hypoglycaemia*. The CSF glucose concentration parallels that of plasma, although there is a delay before changes in plasma glucose levels are reflected in the CSF. Hypoglycaemia may cause coma, and low CSF glucose levels, although there is no primary cerebral abnormality. If there is hyperglycaemia CSF levels will be high. *Both plasma and CSF concentrations should be measured.*
- *widespread malignant infiltration of the meninges* may be associated with low CSF glucose levels, but glucose assay is *not* useful as a diagnostic test for this condition.

Protein

The protein concentration of normal CSF is very low. The ability of individual plasma proteins to cross the normal vascular wall and meninges depends on their molecular weights. The lumbar CSF protein concentration is up to three times higher than that in the ventricles, but even the normal lumbar concentration is below 0.4 g/litre; most of the protein is albumin. In the newborn infant, because of increased vascular permeability, the CSF protein concentration is up to three times that in the adult.

Even in the absence of cerebral disease, changes in the concentration of such CSF proteins as IgG may reflect changes in plasma levels and *results of assays can only be interpreted if those of the two fluids are compared.*

Cerebral disease may change the total concentration of CSF protein, and the proportion of its constituents, for two reasons.

- The vascular and meningeal permeability may be increased, allowing not only more protein, but proteins with molecular weights higher than that of albumin, to enter the CSF.
- Proteins may be synthesized within the cerebrospinal canal by inflammatory or other invading cells.

In some conditions both these factors may be present.

Total protein

Measurement of CSF total protein is a relatively insensitive test for the diagnosis of cerebral disease, because early changes in the concentration of a specific protein do not always cause a detectable rise in total protein concentration. Serial determinations may be used to monitor treatment.

The CSF total protein concentration will be increased:

- in the presence of blood, due to haemoglobin and plasma proteins;
- in the presence of pus, due to cell protein, and to exudation from inflamed surfaces.

If either of these is apparent on visual inspection or by microscopical examination of the specimen *no further information is provided by estimating the total protein concentration and the laboratory staff should not be unnecessarily exposed to potentially dangerous infected material.*

- in non-purulent inflammation of cerebral tissue, when there may be a definite rise in total protein concentration, despite the absence of detectable cells in the CSF. Cells may also be undetectable in some cases of bacterial meningitis, particularly:
  - in children;
  - in immunocompromized patients;
  - if antibiotics have been given before lumbar puncture.
- in blockage of the spinal canal by spinal tumours, by vertebral fractures or due to spinal tuberculosis, increased capillary permeability and fluid reabsorption due to stasis may cause very high protein concentrations distal to the block, often with xanthochromia.

Tests for abnormal CSF protein patterns

Tests for individual CSF proteins are not useful, and may even be misleading, in the presence of blood or pus. They may, however, detect abnormalities when the *total protein concentration is equivocally raised or normal,* and may help to elucidate the cause of a high level.

The concentrations of individual CSF proteins may be collectively assessed by electrophoresis, or individually measured by immunological techniques.

Electrophoresis. Electrophoresis of CSF may yield useful diagnostic information if the pattern is compared with that of serum from blood taken at the same time.

The normal CSF electrophoretic pattern differs from that of serum; there are relatively higher prealbumin and lower γ-globulin concentrations, and a 'tau' fraction, which migrates in the α_2-β region, is formed from degradation products of transferrin.

Increased capillary permeability, with a similar increase in the permeability of the blood-brain barrier, may be demonstrated by finding relatively high molecular-weight plasma proteins in the CSF, which are not normally present in significant amounts. This is a non-specific pattern due to a wide variety of inflammatory conditions, but may sometimes aid diagnosis. Such a pattern may be due to:

- *cerebral tumours*, some of which may also produce abnormal immunoglobulins;
- *acute idiopathic polyneuropathy (Guillain-Barré syndrome)* in which acute-phase proteins (p. 330) and immunoglobulins may also be synthesized locally;
- *local synthesis of immunoglobulins*, particularly IgG and IgA, which may occur within the central nervous system in:
  - multiple sclerosis (the most important indication for the test);
  - encephalitis;
  - neurosyphilis;
  - systemic lupus erythematosus;
  - chronic relapsing Guillain-Barré syndrome;
  - cerebral sarcoidosis;
  - cerebral tumours (rarely);
  - slow virus infection (for example Jakob-Creutzfeldt disease);
  - subacute sclerosing panencephalitis.

Immunoglobulins synthesized within the CSF may be detected by finding a characteristic pattern on electrophoresis, consisting of multiple bands in the γ-globulin region. These bands are rarely monoclonal and the pattern is called '*oligoclonal banding*'. The typical CSF pattern is found in over 90 per cent of patients with multiple sclerosis, *but is not specific for this condition.*

Oligoclonal bands only signify cerebral disease if found in the CSF and not in the serum.

Occasionally intrathecal malignant B lymphocytes produce a local *monoclonal band* detectable by electrophoresis.

Immunological techniques. These techniques are less diagnostically discriminating than electrophoresis.

Concentrations of CSF immunoglobulins, especially of *IgG*, may be measured. One method of improving the diagnostic precision is also to measure the concentration of a lower molecular-weight protein, such as albumin, and to calculate the ratio of the concentrations of CSF IgG to serum IgG, using the ratio of the serum to CSF concentrations of the smaller protein to correct for increased vascular permeability. If high CSF IgG levels are due to the increased permeability of non-specific inflammation the ratio will be low or normal; if permeability is only slightly increased more albumin than IgG will diffuse from the plasma into the CSF and the ratio will be low, but more commonly the permeability is such that albumin and IgG diffuse at almost the same rate, and the ratio is normal. However, in conditions such as *multiple sclerosis* the *ratio is high* because CSF IgG has been synthesized locally.

The recommended procedure for examining the CSF is outlined on p. 432.

Summary

1. CSF is potentially highly infectious.
2. Biochemical analysis of CSF is less useful than simple inspection and bacteriological examination.
3. Estimation of CSF glucose concentration is rarely useful in acute conditions, but more so in cases of suspected tuberculous meningitis.
4. Assessment of the pattern of CSF proteins is the most useful biochemical procedure in chronic cerebral disease.
5. The finding of oligoclonal bands in the CSF may be of diagnostic value in non-purulent cerebral conditions such as multiple sclerosis.

Further reading

Thompsom EJ, Johnson MH. Electrophoresis of CSF proteins. *Br J Hosp Med* 1982; **28**: 600–8.

Hayward RA, Shapiro MF, Oye RK. Laboratory testing on cerebrospinal fluid. *Lancet* 1987; **1**: 1–4.

Procedure for examination of the CSF

If the CSF is:

- **very cloudy,** send for microbiological examination. Chemical estimations are unnecessary;

- **heavily bloodstained** in three consecutive specimens there has probably been a cerebral haemorrhage. Chemical estimations are useless;

- **clear or only slightly turbid,** send for microbiological examination and for estimation of *glucose* and *protein* concentration, and for electrophoresis. *Always send blood for glucose and protein estimations at the same time.*

- **xanthochromic,** send the specimen, with blood, for protein estimation. Examine under the microscope for erythrocytes.

24

Drug monitoring

The concentrations of many drugs can be measured in plasma and other body fluids. Interpretation of results is not simple and measurement of only a few drugs is of proven value for either therapeutic or toxicological monitoring.

Pharmacokinetics is the name given to the study of the fate of drugs after administration and is concerned with their absorption, distribution in body compartments, metabolism and excretion. Absorption depends on whether the drug is taken orally, by intravenous or intramuscular injection, sublingually or rectally. Unless stated otherwise, the drugs considered in this chapter are taken orally.

Possible indications for measuring plasma drug levels are to:

- monitor their therapeutic use by trying to:
  check that the patient is taking the drug as prescribed (*compliance*);
  ensure that the dose is high enough to produce the required effect, but not so high as to be likely to cause toxic effects.
- diagnose obscure conditions;
- elucidate the type of drug taken in overdose and assess the need for treatment. In cases of suspected overdose medicolegal requirements must be taken into account.

In the last two groups screening of the urine or gastric aspirate for drugs or their metabolites is often helpful.

Each of the indications will be discussed briefly.

Monitoring drug treatment

The best way of assessing whether the dose of a drug is optimal is to measure, either clinically or by laboratory assays, the desired effect. For example, the effect of antihypertensive drugs is monitored by measuring the blood pressure. Examples of biochemical effects, such as the monitoring of plasma calcium levels and sometimes alkaline phosphatase activities during calciferol treatment, plasma potassium and $T\text{CO}_2$ levels during potassium supplementation and plasma thyroxine and TSH levels during the treatment of thyroid disease are discussed in the relevant chapters.

The plasma drug level may not parallel cellular effects and its measurement can only be a second best to direct assessment. Measurement of plasma levels may be indicated if the desired result cannot be measured precisely; for example, the incidence of epileptic fits is a poor indicator of the optimal dosage of anticonvulsants, both because the frequency may vary even without treatment and because only partial control may be possible, even with the best treatment. Chemical monitoring is likely to be useful only if the range of plasma levels most effective in producing the desired result without toxic side-effects (the *therapeutic range*) has been defined; this is particularly true if there is a narrow margin between therapeutic and toxic drug levels, such as in the case of lithium (p. 438).

The effect of pharmacokinetics on plasma drug levels will be discussed so that the limitations of chemical drug monitoring may be understood.

Factors affecting plasma levels

The total amount of drug in the extracellular fluid depends on the balance between that entering and that leaving the compartment; the plasma *concentration* also depends on the volume of fluid through which the retained drug is distributed.

Assay of plasma levels of drugs *may* be useful to assess compliance and, for those drugs with a known therapeutic range of plasma concentration, if there is doubt about any of the following steps.

Compliance

It has been shown that very few in-patients receive a drug exactly as prescribed; compliance with instructions is likely to be even worse outside hospital. The patient may fail to take the drug at all, may not take the prescribed dose, or may take it irregularly. The more drugs that are prescribed the more confused he may become about the timing and dose of each. Unless the doctor realizes this he may attribute the poor clinical response, or symptoms of toxicity, to inappropriate dosage. A regular review of therapy and careful explanation to the patient often works wonders, and is more rewarding than changing the dosage without further questioning.

Assay of plasma levels is a fairly crude method of assessing compliance and only tests the situation at the time of seeing the patient. If levels are very low, and certainly if no drug is detectable, there is strong evidence that the patient is not taking the drug, or is taking inadequate amounts: if levels are very high he may be regularly taking too high a dose, or, especially if there are no signs of toxicity, he may have taken a large dose just before seeing the doctor, to 'catch up' because of previous undercompliance.

Timing of the sample. If blood is taken before absorption of the drug is complete a falsely low level may be detected: if it is taken after absorption, but before the drug has been completely distributed through the extracellular compartment, the value may be falsely high. Blood samples must be taken at a standard time after ingestion of the drug, the exact time varying with the known differences in rate of absorption, metabolism and excretion of different drugs.

There may also be intersubject, or even intrasubject, variations in these factors, perhaps due to disorders of the organs responsible for these processes, and sometimes only serial sampling can determine whether a reasonably steady plasma level is being maintained. These factors are discussed below.

Entry of the drug into, and distribution through, the extracellular fluid

Absorption. *Lipid-soluble* drugs can pass cell membranes more readily, and are therefore absorbed more rapidly, than water-soluble ones; they reach the highest plasma concentrations at between 30 and 60 minutes after ingestion. Drugs such as phenoxymethyl penicillin (penicillin V) may be degraded within the gastrointestinal lumen, and allowance must be made for this when the dose is calculated.

The rate of absorption in an individual patient may be affected by the timing of ingestion in relation to meals.

The rate of gastric emptying affects that of absorption and must be allowed for, for example, during treatment with drugs affecting gut motility or after gastric surgery. Vomiting, diarrhoea and the malabsorption syndromes are obvious causes of impaired absorption.

Volume of distribution. The final plasma concentration reached after a standard amount of drug has been absorbed is affected by the volume through which it has been distributed. For example, it may be difficult to predict the appropriate dose in patients with a larger than normal volume of distribution because of oedema or obesity, in whom unexpectedly low plasma levels may be found, or in small children in whom there is danger of overdosage. The dose may have to be calculated allowing for the weight or surface area of the patient.

Binding to plasma albumin. Many drugs, like many endogenous substances, are partly inactivated by protein-, usually albumin-, binding. Most assays estimate the total concentration of the free- plus protein-bound drug. Biological feedback mechanisms do not control free drug levels as they do those of plasma calcium and hormones, and the method of interpretation to allow for altered protein binding is therefore different. *Measured* plasma levels may be little changed by a reduction in protein binding, but a larger *proportion* of the measured drug will be active; unless this is realized dangerously high free levels may be interpreted as being within, or even below, the therapeutic range if the plasma albumin concentration is very low. The proportion bound varies at different plasma albumin levels and there is no valid correction factor which allows for protein abnormalities (compare calcium, p. 194).

About 90 per cent of phenytoin, 70 per cent of salicylic acid, 50 per cent of phenobarbitone and 20 per cent of digoxin is protein-bound if binding is not affected by:

- *abnormalities in plasma albumin concentration.* Blood should be taken without stasis to minimize a possible *rise* in plasma albumin, and albumin-bound drug, levels. A rise may also be due to loss of protein-free fluid from the vascular compartment, causing haemoconcentration. *Low* levels, such as those found in hepatic cirrhosis or the nephrotic syndrome may reduce the proportion of protein-bound drugs;

- *competition for binding sites on protein.* Many drugs, unconjugated bilirubin, and hydrogen ions compete with each other for binding sites. If more than one drug is being taken the free level of each is likely to be higher than the measured concentration would suggest. *Plasma drug levels in patients taking many drugs, or who are jaundiced or acidotic, are difficult to interpret.*

In *renal failure* the use of drug assays to detect reduced excretion may be partially invalidated by acidosis and by retention of other competing ions. Similarly, in *hepatic disease*, low plasma albumin levels or competition for binding sites by unconjugated bilirubin limits the value of monitoring protein-bound drugs.

Metabolism and excretion of drugs

The time taken for the plasma drug concentration to fall to half its original level is called its *elimination half-life (T½).* A drug with a short T½ should be taken more often than that with a long one in order to maintain a steady plasma concentration. *After starting treatment* a steady state is usually only reached after between four and six half-lives have elapsed, and the first specimen of blood for monitoring should not be taken earlier than this.

The rate at which the plasma drug concentration falls after it has reached peak concentration depends on:

- the rate of distribution through the extracellular fluid (see above);
- the rate at which it is metabolized and excreted, whether in urine or bile;
- the rate of entry into cells.

The rate of elimination of most drugs depends on their plasma level. A small increase in the dose of a few, such as phenytoin, may saturate the metabolic or excretory pathways, and so cause a disproportionate increase in plasma levels, which should be monitored. The rate at which this occurs is controlled by a phenomenon known as *saturation kinetics.*

Metabolic conversion to active or inactive metabolites. Drugs such as calciferol (p. 176) are only active after metabolic conversion: others are inactivated, usually by conjugation in the liver. Ideally a drug assay should measure all the active forms, whether the parent compound or its active metabolite, and none of the inactive forms.

Some drugs given orally, and absorbed into the portal system, are inactivated by the liver before reaching the systemic circulation. Changes in hepatic function affect this process. For instance, hepatic dysfunction impairs such metabolism, and allows more active drug than usual to enter the systemic circulation: induction of enzymes which inactivate drugs, such as some anticonvulsants, reduce the amount leaving the liver.

Relationship between plasma levels and cellular effects

Plasma concentrations are measured in an attempt to predict cellular effects. They only do so if they parallel the concentrations at the active sites and if the cells respond predictably to a given concentration at that site. Neither of these conditions is always fulfilled.

Drug concentrations at the active site. Drugs that are lipid-soluble are more rapidly distributed across cell membranes than those that are water-soluble (p. 435).

In the steady state the plasma concentration may not be the same as that at the active site within the cell, but usually parallels it. This may *not* be true if:

- *treatment has just begun.* Plasma levels may be relatively high compared with those in the cell;
- *treatment has just stopped.* Plasma levels may fall more rapidly than those in the cell;
- *the drug is taken irregularly or at intervals inappropriate to the half-life of the drug.* Plasma levels then fluctuate and equilibrium may not be reached with those in the cell;
- *there is a change in pH, or change in other substances competing for binding to the cell membrane or active site.* This adds to the problem of interpreting plasma levels in patients with renal or hepatic dysfunction (p. 436).

Tissue sensitivity

Intersubject variation. Some patients are more sensitive to drugs than others.

The rate of metabolism of some drugs depends on the age of the patient. *Premature infants*, with immature detoxifying mechanisms, metabolize and excrete some drugs more slowly than adults, while *children*, who have a high metabolic rate may eliminate them more rapidly. The *elderly* may metabolize drugs slowly and are particularly sensitive to digoxin.

The rate of metabolism of some drugs may be affected by genetic factors. For example, the inactivation of *isoniazid* by acetylation depends on whether the patient is genetically able to carry out this process at a normal rate, and toxicity is likely at lower plasma levels in 'slow acetylators' than in 'fast acetylators'.

Intrasubject variation. The sensitivity of an individual to a drug may vary. Known causes of increased sensitivity to digoxin are hypokalaemia and hypercalcaemia.

Tolerance. Drugs may, for example, induce the synthesis of enzymes that inactivate them. Much higher levels than normal may be needed to produce the desired effect in patients who have become tolerant.

Additive or antagonistic effects at the active site. Some drugs and endogenous substances, often the same ones as those that compete for plasma albumin-binding sites, also compete for binding at the active site within the cell. Some of these have been discussed above. By contrast, some drugs have synergistic effects. These factors further complicate the interpretation of plasma drug levels in patients taking many drugs.

When to measure drug levels to monitor treatment

Only a few drug assays are of proven clinical value. The clinical picture must be taken into account and all the above factors must be allowed for when interpreting results. The following are some of the drug measurements which have been claimed to be useful for therapeutic monitoring.

Digoxin. Estimation of plasma digoxin levels may be useful:

- to assess compliance. The elderly, for whom digoxin is frequently prescribed, are often on multiple therapy and are liable to be confused about which tablets to take and at what time;
- in patients, such as the elderly and premature infants, who are likely to have a low threshold for toxicity;
- if it is difficult to calculate the appropriate dose because of the size of the patient (for example, in infants and the obese), or because of an increased volume of distribution due to oedema;
- because of impaired excretion, perhaps due to renal dysfunction. The problems of interpreting results in such cases has been discussed above.

Plasma digoxin levels, even within the normal range, are very difficult to interpret in the presence of conditions which may alter receptor sensitivity, such as:

- hypokalaemia, hypercalcaemia or hypomagnesaemia;
- hypoxia;
- thyroid disease.

Plasma levels should be measured at least 6 hours after the last dose, when absorption and distribution are complete.

Anticonvulsants. Once the dose which gives stable plasma levels within the therapeutic range has been determined, monitoring is probably only necessary:

- to assess compliance;
- if the frequency of fits increases in a previously well-controlled patient;
- if the clinical picture suggests toxicity.

Levels in children and in pregnant women should be monitored more often.

Measurement of plasma *phenytoin*, *phenobarbitone* and *possibly sodium valproate and carbamazepine* may be of clinical value. The fact that phenytoin exhibits saturation kinetics (p. 436) means that monitoring of this drug is of the most importance.

Lithium and high-dose salicylate therapy. The margin between therapeutic and toxic levels is narrow for both these drugs. Lithium is given to psychiatric patients, who are especially unlikely to comply with instructions. Plasma levels of both lithium and salicylate given at high dosage should be monitored regularly. Lithium is not protein-bound and interpretation of results of assays is *not* complicated by protein abnormalities.

Plasma levels of lithium should be measured 12 hours after the evening dose.

Theophylline assays may be helpful, especially in *acute asthmatic attacks which are not responding clinically*; this helps to ensure that the poor response is not due to underdosage, and, if it is not, that increasing the dosage will not lead to toxic plasma levels.

Theophylline is used to manage recurrent apnoea in the *newborn*; clearance of the drug from the body is slow at this age and it is metabolized to *caffeine*, an active metabolite. *Both* plasma theophylline and caffeine levels may need to be assayed to assess therapeutic or toxic effects in the newborn.

Samples for theophylline assays should be taken at between two and four hours after the last dose.

Antiarrhythmic drugs, such as procainamide and quinidine, can be measured in special cases. N-acetylprocainamide is a less potent metabolite of procainamide and is only present after oral administration of the parent drug, and need not be assayed.

Aminoglycosides such as gentamicin, in contrast to many other antibiotics, have a narrow margin between the therapeutic range and toxic levels; at toxic levels they may cause ototoxicity and nephrotoxicity. They are given intramuscularly or intravenously.

Measurement of plasma levels is indicated if:

- there is serious infection, or if the treatment is prolonged, which increases the risk of toxicity;
- renal function is impaired;
- the patient is receiving other drugs, such as frusemide, which may potentiate aminoglycoside toxicity.

It is usually necessary to measure aminoglycoside levels at two different times. One specimen should be taken immediately before injection, to ensure that levels are not already high, and that further administration is not likely to cause toxicity, or that they are not so low as to be ineffective. The second should be taken about an hour after injection, at the time of the anticipated peak level, to ensure that an adequate concentration has been achieved for antibacterial action.

Monitoring possible side-effects of drug treatment

Some drugs have harmful side-effects. For example, plasma thyroxine and TSH levels should be measured during lithium treatment because of the danger of hypothyroidism (p. 166), and plasma potassium levels during diuretic therapy. Measurement of plasma transaminase activities, or plasma urea or creatinine levels, may be indicated during treatment with potentially hepatotoxic or nephrotoxic drugs respectively.

Diagnosis of obscure conditions

Drug overdosage must always be excluded as a cause of coma of obscure origin. This subject will be dealt with in the next section.

Clinical findings may suggest effects due to a drug or alcohol despite denial by the patient.

Unexplained laboratory findings may also be due to drug or alcohol ingestion. For example, hypokalaemia may be due to purgative abuse (p. 60) and hypoglycaemia may be a presenting finding in alcoholics (p. 219). Raised plasma transaminase activities which are thought to be due to alcoholic liver disease may be investigated by measuring random plasma alcohol levels.

Measurement of plasma drug levels may be supplemented by screening the urine for drugs or their metabolites.

Investigation of known or suspected overdosage

About half the patients attempting suicide take several drugs, sometimes with alcohol. Screening of plasma, urine or gastric aspirate may be needed to confirm the diagnosis and will usually identify the drugs.

It is often unnecessary to measure plasma *levels* of drugs that the patient is known to have taken in overdose. The need for gastric lavage and measures needed to increase urinary excretion of the drug and to maintain adequate respiration and circulation are not usually affected by a knowledge of drug levels. Only measurements of plasma electrolytes and blood gases are needed.

Drugs and other toxins can be considered in three main groups. These are:

- toxins for which there is a specific antidote;
- toxins for which there is no specific antidote, but which are not, or are only partially, protein-bound;
- toxins for which there is no specific antidote and which are strongly protein-bound.

The case for assay is strongest in the first and weakest in the last group.

Poisons for which there is a specific antidote

Paracetamol (acetaminophen). It is *essential* to measure drug levels in suspected paracetamol poisoning because:

- a metabolite of paracetamol is hepatotoxic; the patient may die of liver failure despite recovery from the immediate effects;
- a specific antidote to the hepatotoxic effect is available;
- the antidote is only useful if given within a defined period of time and if defined plasma levels are reached;
- the likelihood of hepatotoxicity cannot be predicted from the clinical picture at presentation.

Cysteamine, methionine and N-acetylcysteine are all effective antidotes, because they enable paracetamol to be converted to non-toxic metabolites. N-acetylcysteine is the least toxic and is often used. Cysteamine may cause nausea, vomiting, abdominal pain and cardiac arrhythmias. Methionine, although less toxic, may cause vomiting. N-acetylcysteine is only effective in clearly defined circumstances and should only be given after the following factors have been taken into account:

- *timing of the specimen.* Plasma levels cannot be interpreted, and should not be measured, until absorption and distribution are nearly complete, at about *4 hours after ingestion.* If treatment is to be effective it should be started as soon as possible following ingestion and is *ineffective after 15 to 20 hours.* Levels should be measured on specimens taken *as early as possible between 4 and 15 hours after ingestion.*
- *plasma paracetamol levels.* As a rough guide, treatment is indicated if the plasma paracetamol concentration is *200 mg/litre (1300 μmol/litre) or more at 4 hours and 30 mg/litre (200 μmol/litre) or more at 15 hours after the dose.* If the specimen is taken between 4 and 15 hours, treatment is indicated if the plasma concentration at the time falls above the solid line in Fig. 24.1. Treatment is certain to be ineffective after 20 hours.

Serial measurement of plasma transaminase activities and of prothrombin time may be needed to detect liver involvement which may become apparent about three days after ingestion.

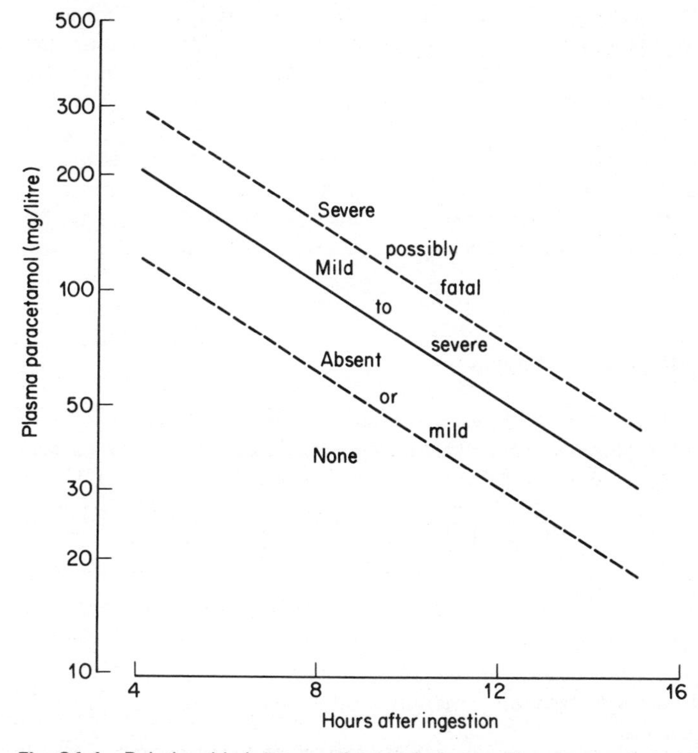

Fig. 24.1 Relationship between plasma paracetamol concentration, time after ingestion and potential severity of liver damage. (Paracetamol in mg/litre × 6.6 = μmol/litre) Reproduced by kind permission from Prescott LF. In Richens A, Marks V, eds. *Therapeutic drug monitoring* Edinburgh: Churchill Livingstone, 1983.

Iron. Iron overdosage is most common in children. Desferrioxamine chelates iron and the chelate is lost in the urine. The level of plasma iron at which treatment is indicated has not been clearly defined; if there is any doubt desferrioxamine should be given. Concentrations above 90 μmol/litre (500 μg/dl) in young children and about 150 μmol/litre (840 μg/dl) in adults are said to be definite indications for treatment.

Poisons for which there is no specific antidote

Weakly protein-bound poisons. Forced diuresis may increase the excretion of some free drugs; induction of alkalosis often further increases the excretion rate, but neither forced diuresis nor induction of alkalosis is without risk and in most cases neither is indicated. If alkaline diuresis is induced plasma potassium levels, as well as blood pH, must be monitored because of the danger of acute hypokalaemia (p. 61). It has been suggested that plasma *lithium* (unbound), *phenobarbitone* (about 50 per cent bound) and *salicylate* (about 70 per cent bound) levels should be measured after they have been taken in overdosage. Serial readings may help the clinician to assess the adequacy of therapy.

Strongly protein-bound poisons. It is agreed that there is no indication to measure levels of strongly protein-bound drugs, such as barbiturates other than phenobarbitone (p. 435); excretion cannot be increased significantly and supportive treatment is based on clinical observation and the results of plasma electrolyte and blood gas estimations.

Monitoring of *methanol* and *paraquat* levels may be desirable. It is wise to contact your laboratory about all cases of suspected overdose; this is especially important in the case of such rare poisons. The local laboratory may not have full facilities and it is then essential to seek specialist advice from a toxicology centre *before* blood is taken.

Medicolegal precautions

Specimens of vomit, gastric washings, urine and plasma from all serious cases of poisoning should be stored in the refrigerator in sealed containers, labelled with the full name of the patient (or, if this is not available, some other form of identification) and the date and time of the specimen. They may be needed later for medicolegal purposes, and should be kept for at least three months.

Summary

1. Assay of drug levels to monitor therapy is only indicated for a few drugs, in defined circumstances.
2. Drug effects are ideally assessed by clinical or laboratory evidence of response.
3. The time at which the specimen is taken after administration is important if results are to be meaningful.

4. A knowledge of the pharmacokinetics of the drug is necessary to interpret results.

5. Screening of plasma or urine to identify toxic substances taken in overdosage may be helpful. It is rarely necessary to measure the concentrations.

6. In most cases of drug overdose plasma, urine and gastric aspirate should be stored for three months in case of medicolegal proceedings.

Further reading

Richens A, Marks V, eds. *Therapeutic drug monitoring.* Edinburgh: Churchill Livingstone. 1981.

Richens A. Monitoring plasma levels of drugs. *Prescribers' J* 1983; **23**: 1–9.

Morgan DB. Plasma digoxin: an overused test? *Ann Clin Biochem* 1984; **21**: 449–52.

25

The clinician's contribution to valid results

Chemical pathology is the study of physiology and biochemistry applied to medical practice and overlaps that of clinical disciplines. An understanding of the basis of chemical pathology is, as we have seen, essential for much diagnosis and treatment.

The medical student and the clinician need only limited knowledge of the technical details of laboratory estimations. However, correct interpretation of results requires some understanding both of achievable analytical accuracy and reproducibility and of physiological variations, and these subjects will be discussed in the next chapter. He should also be aware of laboratory organization in so far as it affects the speed at which tests can be completed. Above all he should realize that *the technique of collecting specimens can affect results drastically*, and should co-operate with the laboratory in its attempt to produce answers rapidly and accurately and identifiable with the relevant patient. To this end he should understand the importance of *accurately completed request forms, of correctly labelled specimens, collected at the appropriate time by the correct technique, and of speedy delivery to the laboratory*. In an emergency *therapy based on technically correct results from a wrongly labelled or collected specimen may be as lethal as faulty surgical technique*: moreover, even if the error is recognized, *time could have been saved by a little thought and care in the first place. An emergency warrants more, not less, than the usual accuracy in collection and identification of specimens.*

Safety recommendations in the UK strongly advise that *potentially dangerous specimens*, such as those from patients with hepatitis and AIDS, should be assayed with special precautions, and should be *clearly identified by the clinician* when he sends them to the laboratory. *Any* such specimen must be sent in a leak-proof, sealed plastic bag, with the request form in a different pocket in the bag. Failure to comply with these guidelines may put many staff, including porters and laboratory staff, at unnecessary risk, and may mean that necessary assays are not performed, or are at best delayed.

If the clinician delegates any part of these tasks to, for example, a nurse or ward clerk he is not absolved from responsibility for their accuracy, any more than senior laboratory staff are absolved from responsibility for the accuracy of analysis and reporting.

Request forms

Clinical information

Most departments take stringent precautions to control the analytical accuracy and precision of results. However, when very large numbers of estimations are being performed under pressure it is impossible to be sure that every result is correct. The clinician can play his part by co-operating with the pathologist in minimizing the chance of error, not only by taking suitable specimens, but also by giving *relevant* clinical information. 'Unlikely' results are checked in most laboratories and if, for example, plasma enzymes had risen from normal to very high activities in a day, estimations would be repeated on both specimens to be sure that they had not been transposed with those of another patient: if, however, it were known that the patient had had a myocardial infarction the result would have been expected and time, money and worry would have been saved. In this example the words 'post MI' would be more informative, and take no longer to write, than any previously stated diagnosis.

Patient identification

Accurate information about the patient, legibly written and including the *surname* and *first names correctly and consistently spelt, and date of birth and hospital case number* are essential for comparing current with previous results on the same patient. A 'cumulative' report includes the results of each patient on successive occasions: this type of form enables the clinician to follow the progress of the patient more easily than by looking at a single result, and the laboratory staff can detect sudden changes, so that the cause can be sought. The system can only work effectively if patient identification is accurate: a surprising number of people have the same names, even when these are apparently uncommon; it is less likely that they will also be of the same age in years, and even less probably will they have the same full date of birth; they should not have the same hospital number. Any of these items may be recorded inaccurately on the form and, unless there is complete agreement with previous details, results may be entered on the wrong patient's record, causing confusion, and even danger to the patient. *If reporting is computerized it is most important to provide accurate information.* Computers cannot think or telephone the ward for elucidation, and if the information fed into them is inaccurate it may be rejected or, worse still, results may be reported as belonging to another patient.

Location of the patient and clinician

It is obvious that if the *ward or department* is not stated it may take time and trouble to determine where the results should be sent. *The requesting doctor must sign the form legibly*, checking at the same time that the information given is correct: this and the consultant's name are necessary, especially if urgent or alarming results are to be notfied rapidly, and advice given about treatment.

The forms designed by pathology and other departments ask only for information essential to ensure the most efficient possible service to the clinician and therefore to the patient. All pathologists have been faced with a form containing only the information 'Smith'; sometimes not even the investigations needed are stated. This is particularly dangerous in an emergency. Unless the pathologist is endowed with psychic powers it is difficult for him to help under these circumstances.

Collection of specimens

Collection of blood

If a clinically improbable result has been checked analytically, and the second agrees well with the first, a fresh specimen should be analysed. Before proceeding further, it is essential to try to determine why the previous one gave a false answer (if it was false). Although contamination of the syringe, needle or specimen tube are obvious possibilities they are relatively rare and should not be accepted until other, more common causes have been excluded.

The errors to be discussed in this chapter arise outside the laboratory and are, in our experience, relatively common. All examples given are genuine.

Effect on results of procedures before venepuncture

Effect of posture. Concentrations of plasma proteins, and of substances bound to them, are lower if the patient is in the supine than in the erect position. It is often difficult to standardize this variable, and the effect must be taken into account when serial results are interpreted. It is discussed more fully on p. 459.

Oral medication. Specimens should not be taken to measure an analyte just after a large oral dose of the same substance has been given. For example, blood should be taken for drug assays at a standard time after the dose; misleadingly high levels may occur at the time of peak absorption (p. 434).

There may be significant hypokalaemia for a few hours after taking potassium-losing diuretics due to rapid clearance of potassium from the extracellular fluid. The plasma concentration returns to its 'true' level as equilibration occurs between cells and extracellular fluid.

Interfering substances. Previous administration of a substance may affect plasma levels for some time. For example, drugs such as salicylates compete with T_4 for binding sites on TBG, and a falsely low plasma T_4 value may be obtained. Other drugs interfere with the chemical reactions used in assays such as some of those for creatinine. The effect of interfering substances is probably widespread and difficult to detect.

Intravenous infusion. A spuriously low plasma sodium concentration may be due to gross hyperlipidaemia or hyperproteinaemia (p. 32). If a lipid solution is being infused it should be replaced by a lipid-free one for at least three hours before sampling. Hyperlipidaemia not only causes spurious results because of the space-occupying effect, but may directly interfere with several assays.

Palpation of the prostate. Palpation of a non-malignant prostate, especially by an inexperienced clinician, may release abnormally large amounts of prostatic (tartrate-labile) acid phosphatase into the circulation, and the falsely elevated plasma levels may persist for several days. Although the effect of this procedure on acid phosphatase results has been questioned, it does, in our experience, sometimes occur; the conflicting reports may be due to the inconsistency of the effect. Very high serum activities are unlikely to be due to this cause, but in such cases the diagnosis of carcinoma of the prostate is usually evident on clinical grounds. Any marginally raised activity should be checked on a specimen taken a few days later.

For example, a specimen was received from a patient who had undergone rectal examination a few hours earlier. The serum tartrate-labile (prostatic) acid phosphatase activity was three times the upper limit of the reference range. Three days later it was still twice this upper limit, but had returned to within the range eight days after the examination.

Effects on results of the technique of venepuncture

Venous stasis. It is usual to apply a tourniquet proximal to the site of venepuncture to ensure that the vein 'stands out' and is easier to enter with the needle. If occlusion is maintained for more than a short time the combined effect of raised intravenous pressure and hypoxia of the vessel wall results in passage of water and small molecules from the lumen into the surrounding interstitial fluid. Large molecules, such as proteins (including lipoproteins), and erythrocytes and other blood cells, cannot pass through the vein wall at the same rate: their plasma concentrations therefore rise. The concentrations of individual proteins, such as immunoglobulins, also increase: day-to day variations in these can often be attributed to the differing amounts of stasis used, as well as to changes in posture.

Many plasma constituents are, at least partly, bound to protein. Prolonged venous stasis can raise the plasma calcium concentration, sometimes to equivocal or slightly high concentrations (Table 25.1). Blood samples for calcium estimation should preferably be taken without stasis, especially if high levels have been found in the previous specimen. Other important protein-bound substances include hormones and drugs.

Prolonged stasis may also cause local hypoxia: consequent leakage of intracellular constituents, such as potassium and phosphate, may cause falsely high plasma levels.

It is sometimes difficult to enter 'bad veins' without applying stasis. A tourniquet may be left on until the needle is in the vein; if it is then released, and at least 15 seconds allowed before withdrawal of blood, a suitable specimen will be obtained.

Table 25.1 Some effects of stasis on plasma concentrations in a normal subject

Stasis for	0	2	4	6	minutes
Calcium	2.38	2.45	2.52	2.58	mmol/litre
Total protein	72	74	77	80	g/litre
Albumin	39	40	42	43	g/litre
Haemoglobin	14.7	14.8	15.1	15.5	g/dl

Site of venepuncture. If the patient is receiving an intravenous infusion, the administered fluid in the veins of the same limb, whether proximal or distal to the infusion site, has not mixed with the total plasma volume; local concentrations will therefore be unrepresentative of those circulating through the rest of the body. Blood taken from the opposite arm will give valid results.

The following example is one of the very many which illustrate this point. The clinical details were given as 'post-op'. On the day before the plasma electrolyte concentrations had been normal, the plasma urea was 16.7 mmol/litre (100 mg/dl), and the plasma total protein 67 g/litre. The relevant plasma results on the day in question were as follows:

sodium	66 mmol/litre
potassium	2.2 mmol/litre
T_{CO_2}	11 mmol/litre
total protein	52 g/litre
urea	8.7 mmol/litre (52 mg/dl)

The patient was 'doing well'. All measured constituents seemed to be diluted, and it was assumed that dextrose (glucose), with or without hypoosmolar saline, was being infused. A plasma glucose concentration of 50 mmol/litre (900 mg/dl) supported this view. A call to the ward confirmed that the specimen had been taken from the arm into which a dextrose infusion was flowing. Analysis of plasma from blood taken from the opposite arm gave results almost identical with those of the day before. Note that glucose infusion is likely to cause some hyperglycaemia, even in plasma from the opposite arm, and that, if the plasma level is above about 11 mmol/litre, there is likely to be glycosuria. Only if the hyperglycaemia persists after the infusion is stopped is the patient likely to have diabetes mellitus.

Such an extreme example is easily detected and, although time has been wasted and the patient subjected to an unnecessary venepuncture, no serious harm has been done. Results might have *appeared* to be correct if isosmolar saline were being infused and, if an earlier plasma urea value had not been available, the wrong treatment might have been given.

If it is impossible to obtain blood from another limb the infusion should be temporarily stopped, the tubing disconnected from the needle, and 20 to 30 ml of blood aspirated through this needle *and discarded*: only then should the sample for analysis be withdrawn and the infusion restarted.

Containers for blood

Most hospital laboratories issue a list of the types of container required for different assays and this list may vary from hospital to hospital. For example, most departments ask that blood for glucose estimation be put into a tube containing fluoride; this inhibits glycolysis in erythrocytes.

Potassium should be estimated on plasma from heparinized blood. Potassium is released from cells, especially platelets, during clotting; serum potassium concentrations are usually higher by a variable amount than those of plasma, and this difference can be clinically misleading. Marked differences may be found in patients with leukaemia, in whom the number of white blood cells is usually greatly increased.

Laboratories should only accept blood in the correct containers, but serious errors can arise if blood is decanted from one container to another. The anticoagulant action of oxalate and of sequestrene (ethylenediamine tetraacetate; EDTA) depends on precipitation or chelation, respectively, of calcium, so invalidating results of calcium estimation. Usually the EDTA is in the form of its potassium salt, and potassium results are also affected. As an example, blood was received, apparently in the correct tube, from a patient whose clinical details were given as 'ureterosigmoidostomy'. The result of plasma calcium assay was 0.4 mmol/litre (1.6 mg/dl), and that of plasma potassium 7.5 mmol/litre: despite this the patient was said to 'feel very well'. Further enquiry confirmed that blood had been taken, at the same time, into EDTA for haematological investigations, and that to bring the blood in the EDTA tube down 'to the mark', some had been tipped into the lithium heparin tube which was sent to the chemical pathology department.

The use of sodium, instead of lithium, heparin will cause a falsely high plasma sodium result. This anticoagulant is often used in specimens taken for 'blood gases'; apparent sodium results of 160 to 170 mmol/litre can result from transferring an aliquot into a lithium heparin tube. This error is particularly likely to be misinterpreted, and the wrong treatment instituted, if the patient is unconscious or confused, and therefore, is a likely candidate for hypernatraemia (p. 45).

Falsely high plasma lithium results may be due to the use of lithium heparin.

Effects of storage and haemolysis of blood

The concentration of many substances is very different in erythrocytes from that in the surrounding plasma. Haemolysis releases the cell contents and false answers will be obtained. Release of haemoglobin imparts a red colour to the plasma, which should be detected in the laboratory, and haemoglobin interferes with some chemical reactions. The chance of haemolysis is minimized if the blood is treated gently. The plunger of the syringe should be drawn back slowly and the blood should flow freely. The needle should be removed from the syringe before the specimen is expelled *gently* into the correct container.

The differential concentrations across cell membranes are maintained by energy derived from glycolysis. *In vitro*, erythrocytes soon use up the available glucose (hence the need for fluoride in specimens taken for glucose assay), and therefore the energy source: concentrations in plasma will then tend to equalize, by passive diffusion across cell membranes, with those in erythrocytes. If plasma is not separated from blood cells within a few hours the effect on plasma levels will be similar to that resulting from haemolysis, with the important difference that, because haemoglobin is not released, the plasma looks normal. If the container is not dated, or wrongly dated, the error may not be detected. Low temperatures *slow* erythrocyte metabolism so that differential concentrations cannot be maintained; refrigeration of whole blood has the same effect on the plasma potassium level *in a shorter time*, as allowing it to stand at room temperature. Plasma must therefore be separated from cells before storing overnight, *even in the refrigerator*.

The following is an example of the effect on plasma potassium and glucose concentrations of leaving unseparated blood, without added fluoride, to stand at room temperature.

Blood separated after	0	4	8	24	hours
Potassium	4.0	4.3	4.8	6.4	mmol/litre
Glucose	4.8	3.9	3.0	1.9	mmol/litre

Bilirubin is one of many plasma constituents which deteriorate even if plasma is correctly separated and stored. Whenever possible, blood should reach the laboratory early in the working day, when the bulk of assays is being performed.

Collection of urine

Urine estimations performed on timed collections are expressed as units/time (for example, mmol/24 h); this figure is calculated by multiplying the concentration by the volume collected during the timed period. The accuracy of the final result depends largely on that of the urine collection and this is surprisingly difficult to ensure. Sometimes, if the patient is incontinent, or has prostatic hypertrophy or neurological lesions which render him incapable of complete bladder emptying, accurate collection is only possible if a catheter is inserted; this is undesirable, unless indicated for clinical reasons, because of the risk of urinary infection. However, most collection errors are due to misunderstanding on the part of the nurse, doctor or patient.

For example, a 24 hour specimen may be collected between 08.00 h on Monday and 08.00 on Tuesday. The volume *secreted by the kidneys* during this period is the crucial one. Urine already in the bladder at 08.00 h on Monday was secreted earlier and should *not* be included; that in the bladder at 08.00 on Tuesday was secreted during the relevant 24 h, and *should* be included in the collection. The procedure is therefore as follows:

08.00 on Monday. The bladder is emptied completely, whether or not the patient feels the need. *Discard the specimen.*
Collect all urine passed until:
08.00 on Tuesday. The bladder is emptied completely, whether or not the patient feels the need. *Add the specimen to the collection.*

The shorter the period of collection, the greater the error if this procedure is not followed.

Before the collection is started a bottle should be obtained from the laboratory, containing a preservative to inhibit growth of bacteria which may destroy the substance being estimated, but which does not interfere with the relevant assay.

Collection of faeces

Rectal emptying is usually erratic and, unlike that of the bladder, can rarely be performed to order. Results of faecal estimations, of for example fat, may vary by several hundred per cent in consecutive 24 h collections. If the collection period lasted for weeks, the *mean* 24-hourly output would be very close to the true daily loss from the body into the intestinal tract. As a compromise most laboratories collect for periods of either 3 or 5 days. To render collection more accurate, many departments use orally administered 'markers' (p. 268).

Faecal collections and estimations are time-consuming and unpleasant for all concerned. Only if *every* specimen passed during the collection period is sent to the laboratory will the timed result be reasonably accurate. Administration of purgatives, enemas or barium before or during the collection period alters conditions and invalidates the answer.

Labelling specimens

Every specimen must be labelled accurately, and the information should correspond with that on the accompanying form in every detail. The date, and sometimes the time, of collecting the specimen should be included, and should be written *at the time of collection.* If it is done in advance the clinician may change his mind and the information will be incorrect: there is also a very real danger of using a container with one patient's name on it for another patient's specimen.

Blood specimens

Specimens in wrongly labelled tubes may cause danger to one or more patients. The date of the specimen is important both from the clinical point of view and, as discussed on p. 449, to assess the suitability of the specimen for the assay requested. The time when the specimen was taken is especially important if, like plasma glucose, the concentration of the analyte varies during the day. If more than one specimen is sent for the same assay on one day, *each must be timed* to determine in which order they were taken.

Urine and faecal specimens

Timed urine specimens should be labelled with the date and time of starting and completing the collection, so that the volume secreted in unit time can be calculated. Faecal collections are best labelled not only with the date and time, but with the specimen number in the series; the absence of a specimen is then immediately obvious.

Sending the specimen to the laboratory

Many estimations can be performed rapidly if the result is needed urgently. However, batching specimens for assay is more economical in staff time and in reagents, as well as being easier to organize. Most laboratories like to receive non-urgent specimens early in the day. A constant 'trickle' may cause delay in reporting the whole batch, and possibly in noticing a result indicating the need for urgent action.

If a patient arrives late in the day it is preferable to delay taking the specimen until the next morning. If this is not possible, and if the result is not needed immediately, the specimen should be sent with a note to that effect. Plasma can then be separated from cells and stored overnight: this service should not be abused, since

Table 25.2 Some extralaboratory factors leading to false results

Cause of error	Possible consequences
Keeping blood overnight before sending to the laboratory	High plasma K, total acid phosphatase, LD, HBD, AST, phosphate
Haemolysis of blood	As above
Prolonged venous stasis during venesection	High plasma Ca, total protein and all protein fractions, lipids, T_4
Taking blood from arm with infusion running into it	Electrolyte and glucose concentrations approaching composition of infused fluid. Dilution of everything else
Putting blood into wrong container or tipping it from this into chemical pathology container	For example, EDTA or oxalate cause low Ca, with high Na or K
Blood for glucose not put into fluoride	Low glucose
Palpation of prostate by rectal examination, passage of catheter, enema etc in last few days	High tartrate-labile acid phosphatase
Inaccurately timed urine collection	False timed urinary excretion values (for example, per 24 h). False and erratic renal clearance values
Loss of stools during faecal fat collection Failure to collect for long period between markers	False faecal fat results

it takes a long time to centrifuge and separate plasma from many specimens, and the chance of transposition of samples is increased.

In cases of true clinical emergency the department should be notified, preferably before the specimen is taken, indicating the reason for the urgency, so that the laboratory may be ready to deal with it quickly. Usually a specimen not known to need urgent attention, and certainly one not accompanied by any information about clinical details, will be assumed to be non-urgent. *It is the clinician's responsibility to indicate the degree of urgency.* The word 'urgent' should not be used lightly. Misuse of emergency services delays truly urgent results both by increasing the amount of work and by inducing cynicism in laboratory staff.

Table 25.2 summarizes some of the errors discussed in this chapter.

Bedside investigation

Traditionally urine testing has been performed by medical or nursing staff in ward side-rooms: the results have varied in accuracy and usefulness. More recently it has become possible to carry out a large range of blood assays near the wards or clinics: examples include 'blood gases', glucose, bilirubin and sodium and potassium. Although the idea is superficially attractive, especially because of the speed of obtaining the result in an emergency, there are many problems to consider. Some of the more important of these include:

- the possibility of untrained staff handling potentially infected specimens;
- the difficulty of maintaining the quality and use of reagents and machinery;
- the difficulty of obtaining correct results;
- the considerable financial implications associated with duplication of apparatus, and its misuse by such staff.

The subject is too complex to be dealt with fully here, but it should be appreciated that it is safer to have no result than the wrong one. *No type of investigation in wards and clinics should be introduced or continued without full consultation and co-operation with the local laboratory.*

Summary

1. The clinician's responsibility for contributing to the accuracy and speed of reporting results includes:

- taking a suitable specimen of blood:
  - (a) at a time when an earlier procedure will not interfere with the result;
  - (b) from a suitable vein;
  - (c) with as little venous stasis as possible;
  - (d) with precautions to avoid haemolysis.
- putting the specimen in the correct container;
- labelling the specimen accurately;
- completing the form accurately, including *relevant* clinical details and drug therapy, and potential hazards due to infection;
- ensuring that safety regulations have been complied with;
- ensuring that the specimen reaches the laboratory without delay, and that plasma is separated from cells before the specimen is stored;
- collecting accurate and complete timed specimens of urine and faeces.

2. Bedside investigations should only be initiated and carried out in collaboration with the local laboratory.

Further reading

Pannall P. Pitfalls in the interpretation of blood chemistry results. *S Afr Med J* 1971; **45**: 1184–7.

Zilva JF. Is unselective biochemical urine testing cost effective? *Br Med J* 1985; **291**: 323–5.

Anderson JR, Linsell WD, Mitchell FM. Guidelines on the performance of chemical pathology assays outside the laboratory. *Br Med J* 1981; **282**: 743.

26

Requesting tests and interpreting results

Requesting tests

Why investigate?

The clinician has a great many tests at his disposal. If used critically these often provide helpful information: if used without thought the results are at best useless, and at worst misleading and dangerous. Writing a request form should not be considered to cast a magic spell, which, in itself, will benefit the patient.

Investigation should be used to diagnose disease and to monitor treatment, not to show how 'clever' the doctor is. Although this may seem obvious, it is often forgotten. Intellectual satisfaction comes from deciding what one hopes to gain from the results of investigation and by achieving this aim as economically as possible in time and money. One test is not necessarily better than another because it is newer, more expensive or more difficult to perform: if it *is* better it should be used instead of, not as well as, the other one.

Far from helping the patient, overinvestigation may harm him by delaying treatment, causing him unnecessary discomfort or danger, or more insidiously, by using resources which might be spent more usefully on other aspects of his care. Of course, under- is just as undesirable as overinvestigation: the cost to the patient of omitting a necessary test is just as high as carrying out unnecessary ones.

Before requesting an investigation the clinician should ask himself:

- will the result, whether high, low, or normal, *affect my diagnosis?*
- *will the result affect my treatment?*
- *will the result affect* my estimate of the patient's *prognosis*?
- *can the abnormality* I am seeking *exist without clinical evidence of it? If so, is such an abnormality dangerous, and can it be treated?*

If, after careful consideration, the answer to any of these questions is 'yes', the test should be performed. If the answer to *all* of them is 'no', there is no need for the test.

Sometimes the test may be necessary even if the clinical diagnosis is obvious. For example, both the clinical and biochemical features of overt hypothyroidism return towards normal once treatment is started, but it is usually necessary to continue such treatment for life. Later another doctor, seeing the patient for the first time, may only be able to confirm the original diagnosis by stopping treatment to see if the symptoms return. This is undesirable, and unequivocally low plasma T_4, with unequivocally high TSH, levels provide objective documentation.

Why not investigate?

The student is warned against the following unqualified statements:

'*It would be nice to know.*' Will it help the patient?

'*We must document it fully.*' Will the extra documentation make any difference to the management of the patient, or even to the sum of medical knowledge?

'*Everyone else does it.*' If they do they may be right, but do you know their reasons? Perhaps they too do it because everyone else does. Do not accept any statement from anyone (not even this book) uncritically. Reassess dogma continually. If you lack experience, at least examine the reasoning behind what you read or are taught; use the statements of 'experts' as working hypotheses until you are in the position to make up your own mind.

How often should I investigate?

This depends on:

- *how quickly numerically significant changes are likely to occur.* For example, concentrations of the main serum protein fractions are most unlikely to change significantly in less than a week, and the plasma urea concentration will not change significantly during the first 12 hours of oliguria;
- *whether a change, even if numerically significant, will alter treatment.* For example, plasma transaminase activities may alter within 24 hours in the course of acute hepatitis. Once the diagnosis has been made this is unlikely to affect treatment. By contrast, the plasma potassium concentrations may alter rapidly in patients given large doses of diuretics and these *may* indicate the need to instigate or to change treatment.

Investigations are very rarely needed more than once in 24 hours, except in some patients receiving intensive therapy.

When is an investigation 'urgent'?

The only valid reason for asking for an investigation to be performed urgently is that an earlier answer will alter treatment. This is a very rare situation. For example, the clinician should ask himself how often treatment would really change within the next 12 hours if it were known that the plasma urea concentration were 10 mmol/litre (60 mg/dl) or 50 mmol/litre (300 mg/dl).

Interpreting results

Before considering diagnosis or treatment based on an analytical result the clinician should ask himself three questions:

1. If it is the first time the estimation has been performed on this patient, *is it normal or abnormal?*
2. If it is abnormal, *is the abnormality of diagnostic value* or is it a non-specific finding?
3. If it is one of a series of results, *has there been a change and, if so, is this change clinically significant?*

Is the result normal?

Reference ranges

The reference ('normal') range of, for example, plasma urea concentration may be quoted as between 3.3 and 6.7 mmol/litre (20 and 40 mg/dl). It is obviously ridiculous to assume that a result of 6.5 mmol/litre (39 mg/dl) is normal, while one of 6.9 mmol/litre (41 mg/dl) is not. Just as there is no clear cut demarcation between 'normal' and 'abnormal' body weight and height, the same applies to any other measurement which may be made.

Most individuals in a normal population will have a value for any constituent near the mean value for the population as a whole; all the values for this population will be distributed around this mean, the frequency with which any one occurs decreasing as the distance from the mean increases. There will be a range of values in which 'normal' and 'abnormal' overlap (Fig. 26.1): all that can be said with certainty is that the *probability* that a value is abnormal increases the further it is from the mean until, eventually, this probability approaches 100 per cent. For example, there is no reasonable doubt that a plasma urea value of 50 mmol/litre (300 mg/dl) is abnormal, whereas it is possible that one of 7.5 mmol/litre (45 mg/dl) is normal for the individual concerned. Nor does a normal result necessarily exclude the disease sought: a value within the population reference range may be abnormal for that individual.

In order to stress this uncertainty at the borders of the reference range it is better to quote limits between which values for 90 and 95 per cent of the 'reference' population fall than to give a rigid 'normal range'. Statistically the 95 per cent limits are two standard deviations from the mean if there is a Gaussian distribution of values. Although the probability is high that a value outside these limits is abnormal, obviously 5 per cent of the 'normal' population may have such a value – if the distribution is symmetrical (Gaussian) 2.5 per cent at either end. The range of variation for a single subject is usually less than that for the population as a whole.

When assessing a result all factors must be considered, especially the clinical findings, before it is possible to reach some estimate of the probability of whether it is normal.

Physiological differences

Physiological factors affect the interpretation of results. For example, there is a *sex* difference in the reference ranges for plasma urate and iron (pp. 384 and 392), and of course for sex-hormone concentrations. The plasma urea concentration tends to rise with *age*, especially in men, and normal values of many variables are different

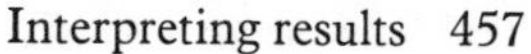

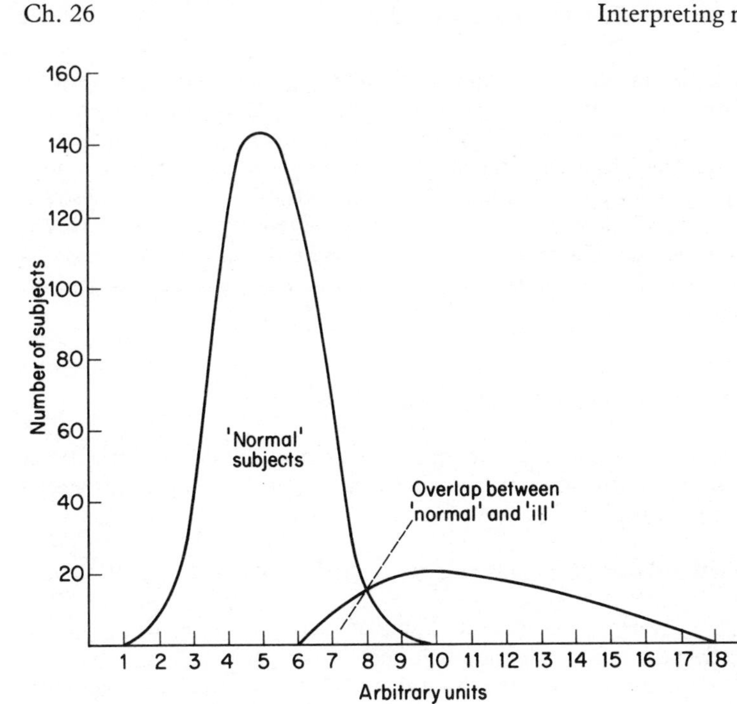

Fig. 26.1 Theoretical distribution of values for 'normal' and 'ill' subjects, showing overlap at upper end of the reference range.

in children from those in adults. There are geographical differences, either because of *racial* or *environmental* factors.

It is therefore clear that *we are not so much expressing 'normal' values as the most usual ones for a given population.* A plasma urea concentration which is 7.5 mmol/litre (45 mg/dl) at the age of 20 suggests mild renal glomerular impairment which *may* progress to clinically severe damage in later life: the same value at the age of 70 suggests the same degree of renal impairment, but, if it is not rising, the subject will usually die of some other disease before this becomes severe. In other words, this rising mean value of urea with age is not strictly normal, and probably does reflect disease. Remember that we are talking of *mean* values for a population at a certain age: in an individual there may be no change with advancing years. By contrast, many of those values found in children and at puberty which differ from those in adults are associated with growth, and are true physiological differences.

Differences between laboratories

It has been pointed out that, even if the same method is used in the same laboratory, it is difficult clearly to define normality. Interpretation may sometimes be even more difficult if results obtained in different laboratories, using different analytical methods, are compared. Agreement between laboratories is close for most constituents. However, for others, such as plasma enzymes, different methods give different results. We have also seen that the definition of 'international units', in which the results of enzyme assays are usually expressed, does not, for example, include the temperature at which the assay is performed. Different laboratories, for various technical reasons, may use temperatures varying from 25°C to 37°C: results on the same specimen at these two temperatures will differ significantly, although apparently expressed in the same units. Even if temperatures were standardized, results would still vary unless substrate, pH and all other variables were the same. If precision is acceptable one method is often no better than another *if the results are compared with the reference ranges for the laboratory in which the estimation was performed, and if serial estimations are carried out by the same method.*

Is the abnormality of diagnostic value?

Results of assays on plasma or serum only express extracellular concentrations. Moreover, some abnormalities are so non-specific as to be of no diagnostic or therapeutic significance.

Relationship between plasma and cellular concentrations

Intracellular constituents are not easily sampled, and plasma concentrations do not always reflect the situation in the whole body; this is particularly true for those constituents, such as potassium, which are at much higher intracellular than extracellular concentrations. A normal, or even high, plasma potassium concentration may be associated with cellular depletion if equilibrium across cell membranes is abnormal, such as in diabetic ketoacidosis; in the short term the aim should be to correct the plasma potassium concentration, whatever the cellular content, but the possible rapid change after treatment of the underlying abnormality should be anticipated (p. 63).

Plasma phosphate levels, like those of potassium, may fall as phosphate moves into cells when, for example, glucose and insulin are being infused. Hypophosphataemia does not necessarily indicate phosphate depletion.

Relationship between extracellular concentrations and total body content

The numerical value for a concentration depends not only on the total amount of the constituent measured, but also on the amount of water through which it is distributed (for example, mmol/litre). Hyponatraemia is not usually due to sodium depletion, but more often to an excess of water: total body sodium may even be increased (p. 51). Conversely, hypernatraemia is much more often due to water deficit than to sodium excess. It is very important to recognize these facts and to adapt treatment accordingly. If stasis is eliminated as a cause of day-to-day variations in plasma protein concentrations, significant short-term changes can be used to assess changes in hydration of the patient: in other words, the amount of protein is not, but the volume of water is, changing significantly.

Non-specific abnormalities

Plasma concentrations of, for example, albumin, calcium and iron vary considerably in diseases unrelated to a primary defect in their metabolism.

The concentrations of all protein fractions, including immunoglobulins, and of protein-bound substances, may fall by as much as 15 per cent after as little as 30 minutes recumbency, possibly due to fluid redistribution in the body. This may at least partly account for the low plasma albumin concentrations found in even quite minor illnesses. Inpatients usually have blood taken early in the morning, while recumbent, and tend to have lower values for these parameters than outpatients.

Most plasma calcium methods measure the total of the protein-bound plus free-ionized concentrations. Changes in albumin levels are associated with changes in those of the calcium bound to it, without alteration of the physiologically important free-ionized fraction; this can occur either artefactually, as discussed on p. 173, or because of true changes in albumin. It is most important not to try to raise the total calcium concentration to normal if there is significant hypoalbuminaemia.

Plasma iron concentrations are very labile, and fall in most types of anaemia, whether due to iron deficiency or not; it is dangerous to give iron to patients with anaemia, especially by the parenteral route, unless there is other, more reliable evidence of iron deficiency (p. 395).

Has there been a clinically significant change?

Before one can interpret day-to-day changes in results, and decide whether the patient's biochemical state has altered, one must know the degree of variation to be expected in results derived from a normal population.

Reproducibility of laboratory estimations

The approximate precision of some assays is given in Table 26.1 at the end of this chapter. Most estimations should give results reproducible to well within 5 per cent: some, such as sodium and calcium, should be even more precise, but variability of, for example, some hormone assays is much greater. Changes of less than the precision of the method are not likely to be clinically significant.

Physiological variations

There are physiological variations of both plasma concentrations and urinary excretion rates of many constituents and false impressions may be gained from the results if this is not taken into account.

Physiological variations may be regular or random.

Regular variations. Regular changes occur throughout the 24 hour period (circadian or diurnal rhythms, like those of body temperature), or throughout the month.

The time of meals affects plasma glucose concentrations: correct interpretation is often only possible if the blood is taken when the patient is fasting, or at a set time after a standard dose of glucose (p. 225). Plasma iron levels may fall by 50 per cent between morning and evening. The circadian variation of plasma cortisol is of diagnostic value (p. 125), but superimposed on this regular variation, 'stress' will cause an acute rise. To eliminate the unwanted effect of circadian variations blood should, ideally, always be taken at the same time of day, preferably in the early morning with the patient fasting. This is not usually possible, and these variations should be taken into account when results are interpreted.

Some constituents vary in monthly cycles, especially in women (again, compare body temperature). These can be very marked in the results of sex hormone assays, which can only be interpreted if the stage of the menstrual cycle is known; plasma iron may fall to very low levels just before the onset of menstruation (p. 392). Some constituents may also vary seasonally.

Although some of these changes, such as the relation between plasma glucose and meals, have obvious causes, many appear to be regulated by the so-called 'biological clock', which may be, but often is not, affected by the alternation of light and dark.

Random variations. Day-to-day variations in, for example, plasma iron concentrations are very large, and may swamp regular changes (p. 392). The causes of these are not clear, but they should be allowed for when serial results are interpreted. The effect of stress on plasma cortisol and other hormone concentrations, and the many factors affecting those of serum proteins, have already been mentioned.

Consultation with laboratory staff

The object of citing the examples given in this and in the preceding chapter is not to confuse the clinician, but to stress the pitfalls of interpreting a laboratory result in isolation. The clinical findings are the lynch-pin of diagnosis and management, and usually, if the specimen has been taken with care, a diagnosis can be made, and treatment instituted *by relating the result to the clinical state of the patient.* If there is any doubt about the type of specimen needed, or about the interpretation of a result, consultation between the clinician and chemical pathologist or biochemist can be mutually helpful.

Inevitably, laboratory errors do occur, even in the best regulated departments, but a discrepant result should not be assumed to be due to this cause before consultation. The estimation may already have been repeated, and, if it has not, the laboratory is usually willing to do so in case of doubt. If it has already been checked, a fresh specimen should be sent to the laboratory, after consultation to determine if and why the first specimen was unsuitable. If the result on this second specimen is the same as that on the first one every effort must be made to find the cause.

Laboratory staff of all grades often take an active interest in the patients they are investigating. On their side they should take trouble to keep the clinician informed of changes needing urgent action, and may suggest further useful tests: the clinician should reciprocate by giving the chemical pathologist or biochemist information relevant to the interpretation of results, and of the clinical outcome of an interesting problem. Such exchange of ideas and information is in the best interests of the patient.

Summary

The clinician should use the laboratory intelligently and selectively, in the best interests of the patient. When interpreting results the following facts should be remembered.

1. The relation of a result to the reference range only indicates the *probability* that it is normal or abnormal.
2. There are physiological differences in reference ranges, and day-to-day physiological variations.
3. There are small day-to-day variations in results due to technical factors and reference ranges may vary with the laboratory technique used.
4. Plasma or serum concentrations reflect those in the extracellular compartment. These may not always reflect intracellular levels. Plasma or serum levels may depend more on the amount of extracellular water than on the body content of the analyte.
5. Changes in the concentration of a given constituent may be unrelated to a defect in the metabolism of that constituent.

Finally, when in doubt, two heads are better than one. Pathologists and clinicians bring different expertise to a problem and full consultation between the two is of great importance.

Further reading

Martin AR, Wolf MA, Thibodeau LA, Dzau V, Braunwald E. A trial of two strategies to modify the test-ordering behaviour of medical residents. *New Engl J Med* 1980; **303**: 1330–6.

Fleming PR, Zilva JF. Work-loads in chemical pathology: too many tests? *Health Trends* 1981; **13**: 46–9.

A very salutary collection of essays on the general subject of common sense in medicine is:

Asher R. *Richard Asher talking sense.* London: Pitman, 1972.

The short section entitled 'Logic and the laboratory' on pp. 166–7 is excellent.

Table 26.1 Reference values

The authors feel strongly that each laboratory should issue its own list of reference values, and that clinicians should consult that list, rather than a textbook, when interpreting results. However, so that students should have some idea of the order of magnitude of reference values, this list gives the MEAN for the authors' laboratories. Students should fill in the blank columns with the values of their own laboratories. Investigations marked with an asterisk indicate those most likely to have different values in different laboratories; this especially applies to plasma enzyme activities.
CV = Coefficient of variation in authors' laboratories at upper reference limit. (This is a measure of the analytical variability if the estimation is repeated on the same specimen several times.) The percentage variation may differ with level, usually being higher at very low levels. It will be near 10–15 per cent for some radioimmunoassays. NOTE THAT THIS IS ANALYTICAL VARIATION ONLY, IT DOES NOT INCLUDE VARIATIONS DUE TO PHYSIOLOGICAL FACTORS OR SPECIMEN COLLECTION, NOR THOSE DUE TO DRUG THERAPY AND INTERFERING SUBSTANCES. These are often greater than analytical variation.

PLASMA, SERUM or BLOOD (Plasma unless otherwise stated)	Approximate MEAN adult reference value for authors' laboratories		CV ± %	Reference range for student's laboratory (fill in)
	SI or other accepted Units	[Other Units]		
Amylase*	200 U/litre at 37°C		5	
Bilirubin	Up to 17 μmol/litre	Up to 1 mg/dl	5	
Calcium (Total)	2.25 mmol/litre	9 mg/dl	1	
Cortisol 09.00 h	440 nmol/litre	16 μg/dl	8	
Cortisol 24.00 h	110 nmol/litre	4 μg/dl		
Creatine kinase*	100 U/litre at 37°C	—	5	
Creatinine	90 μmol/litre	1.0 mg/dl	5	
Electrolytes Sodium	140 mmol/litre	140 mEq/litre	1	
Potassium	4 mmol/litre	4 mEq/litre	2	
Bicarbonate (or T_{CO_2})	24 mmol/litre	24 mEq/litre	5	
Chloride	100 mmol/litre	100 mEq/litre	1	
Gases and pH (whole blood)				
P_{O_2}	12.6 kPa	95 mmHg		
pH ([H$^+$])	7.40 (40 nmol/litre)			
P_{CO_2}	5.3 kPa	40 mmHg		
Glucose (plasma) Fasting	4.5 mmol/litre	80 mg/dl	3	
γ-glutamyltransferase (GGT)*	30 U/litre at 37°C	—	3	

	SI units		Conventional units		
HBD*	125	U/litre at 37°C	—		3
Iron male	21	μmol/litre	120	μg/dl	3
female	14	μmol/litre	80	μg/dl	
Iron-Binding Capacity (total)	54	μmol/litre	300	μg/dl	4
Lipids					
Cholesterol (total)	5.2	mmol/litre	200	mg/dl	5
Triglycerides (fasting)	1.0	mmol/litre	88	mg/dl (as triolein)	2
Magnesium	0.8	mmol/litre	1.6	mEq/litre	5
Phosphatase*					
Acid (serum) (tartrate-labile)	Up to 1.6	U/litre at 37°C	Up to 0.9	KA Units	15
Alkaline	150	U/litre at 37°C	10	KA Units	3
Phosphate	1.0	mmol/litre	3	mg/dl (as P)	2
Proteins (serum)*					
Total	70	g/litre (about 3 g/litre higher for plasma)	7.0	g/dl (about 0.3 g/dl higher for plasma)	
Albumin	40	g/litre	4.0	g/dl	2
Thyroxine (T_4)					
T_4 (Total)	100	nmol/litre	8	μg/dl	8
T_4 (Free)	15	pmol/litre	1.2	ng/dl	8
Transaminases*					
ALT (GPT)	20	U/litre at 37°C	—		6
AST (GOT)	22	U/litre at 37°C	—		5
TSH	1.0	mU/litre	1.0	mU/litre	4
Urate male	0.33	mmol/litre	5.5	mg/dl	2
female	0.27	mmol/litre	4.5	mg/dl	
Urea	4.0	mmol/litre	25	mg/dl	2

Table 26.1 (continued)

URINE or FAECES	Approximate MEAN adult reference values for authors' laboratories		CV	Reference range for student's laboratory
	SI or other accepted Units	Other Units	± %	
URINE				
5-HIAA	30 μmol/day	6 mg/day	11	
HMMA (VMA)	20 μmol/day	4 mg/day	13	
Cortisol	200 nmol/day	70 μg/day	10	
FAECES				
Fat (collected over 5-day period)	Up to 18 mmol/day (as fatty acid)	Up to 5 g/day (as stearic acid)		

The inherent imprecision of collecting a timed specimen usually outweighs analytical variation in these urinary and faecal estimations (and in clearance estimations)

Index

The main page references are in **bold** type.